Advances in Water Security

Series Editor

Ali Fares, College of Agriculture, Food, and Natural Resources, Prairie View A&M University, Prairie View, TX, USA

Water security is vital to a sustainable and secure future of any nation. Addressing water security issues requires: i) a multidisciplinary approach involving highly skilled scientific and technical experts; and ii) substantial long-term funding with little or no-return in the short term. This series has been established as an advanced forum for hydrologists, technologists, policy makers, planners, and other users to discuss the latest innovations, uses and application of new techniques, and policies in dealing with water security in more comprehensive approaches. Topics for volumes in the series include basics of water security; water security and climate change; agriculture and water security; international law and water security; energy security and water security; development and water; analyzing and quantifying the linkage between water and food security; water availability and demand; water and food security considering spatial and temporal variability; analysis of trans-boundary water management and water security; water security adaptation to climate change and variability in land use systems; human dimensions of water security including determinants of water consumption behaviors; big data and water security; water security and technological advances in water sensing technologies such as remote sensing (e.g., LIDAR - Light Detection and Ranging, passive remote sensing, thermal infrared data, passive microwave data, visible and microwave data, visible and near-infrared data), ground penetrating radar, in-situ electromagnetic sensors (e.g., Time domain reflectometry (TDR), Time domain transmission (TDT), frequency domain (e.g., capacitance sensor), the neutron scattering, fiber-optic sensors, and heat dissipative sensors. Proposals for volumes should include as much information as possible, and should be sent to the series editor – Ali Fares (Alfares@PVAMU.Edu) or Senior Publishing Editor - Margaret Deignan (Margaret.Deignan@springer.com).

Sarantuyaa Zandaryaa · Ali Fares · Gabriel Eckstein
Editors

Emerging Pollutants

Protecting Water Quality for the Health
of People and the Environment

Editors
Sarantuyaa Zandaryaa
Division of Water Sciences, Secretariat
of the Intergovernmental Hydrological
Programme
United Nations Educational, Scientific
and Cultural Organization (UNESCO)
Paris, France

Ali Fares
College of Agriculture, Food and Natural
Resources
Prairie View A&M University
Prairie View, TX, USA

Gabriel Eckstein
School of Law
Texas A&M University
Fort Worth, TX, USA

Published jointly by the United Nations Educational, Scientific and Cultural Organization (UNESCO), 7, place de Fontenoy, 75007 Paris, France and Springer Nature
Switzerland AG, Gewerbestrasse 1, 6330 Cham, Switzerland.

© UNESCO 2025.
ISBN UNESCO 978-92-3-100750-7

ISSN 2523-3572   ISSN 2523-3580   (electronic)
Advances in Water Security
ISBN 978-3-031-71757-4   ISBN 978-3-031-71758-1   (eBook)
https://doi.org/10.1007/978-3-031-71758-1

Emerging Water Pollutants:
Advancing Science to Shape Policies

Water is vital for life on Earth. Water quality is declining worldwide, threatening human health, food security, ecosystems, and biodiversity. The United Nations 2030 Agenda for Sustainable Development and the Sustainable Development Goals recognize the crucial need to provide **access to safe water—a fundamental human right**—and protect the world's water resources from pollution.

Water quality management has historically focused on pathogens, nutrients, and heavy metals. A new class of pollutants known as emerging pollutants has garnered increasing attention due to potential risks to human health and ecosystems. Emerging pollutants, which include pharmaceuticals, personal care products, endocrine-disrupting compounds, chemicals, microorganisms, and micro- and nano-plastics, are found in water resources and known to cause chronic toxicity and endocrine disruption in humans and aquatic wildlife, and the development of bacterial pathogen resistance.

Limited knowledge about emerging pollutants hinders the development of appropriate regulatory, monitoring, prevention, and control measures.

UNESCO promotes research, knowledge generation and dissemination, capacity building, and awareness of emerging water pollutants. This book presents state-of-the-art research findings and proposes science-based policy recommendations on managing emerging pollutants toward sustainable water management and healthy ecosystems in the face of global changes and evolving environmental threats.

More than
8 million tonnes
of plastic enter **the ocean** each year
through **rivers** and **waterways**

"Since wars begin in the minds of men and women it is in the minds of men and women that the defences of peace must be constructed"

Foreword

Water, the source of life on Earth, is essential for ecosystems, biodiversity, human well-being, food security, social and economic activities. Protecting and sustainably managing water resources are, therefore, the foundation for the sustainable development of societies. Yet, the world's water resources are increasingly threatened by pollution, unsustainable use and climate change. New and emerging pollutants further excretable water pollution, adversely affecting both natural ecosystems and human health.

This publication is timely and essential on emerging water pollutants. Rapid industrialisation, urban expansion, and intense agricultural activities have introduced a wide and complex array of new contaminants into freshwater bodies, posing human health and ecological risks that often go unnoticed until it is too late.

The Intergovernmental Hydrological Programme (IHP)—UNESCO's international scientific cooperative program in water research, water resource management, water education, and capacity-building, and the only broadly-based science programme of the UN system in this area—has been at the forefront of addressing water-related issues for over five decades now. It fosters scientific collaboration and interdisciplinary research to support countries in enhancing their capacities to monitor, manage and protect water quality, promoting the sustainable use of this vital resource for sustainable development. In particular, IHP has been championing research promotion, knowledge generation, and capacity building on emerging pollutants during the past decade.

This critical book reflects our restless commitment to advancing scientific understanding of emerging pollutants, a crucial resource for scientists, policymakers, and practitioners. By integrating scientific knowledge with practical applications, this publication aims to foster a deeper understanding of the interconnectedness of water, health, and the environment. The chapters within this book explore the sources and effects of emerging water pollutants and highlight the urgent need for decision-makers to fully understand and rapidly exploit these trustful scientific findings to act and design comprehensive strategies to preserve ecosystems and protect people's health.

I invite you to delve into this critical collective work and join us in pleading for urgent action to protect and better manage our water resources for a cleaner, healthier, and more sustainable world.

Lidia Brito
Assistant Director-General for Natural
Sciences of UNESCO

Preface

Pollutants of emerging concern are a wicked problem. New and emerging pollutants are known to be harmful to people and the environment. These substances are used and released continuously into the environment, even in low quantities. Some may cause chronic toxicity and endocrine disruption in humans and aquatic wildlife and contribute to anti-microbial-resistant pathogens. Yet, the sheer number and diversity of these contaminants and the complexities of their origins, transport, toxicity, chronic effects, intermixing, cumulative effects, and elimination make it especially challenging to formulate appropriate policy responses.

Current knowledge of the extent of human and ecosystem health risks posed by emerging pollutants is inadequate. Moreover, most emerging pollutants found in the environment are not regulated through environmental, water quality, wastewater discharge, or health impact regulations. Similarly, regulations for monitoring or tracing the origin or fate of these substances are lacking. Traditional water treatment technologies often fail to remove emerging pollutants, highlighting the need for advanced research and effective monitoring and control measures. As a result, there is an urgent need to implement appropriate measures and policies to reduce emerging pollutants and associated threats to humans and ecosystems. In particular, more research and action are needed to: strengthen scientific knowledge and adopt appropriate technological and policy approaches to monitor emerging pollutants in water resources and wastewater; assess the potential human health and environmental risks posed by emerging pollutants; prevent and control the disposal of emerging pollutants into water resources and the environment; and develop regulations to mitigate and prevent the introduction of these pollutants in the aquatic environment.

Aiming to achieve some of these goals, the United Nations Educational, Scientific and Cultural Organization (UNESCO), through its Intergovernmental Hydrological Programme (IHP) and jointly with the International Water Resources Association, organized an online conference under the theme of *Emerging Pollutants: Protecting Water Quality for the Health of People and the Environment* in January 2023. The conference highlighted research findings, approaches, methodologies, technologies, and policies that communities worldwide can use to advance knowledge and research, identify solutions, and develop policies for managing pollutants of emerging concern in the aquatic environment.

Aligned with the theme of Springer series *Advances in Water Security*, this book focuses on protecting water quality from emerging pollutants. It features selected contributions presented at the conference and proposes science-based policy recommendations. A few exceptional articles were also invited to enrich this compilation of research and scientific studies on emerging water pollutants. The book emphasizes the importance of interdisciplinary scientific collaboration, innovative policies, and adaptive governance. It is an important, comprehensive knowledge resource for policymakers, researchers, and stakeholders to manage emerging pollutants and mitigate associated risks.

As the series and book editors and the conference's International Scientific Committee co-chairs, we are grateful for the excellent contributions of all authors, as well as for the scientific work of researchers worldwide who dedicate their expertise and research in various disciplines to advancing knowledge in this developing area of crucial importance. Their efforts are indispensable to the global efforts to find solutions to improve and protect water quality and manage emerging pollutants in our changing world. Only through better knowledge, scientific evidence, and collaborative efforts will we be able to identify sustainable solutions to the wicked problem of pollutants of emerging concern.

Lead editor of this book Dr Sarantuyaa Zandaryaa, who coordinated UNESCO IHP Theme on Water Quality during IHP-VII and IHP-VIII phases, recalls UNESCO's pioneering role in research promotion and knowledge generation on emerging water pollutants, which began as a case study back in 2011 and evolved into the implementation of a comprehensive UNESCO project in 2015–2018 dedicated to scientific cooperation, research, policy development, capacity building, and awareness raising on emerging pollutants and microplastics in freshwater resources. UNESCO's work on emerging pollutants catalyzed and fostered a rich landscape of research in various scientific disciplines across world's regions investigating different aspects of emerging pollutants and exploring solutions for addressing this global concern. As this book elucidates, the scientific advancements in understanding human and environmental health effects of emerging pollutants reflect not only the dedication of researchers worldwide but also the critical role that international scientific cooperation plays in tackling global environmental challenges. Through the concerted efforts facilitated by UNESCO, a dynamic field of study has emerged, positioning the global scientific community at the forefront of addressing the complexities posed by these new and emerging pollutants. As the world is facing increasingly acute challenges associated with climate change, population growth, and pollution, the insights provided in this book are more crucial than ever. Readers will discover not only state-of-the-art research findings but key policy recommendations guiding action.

Paris, France   Sarantuyaa Zandaryaa
                                               UNESCO-IWRA Conference's International
                                                       Scientific Committee Co-chair

Prairie View, USA   Ali Fares
                                               UNESCO-IWRA Conference's International
                                                       Scientific Committee Co-chair

Fort Worth, USA   Gabriel Eckstein
                                               UNESCO-IWRA Conference's International
                                                       Scientific Committee Co-chair

Acknowledgements

UNESCO, IWRA, and the Editors of this book gratefully acknowledge the invaluable contributions provided by the following individuals who served as external peer-reviewers for the chapters included in the book. Their expertise and advice have greatly enhanced the quality and depth of this work.

Peer-Reviewers

Ripendra Awal, Prairie View A&M University, United States
Suryakanta Acharya, International Water Resources Association
Sougata Bardhan, Lincoln University, United States
Heather Bond, European Open Rivers Programme, The Netherlands
Daniele Cocca, Università degli Studi di Torino, Italy
Bassel Daher, Texas A&M University, United States
Rojano Fernando, West Virginia State University, United States
Malcolm Gander, Naval Base Kitsap-Bangor, United States
Piero Gardinali, Florida International University, United States
Addisie Geremew, Prairie View A&M University, United States
Suat Irmak, Penn State University, United States
Anish Janatrania, Texas A&M University, United States
Davie Kadyampakeni, University of Florida, United States
Binod Khanal, Prairie View A&M University, United States
Marijn Korndewal, Organisation for Economic Co-operation and Development, France
Dorina Murgulet, Texas A&M University-Corpus Christi, United States
Atikur Rahman, Prairie View A&M University, United States
Nikola Rakonjac, Wageningen University, The Netherlands
Dahlia Sabri, International Water Resources Association
Hatim Sharif, University of Texas San Antonio, United States
Zina Souaissi, Université du Québec à Montréal, Canada
Binita Thapa, Prairie View A&M University, United States
T. Nelson Thompson, former Maritime Environmental and Energy Technical Adviser to US Transportation Secretary, United States
Geoffrey R. Tick, The University of Alabama, United States
Jesse Traller Ojeda, Algae Foundation, United States
Anoop Veettil, Prairie View A&M University, United States

Acknowledgements

Contents

Contributors

Tanvir Ahmed Department of Civil Engineering, Bangladesh University of Engineering and Technology, Dhaka, Bangladesh

Chayma Alaya Laboratory of Agrobiodiversity and Ecotoxicology, Higher Institute of Agronomy, University of Sousse, Sousse, Tunisia

Mohamed Banni Laboratory of Agrobiodiversity and Ecotoxicology, Higher Institute of Agronomy, University of Sousse, Sousse, Tunisia;
Higher Institute of Biotechnology, University of Monastir, Monastir, Tunisia

Farhad Bolouri Civil Engineering Department, Energy, Environment, and Water Research Center, Faculty of Civil and Environmental Engineering, Near East University, Nicosia, Türkiye

Regina M. Buono Environmental Law Institute, Washington, DC, USA;
International Water Resources Association, Paris, France

Zehao Chen Water Management and Hydrological Science, Texas A&M University, College Station, TX, USA

Mary Chibwe Institute for Water Research, Rhodes University, Grahamstown, South Africa

Edith Nwakaego Chima Department of Civil Engineering, College of Engineering, Kwame Nkrumah University of Science and Technology, Kumasi, Ghana;
Regional Centre for Integrated River Basin Management, under the auspices UNESCO, Abuja, National Water Resources Institute, Kaduna, Nigeria

Burcu Cömert Ministry of Agriculture and Forestry, Regional Directorate of State Hydraulic Works, Aydın, Türkiye

Carolina Cuchimaque Lugo Instittute of Environment, Florida International University, Miami, FL, USA;

Department of Chemistry and Biochemistry, Florida International University, North Miami, FL, USA

Thiago de A. Neves Department of Environmental and Sanitary Engineering, School of Engineering, Federal University of Minas Gerais, Belo Horizonte, Minas Gerais, Brazil

Robert Michael Di Filippo National Institute of Geological Sciences, University of the Philippines, Diliman, The Philippines

Luciana Teresa Dias Cappelini Instittute of Environment, Florida International University, Miami, FL, USA;
Department of Chemistry and Biochemistry, Florida International University, North Miami, FL, USA

Mariana A. Dias Environmental Chemistry Laboratory, Institute of Chemistry, University of Campinas, São Paulo, Brazil

Vinicius Diniz Department of Analytical Chemistry, Institute of Chemistry, University of Campinas, São Paulo, Brazil

Armin Dolatimehr Independent Researcher, Berlin, Germany

Ramon Domingues Environmental Chemistry Laboratory, Institute of Chemistry, University of Campinas, São Paulo, Brazil

Gabriel Eckstein Texas A&M University School of Law, Fort Worth, TX, USA; International Water Resources Association, Paris, France

Regina E. Edziyie Department of Fisheries and Watershed Management, Kwame Nkrumah University of Science and Technology, Kumasi, Ghana

Helen M. K. Essandoh Department of Civil Engineering, College of Engineering, Kwame Nkrumah University of Science and Technology, Kumasi, Ghana

Ali Fares College of Agriculture, Food and Natural Resources, Prairie View A&M University, Prairie View, TX, USA

Hajar Farzaneh Hamad Bin Khalifa University, Doha, Qatar

Anderson Fozina Krüger Economic Development and Environment, Ibirama City Hall, Brazil

Piero R. Gardinali Institute of Environment and Department of Chemistry and Biochemistry, Florida International University, Miami, FL, USA

Willian Jucelio Goetten Santa Catarina State University, Ibirama, Brazil; Intermunicipal Sanitation Regulatory Agency (ARIS), Florianópolis, Brazil

Hüseyin Gökçekuş Civil Engineering Department, Energy, Environment, and Water Research Center, Faculty of Civil and Environmental Engineering, Near East University, Nicosia, Türkiye

Sibel Mine Güçver Ministry of Agriculture and Forestry, General Directorate of Water Management, Ankara, Türkiye

Maria Guerra de Navarro Instittute of Environment, Florida International University, Miami, FL, USA;
Department of Chemistry and Biochemistry, Florida International University, North Miami, FL, USA

Brandon Hardiman Water Management and Hydrological Science, Texas A&M University, College Station, TX, USA

Sabrine Hattab Laboratory of Agrobiodiversity and Ecotoxicology, Higher Institute of Agronomy, University of Sousse, Sousse, Tunisia;
Regional Research Centre in Horticulture and Organic Agriculture, Sousse, Tunisia

Courtney Heath Instittute of Environment, Florida International University, Miami, FL, USA;
Department of Chemistry and Biochemistry, Florida International University, North Miami, FL, USA

Samkelisiwe Hlophe-Ginindza Water Research Commission, Pretoria, South Africa

Nonhlanhla Kalebaila Water Research Commission, Pretoria, South Africa

Aybala Koç Orhon Ministry of Agriculture and Forestry, General Directorate of Water Management, Ankara, Türkiye

Marijn Korndewal Organisation for Economic Co-operation and Development, Paris, France

Raisibe Florence Lehutso Water Research Centre, Council for Scientific and Industrial Research, Pretoria, South Africa

Camila Leite Madeira Environmental Chemistry Laboratory, Institute of Chemistry, University of Campinas, São Paulo, Brazil;
Department of Civil Engineering, College of Engineering, University of Texas at El Paso, Texas, USA

Wifag Hassan Mahmoud Regional Center for Capacity Building and Research in Water Harvesting, under the auspices of UNESCO, Khartoum, Sudan;
Water Harvesting Center, Faculty of Engineering Science, University of Nyala, Nyala, Sudan

Gordon McKay Hamad Bin Khalifa University, Doha, Qatar

Itza Mendoza-Sanchez Zachary Department of Civil and Environmental Engineering, Texas A&M University, College Station, TX, USA

Nwude O. Micheal Regional Centre for Integrated River Basin Management, under the auspices UNESCO, Abuja, National Water Resources Institute, Kaduna, Nigeria

Mbuyiselwa Shadrack Moloi Centre for Environmental Management, University of the Free State, Bloemfontein, South Africa;
Water Research Centre, Council for Scientific and Industrial Research, Pretoria, South Africa

Cassiana C. Montagner Environmental Chemistry Laboratory, Institute of Chemistry, University of Campinas, São Paulo, Brazil

Dorina Murgulet Department of Physical and Environmental Sciences, Center for Water Supply Studies, Texas A&M University-Corpus Christi, Corpus Christi, TX, USA

Ioana Murgulet Department of Biosciences, Rice University, Houston, TX, USA

Muna Mohammed Musnad UNESCO Chair in Water Resources, Omdurman Islamic University, Khartoum, Sudan

Thabiso Mzinyati Department of Chemical Sciences, University of Johannesburg, Johannesburg, South Africa

Kei Namba Technische Universität Berlin, Berlin, Germany

Roya Narimani Department of Physical and Environmental Sciences, Center for Water Supply Studies, Texas A&M University-Corpus Christi, Corpus Christi, TX, USA

James E. Nickum International Water Resources Association, Paris, France

Chika Felicitas Nnadozie Institute for Water Research, Rhodes University, Grahamstown, South Africa

Vahid Nourani Civil Engineering Department, Energy, Environment, and Water Research Center, Faculty of Civil and Environmental Engineering, Near East University, Nicosia, Türkiye;
Center of Excellence in Hydroinformatics, Faculty of Civil Engineering, University of Tabriz, Tabriz, Iran

Paul J. Oberholster Centre for Environmental Management, University of the Free State, Bloemfontein, South Africa

Joshua Omaojo Ocheje Instittute of Environment, Florida International University, Miami, FL, USA;
Department of Chemistry and Biochemistry, Florida International University, North Miami, FL, USA

Oghenekaro Nelson Odume Institute for Water Research, Rhodes University, Grahamstown, South Africa

Olutobi Daniel Ogunbiyi Instittute of Environment, Florida International University, Miami, FL, USA;
Department of Chemistry and Biochemistry, Florida International University, North Miami, FL, USA

Gülnur Ölmez Ministry of Agriculture and Forestry, General Directorate of Water Management, Ankara, Türkiye

Natalia Quinete Institute of Environment, Florida International University, Miami, FL, USA;
Department of Chemistry and Biochemistry, Florida International University, North Miami, FL, USA

Md. Mezanur Rahaman TA Consultant, Asian Development Bank, Bangladesh Residence Mission, Dhaka, Bangladesh

Atikur Rahman Cooperative Agricultural Research Center, College of Agriculture, Food, and Natural Resources, Prairie View A&M University, Prairie View, TX, USA

Susanne Rath Department of Analytical Chemistry, Institute of Chemistry, University of Campinas, São Paulo, Brazil

Jarbas José Rodrigues Rohwedder Department of Analytical Chemistry, Institute of Chemistry, University of Campinas, São Paulo, Brazil

Jayaprakash Saththasivam Hamad Bin Khalifa University, Doha, Qatar

Camila Schwarz Pauli Department of Environment, Ibirama City Hall, Brazil

Maria Pilar Serbent Santa Catarina State University, Ibirama, Brazil;
Chemical Research and Technology Center, UTN-CONICET, Córdoba, Argentina

Esra Şıltu Ministry of Agriculture and Forestry, General Directorate of Water Management, Ankara, Türkiye

Maria Clara V. M. Starling Department of Environmental and Sanitary Engineering, School of Engineering, Federal University of Minas Gerais, Belo Horizonte, Minas Gerais, Brazil

Melusi Thwala Centre for Environmental Management, University of the Free State, Bloemfontein, South Africa;
Science Advisory and Strategic Partnerships, Academy of Science of South Africa, Pretoria, South Africa

Mary Trudeau Envirings Inc, Ottawa, Canada;
International Water Resources Association, Paris, France

Anoop Veettil Prairie View A&M University, Prairie View, TX, USA

Sabry Zagloul Wahba Water Pollution Research Department, Institute of Environmental and Climate Change Research, National Research Center, Cairo, Egypt

Chunmiao Wang Research Center for Eco-Environmental Sciences, Chinese Academy of Sciences, Beijing, China

Yinuo Wang Department of Geology and Geophysics, Texas A&M University, College Station, TX, USA

Xinghui Xia School of Environment, Beijing Normal University, Beijing, China

Min Yang Research Center for Eco-Environmental Sciences, Chinese Academy of Sciences, Beijing, China

Omagbemi Omoloju Yaya Regional Centre for Integrated River Basin Management, under the auspices UNESCO, Abuja, National Water Resources Institute, Kaduna, Nigeria

Jianwei Yu Research Center for Eco-Environmental Sciences, Chinese Academy of Sciences, Beijing, China

Sarantuyaa Zandaryaa Division of Water Sciences, Secretariat of the Intergovernmental Hydrological Programme, United Nations Educational, Scientific and Cultural Organization (UNESCO), Paris, France

Hongbin Zhan Water Management and Hydrological Science, Texas A&M University, College Station, TX, USA;
Department of Geology and Geophysics, Texas A&M University, College Station, TX, USA

Junyuan Zhang Department of Geology and Geophysics, Texas A&M University, College Station, TX, USA

Chapter 1
Introduction—Emerging Pollutants in Water: Threats, Challenges, and Research Needs

Sarantuyaa Zandaryaa, Ali Fares, and Gabriel Eckstein

Abstract Water is indispensable for life, health, and environmental sustainability, as underscored by the United Nations Sustainable Development Goals. Traditional water quality assessments have historically focused on pathogens, nutrients, and heavy metals. However, recent decades have witnessed growing concerns over Contaminants of Emerging Concern (CECs), a diverse class of pollutants with potential risks to human health and ecosystems. CECs encompass pharmaceuticals, personal care products, endocrine-disrupting compounds, and microplastics, entering water bodies via wastewater, industrial discharge, and agricultural runoff. Their persistence and adverse effects pose significant challenges to water treatment technologies. This manuscript explores the scientific understanding, environmental fate, and societal implications of emerging pollutants globally. It highlights gaps in regulatory frameworks, monitoring data scarcity, and the urgent need for innovative strategies to mitigate and manage impacts of emerging pollutants in water. Drawing from international collaborations and advancements in scientific research, technology and policy, it advocates for integrated approaches to monitor, assess, and manage emerging pollutants, ensuring sustainable water resource management and safeguarding human and environmental health in a changing world.

Keywords Water · Aquatic ecosystems · Health · Emerging pollutants · Contaminants of emerging concern

S. Zandaryaa (✉)
Division of Water Sciences, Secretariat of the Intergovernmental Hydrological Programme (IHP), UNESCO, Paris, France
e-mail: s.zandaryaa@unesco.org

A. Fares
College of Agriculture, Food and Natural Resources, Prairie View A&M University, Prairie View, TX, USA

G. Eckstein
Texas A&M University School of Law, Fort Worth, TX, USA

International Water Resources Association, Paris, France

© UNESCO 2025
S. Zandaryaa et al. (eds.), *Emerging Pollutants*, Advances in Water Security,
https://doi.org/10.1007/978-3-031-71758-1_1

Water is a fundamental resource for life, health, and the environment. The United Nations 2030 Agenda for Sustainable Development *"Transforming Our World"* and its Sustainable Development Goals (SDGs) recognize the crucial importance of access to safe water and the urgent need to protect the quality of the world's water resources and reduce water pollution. Good ambient water quality is essential for access to safe water and a prerequisite for achieving many SDGs, such as the goals related to health, food security, poverty reduction, gender equality, ecosystems, and biodiversity.

Water quality evaluations have historically focused on pathogens, nutrients, and heavy metals. However, over the past two decades, a new class of pollutants known as emerging pollutants, also called Contaminants of Emerging Concern (CECs), has garnered increasing attention due to their potential risks to human health and the environment. Emerging pollutants comprise a broad group of contaminants found in water resources with known or unknown effects on human health and ecosystems. UNESCO broadly defines emerging pollutants as synthetic or naturally occurring pollutants, chemicals, and microorganisms not commonly monitored or regulated in the environment but with potentially known or suspected adverse ecological and human health effects. Furthermore, the Organisation for Economic Cooperation and Development (OECD) defines contaminants of emerging concern as a vast array of contaminants that have only recently appeared in water or that are of recent concern because they have been detected at concentrations significantly higher than expected or their risk to human and environmental health may not be fully understood.

Emerging pollutants encompass a wide array of substances, including pharmaceuticals, personal care products, endocrine-disrupting compounds, household and industrial chemicals, microorganisms, and microplastics. These contaminants are not routinely monitored in water systems, or in ambient waters, but have been detected in various water bodies worldwide. They enter the environment from various sources through multiple pathways, such as municipal wastewater effluents, industrial discharges, agricultural runoff, and leachates from landfills. New and emerging pollutants present a global water quality challenge and potentially severe threats to human health and ecosystems.

Emerging pollutants are a human health and ecological concern because some are known to cause chronic toxicity and endocrine disruption in humans and aquatic wildlife, as well as the development of environmental antimicrobial resistance. The potential human health risks of emerging pollutants through exposure via drinking water require special attention and further scientific research because conventional water purification and wastewater treatment facilities are ineffective in removing them (Lei et al., 2015). Research studies stress that various pharmaceuticals and household and industrial chemicals are continuously used and released into the environment, even in meager quantities (Gavrilescu et al., 2015).

Several serious challenges related to CECs have been identified (Wilkinson et al., 2017), including limited scientific understanding and knowledge of the potential human and ecosystem health risks posed by emerging pollutants. Additionally, monitoring data on emerging pollutants in freshwater resources and wastewater, as well as on their pathways and accumulation in the aquatic environment, are scarce. With few exceptions, emerging pollutants are rarely regulated in national environmental, water quality, and wastewater discharge regulations.

There is an urgent need to improve scientific understanding and information on emerging pollutants to develop and adopt science-based appropriate technological and policy approaches to monitor emerging pollutants in water resources and wastewater, assess their potential human health and environmental risks, and prevent and control their disposal to the environment (Lei et al., 2015). Scientific knowledge and understanding of potential human and ecosystem health risks posed by emerging pollutants still need to be explored, as well as of their presence in water resources and wastewater and their pathways and accumulation in the environment. Consequently, more research is needed to investigate these substances' occurrence, sources, fate, and transport in wastewater (Gosh et al., 2024).

For over a decade, UNESCO has promoted research, knowledge generation and dissemination, capacity building, and awareness of emerging pollutants in water resources and wastewater. The ninth phase of UNESCO's Intergovernmental Hydrological Programme (IHP-IX, 2022–2029), "Science for a water-secure world in a changing environment," aims at promoting scientific research and innovation (Priority Area 1) and Integrated Water Resources Management under conditions of global change (Priority Area 4). These priorities aim to support UNESCO Member States in developing the knowledge, scientific and research capacity, new and improved technologies, and management skills that support water resources management for human development and healthy ecosystems for sustainable development. In particular, the IHP-IX Strategic Plan focuses explicitly on enhancing understanding and knowledge of pollutant sources, behavior, fate, and movement in freshwater systems, such as rivers, lakes, wetlands, and groundwater. In addition, the international scientific community and the UNESCO Water Family—comprising UNESCO water-related category II centres, UNESCO Water-related Chairs, and experts' networks—collaborate to improve knowledge on emerging pollutants to help prevent and reduce water pollution globally. UNESCO supports effective water resource management strategies (UNESCO, 2023) by promoting knowledge generation and dissemination, research, and international scientific cooperation on emerging pollutants in freshwater systems to help develop solutions for pollution prevention and control. To implement these IHP-IX priorities, and also in the framework of IHP's International Initiative on Water Quality (IIWQ), UNESCO partnered with the International Water Resources Association (IWRA) on the organization of the UNESCO-IWRA Online Conference on "Emerging Pollutants: Protecting Water Quality for the Health of People and the Environment," held on 17–19 January 2023.

The conference's purpose was to disseminate state-of-the-art knowledge and draw science-based solutions and policy recommendations upon which the world can advance progress towards managing emerging pollutants to improve and protect water quality in a changing world for the health of people and ecosystems. The main objectives were to: (a) explore the relations between water and emerging pollutants, bearing in mind increasing demand on freshwater resources and global climatic and demographic changes, as well as considering technical and policy solutions; (b) determine the appropriate water policies and technologies to motivate change in business and political behaviour to address challenged posed by emerging pollutants; (c) high-light the bold policy and institutional change needed to support the transformation to better and more reliable water quality; (d) showcase successful efforts of nature posi-tive as well as technological solutions for improved water quality; and (e) examine the life cycle management of emerging pollutants to find solutions in collaboration with industry, regulators, and local communities (including under-represented groups.

The conference focused on five main themes: (i) Emerging pollutants in aquatic ecosystems; (ii) Emerging pollutants and groundwater; (iii) Emerging pollutants and managing wastewater and waste; (iv) A circular economy approach: Lifecycle management of emerging pollutants; and (v) "Priority" emerging pollutants in the hydro-cycle. Research developments and findings, success stories, and case studies were globally disseminated and shared in 14 thematic sessions under these five themes over three days, featuring 70 oral presentations, 100 poster presentations, two high-level science-policy panel discussions, and two plenaries (Opening and Closing Ceremonies). The conference showcased the work and scientific advances of some of the world's leading experts on emerging pollutants. Featuring more than 90 speakers, 27 moderators, co-moderators, and rapporteurs, and several high-level representatives of UN, international and national scientific organizations, the confer-ence successfully provided a global platform for scientific exchange and cooper-ation among scientists, experts, researchers, practitioners, young researchers, and UN, international and national scientific organizations from different sectors, who presented their work and explored the science, policy, law, economics, and other aspects of emerging pollutants. With 170 research studies disseminated to more than 2700 registered participants from over 120 countries worldwide, the confer-ence provided a platform for stakeholders to engage in these productive scientific discussions and science-policy dialogues.

This book is one of the outputs of the conference. It comprises 21 chapters, including selected contributions presented at the conference, along with introductory and concluding chapters. The chapters are distributed across five sections (Parts I to V), corresponding to the conference's themes. The introductory Chapter 1 provides comprehensive insights into the topic's significance and the conference itself and previews the content of the subsequent chapters within the book.

Part I, *Emerging Pollutants in Aquatic Ecosystems*, features six chapters exploring various facets of environmental occurrences of emerging pollutants. Topics include the impact of pesticides and emerging contaminants on the UNESCO Heritage Site of Pampulha Lake in Brazil and a detailed examination of emerging chemical pollutants in aquatic ecosystems using Turkey as a case study. The chapter on *Campylobacter* highlights its emergence as a pollutant in aquatic environments. In contrast, another chapter investigates the presence of odor-causing compounds in drinking water across major Chinese cities. Furthermore, assessing harmful algae as an emerging pollu-tant in domestic water supply from rainwater harvesting facilities in Sudan under-scores the global scope and varied nature of emerging aquatic pollutants. Lastly, the evolution of water research in South Africa, focusing on the transition from legacy pollutants to newer contaminants of concern found in the aquatic environment, is discussed in another chapter, outlining successful strategies and opportunities for future research advancements on emerging pollutants.

Part II, *Emerging Pollutants and Groundwater* presents three focused chapters that delve into critical aspects of groundwater contamination by emerging pollutants. The first chapter examines the origins, transport pathways, remediation techniques, and ongoing challenges associated with emerging pollutants in groundwater. It provides a comprehensive overview of how these contaminants enter and move through aquifers, emphasizing the complexities involved in their cleanup and mitigation. The second chapter focuses on the broader challenges, management strategies, and policy perspectives related to emerging contaminants in groundwater. It explores the regulatory frameworks and governance structures necessary to monitor and address these pollutants effectively, highlighting the intersection of scientific research with policy formulation and implementation. A significant concern addressed in this Section is the implications of antibiotics and antibiotic resistance in aquifer recharge practices, detailed in the third chapter. This chapter scrutinizes how agricultural and medical practices contribute to the presence of antibiotics in groundwater, posing risks to environmental and human health. It emphasizes the need for sustainable water management practices considering the broader impacts of pharmaceutical contaminants on aquifer systems.

Part III, *Emerging Pollutants and Managing Wastewater and Waste*, comprises three chapters that tackle various challenges associated with wastewater treatment and waste management. The first chapter focuses on the complexities of removing emerging pollutants from wastewater, illustrated through a detailed case study. It examines the pitfalls and challenges encountered in conventional wastewater treatment methods when dealing with persistent and emerging contaminants, emphasizing the need for innovative approaches to enhance removal efficiency. The second chapter discusses using ceramic membrane filters to remove pharmaceutical contaminants from wastewater. It explores the application of advanced filtration technologies as a promising solution to mitigate the presence of pharmaceuticals in treated wastewater, contributing to the protection of aquatic ecosystems and public health. In the third chapter, the focus shifts to the emergence of new pollutants in wastewater and their implications for water reuse strategies, highlighting technological advancements and regulatory hurdles in achieving safe and sustainable water recycling practices.

Part IV, *A Circular Economy Approach: Lifecycle Management of Emerging Pollutants,* includes four chapters that explore innovative, sustainable approaches to managing emerging pollutants throughout their lifecycle using the circular economy approach. The second chapter outlines a circular economy approach to managing emerging pollutants, drawing lessons from developed countries and proposing strategies for implementing similar frameworks in Northern Cyprus. The second chapter focuses on agricultural practices and their socio-environmental responsibilities, particularly in rural communities in Brazil. It examines educational initiatives promoting sustainable agricultural practices to minimize the environmental impact of agrochemicals and emerging pollutants in water resources. It emphasizes the importance of resource efficiency and waste reduction in mitigating the environmental footprint of emerging pollutants. Bio-based and circular solutions for addressing harmful algal blooms (HABs) and challenges related to water quality and climate change are discussed in the third chapter through the case of Lake Tegel in Berlin-Brandenburg, Germany. The chapter underscores the role of innovative technologies and sustainable practices in combating HABs and enhancing water resilience in the face of climate variability. Another chapter assesses the spatio-temporal distribution of chlorine residuals in the water distribution system of Dhaka City, Bangladesh, and its potential links to emerging pollutants, highlighting challenges and strategies for optimizing water quality management in urban settings.

Part V, *Priority Emerging Pollutants in the Hydro-cycle*, consists of three chapters that prioritize key emerging pollutants and propose strategies for monitoring and managing their impact on the hydrological cycle. The first chapter focuses on curbing the environmental implications of emerging nano-pollutants, highlighting recent advancements in preventing their environmental exposure and understanding their adverse effects. The second chapter advocates for direct potable reuse as a strategy for sustainable water management, emphasizing the prioritization of emerging contaminants for monitoring and advanced treatment at pilot scales. It underscores the importance of ensuring water safety and reliability through rigorous monitoring and treatment protocols. Lastly, the third chapter discusses prioritizing emerging pollutants for fingerprinting specific water sources, emphasizing their role in tracing contamination pathways and informing targeted mitigation strategies. It illustrates how advanced analytical techniques can enhance our understanding of pollutant sources and facilitate effective water resource management and protection.

The concluding chapter distills policy-relevant practical insights, spanning the five themes, gleaned from the comprehensive exploration of the thematic chapters of the book and the findings of other research studies presented at the conference. It synthesizes key findings and actionable science-based policy recommendations emerging from the discussions on emerging pollutants in aquatic ecosystems, groundwater contamination, wastewater management, circular economy approaches to pollutant lifecycle management and prioritization of pollutants in the hydro-cycle. By highlighting scientific findings, successful case studies, innovative technologies, and policy frameworks, the concluding chapter emphasizes the importance of interdisciplinary scientific collaboration and adaptive governance in addressing complex environmental challenges like emerging water pollutants. Putting forward science-based policy recommendations and identifying research and policy gaps, this concluding chapter serves as a roadmap for policymakers, researchers, and other stakeholders alike, offering strategic pathways to managing emerging pollutants and associated human health and ecological risks toward sustainable water management practices and resilient ecosystems in the face of global changes and evolving environmental threats.

References

Gavrilescu M, Demnerová K, Aamand J, Agathos S, Fava F (2015) Emerging pollutants in the environment: present and future challenges in biomonitoring, ecological risks and bioremediation. New Biotechnol 32(1):147–156. https://doi.org/10.1016/j.nbt.2014.01.001

Ghosh R, Parde D, Bhaduri S, Rajpurohit P, Behera M (2024) Occurrence, fate, transport, and removal technologies of emerging contaminants: A review on recent advances and future perspectives. Clean Soil Air Water. https://doi.org/10.1002/clen.202300259

Lei M, Zhang L, Lei J, Zong L, Li J, Wu Z, Wang Z (2015) Overview of emerging contaminants and associated human health effects. Biomed Res Int 404796. https://doi.org/10.1155/2015/404796

UNESCO (2023) Proceedings of the 41st session of the General Conference. pp 412. UNESCO

Wilkinson J, Hooda PS, Barker J, Barton S, Swinden J (2017) Occurrence, fate, and transformation of emerging contaminants in water: An overarching review of the field, Environmental Pollution. V 231. Part 1:954–970. https://doi.org/10.1016/j.envpol.2017.08.032

Part I
Emerging Pollutants in Aquatic Ecosystems

Chapter 2
Occurrence of Pesticides and Emerging Contaminants in the Pampulha Lake: Anthropic Pollution of a UNESCO Heritage Site

Ramon Domingues⊚, Mariana A. Dias⊚, Camila Leite Madeira⊚,
Maria Clara V. M. Starling⊚, Thiago de A. Neves⊚,
and Cassiana C. Montagner⊚

Abstract Pampulha Lake is an urban lake located in Belo Horizonte, Brazil, that is part of a modern ensemble considered one of the main tourist attractions in the city. The ensemble was nominated as a UNESCO Cultural World Heritage Site in 2016. Despite its cultural and recreational relevance to the local population, Pampulha Lake has suffered from anthropogenic pollution, which has led to water quality impairment over time. Once used for water supply, the lake became eutrophic due to domestic sewage and industrial wastewater discharge in its main tributaries, thus showing recurrent algae blooms and contamination by potentially toxic metals. This study provides the first broad assessment of the occurrence of pesticides and emerging contaminants such as pharmaceutical drugs, personal care products, and hormones in different regions of the lake. Seasonality affected the concentration of several contaminants, such as atrazine, simazine, and caffeine. Spatial variations were also observed for target contaminants, indicating possible sources related to anthropic activities. Risk assessment indicated that imidacloprid and fipronil detected in Pampulha Lake have the potential to threaten aquatic life. Hence, a monitoring plan is needed to identify the main sources of pollution and provide information to support decision-making on remediation strategies.

R. Domingues · M. A. Dias · C. L. Madeira · C. C. Montagner (✉)
Environmental Chemistry Laboratory, Institute of Chemistry, University of Campinas, Campinas, São Paulo, Brazil
e-mail: ccmonta@unicamp.br

C. L. Madeira
Department of Civil Engineering, College of Engineering, University of Texas at El Paso, Texas, USA

M. C. V. M. Starling · T. de A. Neves
Department of Environmental and Sanitary Engineering, School of Engineering, Federal University of Minas Gerais, Belo Horizonte, Minas Gerais, Brazil

© UNESCO 2025
S. Zandaryaa et al. (eds.), *Emerging Pollutants*, Advances in Water Security,
https://doi.org/10.1007/978-3-031-71758-1_2

Keywords Urban lake · Eutrophication · Pesticides · Caffeine · Surface runoff · Ecological risks

2.1 Introduction

Pampulha Lake was created in 1938 when a dam was built in the Pampulha River to supply water to the city of Belo Horizonte in Minas Gerais, Brazil (Resck et al. 2007). Additionally, the lake was used for recreation (e.g., canoeing, rowing, swimming), fishing, and flood control. Later, in 1943, it was incorporated into the Pampulha Modern Ensemble, a complex of buildings designed by Oscar Niemeyer, a key figure in the development of modern architecture in Brazil. The Ensemble includes a church, a museum, a dance hall, a soccer stadium, a sports center, and gardens (Fig. 2.1).

Since its construction, Pampulha Lake's main tributaries have received untreated wastewater and surface runoff from the neighborhoods in the watershed and houses surrounding the lake, leading to water quality impairment. Consequently, its use as a water supply was banned in 1972 (Lopes et al. 2019). In the 1980's, recreational activities in the lake were forbidden. Nonetheless, due to the uniqueness of its architecture

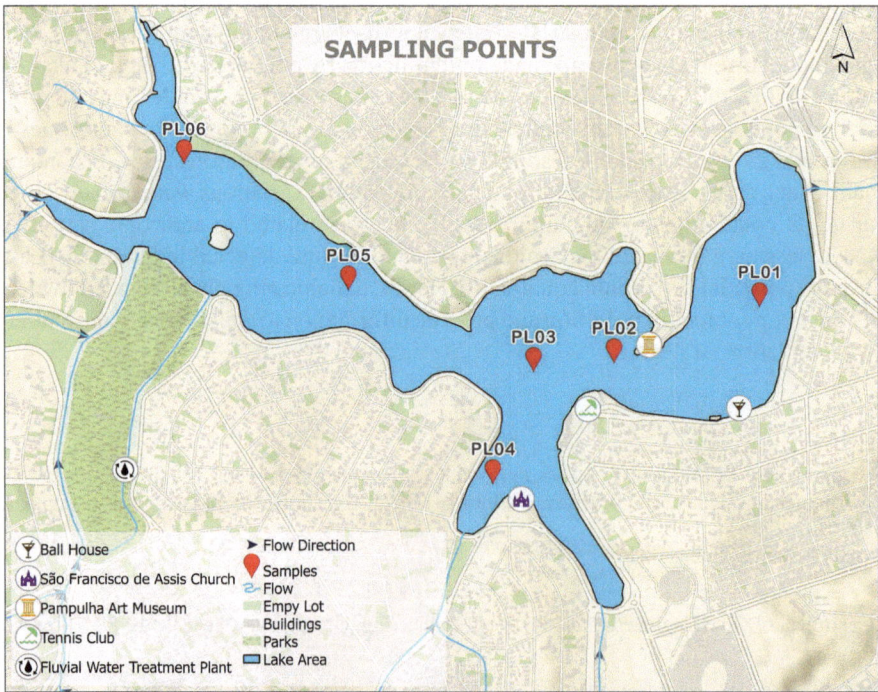

Fig. 2.1 Map of the study area showing the location of the six sampling points, main tributaries, and buildings forming the Pampulha Modern Ensemble. *Source* Authors

and landscape concepts, the Pampulha Lake Modern Ensemble was nominated as a UNESCO's Cultural World Heritage Site by the United Nations Scientific and Cultural Organization (UNESCO) in 2016.

Pampulha Lake was first reported as an eutrophic environment in 1988 by Giani et al. (1988). Samples collected in 1984 and 1985 revealed oxygen concentrations close to zero in the hypolimnion and high electrical conductivity, turbidity, and nitrogen concentrations across the lake. The authors identified the tributaries as significant sources of nutrients to the lake as they receive domestic and industrial wastewater. Evidence of eutrophication remained even after the implementation of an extensive program for macrophyte removal (Pinto-Coelho 1998). Furthermore, Costa (2015) reported the occurrence of cyanobacteria blooms (*Cylindrospermopsis*, *Planktothrix*, and *Microcystis*) in the lake, which can produce toxins that may harm human health.

Although several studies have investigated water quality and its effects on aquatic organisms in Pampulha Lake, only a few have focused on the occurrence of pesticides and emerging contaminants. A study performed by Rietzler et al. (2018) using a thorough toxicity identification and evaluation approach indicated that nonpolar organic and filterable compounds, such as cyanotoxins and oxidants, as well as metals, were possibly the main compounds responsible for the toxicity to water fleas. For instance, the presence of heavy metals such as Fe, Cr, Ni, Sb, Cd, Cu, Sn, and Zn was reported by Friese et al. (2010). The authors suggested that these compounds had an anthropic origin, possibly related to the discharge of iron and steel industries and landfill leachate in Pampulha Lake tributaries.

Several pesticides and emerging contaminants have been reported to occur in artificial lakes located in urban areas in Brazil. A study assessing the occurrence of 35 contaminants in Paranoá Lake, Brasilia, detected caffeine, bisphenol A, and the insect-repellent DEET (3-Methyl-N,N-diethylbenzamide) in all samples, and atrazine, carbamazepine, paraxanthine, mefenamic acid, and nicotine were observed in at least 57% of the samples (Sodré & Sampaio 2020). Similarly, mefenamic acid, atenolol, simazine, lidocaine, and ibuprofen were reported to occur in a high frequency in Guaíba Lake, located in a highly populated area containing industries and agricultural fields (Perin et al. 2021). In both studies, risk assessment indicated the potential harm of these compounds to aquatic organisms.

Due to its importance as a UNESCO Heritage Site and one of the main tourist attractions in town and its integration with the Pampulha Ecological Park, an area known for its high diversity of plants, mammals, amphibians, reptiles, and birds, it is critical to understand the dynamics and fate of pesticides and emerging contaminants in Pampulha Lake to support the development of monitoring plans and cleanup strategies to promote a healthier ecosystem.

2.1.1 Study Area: Pampulha Lake and Architectural Complex - a UNESCO Cultural Heritage Site

Pampulha Lake Watershed comprises 90 km^2 and is part of the Velhas Basin, a sub-basin of the São Francisco Basin, one of the main watersheds in Brazil. The lake has a surface area of 1.8 km^2 with a total water volume of 10 Mm^3, maximum and mean depths of 16.0 m and 5.1 m, respectively. The climate in the location (19°51′05.7″ S 43°58′46.4″ W, altitude 801 m) is classified as high-subtropical (Köppen Cwa), which is characterized by dry winters (April-August) and wet summers (October–March). Pampulha Lake has eight tributaries. The two main tributaries that contribute with more than 70% of the total inflow are Sarandi (sub-basin area: 41 km^2) and Ressaca (sub-basin area: 20.38 km^2) streams (Lopes et al. 2020). The watershed encompasses political boundaries corresponding to the cities of Belo Horizonte and Contagem, a highly urbanized area (nearly 500,000 inhabitants). Both cities are part of the Metropolitan Area of Belo Horizonte, home to almost 6 million people.

The Water Quality Index (WQI) reflects water contamination related to organic and fecal matter, solids, and nutrients by assembling results of 9 parameters (dissolved oxygen, *Escherichia coli*, pH, biochemical oxygen demand, nitrate, total phosphate, temperature, turbidity, and total solids). WQI values vary between 0 and 100, and quality levels are classified as Very Bad ($0 \leq$ WQI ≤ 25), Bad ($25 <$ WQI ≤ 50), Fair ($50 <$ WQI ≤ 70), Good ($70 <$ WQI ≤ 90) and Excellent ($90 <$ WQI ≤ 100) (CETESB 2021; Chidiac et al. 2023). The WQI in the lake varies from bad to fair (IGAM 2018). Anoxic conditions, algae blooms, and fish death are frequent in the lake.

Despite its eutrophic state, Pampulha Lake still provides several ecosystem services, such as landscape improvement, leisure activities, and tourism. Although activities involving primary contact with water were prohibited in the 1980's, fishing is still performed by a small community.

Considering the relevance of the lake to the city of Belo Horizonte, the Program to Recover and Develop Pampulha Lake Watershed (PROPAM) was created in 1998 with three main divisions: sanitation, lake recovery and environmental management. Since then, the program has undertaken several initiatives to restore water quality and ecosystem services provided by the lake. From 2000 to 2006, 1.8 × 10^6 m^3 of sediments and floating macrophytes were dragged from the lake. Also, a Fluvial Water Treatment Plant (flow rate of 750 L s^{-1}) was installed in 2003 to treat water coming through Ressaca and Sarandi Streams, which receive raw wastewater. The treatment plant encompasses a screen for solids retention, followed by a grit chamber, coagulation, flocculation, and flotation. The plant was designed to treat 100% of the inflow in the dry season, thus reducing the load of contaminants reaching Pampulha Lake. In addition, the sanitation facility improved sewage infrastructure to 95% of total flow in the cities of Belo Horizonte and Contagem and a system to retain sediments was installed at the points of discharge of these tributaries into the lake.

Besides sediment dragging and sewage discharge control, corrective actions to manage eutrophication have also been applied to the lake, such as the use of Phoslock®, a lanthanum-modified bentonite clay that binds with dissolved phosphate ions forming $LaPO_4$. This compound binds to phosphorus leading to its precipitation in the sediment (Douglas 2002). The application of this product started in 2016 and took place monthly for 23 months (1:1 molar ratio of La/P; 5-day application schemes per month, a total of 33–47 tons of Phoslock® were applied to the lake) (Barçante 2020). After two years of Phoslock® application, new contracts with the company were signed to continue its use in the following years. Besides, another product named ENZILIMP® is periodically applied to the lake for disinfection and organic matter degradation. However, as of now, there are no academic studies related to the effects of these products on the occurrence and environmental fate of pesticides and emerging contaminants.

Thus, this work aimed to assess the occurrence of several pesticides and emerging contaminants in Pampulha Lake and their spatiotemporal variation to identify potential sources of anthropogenic pollution and assess the ecological risks imposed by target contaminants.

2.2   Sampling, Materials and Methods

2.2.1   Sampling Sites

Sampling points were defined to cover multiple areas within the lake, including areas near the discharge of tributaries and close to touristic sites of the Pampulha Modern Ensemble. Point PL01 (-19.8503320, -43.9691340) is the closest to the dam, with an average depth of 14 m. Points PL02 (-19.8526900, -43.9752020) and PL03 (-19.8530610, -43.9784950) are in a transition zone with a 6-m depth. The remaining points are located near the margins: LP04 (-19.8577600, -43.9801640) is close to the São Francisco de Assis Square (average depth of 1.5 m), while PL05 (-19.5496260, -43.9861180) and PL06 (-19.8443500, -43.9929600) are close to the entrance of the main tributaries (average depth of 1.5 m) (Silva et al. 2016). The six sampling points, the main tributaries (> 70% of inflow), and the buildings forming the Pampulha Modern Ensemble are presented in Fig. 2.1.

The tributaries of Pampulha Lake cross different sub-basins and neighborhoods of the metropolitan city of Belo Horizonte (Fig. 2.2). Land use and occupation in the drainage basin of each tributary can affect the type and concentration of pesticides and emerging contaminants that flow into the lake as they influence diffuse (stormwater runoff) and point sources. Tributaries that run through green areas (such as parks and gardens) can receive contaminants, especially pesticides, through surface runoff. This could also be a significant source of pesticides to the lake. As the lake is in an urban area, the land around the important tributaries designated for agriculture is not

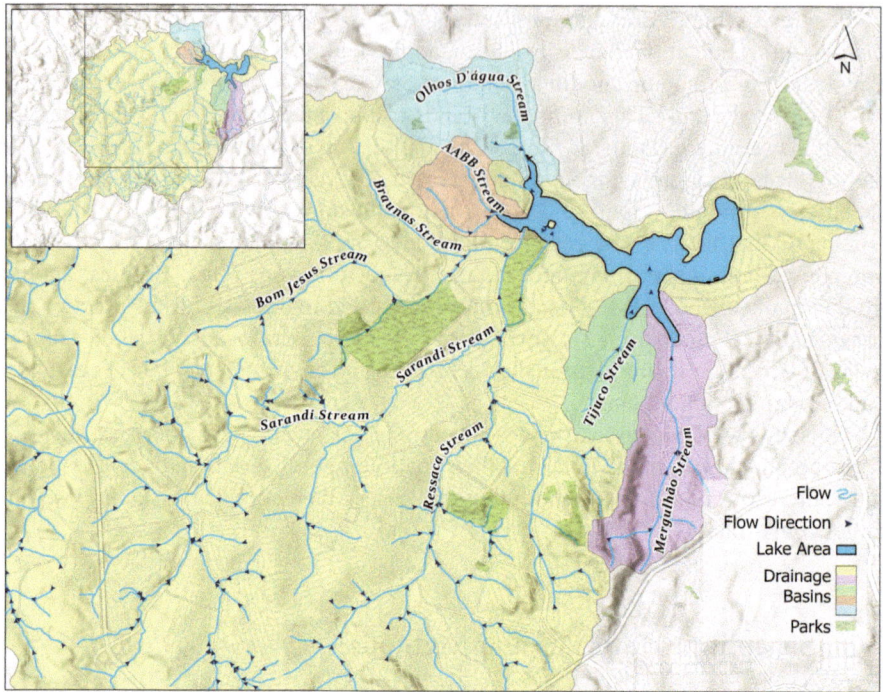

Fig. 2.2 Map of the drainage basins showing the tributaries forming Pampulha Lake. *Source* Authors

substantial. A complete map of the location of the Pampulha Lake in the sub-basins and the drainage basins associated with each of the tributaries is presented in Fig. 2.2.

Considering the proximity of each sampling point to the tributaries and land use and occupation in the watershed (e.g., tourist attractions, marketplaces, sports, and other activities), a higher concentration of a target contaminant in a specific lake area could indicate a potential source. For instance, PL06 is the most impacted by the main tributaries (Sarandi and Ressaca streams), which receive raw wastewater from surrounding neighborhoods, and is also influenced by smaller tributaries, such as the AABB (Associação Atlética Banco do Brasil) stream and the Olhos d'água stream. The latter had a poor and moderate rating for the WQI used to assess the degree of pollution of aquatic bodies. WQI values calculated for the Sarandi and Ressaca streams classify them as "bad" ($25 < WQI \leq 50$), which may negatively affect the water quality at PL06. Ressaca stream receives effluents from food industries and domestic effluents from several neighborhoods in the metropolitan region of Belo Horizonte. Likewise, the Sarandi stream receives a contribution from the industrial and domestic sectors produced by the city of Contagem, which is located within the same metropolitan region (IGAM 2018). PL04 is influenced by the Mergulhão stream, which has a fair WQI. This stream is affected by the discharge of domestic effluents from many neighborhoods in the city of Belo Horizonte. Sampling points

PL02, PL03 and PL05 are located in transition zones inside Pampulha Lake. These points are mainly influenced by the effects of local land use and occupation (e.g., surface runoff from paved streets and gardens, urban drainage, and soil contamination). However, they can be influenced by the geomorphology of the watershed (residence time, water body stratification and flow), microclimate, and water level in the system. PL01 is the closest point to the Pampulha reservoir spillway.

Finally, the Trophic State Index (TSI) was considered for choosing the location of sampling points in the Pampulha reservoir (Cunha 2013). The TSI aims to classify water bodies according to their primary productivity. As primary productivity is influenced by the concentration of nutrients (nitrogen and phosphorus), TSI reflects water quality according to nutrient enrichment and its effect on excessive algal growth (eutrophication). As a result of the eutrophication process, the aquatic ecosystem can undergo significant changes in physicochemical characteristics such as the concentration of dissolved oxygen, pH, turbidity, and total suspended solids, primarily total suspended solids. All the main tributaries of Pampulha Lake present a hypereutrophic state, mainly due to the input of domestic and industrial effluents. The concentration of suspended solids can influence the partition dynamics of pesticides and emerging compounds in aquatic bodies (e.g., the Octanol/water partition coefficient—K_{ow} and the sorption coefficient—K_d), affecting the distribution of these compounds in the liquid and solid phases. A higher presence of suspended solids could increase the sorption of the target analytes, reducing their concentration in the aqueous phase.

2.2.2 Sampling

Before sampling, 1 L amber glass bottles were washed with a 10% (v/v) Extran alkaline solution (Merck MA01), rinsed with tap water and distilled water, and finally with ethanol and acetone. After drying, flasks were heated at 400 °C for 4 h and covered with aluminum foil.

Four sampling campaigns were performed to assess the occurrence of 22 emerging contaminants in Pampulha Lake. Sampling took place in two different seasons throughout the year to assess the seasonal variation of contaminants in the lake. The first samplings were carried out in June and August/2022, which showed total monthly precipitation of 0 and 4.0 mm and average temperatures of 20.6 and 21.0 °C, respectively. The two remaining sampling campaigns were performed in January and February/2023, during the wet season, as verified by total monthly precipitation of 460.8 and 116.3 mm and average temperatures of 23.7 and 25.6 °C, respectively, according to data published by the Brazilian National Institute of Meteorology (INMET). In each sampling event, samples were collected at the six sampling points. All campaigns were conducted during the morning from 8 to 12 am. This is relevant considering higher primary productivity during the day due to light incidence in the water surface that can affect dissolved oxygen, pH, solids and other physicochemical parameters.

For each sample, 1 L of water was collected using a clean amber flask which was sealed, and transported in thermal boxes to the laboratory. Surface water quality in sampling points was characterized in situ for temperature (°C), pH, conductivity (μS cm^{-1}), total suspended solids (TSS, in mg L^{-1}), chlorophyll (μg L^{-1}), turbidity (FNU), Optical Dissolved Oxygen (DO, mg L^{-1}) and Fluorescent Dissolved Organic Matter (fDOM, QSU), by using an EXO 2 (YSI) multi-parameter probe.

2.2.3   Detection of Pesticides and Emerging Contaminants

A list of target pesticides and emerging contaminants analyzed for surface water samples from Pampulha Lake is shown in Table 2.1. Pesticides were classified into three classes (herbicide, insecticide, and fungicide) while emerging contaminants were either pharmaceuticals/personal care products (PCP) or hormones.

Water samples (500 mL) were vacuum-filtered in glass fiber filters (47 mm, grade 13,400, Sartorius) prior to solid phase extraction performed with cartridges (Waters Oasis HLB 500 mg) previously conditioned with methanol and ultrapure water. Quantification was performed by liquid chromatography coupled to mass spectrometry (LC–MS/MS, Agilent Technologies) using Selected Reaction Monitoring mode and internal calibration with atrazine-d$_5$ (CAS #163,165–75-1), fipronil-^{13}C$_4$ (CAS #2,140,327–54-2), 2,4-D-^{13}C$_6$ (CAS #150,907–52-1), and diuron-d$_6$ (CAS #1,007,536–67-5) as internal standards.

Instrumental parameters of the method used to quantify pesticides and emerging contaminants are shown in Table 2.2 and were set as developed and validated by Santos et al. (2022). The instrumental limits of detection (LOD) and limits of quantification (LOQ) were determined by the lowest concentration at the calibration curve for which the signal-to-noise ratio (S/N) was adequate for each of the parameters, and calculated as $3 \times \frac{S}{N}$ for the LOD and $10 \times \frac{S}{N}$ for the LOQ. An approximation of the LOD and LOQ for the method is obtained by multiplying the instrumental values presented by the sample concentration factor (1000 times).

2.2.4   Risk Assessment

A preliminary risk assessment was performed to assess risks to aquatic life considering the detected concentrations of target contaminants in Pampulha Lake. A Risk Quotient was calculated for each of the detected compounds as the ratio between measured environmental concentrations (MEC) and respective Predicted No-Effect Concentrations (PNEC) derived for freshwater reported in the NORMAN Ecotoxicology Database (NORMAN 2023). A risk quotient greater than 1 represents a potential risk to aquatic life.

Table 2.1 List of pesticides and emerging contaminants analyzed in surface water samples from Pampulha Lake

Class		EC
Pesticide	Herbicide	2,4-D (CAS #94–75-7)
		Ametryn (CAS #2163–68-0)
		Atrazine (CAS #1912–24-9) and its degradation products: DIA* (CAS #1007–28-9) and DEA** (CAS #6190–65-4)
		Diuron (CAS #330–54-1)
		Hexazinone (CAS #51,235–04-2)
		Simazine (CAS #122–34-9)
		Tebuthiuron (CAS #34,014–18-1)
	Insecticide	Carbofuran (CAS #1563–66-2)
		Imidacloprid (CAS #138,261–41-3)
		Fipronil (CAS #120,068–37-3)
	Fungicide	Azoxystrobin (CAS #131,860–33-8)
		Carbendazim (CAS #10,605–21-7)
		Tebuconazole (CAS #107,534–96-3)
Pharmaceutical/PCP		Acetaminophen (CAS #103–90-2)
		Caffeine (CAS #58–08-2)
		Triclosan (CAS #3380–34-5)
Hormone		Estrone (CAS #53–16-7)
		17β-Estradiol (CAS #50–28-2)
		Estriol (CAS #50–27-1)
		17α-Ethinylestradiol (CAS #57–63-6)

Source Authors
* DIA—Deisopropylatrazine, ** DEA—Desethylatrazine

2.3 Results

2.3.1 Sample Characterization

Surface water quality parameters were analyzed in situ by a multi-parameter probe and results obtained at each sampling point are presented in Table 2.3.

Higher surface water temperatures were observed during the summer (January and February 2023). Differences between sampling points in a single sampling campaign were not greater than 2.4°C. pH values were stable between 7 and 9 throughout the sampling period and slightly higher values were measured in PL01-04 during the wet season. Conductivity was somewhat higher during the dry weather for most points, except for PL05 and PL06. Total suspended solids ranged from 2 to 13 mg L^{-1}, except for PL05 in February/2023 (585 mg L^{-1}), and PL06 in January/2023 (901 mg L^{-1}). Higher turbidity was also observed in PL05 and PL06, especially during

Table 2.2 Instrumental parameters of the method used for liquid chromatography analysis of samples aiming for the detection of pesticides and emerging contaminants in surface water samples from Pampulha Lake. LOD = Instrumental Limit of Detection; LOQ = Instrumental Limit of Quantification

Compound	Linear range (μg L^{-1})	Linearity (r^2)	LOD (μg L^{-1})	LOQ (μg L^{-1})
2,4-D	1–300	0.997	0.5	1
Ametryn	1–300	0.999	0.5	1
Atrazine	1–300	0.998	0.5	1
DEA	5–300	0.996	1	5
DIA	5–300	0.999	1	5
Diuron	5–300	0.998	1	5
Hexazinone	1–300	0.999	0.5	1
Simazine	5–300	0.996	1	5
Tebuthiuron	1–300	0.999	0.5	1
Carbofuran	1–300	0.999	0.5	1
Imidacloprid	2.5–300	0.996	1	2.5
Fipronil	1–200	0.995	0.5	1
Azoxystrobin	1–300	0.995	0.5	1
Carbendazim	5–300	0.998	1	5
Tebuconazole	5–300	0.999	1	5
Caffeine	25–300	0.999	10	25
Acetaminophen	25–300	0.997	10	25
Triclosan	10–300	0.979	5	10
Estrone	10–300	0.998	5	10
17β-estradiol	5–300	0.994	5	10
Estriol	25–200	0.995	10	25
17α-Ethinylestradiol	50–300	0.988	10	50

Source Authors

January 2023, probably because these sampling points are close to the discharge of Ressaca and Sarandi streams and water flowing through these tributaries receives raw wastewater.

In general, chlorophyll levels were higher during the wet season. This season presents higher temperatures, higher daily exposure to sunlight, and surface water and stormwater runoff flowing into the lake. This scenario is expected to favor microbial and algae growth in the lake.

Dissolved oxygen levels ranged from 4 to 11 mg L^{-1} for most of the samples. PL05 and PL06 were also exceptions for this parameter for measurements performed in February/2023 (0.58 mg L^{-1}) and January/2023 (2.92 mg L^{-1}), respectively. A similar pattern was observed for fDOM, with values ranging between 22 and 33 QSU for almost all samples from points PL 01–05. PL05 presented a slightly higher value

Table 2.3 Range of values obtained in situ for surface water quality parameters at Pampulha Lake: temperature (T, °C), pH, conductivity (Cond., µS cm^{-1}), total suspended solids (TSS, mg L^{-1}), Chlorophyll (µg L^{-1}), turbidity (FNU) and optical Dissolved Oxygen (DO, mg L^{-1})

Sampling point	T (°C)	pH	Cond. (µS cm^{-1})	TSS (mg L^{-1})	Chlorophyll (µg L^{-1})	Turbidity (FNU)	DO (mg L^{-1})
PL01	20.3–26.5	7.62–8.74	179.3–315.1	02–08	11.50–43.40	5.99–14.66	4.30–7.98
PL02	20.6–27.9	7.75–8.91	184.8–321.7	11–12	14.12–37.39	12.56–17.86	6.78–9.99
PL03	20.7–27.9	7.74–8.73	187.7–322.0	8–12	16.99–34.87	13.45–22.02	6.70–9.8
PL04	20.8–28.4	7.70–8.85	186.0–323.6	11–12	20.83–41.88	9.96–16.66	6.42–10.78
PL05	19.6–26.5	7.26–7.80	222.5–343.2	13–585	9.73–50.56	28.12–75.56	0.58–7.95
PL06	20.2–27.0	7.42–7.78	165.3–340.0	12–901	5.14–59.85	18.60–78.47	2.92–7.25

Source Authors

in February/2023, reaching 41.54 QSU. On the other hand, values obtained for PL06 were consistently higher for all sampling campaigns, ranging from 34.31 in June/2022 to 52.52 in January/2023.

Patterns observed for PL05 and PL06, especially during the wet season, are consistent with the influence of the Ressaca and Sarandi tributaries, as raw sewage is discharged into these tributaries. This may cause higher turbidity, lower dissolved oxygen levels, and higher organic matter content. Seasonality observed by higher concentrations of solids and organic matter as well as turbidity in the wet season is associated with increased surface water runoff to the lake, the contribution of tributaries that carry significant loads of sewage, and the reduced capacity of the surface water treatment plant, which operates at a maximum capacity of 50% in the wet season. Furthermore, it may also be related to the effect of mixing of the water column during heavy rains.

In general, the physicochemical characterization of surface water was consistent with monitoring data from other studies performed in the study area (IGAM 2017; IGAM 2018; IGAM 2022). Data presented in Table 2.3 was compared with the data from IGAM (Institute of Water Management of Minas Gerais), showing that at least one of the physicochemical parameters was above the recommended values set for Pampulha Lake at all sampling points. Among all parameters evaluated, chlorophyll-a concentration presented values above the state reference value ($30 \mu g L^{-1}$) for all sampling points. In the historical series of data gathered by IGAM for the Pampulha Lake, other parameters such as total phosphorus and Chemical Oxygen Demand also presented values higher than their local standards.

According to the CONAMA Resolution 357/05 (CONAMA 2005), the Pampulha Stream basin and its tributaries are classified as Class 2, while the Pampulha Stream section downstream from the dam is classified as Class 3, thus being indicated for more restrictive use. For instance, when it comes to leisure activities, Class 2 indicates that the water can be used for indirect contact recreation and fishing, while Class 3 use is limited to boating and landscaping.

2.3.2 Pesticides and Emerging Contaminants in Pampulha Lake

The concentrations of all target analytes are presented in Fig. 2.3, as well as their frequency of occurrence (%) in samples. The risk quotient was calculated for each analyte and is shown in Fig. 2.4.

Five out of the 22 target analytes were not detected above their detection limit in any of the samples: 17β-estradiol, estriol, estrone, 17α-ethinylestradiol, and acetaminophen. For the hormones, the LOD presented in this study is fairly above the appropriate value for analyzing hormones in a system where raw sewage is not the major component of the lake. PNEC values for 17α-ethinylestradiol, 17β-estradiol, estrone, and estriol are 0.037, 0.4, 3.6, and $60 ng L^{-1}$, respectively (NORMAN 2023).

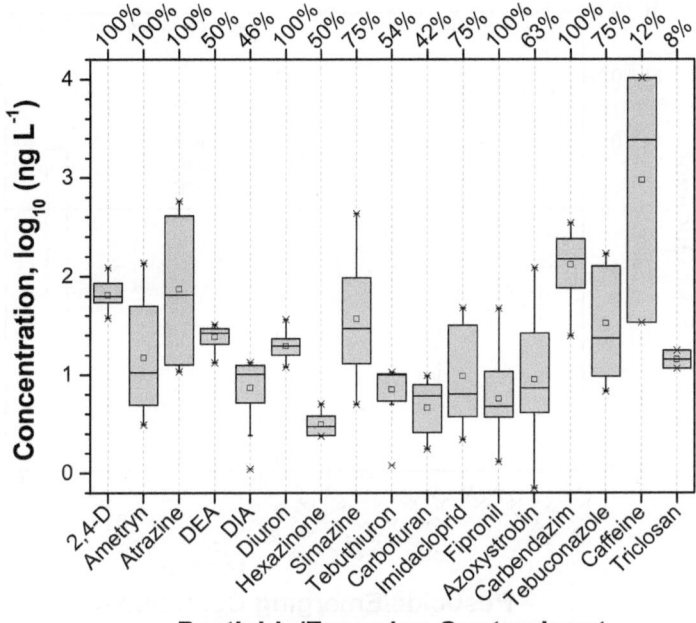

Fig. 2.3 Box-plot of the concentration of pesticides and emerging contaminants detected in Pampulha Lake throughout the four sampling campaigns, in logarithmic scale, and their frequency of occurrence (%). *Source* Authors

Estriol is the only compound that presented a LOD lower than its PNEC. So, besides the fact that these contaminants were not detected in samples, the data could not be used to properly assess the environmental risk posed by hormones for the lake ecosystem.

On the other hand, pesticides 2,4-D, ametryn, atrazine, diuron, fipronil, and carbendazim were found in all sampling points throughout the four sampling campaigns in concentrations varying between 1.3 and 575.7 ng L^{-1}.

Caffeine was only quantified in three sampling points in samples obtained during the wet season, with concentrations ranging from 33 to 10,223.4 ng L^{-1}, indicating a moderate potential risk to aquatic biota. This contaminant was only quantified during the wet seasons in the areas close to the discharge of Ressaca and Sarandi tributaries. This is in accordance with the use of caffeine as an excellent domestic sewage indicator (Gonçalves 2017). As mentioned previously, the lake is reported to receive wastewater throughout the year from industrial and domestic sources. However, wastewater input is higher in the wet season as the surface water treatment plant operates at a lower capacity.

Triclosan was quantified in PL01 and PL02 during the dry season (August 2022) in concentrations between 11.6 and 17.6 ng L^{-1}. Triclosan is a biocide widely used in personal care products and a well-known anthropic pollutant. Sodré & Sampaio

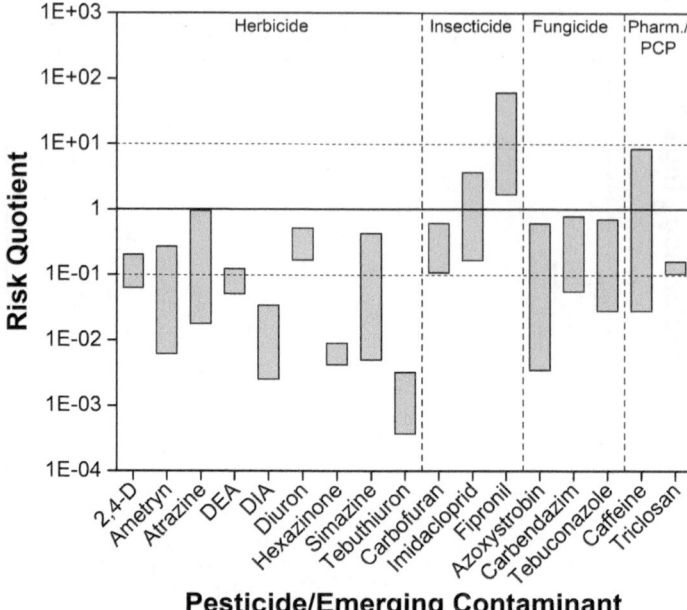

Fig. 2.4 Risk quotient obtained for each of the detected target analytes in Pampulha Lake considering the Lowest Predicted No-Effect Concentrations (PNEC) for Freshwater from NORMAN Ecotoxicology Database. *Source* Authors

(2020) quantified its occurrence in Paranoá Lake, in the capital of Brazil. The analyte was only found in two of the eight analyzed samples, also during the dry season and with at an average concentration of 2.36 ng L^{-1}. In this study, the LOQ for the compound was higher than the one used at Paranoá Lake, thus not appropriate to quantify triclosan at concentrations near the ones reported there. Therefore, the pollutant could be present in other lake areas but at concentrations below the LOQ. Considering the risk assessment results in Fig. 2.4, a concentration lower than 10 ng L^{-1} is not likely to pose a threat to aquatic organisms.

Due to seasonal or spatial factors, most contaminants significantly varied throughout the sampling period. In terms of concentration, considering all samples analyzed in this study, the most abundant contaminants in Pampulha Lake, indicating an imminent risk for aquatic organisms, were atrazine (concentrations ranging from 10.1 to 575.7 ng L^{-1}) > simazine (< LOD to 431.2 ng L^{-1}) > carbendazim (24.7 to 347.2 ng L^{-1}) > tebuconazole (< LOD to 169.9 ng L^{-1}) > ametryn (3.1 to 136.7 ng L^{-1}) > 2,4-D (37.6 to 122.8 ng L^{-1}) > azoxystrobin (<LOD to 122.1 ng L^{-1}).

Both atrazine metabolites (DEA and DIA), diuron, hexazinone, tebuthiuron, carbofuran, imidacloprid, fipronil, and triclosan were detected in low concentrations in all analyzed samples (< 50 ng L^{-1}). As they were detected at low concentrations, their occurrence was less influenced by seasonal and spatial effects when compared to the analytes listed in the previous paragraph.

Even at low concentrations, the risk assessment indicated a moderate risk related to the occurrence of insecticides fipronil and imidacloprid. In addition, special attention should be given to diuron and carbofuran, quantified in 100% and 42% of the samples, respectively, and associated with risk quotients close to 1.

2.3.3   Seasonal Effect

To better understand the spatiotemporal variation of target analytes in Pampulha Lake, Fig. 2.5 presents the total number of contaminants detected in all sampling points for each of the four sampling campaigns. The first two columns for each point correspond to the sampling campaign carried out during the dry season and the last two in the wet season. Caffeine, observed in a significantly higher concentration in only three samples, was excluded from the figure for better visualization.

The total concentration of target contaminants was higher in PL01 and PL02 in August 2022 (second campaign, dry season), followed by PL05 and PL06 in the same sampling campaign. On the other hand, sampling points PL05 and PL06 showed the lowest total sum during the wet season, yet similar to the sum observed at the other four points of the lake for this season.

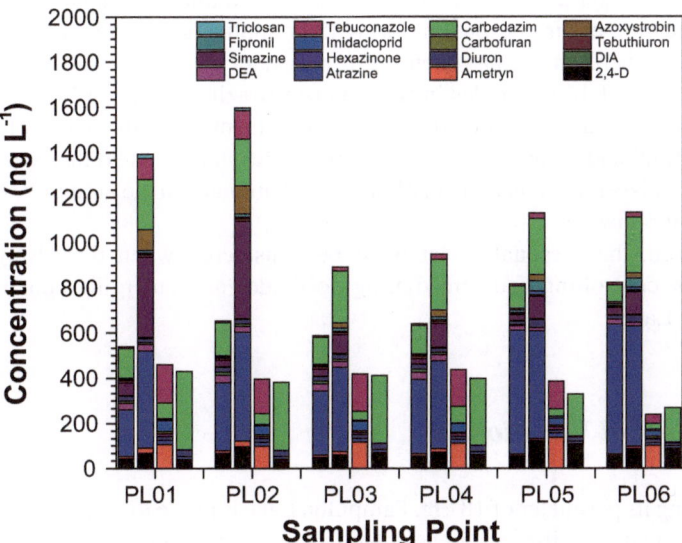

Fig. 2.5 Sum of pesticides and emerging contaminants (ng L^{-1}) quantified in samples obtained in each of the different sampling campaigns and sampling points. The first two columns for each point correspond to the sampling carried out during the dry season and the last two to the wet season. *Source* Authors

In Fig. 2.5 it is also possible to observe that both sampling campaigns performed during the dry season (first two columns of each sampling point) presented a higher total amount of contaminants in all areas of the lake when compared to the two campaigns during the wet season. The general observation of lower levels of contamination during the wet season is associated with the high flow of stormwater to the lake which causes a dilution effect (Montagner and Jardim 2011; Sodré et al. 2018). Another factor influencing the concentration during the wet season may be the higher level of cyanobacteria and algae in the lake, evidenced by the higher levels of chlorophyll-a found during sample characterization as target contaminants may be adsorbed to suspended solids.

Considering each compound separately, no seasonal variations were observed for 2,4-D, diuron, and carbendazim. Their concentrations differed by less than 10% between seasons. Compounds like atrazine, DEA, DIA, hexazinone, tebuthiuron, simazine, fipronil, and azoxystrobin presented lower levels during the wet season due to the dilution effect. Both atrazine degradation products were mainly detected during the dry season when higher levels of atrazine were also observed.

On the other hand, ametryn, carbofuran, imidacloprid, and tebuconazole showed higher levels in the wet season. A higher level of contamination during the wet period can be linked to a diffuse source of pollution, indicating that these contaminants may be transported to the lake through urban drainage canals and surface water runoff from roads, residential areas, parks and gardens. These contaminants could also be present in rainwater due to processes such as evaporation from plants, soil, and water after application or air dissipation during their application. Pesticides can be found in the atmosphere and, consequently, in rainwater. Moreira et al. (2012) studied the occurrence of 16 pesticides in rainwater in Brazil. In total, 56% of the samples were positive for at least three different contaminants, and the levels observed for an agricultural and an urban area were similar. Results indicated that, even in urban centers, the dispersion mechanism of these contaminants in the air may result in their presence in rainwater.

Therefore, the seasonal effect must be considered when discussing possible actions for controlling and remediating pesticides and emerging contaminants in Pampulha Lake.

2.3.4 Spatial Variation

Considering its perimeter of 18 km, Pampulha Lake is prone to variations in the levels of contaminants in different areas across the lake. When considering the primary contribution observed for each region, it is possible to infer the potential sources of contaminants based on their distribution pattern throughout the lake.

During the dry season, the concentrations of atrazine, simazine, fipronil and azoxystrobin presented a significant variation between the different parts of the lake. A graphical representation of the spatial distribution of these compounds during the dry season is shown in Fig. 2.6.

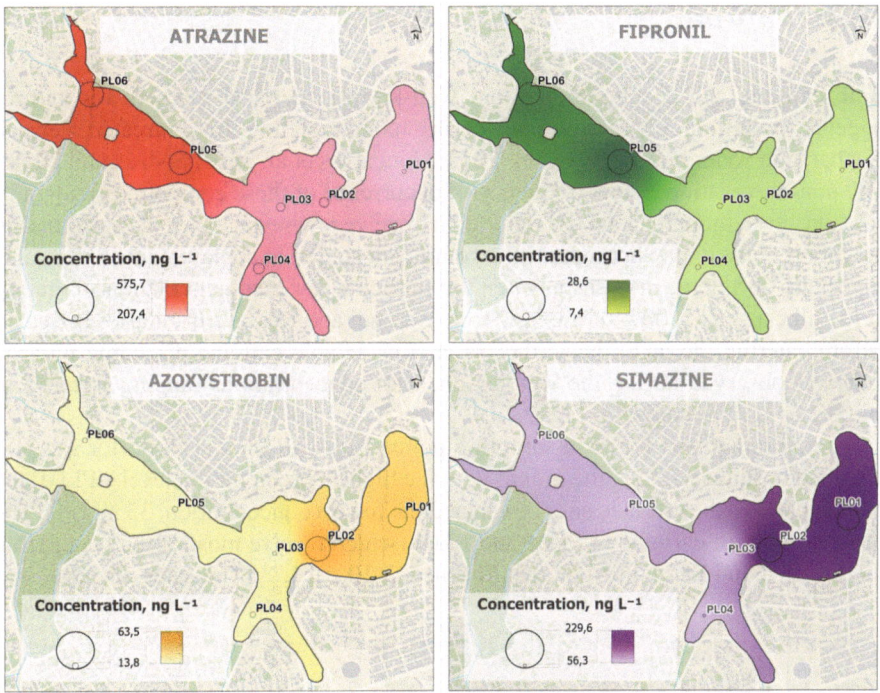

Fig. 2.6 Spatial distribution of atrazine, azoxystrobin, simazine, and fipronil in Pampulha Lake surface water during the dry season (represented by the color gradient and the circle size). *Source* Authors

Pesticides with distribution patterns like atrazine and fipronil—higher concentrations in sampling points PL05 and 06—indicate that Ressaca and Sarandi tributaries and urban drainage canals are likely to be significant sources of these compounds to the lake. On the other hand, azoxystrobin and simazine showed higher concentrations in sampling points PL01 and PL02, which may be associated with their application in the maintenance of the gardens around the lake as part of the Pampulha Modern Ensemble.

Hence, it is essential to identify the primary pollution pathways to promote landscaping practices aiming at the conservation of the lake and water quality improvement. These results indicate that the pesticides applied in the gardens and recreational sites within the watershed may reach the lake through the tributaries, as well as chemicals used for the maintenance of the landscape at touristic sites, thus contributing to the presence of contaminants in Pampulha Lake.

2.3.5   Caffeine

Both caffeine and acetaminophen are well-known chemical markers for anthropogenic pollution, especially considering contamination by untreated sewage discharge in water bodies (Wu et al. 2012; Gonçalves et al. 2017; Sodré et al. 2018). The non-observation or low frequency of occurrence of these compounds may indicate that, even though the lake is reported to receive raw sewage from its surrounding neighborhoods, it may not represent a significant source of contaminants to the lake, especially during the dry season when the surface water treatment plant is operating at its full capacity (100% of the flow from Sarandi and Ressaca tributaries is treated before entering the lake). Nevertheless, a considerable caffeine discharge into the lake was observed during the wet season, with concentrations ranging from 33 to 10,000 ng L^{-1}, as shown in Fig. 2.7.

The pollutant was found at sampling points LP04, PL05 and PL06 for samples obtained in January/2023, the month with the highest precipitation levels. PL04, the closest point to the touristic sites, presented the lowest concentration, 33 ng L^{-1}. The concentration of caffeine in PL05 and PL06, which receive most of the water from the Ressaca and Sarandi tributaries, reached 2–10 µg L^{-1}. The same scenario was

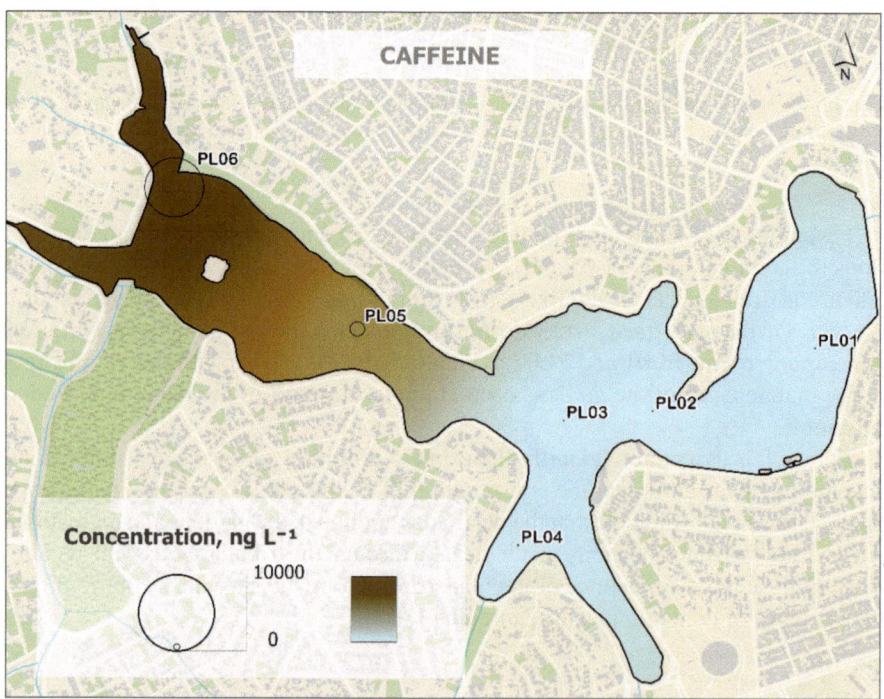

Fig. 2.7 Spatial distribution of the caffeine discharge quantified at the sampling campaign of January/2023 (wet season) in Pampulha Lake. *Source* Authors

not observed in the following sampling campaigns, indicating that the discharge of caffeine that reached the lake underwent a process of degradation and dilution during that period. Caffeine can undergo different environmental degradation pathways, especially biodegradation and photolysis (Korekar et al. 2020).

A possible explanation for the observed values is that although caffeine presents high solubility and a low K_{ow}, indicating that it is preferably found dissolved in the aqueous phase, its distribution in aquatic bodies is a recurring discussion in the literature. Several studies show that higher occurrences of caffeine are frequently closer to contamination sources. Other studies indicate that caffeine is rapidly degraded or sorbed onto solids and transported to groundwater. This information corroborates with data obtained in this study as higher occurrences and concentrations of caffeine were detected in the sampling points close to the main tributaries of Pampulha Lake. The lake has an area of approximately 197 ha, a volume of around 10 million m^3, a wet perimeter of 14.89 km, and maximum and average depths of 16.2 and 5.1 m, respectively. This means that between sampling points PL01 and PL06, there is a difference in depth that can promote the formation of mixing gradients and significant dilution effects due to the flow control of the spillway into the Pampulha stream. Because Pampulha Lake is a reservoir with lotic and lentic characteristics, the residence time in the system may play an important role in caffeine degradation and sorption routes in the environment. This highlights the need for new studies to understand this pollutant's behavior, transport, and environmental fate since it presents interesting properties for studies focusing on the identification of pollutant sources (Canela et al. 2014).

2.4   Conclusion

Understanding the fate, behavior, and toxicity of contaminants in aquatic bodies in large urban centers requires a detailed knowledge of the history and the basin's characteristics. Despite major efforts to recover water quality in Pampulha Lake, it still presents levels of chlorophyll and suspended solids above state reference values, and which confirm its eutrophic state. This is probably associated with the discharge of raw wastewater and stormwater runoff onto the lake, which are likely to be the sources of pesticides and emerging contaminants. Furthermore, among the main pressure factors, the supply of nutrients and sediments to the lake is a challenge, which implies emergency investments in chemical products such as Phoslock® and ENZILIMP®, as well as a significant increase in the inspection of waste discharges.

Inadequate solid waste disposal and illegal dumping, degradation of waterways, and release of chemicals, food, and textiles into the lake's tributaries contribute to the lake's low WQI. It is essential to improve and expand domestic sewage collection and treatment services in the sub-basin, in the same way that industrial effluents require specific treatment before returning to water bodies. Furthermore, it is important to control diffuse sources of pollution and efforts to restore and preserve the green areas of the lake. However, encouraging environmental education among the population is extremely important to make them aware of the importance of preserving the entire basin and also the risks associated with fishing and primary contact with the lake's water.

Pesticides 2,4-D, ametryn, atrazine, diuron, fipronil, and carbendazim were found in 100% of the samples from Pampulha Lake. Most of the target contaminants presented higher concentrations during the dry season. No significant seasonal difference was observed for 2,4-D, diuron, and carbendazim. Few compounds, such as ametryn, carbofuran, imidacloprid and tebuconazole, presented higher concentrations during the wet season, indicating that these pesticides might reach the lake through a diffuse source. From a chemical point of view, this study brought unprecedented information about the occurrence of pesticides and emerging contaminants in Pampulha Lake.

It was also possible to assess the spatial distribution patterns for atrazine, fipronil, azoxystrobin and simazine. These compounds show a gradient with increasing concentrations towards the shallower area of the lake (atrazine and fipronil), thus indicating that tributaries are potential sources for these compounds. On the other hand, a higher concentration of simazine and azoxystrobin near the recreational areas and gardens indicates stormwater runoff from streets and parks in the Pampulha Modern Ensemble as possible sources of contaminants.

Combining results found for the characterization of the sampling sites and considering caffeine as a chemical marker for anthropic pollution, sewage discharge was identified as one of the main sources of target analytes to the lake during the wet season.

Finally, the environmental risk assessment, based on the NORMAN Ecotoxicological Database, shows that the concentrations found for fipronil and imidacloprid could threaten the aquatic ecosystem of Pampulha Lake. In the field, we were able to observe that the impacts on the water quality of the lagoon are beyond the scenario reported in this study. Despite being recalcitrant, actions to mitigate these trace contaminants must occur within a broader scope, with prioritization for the elimination of point sources, such as the disposal of untreated effluents, thus reducing eutrophication events, in order to restore the microfauna of the lake and allow natural processes, such as self-debugging, to be considered the main degradation routes for pesticide residues and emerging contaminants also in urban aquatic systems.

Acknowledgements The authors acknowledge the financial support from the National Institute of Advanced Analytical Science and Technology (INCTAA). R.D. acknowledges the São Paulo Research Foundation (FAPESP) for his scientific initiation fellowship (2022/09112-5). C.L.M. acknowledges FAPESP for her postdoctoral fellowship (2021/12484-9). This study was financed in part by the Coordenação de Aperfeiçoamento de Pessoal de Nível Superior-Brasil 2015/18790-3 (CAPES)-Finance Code 001. The city of Belo Horizonte also collaborated by allowing access to the lake and providing the boat for sampling.

References

Barçante B, Nascimento NO, Silva TFG, Reis LA, Giani A (2020) Cyanobacteria dynamics and phytoplankton species richness as a measure of waterbody recovery: Response to phosphorus removal treatment in a tropical eutrophic reservoir. Ecol Indic 117:106702. https://doi.org/10.1016/j.ecolind.2020.106702

Canela MC, Jardim WF, Sodré FF, Grassi MT (eds) (2014) Cafeína em águas de abastecimento públicos no Brasil [Caffeine in public water supplies in Brazil]. São Carlos, São Paulo, Brasil

CETESB—Companhia Ambiental do Estado de São Paulo (2021) Apêndice E Índices de Qualidade das Águas, Critérios de Avaliação da Qualidade dos Sedimentos e Indicador de Controle de Fontes [Appendix E Water Quality Indexes, Sediment Quality Assessment Criteria ans Source Control Indicatior]. https://cetesb.sp.gov.br/aguas-interiores/wp-content/uploads/sites/12/2022/11/Apendice-E-Indices-de-Qualidade-das-Aguas.pdf. Accessed 15 July 2023

Chidiac SE, Najjar P, Ouani N et al (2023) A comprehensive review of water quality indices (WQIs): history, models, attempts and perspectives. Rv Environ Sci Biotechnol. https://doi.org/10.1007/s11157-023-09650-7

CONAMA—Conselho Nacional do Meio Ambiente, Resolution n° 357 (2005). Dispõe sobre a classificação dos corpos de água e diretrizes ambientais para o seu enquadramento, bem como estabelece as condições e padrões de lançamento de efluentes, e dá outras providências [Provides guidelines for classifying and releasing effluents from water bodies, as well as other measures]. https://conama.mma.gov.br/?option=com_sisconama&task=arquivo.download&id=450. Accessed 17 July 2023.

Costa NB (2015) Dinâmica temporal das cianobactérias em um reservatório urbano hipereutrófico: uma abordagem morfológica e molecular [Temporal dynamics of cyanobacteria in a hyper-eutrophic urban reservoir: a morphological and molecular approach]. Dissertation, Federal University of Minas Gerais.

Cunha DGF, Calijuri MC, Lamparelli MC (2013) A trophic state index for tropical/subtropical reservoirs (TSI$_{tsr}$). Ecol Eng. https://doi.org/10.1016/j.ecoleng.2013.07.058

Douglas GB (2002) Remediation material and remediation process for sediments. US Patent 6350383, 26 Feb 2002

Friese K, Schmidt G, de Lena JC, Arias Nalini H, Zachmann DW (2010) Anthropogenic influence on the degradation of an urban lake—The Pampulha reservoir in Belo Horizonte, Minas Gerais, Brazil. Limnologica 40:114–125. https://doi.org/10.1016/j.limno.2009.12.001

Giani A, Pinto-Coelho R, Oliveira S, Pelli A (1988) Ciclo sazonal de parâmetros físico-químicos da água e distribuição horizontal de nitrogênio e fósforo no Reservatório da Pampulha (Belo Horizonte, MG, Brasil) [Seasonal cycle of physical-chemical parameters of water and horizontal distribution of nitrogen and phosphorus in the Pampulha Reservoir (Belo Horizonte, MG, Brazil)]. Ciência e Cultura 40(1):69–77

Gonçalves ES, Rodrigues SV, da Silva-Filho EV (2017) The use of caffeine as a chemical marker of domestic wastewater contamination in surface waters: seasonal and spatial variations in Teresópolis, Brazil. Rev Ambient Água 12:192–202. https://doi.org/10.4136/ambi-agua.1974

IGAM – Instituto Mineiro de Gestão das Águas (2017) Monitoramento da qualidade das águas superficiais da sub-bacia do Ribeirão Pampulha [Monitoring the quality of surface waters in the Pampulha Stream sub-basin]. http://repositorioigam.meioambiente.mg.gov.br/jspui/handle/123456789/2364. Accessed 13 July 2023

IGAM – Instituto Mineiro de Gestão das Águas (2018) Monitoramento da qualidade das águas superficiais da sub-bacia do Ribeirão Pampulha [Monitoring the quality of surface waters in the Pampulha Stream sub-basin]. http://repositorioigam.meioambiente.mg.gov.br/handle/123456789/2392. Accessed 13 July 2023

IGAM – Instituto Mineiro de Gestão das Águas (2022) Relatório de avaliação da qualidade das águas na sub-bacia do Ribeirão Pampulha [Water quality assessment report in the Pampulha Steam sub-basin]. http://repositorioigam.meioambiente.mg.gov.br/jspui/handle/123456789/4409. Accessed 13 July 2023

Korekar G, Kumar A, Ugale C (2020) Occurrence, fate, persistence and remediation of caffeine: a review. Environ Sci Pollut Res 27:34715–34733. https://doi.org/10.1007/s11356-019-06998-8

Lopes FA, Silveira JS, Leite AC, Piazi J, Lopes NI de A (2019) Recreação de contato secundário em lagos urbanos: o caso da Lagoa da Pampulha [Secondary contact recreation in urban lakes: the Pampulha Lake case]. Revista Geografias 15:42–60. https://doi.org/10.35699/2237-549X.2019.19887

Lopes FA, Davies-Colley R, Piazi J, Silveira JS, Leite AC, Lopes NIA (2020) Challenges for contact recreation in a tropical urban lake: assessment by a water quality index. Environ Dev Sustain 22:5409–5423. https://doi.org/10.1007/s10668-019-00430-4

Montagner CC, Jardim WF (2011) Spatial and seasonal variations of pharmaceuticals and endocrine disruptors in the Atibaia River, São Paulo State (Brazil). J Braz Chem Soc 22:1452–1462. https://doi.org/10.1590/S0103-50532011000800008

Moreira JC, Peres F, Simões AC, Pignati WA, Dores E de C, Vieira SN, Strüssmann C, Mott T (2012) Contaminação de águas superficiais e de chuva por agrotóxicos em uma região do estado do Mato Grosso [Contamination of surface water and rainwater by pesticides in a region of Mato Grosso state]. Ciênc saúde coletiva 17:1557–1568. https://doi.org/10.1590/S1413-812320120 0600019

NORMAN Ecotoxicology Database. https://www.norman-network.com/. Accessed 28 Aug 2023

Perin M, Dallegrave A, Suchecki Barnet L, Zanchetti Meneghini L, Araújode GA, Pizzolato TM (2021) Pharmaceuticals, pesticides and metals/metalloids in Lake Guaíba in Southern Brazil: Spatial and temporal evaluation and a chemometrics approach. Sci Total Environ 793:148561. https://doi.org/10.1016/j.scitotenv.2021.148561

Pinto-Coelho RM (1998) Effects of eutrophication on seasonal patterns of mesozooplankton in a tropical reservoir: a 4-year study in Pampulha Lake, Brazil. Freshw Biol 40:159–173. https://doi.org/10.1046/j.1365-2427.1998.00327.x

Resck RP, Neto JFB, Coelho RMP (2007) Nova batimetria e avaliação de parâmetros morfométricos da Lagoa da Pampulha (Belo Horizonte, Brasil) [New bathymetry and evaluation of morphometris parameters of Pampulha Lake (Belo Horizonte, Brazil)]. Revista Geografias 24–37. https://doi.org/10.35699/2237-549X..13230

Rietzler AC, Botta CR, Ribeiro MM, Rocha O, Fonseca AL (2018) Accelerated eutrophication and toxicity in tropical reservoir water and sediments: an ecotoxicological approach. Environ Sci Pollut Res 25:13292–13311. https://doi.org/10.1007/s11356-016-7719-5

Santos VS, Anjos JSX, de Medeiros JF, Montagner CC (2022) Impact of agricultural runoff and domestic sewage discharge on the spatial-temporal occurrence of emerging contaminants in an urban stream in São Paulo. Brazil. Environ Monit Assess 194:637. https://doi.org/10.1007/s10 661-022-10288-1

Silva TF das G, Vinçon-Leite B, Giani A, Figueredo CC, Petrucci G, Lemaire B, Sperling EV, Tassin B, Seidl M, Khac VT, Viana PS, Viana VFL, Toscano RA, Rodrigues BHM, Nascimento N de O (2016) Modelling Lake Pampulha: a tool for assessing the catchment area impacts on the phytoplankton dynamics. Eng Sanit Ambient 21:95–108. https://doi.org/10.1590/S1413-415 20201600100125692

Sodré F, Santana J, Sampaio T, Brandão C (2018) Seasonal and spatial distribution of caffeine, atrazine, atenolol and DEET in surface and drinking waters from the Brazilian federal district. J Braz Chem Soc. https://doi.org/10.21577/0103-5053.20180061

Sodré FF, Sampaio TR (2020) Development and application of a SPE-LC-QTOF method for the quantification of micropollutants of emerging concern in drinking waters from the Brazilian capital. Emerg Contam 6:72–81. https://doi.org/10.1016/j.emcon.2020.01.001

Wu S, Zhang L, Chen J (2012) Paracetamol in the environment and its degradation by microorganisms. Appl Microbiol Biotechnol 96:875–884. https://doi.org/10.1007/s00253-012-4414-4

The opinions expressed in this chapter are those of the author(s) and do not necessarily reflect the views of the UNESCO: United Nations Educational, Scientific and Cultural Organization, its Board of Directors, or the countries they represent.

Chapter 3
Campylobacter—an Emerging Pollutant of Aquatic Environments

Mary Chibwe, Oghenekaro Nelson Odume, and Chika Felicitas Nnadozie

Abstract *Campylobacter* species are primarily known for causing gastrointestinal infections in humans. Globally, *Campylobacter* species are increasingly recognised as emerging pollutants in aquatic environments. This chapter reviews the emergence of *Campylobacter* in aquatic environments, exploring its sources, factors driving its emergence, public health implications, and monitoring and management interventions for preventing its emergence. The information presented in this chapter shows that the emergence of *Campylobacter* in aquatic environments is a concern in many parts of the world and is driven by several complex factors which may be biological, social, and environmental. This chapter shows evidence of increasing occurrence of *Campylobacter spp.* and its resistance determinants in different aquatic environments, the detection of *Campylobacter spp.* in aquatic environments, and factors that may drive its emergence in aquatic ecosystems. Given that aquatic environments serve as sources of water for recreation and domestic consumption, the emergence of *Campylobacter spp.* presents a considerable risk to public health. The presence of *Campylobacter* and its antibiotic-resistance genes in aquatic environments poses a high risk of transmission to humans and animals. Furthermore, the chapter highlights the need for comprehensive monitoring, mitigation strategies, and further research to address this emerging environmental concern.

3.1 Introduction

Aquatic environments such as lakes and rivers are vital for sustaining life and providing ecosystem benefits that are critical for the survival of mankind and the well-being of communities. They provide ecosystem benefits such as drinking water, food, and water for agriculture. Additionally, aquatic environments provide water for livestock, wild animals, and birds and are natural habitats for numerous organisms (Hodgson and Manus 2006; Njage and Buys 2017; Makaya et al. 2020). Other ecosystem benefits include recreation and tourism, and spiritual and cultural

M. Chibwe (✉) · O. N. Odume · C. F. Nnadozie
Institute for Water Research, Rhodes University, Grahamstown, South Africa
e-mail: mchibwe@evelynhone.edu.zm

© UNESCO 2025

S. Zandaryaa et al. (eds.), *Emerging Pollutants*, Advances in Water Security,
https://doi.org/10.1007/978-3-031-71758-1_3

purposes for many communities (Mboweni and De Crom 2016; Sekwadi et al. 2018). Despite being providers of these ecosystem benefits, over the past decades, there has been an increase in pollution of aquatic environments mainly driven by anthropogenic activities (Almeida et al. 2021). Aquatic environments have become more prone to pollution, especially to emerging pollutants of chemical and biological origin. Emerging pollutants in aquatic environments refer to pollutants of microbiological or chemical (natural or synthetic) origin that have only recently been detected or have become of concern due to their potential to cause adverse effects on human/ecosystem health, and whose health effects may not be well understood (Radwan et al. 2023). Emerging pollutants also refer to pollutants not yet included in routine environmental monitoring programs (Bunke et al. 2019). Emerging pollutants include endocrine-disrupting chemicals, flame retardants, plasticizers, artificial sweeteners, industrial chemicals, surfactants, some pesticides, engineered nanoparticles, analgesics, antibiotics, hormones, anti-inflammatory, antidiabetic, and antiepileptic drugs, personal care products (PCPs), some disinfection by-products (DBPs), microplastics, antibiotic-resistant genes (ARGs) and some pathogenic organisms (Radwan et al. 2023). There have been growing concerns over emerging pollutants in aquatic environments, and many of these pollutants are being detected in the environment (Radwan et al. 2023).

Campylobacter was first reported as early as 1886 by Theodore Escherich. *Campylobacter* species are one of the main pathogens responsible for gastroenteritis in humans. Infections are characterised by watery, non-bloody, non-inflammatory diarrhoea, abdominal pain, and fever (Shobo et al. 2016). The infections are usually self-limiting but more severe cases can result in Guillain–Barre syndrome, reactive arthritis, and abortion (Aksomaitiene et al. 2019). However, infection can be severe in immunocompromised individuals, children, elderly people, or pregnant women (Kashoma et al. 2016; Bolinger and Kathariou 2017). When symptoms persist and in severe cases, antibiotics that are often prescribed include tetracycline, macrolides (erythromycin), and fluoroquinolones (ciprofloxacin). Systemic infections are treated with gentamicin, clindamycin, and ampicillin. Nevertheless, *Campylobacter* species have demonstrated resistance to these antibiotics of choice (Ghunaim et al. 2015; Kashoma et al. 2016; Pillay et al. 2020). Globally, it is estimated that *Campylobacter* is responsible for over 150 million cases of acute gastroenteritis per annum and over 400–500 million infection cases each year (Igwaran and Okoh 2019; Goddard et al. 2022). In 2019, over 220,000 confirmed cases of gastrointestinal were linked to *Campylobacter* in the European Union (EU) (Goddard et al. 2022). In the United Kingdom (UK), *Campylobacter* is responsible for acute gastroenteritis in about 600,000 people a year while *Campylobacter* infections affect approximately 1.5 million people in the USA annually (Goddard et al. 2022). Similarly, high cases of *Campylobacter* infections are also recorded in Canada (about 447 per 100,000 people) (Smith et al. 2019). *Campylobacter* is also one of the major causes of gastroenteritis in developing countries. In Low and Middle-Income Countries (LMICs), *Campylobacter* outbreaks are often waterborne or foodborne. Waterborne outbreaks are more common in LMICs, especially among communities that use raw river water for domestic consumption (Igwaran and Okoh 2019).

There is growing evidence of *Campylobacter* as an emerging pollutant (pathogenic) of aquatic sources (Saifur and Gardner 2021). Recognition of *Campylobacter* as an emerging pollutant of aquatic environments is largely due to its increasing incidence in aquatic environments (Rodríguez-Martínez et al. 2013). These pathogenic bacterial species are now being detected at higher concentrations in aquatic environments and are being linked to outbreaks of diseases in humans (Vouga and Greub 2016; Kovanen et al. 2016; Hassen et al. 2020; Mpondo et al. 2021). This chapter provides evidence of increasing occurrence of *Campylobacter spp.* and its resistance determinants in different aquatic environments, the detection of *Campylobacter spp.* in aquatic environments, and factors that may drive its emergence in aquatic ecosystems. Given that aquatic environments serve as sources of water for recreation and domestic consumption, the emergence of *Campylobacter spp.* presents a considerable risk for waterborne exposure. The presence of *Campylobacter* and its antibiotic-resistance genes in aquatic environments poses a high risk of transmission to humans and animals. Additionally, the public health implications of the emergence of *Campylobacter* in aquatic environments are discussed. Furthermore, the monitoring and intervention measures for preventing microbial pollution of aquatic environments and improved management of aquatic environments are proposed.

3.2 General Biology of the Genus *Campylobacter*

This section discusses the general biology of the genus *Campylobacter,* which is important for a thorough comprehension of the factors influencing its occurrence, as elaborated in the subsequent section.

3.2.1 Classification

Theodore Escherich observed non-culturable spiral-shaped bacteria which were later on identified as *Campylobacter* in 1906 (Silva et al. 2011). These bacteria were also isolated from aborted bovine foetuses in 1913 and from cattle faecal matter in 1927 and named *Vibrio jejuni* by Smith and Orcutt. Several years later (1944), Doyle isolated a bacteria from porcine faecal matter and classified them as *Vibrio coli*. Unlike the *Vibrio* species, these bacteria were microaerophilic and did not ferment carbohydrates hence they were distinguished from the true *Vibrio* species and named *Campylobacter* in 1963 by Sebald and Véron (Silva et al. 2011). The genus *Campylobacter* falls under the phylum Proteobacteria, class Epsilonproteobacteria, order Campylobacterales, and family Campylobacteraceae (Kaakoush et al. 2015). Other genera in the family Campylobacteraceae are *Arcobacter* and *Sulfurospirillum*. These genera have close phenotypic similarity and growth requirements (On et al. 2017). There are 16 species and six subspecies of the genus *Campylobacter* (Silva et al.

2011). The well-studied species are *Campylobacter jejuni* and *Campylobacter coli* which are the main etiological agents for campylobacteriosis (Asuming-Bediako et al. 2019). Other *Campylobacter* species include *Campylobacter concisus*, *Campylobacter ureolyticus*, *Campylobacter upsaliensis*, and *Campylobacter lari* which are said to be "emerging *Campylobacter* species," due to their poorly understood roles in human and animal diseases (Kaakoush et al. 2015).

3.2.2 Physiology

Campylobacter species show similarities in physiological characteristics. Except for *C. gracillis* and some strains of *C. showae*, most *Campylobacter* species produce oxidase (Silva et al. 2011; On et al. 2017). Additionally, they neither ferment nor oxidize carbohydrates, but are chemoorganotrophs that obtain energy from amino acids, or tricarboxylic acid cycle intermediates. *Campylobacter jejuni* hydrolyses hippurate, and indoxyl acetate and reduces nitrate (Silva et al. 2011; Kaakoush et al. 2015). *Campylobacter* species grow well under micro-aerobic conditions (less than 8% oxygen) but some strains are also able to grow under aerobic and anaerobic conditions. *Campylobacter* species such as *C. concisus*, *C. curvus*, *C. rectus*, *C. mucosalis*, *C. showae*, *C. gracilis, and C. hyointestinalis* utilise hydrogen or formate as an electron donor for micro-aerobic growth (Kaakoush et al. 2015; Asuming-Bediako et al. 2019). Other species such as *Campylobacter jejuni* are obligate microaerophilic (requires oxygen levels within a range of 3 to 15%). (Shagieva et al. 2021). The growth temperature ranges from 31 °C to 36 °C, with the optimum temperature at 42 °C (Khan et al. 2014; Shagieva et al. 2021). *Campylobacter* species are sensitive to saline conditions and cannot withstand sodium chloride concentrations greater than 2% w/v. Furthermore, *Campylobacter* does not grow in environments with a water activity (aw) lower than 0.987 but their optimal growth occurs at aw = 0.997 (approximately 0.5% w/v NaCl). *Campylobacter* species cannot survive below a pH of 4.9 and above pH 9.0 and their optimum growth is at pH 6.5–7.5. All *Campylobacter* species are non-spore-forming, and have low G + C content (between 29 and 47 mol%) (Silva et al. 2011).

3.2.3 Morphology

Campylobacter species are gram-negative, non-spore-forming, curved, or spiral bacteria. Some species, such as *C. showae and* strains of *C. jejuni* may exist as straight rods. Most species have flagella (polar or non-polar) while a few lack flagella (Kaakoush et al. 2015; On et al. 2017; Asuming-Bediako et al. 2019). They are non-spore-forming, with an average size of 0.2 to 0.8 by 0.5–5 um (Kaakoush et al. 2015). *Campylobacter* species are motile except for *Campylobacter gracilis*, which is non-motile and lacks a flagella (Silva et al. 2011).

3.2.4 Ecology

Campylobacter species are primarily found in the intestines of mammals and aves (Magana-Arachchi and Wanigatunge 2020). In general, this colonization either occurs as a commensal, as observed in aves, or as an asymptomatic transient infection pathogen as in livestock and humans, or as pathogens in mammals with lower body temperatures (Magana-Arachchi and Wanigatunge 2020). The microaerophilic nature and temperature sensitivity of these organisms prevent their growth outside intestinal environments. Nonetheless, they can still be isolated from environmental sources contaminated with feces, such as surface water, and animal products like milk and meat, where they can remain viable for extended periods (Vereen et al. 2013; Igwaran and Okoh 2020). In such conditions, the organisms can adapt to various hostile environmental stresses, such as fluctuations in oxygen levels, temperature extremes, osmolarity variations, biotic interactions, and nutrient deprivation (Pitkänen 2013). *Campylobacter* bacteria respond to hostile environmental conditions through various physiological, morphological, and biochemical adaptations (Enany et al. 2021). These responses are generally reversible to some extent, but prolonged or cumulative stress can ultimately lead to bacterial death. One notable response to stress in *Campylobacter* is a transition from a spiral to a coccal morphology. However, this change in morphology does not necessarily correlate with loss of culturability and both properties may occur independently. Furthermore, *Campylobacter*'s survival outside the host is influenced by its ability to form biofilms and interact with other bacteria (Enany et al. 2021). In conditions such as high oxygen levels, limited nutrient availability, heat, acidic pH, temperature fluctuations, and exposure to antimicrobials, *Campylobacter* responds by forming biofilms and transitioning to a viable but nonculturable (VBNC) state, a trait that varies among strains (Enany et al. 2021).

The emergence of *Campylobacter* species in aquatic environments involves several mechanistic aspects at cellular and genomic levels. For instance, *Campylobacter* species express virulence factors in response to environmental stress. *Campylobacter* species express virulence factors responsible for surviving under high temperature, hyperosmotic shock, carbon starvation, oxidative and aerobic (O2) stress) (Hung et al. 2011; Kim et al. 2015; Otigbu et al. 2018; Reddy and Zishiri 2018). Virulence factors expressed in response to environmental stress include Caseinolytic proteases (*clpP*), *dnaJ*, Carbon starvation regulator gene (*csrA*), and High-temperature requirement gene B (*htrB*) (Hung et al. 2011; Kim et al. 2015; Otigbu et al. 2018; Reddy and Zishiri 2018). Additionally, horizontal gene transfer (HGT), also plays an important role in the survival and emergence of *Campylobacter* species in aquatic environments. Horizontal gene transfer aids the sharing and acquisition of genes responsible for the expression of antibiotic resistance, virulence factors, and environmental adaption traits in *Campylobacter* species (Laffite et al. 2016; Iwu et al. 2020). Furthermore, the emergence of *Campylobacter spp.* in aquatic environments may be enhanced by mechanisms such as *Campylobacter*'s aerotolerance properties, ability to form biofilms and transform into a viable but non-culturable

(VBNC) (Elmonir et al. 2022). For instance, biofilms may act as habitats for *Campylobacter* species in water (Banting et al. 2016; Magana-Arachchi and Wanigatunge 2020). Additionally, biofilms also protect the bacteria from the cidal effects of antibiotics enhancing their survival in aquatic environments (Imran et al. 2019). Moreover, when conditions become unfavorable (low nutrients, temperature, and aerobic conditions), *Campylobacter* is transformed into a viable but non-culturable (VBNC). This enables the survival and emergence of *Campylobacter* under changing environmental conditions (Okada et al. 2022).

Overall, *Campylobacter* are not indigenous inhabitants of environments outside host organisms. *Campylobacter* occurrence in aquatic environments is due to fecal contamination. The species have developed adaptive strategies to survive in diverse environmental conditions. Factors such as biofilm formation, adaptation to stressors like high oxygen tension and limited nutrient availability, and the ability to enter a viable but non-culturable state contribute to their persistence outside the host. These factors are pertinent for *Campylobacter* to adapt to and survive in various environmental conditions outside the host (Banting et al. 2016; Magana-Arachchi and Wanigatunge 2020; Okada et al. 2022).

3.3   Occurrence of *Campylobacter* in Aquatic Environments: Sources and Patterns

Campylobacter occurrence in aquatic environments signifies recent fecal contamination. The increase in the occurrence of *Campylobacter* is often attributed to the discharge of raw or untreated wastewater, agricultural run-off, livestock waste, and feacal matter from wild animals or birds (Rodríguez-Martínez et al. 2013). *Campylobacter* has been isolated from various aquatic environments (Table 3.1).

Globally, *Campylobacter* species has been isolated from different aquatic environments, such as rivers, lakes, streams, and coastal waters (Table 3.1). This shows that aquatic environments are prone to *Campylobacter* contamination. Studies have reported the occurrence of *Campylobacter* in Africa (Profitós et al. 2014; Karikari et al. 2016; Elfadaly et al. 2018; Otigbu et al. 2018; Chukwu et al. 2019; Tapela and Rahube 2019; Chala et al. 2021; Taviani et al. 2022; Yitayew et al. 2022), Australia/New Zealand (Devane et al. 2014; Meng et al. 2018; Gilpin et al. 2020), China (Yuan et al. 2019), Europe (Szczepanska et al. 2017; Nilsson et al. 2018; Mulder et al. 2021; Andrzejewska et al. 2022; Strakova et al. 2022) and USA (Vereen et al. 2013; Meinersmann et al. 2023). The most common isolated species are *C. jejuni and C. coli.* (Table 3.1). Other species such as *C. fetus C. lari, C. upsaliensis* have also been detected in aquatic environments (Karikari et al. 2016; Otigbu et al. 2018; Chukwu et al. 2019; Chala et al. 2021; Yitayew et al. 2022). The common sources of contamination leading to *Campylobacter* presence in aquatic environments include wastewater (Vereen et al. 2013; Devane et al. 2014; Karikari et al. 2016; Otigbu et al. 2018; Tapela and Rahube 2019; Yuan et al. 2019; Chala et al. 2021; Yitayew et al.

Table 3.1 Occurrence of *Campylobacter* in different aquatic environments

Campylobacter species	Antibiotic-resistant genes detected	Prevalence	Source(s) of contamination	Type of water source	Country	Reference
C. jejuni C. coli C. upselensis	cmeABC active efflux pump 72.6% (69/95), cmeA 55% (11/20) cmeB 90% (18/20) cmeC 70% (14/20)	C. jejuni 38.9% (95/244) C. coli 7.4% (18/244) C. upselensis 2.9% (7/244)	Industrial effluent, urban run-off, migratory birds, fishing activities	Estuarine Water River water	South Africa	Otigbu et al. (2018)
C. jejuni C. coli C. upselensis	gyrA 25% (5/20) tetO 40% (8/20) Mutation at A2074C/A2075G 75% (15/20)	Campylobacter spp. 21.7% (20/92) C. jejuni 55% (11/20) C. coli, 40% (8/20) C. upsaliensis 5% (1/20)	Recontamination in storage vessels at household level	Municipal taps, wells, rainwater, and river water	South Africa	Chukwu et al. (2019),
C. fetus, C. jejuni C. coli	catII 95%, tetA 88.71% tetB 7.42%, tetM 32.26% ermB 15.38%, gyrA 39.13% ampC 81.54% aac(3)-IIa-(aacC2)[a] 84.85%	Campylobacter spp. 58.9% (33/56) C. fetus 5.8% C. jejuni 38.8% C. coli 7.8%	No source attributed	Rivers, ponds, and dams	South Africa	Igwaran and Okoh (2020)
C. jejuni C. coli	tetA, mphA, strB, sul1, dfr	7% (37/526)	Wastewater	River water	Botswana	Tapela and Rahube (2019)
Campylobacter spp.	tetQ	12% (n = 8)	Livestock	Drinking water, Wells	Cameroon	Profitós et al. (2014)
Campylobacter spp.	Not tested/Detected	36% (n = 29)	Poultry production	River water, drinking water supply	Mozambique	Taviani et al. (2022)

(continued)

Table 3.1 (continued)

Campylobacter species	Antibiotic-resistant genes detected	Prevalence	Source(s) of contamination	Type of water source	Country	Reference
C. coli, C. upsaliensis	aac(6)-Ib-cr, aadA1 bla_{OXA}-10 qnrS, mefA, tetA	Not reported	Agriculture, hospital, and industrial effluent, urban run-off	River water	Ethiopia	Yitayew et al. (2022)
C. jejuni	No ARGs detected	28/245 (11.4%)	Poor sanitary conditions	Ground water	Egypt	Elfadaly et al. (2018)
C. coli, C. lari C. jejuni	Not detected	Overall 42/188 (22.3%) 35.7%-Rivers 26.2%–Streams 21.4%–Wells 9.5% -Ponds 7.1% -Boreholes	Hospital and pharmaceutical effluent, Open defecation, Livestock production, Wastewater	Rivers, streams, wells, ponds and boreholes	Ghana	Karikari et al. (2016)
Campylobacter spp.	Not detected	10/30 (33.3%)	Sewage	River water	New Zealand	Devane et al. (2014)
C. jejuni	Not detected	Not reported	Sheep faecal matter	Groundwater, Livestock drinking water source	New Zealand	Gilpin et al. (2020)
Campylobacter spp.	Not detected	Not reported	Waterfowls	Stormwater constructed wetlands	Australia	Meng et al. (2018)

(continued)

Table 3.1 (continued)

Campylobacter species	Antibiotic-resistant genes detected	Prevalence	Source(s) of contamination	Type of water source	Country	Reference
C. jejuni	Not detected	4/39 (4%)	is largely due to its increasing incidence in aquatic environments	Surface water	China	Yuan et al. (2019)
C. jejuni, C. coli	Not detected	Not reported	Not attributed to any source	Freshwater lake, well	Sweden	Nilsson et al. (2018)
C. jejuni, C. coli	Not detected	21/56 (38%)	Wastewater, wild birds, and run-off from farms	Wastewater, ponds and lakes	Czech Republic	Strakova et al. (2022)
C. jejuni, C. coli	Not detected	42/250 (16.8%)	Birds, Pets, and wild animals	Ponds and ornamental lakes, freshwater beaches, rivers, fountains	Northern Poland	Szczepanska et al. (2017)
C. Jejuni, C. fetus C. coli	No ARGs detected	10.5% (n = 172)	Agricultural activities, livestock production, wastewater	Surface water, piped water, groundwater, and stored water	Ethiopia	Chala et al. (2021)
C. jejuni, C.coli	Not detected	66% (230/348)	Poultry, ruminants and Pigs	Recreational water, drainage ditches, irrigation canals, surface water, wastewater	Netherlands	Mulder et al. (2021)

(continued)

Table 3.1 (continued)

Campylobacter species	Antibiotic-resistant genes detected	Prevalence	Source(s) of contamination	Type of water source	Country	Reference
Campylobacter spp.	Not detected	Not reported	Animal and human faecal contamination	Rainwater collected in a water plaza	Netherlands	Sales-ortells and Medema (2015)
Campylobacter spp.	Not detected	Not reported	Wastewater, Pets	Urban river, lake, rainwater, pond,	Netherlands	Sales-ortells et al. (2015)
C. jejuni	Not detected	6.3% (5/80)	Not reported	Natural waters	Finland	Raulo et al. (2015)
C. jejuni, C. coli	Not detected	Not reported	Wild birds, Poultry, ruminants, and pigs	Rivers, ponds/ lakes, streams/ canals/ditches, and wastewater	Netherlands and Luxembourg	Mughini-Gras et al. (2016)
C. jejuni	Not detected	15/185 (8.1%)	Migratory birds and environmental sources	Rivers, ponds, ornamental lakes, and freshwater beaches	Poland	Andrzejewska et al. (2022)
Campylobacter spp.	Not detected	Prevalence not reported	Wastewater and Livestock production	River water	Georgia, USA	Meinersmann et al. (2019)
Campylobacter spp.	Not detected	62% (96/156)	Agriculture, poultry, and wastewater	Irrigation water	Georgia, USA	Vereen et al. (2013),

Source Authors

2022; Meinersmann et al. 2023), industrial effluent (Karikari et al. 2016; Otigbu et al. 2018; Yuan et al. 2019; Yitayew et al. 2022), urban run-off (Otigbu et al. 2018; Yitayew et al. 2022), migratory birds (Szczepanska et al. 2017; Meng et al. 2018; Otigbu et al. 2018; Andrzejewska et al. 2022), wild animals (Szczepanska et al. 2017), agriculture activities (Yuan et al. 2019; Chala et al. 2021; Yitayew et al. 2022), hospital effluent (Karikari et al. 2016; Yitayew et al. 2022) and livestock production (Vereen et al. 2013; Profitós et al. 2014; Karikari et al. 2016; Chala et al. 2021; Mulder et al. 2021; Taviani et al. 2022; Meinersmann et al. 2023).

Furthermore, some *Campylobacter* isolates from aquatic environments may be harboring genes conferring resistance to antibiotics used in treating human infections. However, many studies reporting the occurrence of *Campylobacter* in aquatic environments have not reported detection of antibiotic resistant genes (Table 3.1). In the few studies (Table 3.1) have reported the detection of antibiotic resistant genes in *Campylobacter* isolates from aquatic environments; the most commonly detected genes are those conferring resistance to fluoroquinolones (*gyrA*) (Chukwu et al. 2019; Igwaran and Okoh 2020) and tetracycline resistance (*tetO, tetA, tetB* and *tetM*)(Profitós et al. 2014; Chukwu et al. 2019; Tapela and Rahube 2019; Yitayew et al. 2022).

It is evident that the aquatic environment receives *Campylobacter* from various sources and can therefore act as an effective mode of transmission to humans and animals (Kaakoush et al. 2015). Consequently, *Campylobacter* species are increasingly implicated in both animal (zoonotic) and human diarrhoea or gastroenteritis infections and are a major cause of bacterial foodborne and waterborne infections. The rise in the incidence of *Campylobacter* infections globally, and their tendency to withstand the effects of antibiotics, underscores their serious threat to public health (Kaakoush et al. 2015; Igwaran and Okoh 2019). The ability of *Campylobacter spp.* to acquire resistance to antibiotics is an issue of concern. Due to high antibiotic resistance among *Campylobacter* species, the World Health Organization (WHO) named *Campylobacter* as one of the 12 bacterial species that pose the greatest threat to human health (Sproston et al. 2018; Veltcheva et al. 2022).

There are concerns about the increase in antibiotic-resistant *Campylobacter spp*, and their antibiotic-resistant genes in aquatic environments (Table 3.1). The presence of antibiotic-resistant *Campylobacter* in water is a threat to human and animal health. This problem may be more severe in developing countries where there is overuse and uncontrolled use of antibiotics, poor surveillance, higher contact between humans and animals, and poor water and sanitation (Kashoma et al. 2016; Chala et al. 2021).

This chapter provides evidence of the increasing occurrence of *Campylobacter spp.* and its resistance determinants in different aquatic environments, the detection of *Campylobacter spp.* in aquatic environments and that may drive its emergence in aquatic ecosystems. Given that aquatic environments serve as sources of water for recreation and domestic consumption, the emergence of *Campylobacter spp.* presents a considerable risk for waterborne exposure.

3.4 The Emergence of *Campylobacter* in Aquatic Environments

Emerging and re-emerging of pathogens such as *Campylobacter* in aquatic environments can be attributed to several complex factors which may be of biological, social, and ecological nature (Vouga and Greub 2016). The following sections describe how these factors are leading to the emergence of *Campylobacter* in aquatic environments.

3.4.1 Changes in Environmental Water Quality Due to Anthropogenic Activities

Anthropogenic activities such as urbanisation, industrialisation, agriculture, and many others are resulting in changing of natural aquatic environments (Vouga and Greub 2016). Urban areas in many parts of the world especially in the LMICs are facing challenges such as discharge of raw or untreated wastewater in rivers and poor solid waste management (Bastaraud et al. 2020). Inadequately treated hospital effluent, wastewater, industrial effluent and solid waste from urban areas often carries pollutants such as antibiotic residues, plastics, heavy metals, pathogens, and antibiotic resistant genes into aquatic environments (Bastaraud et al. 2020; Edokpayi et al. 2020). Additionally, agriculture run-off may potentially carry antibiotic resistant bacteria, antibiotic resistant genes, pesticide residues, biocides and other pollutants into aquatic environments (Irfan and Almotiri 2022). Animal excreta from livestock production farms may also contain heavy metals and antibiotic residues such as nor-floxacin, ofloxacin, ciprofloxacin, trimethoprim, sulfamethoxazole and doxycycline. The presence of these pollutants in rivers may create favourable conditions for the survival of *Campylobacter* in aquatic environments. Previous studies have demonstrated high prevalence of *Campylobacter* in rivers impacted by pollution from anthropogenic activities (Henry et al. 2015; Schang et al. 2016; Meng et al. 2018; Otigbu et al. 2018). Furthermore, the pollutants may promote selection pressure that facilitates the emergence and prevalence of antibiotic resistance in aquatic environments (Irfan and Almotiri 2022; Reddy et al. 2022). Pollutants can create stressful conditions for the bacteria community in rivers consequently enhancing the bacteria's survival mechanisms (Rodgers et al. 2019; Hubeny et al. 2021). Additionally, heavy metals, biocides, pesticides, and some persistent organic compounds co-select for antibiotic resistance genes present in river water. Consequently, their presence in aquatic environments can lead to selective pressure increasing the bacteria's resistance to antibiotics (Rodgers et al. 2019; Hubeny et al. 2021). Various studies have recognised urbanisation as a key driver of the emergence of pathogens such as *Campylobacter* in aquatic environments in urban areas. For instance, *Campylobacter* was detected in 92% of stormwater samples discharged into the Yarra River estuary, Australia (Siddiqee et al. 2019). This demonstrates that urbanisation may be an important cause of *Campylobacter* emergence in aquatic environments. Stormwater often carries contaminants from the community hence detection of *Campylobacter* demonstrates the importance of urbanization-linked stormwater in the increasing incidence of *Campylobacter spp* in the estuarine part of this river (Siddiqee et al. 2019). Similarly, a study in Changzhou City of Yangtze River Delta demonstrated that urban rivers were harboring diverse pathogens and reported the concentration of *Campylobacter jejuni*, ranging from 3.30 to 5.85 log10 copies/100 mL (Cui et al. 2019).

3.4.2   Changing Physico-Chemical Characteristics of Aquatic Environments

Anthropogenic activities have also resulted in alteration of the physico-chemical characteristics of aquatic environments. Physico-chemical characteristics of the aquatic environment affect the occurrence, survival, morphological, and physiological factors of *Campylobacter spp.* (Strakova et al. 2022). Pollution of aquatic environments with wastewater, stormwater, agricultural and industrial waste has resulted in high turbidity, low dissolved oxygen, and high concentrations of ions in aquatic environments (Vereen et al. 2013; Henry et al. 2015; Otigbu et al. 2018). High turbidity is linked with *Campylobacter* occurrence in aquatic environments (Henry et al. 2015). The particles in turbid water may promote the survival of *Campylobacter*. This is worsened by the presence of other emerging pollutants such as microplastics. Microplastics act as a substrate for microbial colonisation in aquatic environments. Their small size, rough surface, hydrophobic properties, and persistence are promoting the development of minute ecological niches which are referred to as the "Plastisphere"(He et al. 2022). Microplastics in aquatic environments are being colonised by pathogenic bacteria such as *Vibrio, Campylobacter*, and *Arcobacter* (Kelly et al. 2020). In urban areas eutrophication, heavy metals, biocides, pharmaceuticals, persistent organic pollutants, and wastewater are leading to low dissolved oxygen in aquatic environments (Huang et al. 2017). Low dissolved oxygen favors the survival of *Campylobacter* in aquatic environments (Teh et al. 2017; Otigbu et al. 2018; Shagieva et al. 2021). *Campylobacter* species are microerophilic hence they are likely to survive in polluted aquatic environments characterized by low dissolved oxygen (Teh et al. 2017; Otigbu et al. 2018; Shagieva et al. 2021). Furthermore, nutrient levels also influence the survival of *Campylobacter*. Fecal contamination of aquatic environments by agriculture effluents, wild animals, livestock, or wastewater often creates nutrient-rich environments that enhance the survival of *Campylobacter* (Nilsson et al. 2018).

3.4.3 Seasonality and Survival of Campylobacter Species in Aquatic Environments

Apart from physico-chemical characteristics of the water, environmental factors such as weather conditions, also influence the occurrence and survival of *Campylobacter spp.* (Strakova et al. 2022). Many studies have reported a high occurrence of *Campylobacter* in autumn and winter; and a lower occurrence in summer and spring in aquatic environments (Horman et al. 2004; Abulreesh et al. 2006; Strakova et al. 2022). Low temperatures in aquatic environments (below 18°C) in autumn and winter promote the survival of *Campylobacter spp.* (Wilkes et al. 2011; Strakova et al. 2022). Consequently, *Campylobacter* recovery from aquatic environments is higher during the colder months of autumn and winter due to lower water temperatures, while it declines in spring and summer (Strakova et al. 2022). Higher temperatures in summer and spring may hinder the survival of *Campylobacter* in aquatic environments (Horman et al. 2004; Abulreesh et al. 2006; Strakova et al. 2022). However, a high prevalence of *Campylobacter* has also been observed in hotter months and this may be attributed to feacal contamination by livestock, wild animals, and migratory birds (Mughini-Gras et al. 2016; Andrzejewska et al. 2022; Meinersmann et al. 2023). Furthermore, heavy rainfall contributes to the contamination of aquatic environments with *Campylobacter*. Run-off transports organisms such as *Campylobacter* from human settlements, agriculture, or wastewater treatment plants (WWTPs) into aquatic environments. Studies have reported an increase in *Campylobacter* concentration in rivers after heavy rainfall (Rechenburg and Kistemann 2009; Henry et al. 2015).

3.4.4 Climate Change

The global climate plays a key role in the distribution, emergence, and re-emergence of infectious pathogens (Yang et al. 2012). Increasing temperature, changes in climate patterns, and extreme weather are affecting the occurrence of pathogens in aquatic environments and their transmission to humans. Extreme conditions such as rainfall, storms, and floods may lead to surface water contamination by overflowing sewers and wastewater treatment plants or agricultural runoffs, while changes in temperature and salinity may optimise conditions for pathogen survival in aquatic environments (Edelson et al. 2022). Extreme weather patterns such as higher frequency and intensity of rainfall lead to an increase in the movement of *Campylobacter* from solid waste, sewer lines, WWTW, livestock farms, and the community into aquatic environments. Run-off carries along the faecal matter of livestock and wild animals into water bodies (Sterk et al. 2013). Studies have reported a higher prevalence of *Campylobacter spp.* at sites impacted by wastewater and lower prevalence at reference sites (Rechenburg and Kistemann 2009; Mughini-gras et al. 2016; Meinersmann et al. 2019b). Additionally, extreme weather events can damage wastewater and sewage infrastructure leading to spillage of polluted water in aquatic environments (Magnano San Lio et al. 2023). Furthermore, climate change is leading to droughts in many parts of the world. Droughts affect the quantity of water in aquatic environments resulting in low water levels. Consequently, this may affect the ability of aquatic environments to dilute pollutants introduced from various sources (Yang et al. 2012; Sterk et al. 2013).

Climate change is likely to affect the waterborne disease burden as human exposure to water pathogens through recreation or drinking will increase (Sterk et al. 2013). Furthermore, climate change-induced extreme weather events will also increase the disease burden through the spread of waterborne bacteria by flooding. Changes in temperature and salinity may also lead to optimisation conditions for pathogens and the expansion of species to new geographic areas (Edelson et al. 2022). This subsequently leads to an increase in the waterborne disease burden. increased pressure on health systems (Edelson et al. 2022; Magnano San Lio et al. 2023). The high disease burden may consequently increase the use of antibiotics in humans, animals, and plants leading to an increase in antibiotic resistance among pathogens such as *Campylobacter* (Magnano San Lio et al. 2023).

Temperature change also influences bacterial growth, survival, and adaptation in the environment. The increased bacterial growth rates associated with higher temperatures in some bacterial species may lead to an increase in horizontal gene transfer (HGT) (Magnano San Lio et al. 2023). Additionally, warmer climatic conditions influence the concentration of heavy metals or biocides in water (Magnano San Lio et al. 2023). These pollutants may be introduced into aquatic environments during flooding (Burnham 2021). Consequently, heavy metals may trigger antibiotic resistance through co-resistance mechanisms (Magnano San Lio et al. 2023). The increase in antibiotic resistance in the environment will necessitate the progressive use of broad-spectrum antibiotics in both human and animal medicine (Burnham 2021).

3.4.5 Contribution of Livestock Production and Wild Animals on the Emergence of Campylobacter Spp. *In Aquatic Environments*

Livestock production and wild animals also contribute to the emergence of *Campylobacter* in aquatic environments. Pathogens such as *Campylobacter* are excreted in large concentrations in animal excreta and consequently into the environment (Khan et al. 2014). Grazing animals, applications of manure on agricultural land, run-off from livestock rearing farms, and effluent from abattoirs often contaminate aquatic environments with bacteria (Vereen et al. 2013; Khan et al. 2014; Rukambile et al. 2019). A high prevalence of *Campylobacter* has been observed in aquatic environments close to farms, grazing fields, or fields where manure is applied as fertiliser (Mulder et al. 2021). Other studies have traced the origin of *Campylobacter* species in river water using molecular methods such as Multi-locus Sequence Typing (MLST)

(Mughini-Gras et al. 2016; Kobayashi et al. 2022). *Campylobacter* strains from water have been attributed to wild birds and livestock sources (Table 3.2). Furthermore, high antibiotic use in livestock production may lead to the introduction of *Campylobacter* and antibiotic-resistant genes in aquatic environments (Pillay et al. 2020; Sithole et al. 2021). It is estimated that about 30 – 90% of the antibiotics administered in animals are excreted in feacal matter and urine (Huygens et al. 2021). The antibiotics often pass through the animal bodies unaltered or they are excreted as active metabolites. Contamination of aquatic environments with animal excreta may therefore introduce antibiotic residues and antibiotic-resistant bacteria (Huygens et al. 2021).

Animal sources are therefore playing a role in the spread of *Campylobacter* in aquatic environments (Table 3.2). Additionally, there is excessive use of antibiotics in livestock production where antibiotics are used for therapeutic or prophylactic purposes and/or as growth promoters (Igwaran and Okoh 2020; Pillay et al. 2020). The antibiotics used in humans and animals are very similar in action as they are usually from the same classes (Finley et al. 2013; Pillay et al. 2020; Sithole et al. 2021). Consequently, antibiotics such as the third and fourth-generation cephalosporins, fluoroquinolones, and macrolides are listed by WHO as critically important antibiotics in both human and animal medicine but bacteria pathogens in aquatic environments are demonstrating high resistance to the classes of antibiotics (Finley et al. 2013).

3.4.6 Influence of Raw or Poorly Treated Wastewater Discharge on the Emergence of Campylobacter Species in Aquatic Environments

Wastewater treatment plants (WWTPs) are also another key driver of the emergence of *Campylobacter* in aquatic environments. Aquatic environments near wastewater are at high risk of *Campylobacter* contamination if the treatment process is not efficient at removing pathogens (Karikari et al. 2016; Tapela and Rahube 2019; Chala et al. 2021). A high prevalence of *Campylobacter* has been reported at some wastewater discharge points indicating that some WWTPs may not effectively completely remove bacteria pathogens (Mulder et al. 2021; Magnano San Lio et al. 2023). Furthermore, studies have demonstrated genetic similarities between isolates from wastewater and aquatic sources (Rechenburg and Kistemann 2009; Pitkanen and Hanninen 2017; Meng et al. 2018; Meinersmann et al. 2019b; Mulder et al. 2021). For example 15 of 54 *Campylobacter* isolates from recreational waters were observed to have known STs (sequence types), which were each detected in isolates from WWTP discharge points (T2654) and agricultural waters (ST137) (Mulder et al. 2021). Furthermore, effluent from WWTPs may contain residual nutrients such as phosphates and nitrates which may promote the growth of Campylobacter (Henry et al. 2015; Zaman and Sizemore 2017; Phungela et al. 2022). Additionally, residual

Table 3.2 Attribution of *Campylobacter* in aquatic environments to zoonotic sources

Attribution of livestock/wildlife as a source of *Campylobacter* in aquatic environments	Reference
A high prevalence of *Campylobacter* (77%) in agricultural run-off in the Netherlands was reported. *Campylobacter jejuni* and *Campylobacter coli* isolates from aquatic environments were attributed to wild birds (*C. jejuni*–60.0%; *C. coli*–93.7%)	Mulder et al. (2021)
Campylobacter strains from water samples in Luxembourg were attributed to wild birds (61.0%), poultry (18.8%), ruminants (15.9%), and pigs (4.3%). Similarly, *Campylobacter* strains from Danish aquatic environments, and were attributed to poultry (51.7%), wild birds (37.3%), ruminants (9.8%), and pigs (1.2%)	Mughini-Gras et al. (2016)
Campylobacter occurrence in a freshwater lake in Japan was linked to migratory geese. Isolates from water and geese fecal samples were genetically identical demonstrating *Campylobacter* originated from the geese	Kobayashi et al. (2022)
An increase in *Campylobacter* concentration in rivers was linked to wild animals migrating wild birds	Fravalo et al. (2011) Strakova et al. (2021) Andrzejewska et al. (2022)
Campylobacter strains in aquatic environments were attributed to livestock and wild animals	Rodríguez-Martínez et al. (2013) Kobayashi et al. (2022)
C. jejuni, *C. coli*, and *C. lari* were detected in river water samples from river basins where there is intensive livestock production. *Campylobacter spp.* detection was higher at sites impacted by livestock production compared to the reference sites in the river	Khan et al. (2014)
Campylobacter jejuni from water samples collected from the Salmon River in British Columbia were genetically identical to *Campylobacter jejuni* from faecal samples of wildlife and domestic animals	Jokinen et al. (2010)
A higher prevalence (58%) of *Campylobacter* was reported in streams near cattle farms than in forested land (30%)	Meinersmann et al. (2019)
Detection of *Campylobacter* was high at sites influenced by poultry production as compared to reference sites on the river	Vereen et al. (2013)

Source Authors

antibiotics from WWTPs discharged into surface waters may act as selective pressure on bacteria present, potentially enabling the survival and propagation of antibiotic *Campylobacter* strains in the water (Li et al. 2015; Andleeb et al. 2020). This explains the high level of resistance found in *Campylobacter* isolated from aquatic environments in many parts of the world (Table 3.1). However, it is important to note that a high level of resistance in *Campylobacter* from aquatic environments can occur even in the absence of exposure to antibiotics (Luangtongkum et al. 2009; Abraham et al. 2020). Residual antibiotic resistance genes contained in discharged effluents can be picked up by *Campylobacter* in the environment, through horizontal gene transfer (Koch et al. 2021; Meinersmann et al. 2023). The continual release of ARGs from WWTPs effluent can lead to the proliferation of antibiotic-resistant bacteria in river water (Henry et al. 2015; Zaman and Sizemore 2017; Phungela et al. 2022). Lastly, wastewater discharge into aquatic environments has the potential to alter the composition and diversity of microbial communities in receiving waters (Bengtsson-Palme et al. 2019). Shifts in the structure of these microbial communities can establish environmental conditions that support the survival and proliferation of *Campylobacter spp* (Bronowski et al. 2014; Bengtsson-Palme et al. 2019; Kreling et al. 2020).

Generally, the emergence of *Campylobacter* species in aquatic environments is influenced by several interconnected factors as demonstrated in Fig. 3.1.

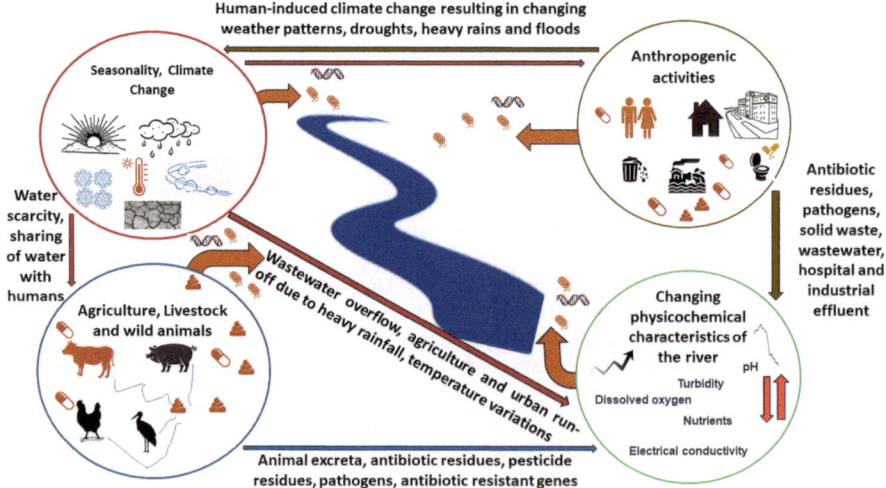

Fig. 3.1 Relationship between factors that influence the occurrence of *Campylobacter* species in aquatic environments. *Source* Authors

3.5 Public Health Implications of the Emergence of *Campylobacter Spp.* In Aquatic Environments

Campylobacter species are one of the leading bacterial causes of human gastroenteritis. Due to the increasing rates of human campylobacteriosis, *Campylobacter* is considered a serious public health concern. In high-income countries (HICs), human *Campylobacter* infections are mainly due to consumption of the contaminated meat or meat products while in low- and middle-income countries (LMICs), human infections are mainly attributed to contaminated environments, food, water, and living in close proximity with livestock (Rukambile et al. 2019). However, the emergence of *Campylobacter* in aquatic environments as demonstrated in the previous sections shows that aquatic environments may serve as routes of exposure for humans (Mughini-Gras et al. 2016). This is a public health concern as outbreaks of *Campylobacter* have previously been reported in many parts of the world (Table 3.3). Outbreaks have rarely been reported in Asia and the Middle East. The incidences are also said to be underreported in these regions but local prevalence studies suggest that *Campylobacter* infections are endemic in these regions (Kaakoush et al. 2015). Sporadic cases of *Campylobacter* due to recreational exposure have also been reported in the USA, Taylor et al. (2013), in Denmark, Kuhn et al. (2022), and in Canada, Ravel et al. (2017). Aquatic environments serve as sources of water for domestic purposes, recreation, irrigation, livestock watering, and spiritual/cultural purposes. The emergence of *Campylobacter in* aquatic environments is a public health concern because *Campylobacter* in aquatic environments can be transmitted to humans through drinking contaminated river water, recreational exposure, consumption of crops irrigated with contaminated water, and zoonotic exposure (Mughini-gras et al. 2016).

3.6 Monitoring and Management Strategies for Preventing the Emergence of *Campylobacter Spp.* In Aquatic Environments

To counteract the emergence of *Campylobacter* in aquatic environments, proposed monitoring and intervention measures are presented herewith. Monitoring the prevalence and concentration of *Campylobacter* in surface water reservoirs will enable prompt interventions to protect the environment and public health. Routine monitoring of *Campylobacter spp.* in the aquatic environment is among the keys to mitigating their emergence in aquatic environments. However, routine monitoring of water quality using indicator organisms, such as *Escherichia coli*, intestinal *Enterococci* (faecal *Streptococci*), and *Clostridium perfringens* is often used to indicate the presence of *Campylobacter* (Pitkänen 2013). Contrary, the presence of indicator organisms in water does not always correlate with the presence of *Campylobacter spp.* High numbers of *Campylobacter* have been detected in water samples even

Table 3.3 Outbreaks of waterborne diseases linked to *Campylobacter* species

Year	Nature of outbreak	Pathogen implicated	Country	Reference
1998–2011	*Campylobacter* was linked to 19% of the recorded waterborne outbreaks between during this period	*Campylobacter jejuni*	Finland	Revez et al. (2014)
2017	An outbreak involving 39 cases was linked to contaminated municipal water supply	*Campylobacter jejuni*	Nebraska, USA	Liu et al. (2022)
2019	An outbreak of waterborne campylobacteriosis resulted in over 2000 sick people, 67 hospitalised cases, and two deaths	*Campylobacter jejuni*	Norway	Mortensen et al. (2021)
1997–2008	13 outbreaks involving 276 cases were recorded in this period. These outbreaks were linked to drinking water and recreational water	*Campylobacter spp.*	USA	Taylor et al. (2013)
2004–2012	25 waterborne outbreaks were recorded. Among these outbreaks, 76% (21/25) were linked to consumption of contaminated drinking water and 24% (4/25) were due to recreational water exposure	*Campylobacter spp.*	USA	Geissler et al. (2017)
1970–2014	22 outbreaks linked to small non-community drinking water systems were recorded during this period	*Campylobacter jejuni*	Canada and USA	Pons et al. (2015)

(continued)

Table 3.3 (continued)

Year	Nature of outbreak	Pathogen implicated	Country	Reference
2016	An outbreak associated with contaminated water supply was linked to 6260 and 8320 cases of illness. 953 cases were reported by physicians while 42 people were hospitalised, 4 people died and 3 people developed Guillain-Barré syndrome	*Campylobacter spp.*	Havelock North, New Zealand,	Gilpin et al. (2020)
2000	Water outbreak associated with contaminated water from a well. Run-off after heavy rainfall contaminated the shallow well with livestock faecal matter. The outbreak led to 6 deaths, 27 cases of the haemolytic uraemic syndrome, and about 2300 sick people	*E. coli* O157:H7 and *Campylobacter spp.*	Walkerton, Ontario, Canada	Garg et al. (2008)
1998	An outbreak of campylobacteriosis linked to drinking non-chlorinated municipal tap resulted in 2700 illnesses	*Campylobacter jejuni*	Finland	Kuusi et al. (2005)
2019	Gastroenteritis outbreak due to contaminated drinking water supply resulted in 638 cases	Norovirus, *Campylobacter jejuni, E. coli*	Northern Greece	Tzani (2020)

Source Authors

when *E. coli* was not detected, or low concentrations of *E. coli* were detected. Therefore, pathogens such as *Campylobacter* should be included in the routine monitoring of water (Pitkänen 2013).

The nature and characteristics of *Campylobacter spp.* have contributed to difficulty in monitoring. Culture methods are the benchmark (Standardization ISO, 2005) for the detection of *Campylobacter* in samples. For instance, *Campylobacter* is a fastidious microorganism. It is microaerophilic, has a slow growth rate, and requires special growth requirements and this has made its isolation laborious and costly (Jokinen et al. 2012). In addition to this, the samples have to be concentrated and undergo pre-enrichment before culturing. Although enrichment significantly improves recovery efficiencies, it makes culture-based methods time-consuming (about 9 days) and expensive (Henry et al. 2015). Furthermore, the concentration of *Campylobacter* in aquatic environments is usually lower compared to high numbers of *Campylobacter* detected in clinical (stool) samples and wastewater (Jokinen et al. 2012; Henry et al. 2015; Ferrari et al. 2019). Therefore, large volumes of water (> 1000 ml) need to be filtered to detect *Campylobacter* in aquatic environments (Pitkänen 2013; Henry et al. 2015). Additionally, culture-based methods are also unable to detect viable but non-culturable (VBNC) forms of *Campylobacter*. The VBNC are usually present in environmental water due to environmental stress and if they are not detected, this can lead to false-negative results (Pitkänen 2013).

Fortunately, molecular methods such as conventional polymerase chain reaction (PCR) and quantitative PCR may overcome the limitations of culture-based methods, quantify the concentration, and even detect the presence of VBNC form (Cangelosi and Meschke 2014; Yuan et al. 2018). Additionally, microbial source tracking and characterisation can be used to identify the potential point and diffuse sources of *Campylobacter* contamination, and host-specific *Campylobacter* strains in aquatic environments (Ahmed et al. 2018; Gitter et al. 2020). Microbial source tracking has been used to detect host-associated genetic markers consequently determining the origins of *Campylobacter spp.* in aquatic environments (Paruch et al. 2020).

Furthermore, employing management intervention strategies to prevent contamination of aquatic environments with *Campylobacter* is essential in preventing the emergence of *Campylobacter* in aquatic environments. Additionally, adopting sound livestock management practices and promoting prudent use of antibiotics in livestock production will play a role in preventing the emergence of *Campylobacter* and antibiotic-resistant genes in aquatic environments (Donkor et al. 2012). Improving wastewater treatment through efficient filtration, disinfection, and UV irradiation techniques and employing nature-based solutions can help in removing *Campylobacter*, antibiotic-resistant genes, and antibiotic residues from wastewater and subsequently the aquatic environment (López-Gálvez et al. 2018; Andleeb et al. 2020; Bai et al. 2022).

Other intervention strategies include proper management of solid waste, and strict enforcement of regulations for discharge of wastewater (Mbanga et al. 2020). Increasing public education/awareness on the adverse effects of microbial pollution, and more research/funding are also key for mitigating the emergence of *Campylobacter* in aquatic environments and safeguarding public health. Overall interdisciplinary partnerships among public health authorities, environmentalists, stakeholders in agriculture and livestock production, and local communities are key for developing comprehensive strategies for water quality management in aquatic environments. Management and intervention strategies adopting a holistic approach involving policy and regulatory measures, and practical interventions are key for preventing the emergence of *Campylobacter spp.* in aquatic environments.

3.7 Conclusion and Future Perspectives

Campylobacter spp. is an emerging pathogen in aquatic environments. Its emergence is mainly driven by anthropogenic activities in the catchment, climate change, and changing physico-chemical conditions of aquatic environments. These complex interrelated factors are promoting the emergence of *Campylobacter* in aquatic environments and have implications on public health. Therefore, monitoring of *Campylobacter* species and interventions for preventing the emergence of *Campylobacter* in aquatic environments need to involve a multi-sectoral and interdisciplinary approach to safeguard public health. Effective intervention strategies and collaborative efforts from stakeholders involved in wastewater treatment, solid waste management, environment awareness, livestock production, environmental monitoring, and surveillance need to be strengthened. Additionally, communities need to be educated about the health risks associated with contamination of aquatic environments. The findings from this study offer valuable insight into the intricate factors influencing the emergence of *Campylobacter* in surface water environments. This key finding advances knowledge by illuminating the interconnected processes and drivers of *Campylobacter* emergence in aquatic environments. Various intervention strategies for reducing *Campylobacter* contamination of aquatic environments have been suggested. These include implementing prevention of microbial pollution at point and non-point sources, improving wastewater treatment processes, and prudent use of antibiotics in human/veterinary medicine. Future research to investigate the effectiveness of these strategies and explore additional interventions for controlling *Campylobacter* in aquatic environments is required.

References

Abraham S, Sahibzada S, Hewson K, et al (2020) Emergence of fluoroquinolone-resistant *Campylobacter jejuni* and *Campylobacter coli* among Australian chickens in the absence of fluoroquinolone use Sam. Appl Environ Microbiol 8. https://doi.org/10.1128/AEM.02765-19

Abulreesh HH, Paget TA, Goulder R (2006) *Campylobacter* in waterfowl and aquatic environments: Incidence and methods of detection. Environ Sci Technol 40:7122–7131. https://doi.org/10.1021/es0603271

Ahmed W, Hamilton KA, Lobos A et al (2018) Quantitative microbial risk assessment of microbial source tracking markers in recreational water contaminated with fresh untreated and secondary treated sewage. Environ Int 117:243–249. https://doi.org/10.1016/j.envint.2018.05.012

Aksomaitiene J, Ramonaite S, Tamuleviciene E et al (2019) Overlap of antibiotic resistant *Campylobacter jejuni* MLST genotypes isolated from humans, broiler products, dairy cattle and wild birds in Lithuania. Front Microbiol 10:1–8. https://doi.org/10.3389/fmicb.2019.01377

Almeida AR, Tacão M, Soares J, et al (2021) Tetracycline-resistant bacteria selected from water and Zebrafish after antibiotic exposure. Int J Environ Res Public Health 18:. https://doi.org/10.3390/ijerph18063218

Andleeb S, Majid M, Sardar S (2020) Chapter 18—Environmental and public health effects of antibiotics and AMR/ARGs. Elsevier Inc.

Andrzejewska M, Grudlewska-Buda K, Spica D, et al (2022) Genetic relatedness, virulence, and drug susceptibility of *Campylobacter* isolated from water and wild birds. Front Cell Infect Microbiol 12:1–10. https://doi.org/10.3389/fcimb.2022.1005085

Asuming-Bediako N, Kunadu AP-H, Abraham S, Habib I (2019) *Campylobacter* at the human – food Interface: The African Perspective. Pathogens 8:1–30. https://doi.org/10.3390/pathogens 8020087

Bai S, Wang X, Zhang Y et al (2022) Constructed wetlands as nature-based solutions for the removal of antibiotics: performance, microbial response, and emergence of antimicrobial resistance (AMR). Sustainability 14:1–12. https://doi.org/10.3390/su142214989

Banting GS, Braithwaite S, Scott C et al (2016) Evaluation of various *Campylobacter*-specific quantitative PCR (qPCR) assays for detection and enumeration of Campylobacteraceae in irrigation water and wastewater via a miniaturized most-probable-number-qPCR assay. Appl Environ Microbiol 82:4743–4756. https://doi.org/10.1128/AEM.00077-16

Bastaraud A, Cecchi P, Handschumacher P, et al (2020) Urbanization and waterborne pathogen emergence in Low-Income Countries : Where and how to conduct surveys? Int J Environ Health Res 17:1–19. https://doi.org/10.3390/ijerph17020480

Bengtsson-Palme J, Milakovic M, Svecov H et al (2019) Industrial wastewater treatment plant enriches antibiotic resistance genes and alters the structure of microbial communities a. Water Res 162:437–445. https://doi.org/10.1016/j.watres.2019.06.073

Bolinger H, Kathariou S (2017) The current state of macrolide resistance in *Campylobacter spp*: Trends and impacts of resistance mechanisms. Appl Environ Microbiol 83. https://doi.org/10.1128/AEM.00416-17

Bronowski C, James CE, Winstanley C (2014) Role of environmental survival in transmission of *Campylobacter jejuni*. FEMS Microbiol Lett 356:8–19. https://doi.org/10.1111/1574-6968.12488

Bunke D, Moritz S, Brack W et al (2019) Developments in society and implications for emerging pollutants in the aquatic environment. Environ Sci Eur. https://doi.org/10.1186/s12302-019-0213-1

Burnham JP (2021) Climate change and antibiotic resistance : a deadly combination. Ther Adv Infect Dis 8:1–7. https://doi.org/10.1177/204993

Cangelosi GA, Meschke JS (2014) Dead or alive: Molecular assessment of microbial viability. Appl Environ Microbiol 80:5884–5891. https://doi.org/10.1128/AEM.01763-14

Chala G, Eguale T, Abunna F et al (2021) Identification and characterization of *Campylobacter* species in livestock, humans, and water in livestock owning households of peri-urban Addis Ababa, Ethiopia : A One Health approach. Front Public Heal 9:1–11. https://doi.org/10.3389/fpubh.2021.750551

Chukwu MO, King Abia AL, Ubomba-Jaswa E et al (2019) Characterization and phylogenetic analysis of *Campylobacter* species isolated from paediatric stool and water samples in the northwest province, south africa. Int J Environ Res Public Health 16:1–22. https://doi.org/10.3390/ijerph16122205

Cui Q, Huang Y, Wang H, Fang T (2019) Diversity and abundance of bacterial pathogens in urban rivers impacted by domestic sewage. Environ Pollut 249:24–35. https://doi.org/10.1016/j.envpol.2019.02.094

Devane ML, Moriarty EM, Wood D et al (2014) The impact of major earthquakes and subsequent sewage discharges on the microbial quality of water and sediments in an urban river. Sci Total Environ 485–486:666–680. https://doi.org/10.1016/j.scitotenv.2014.03.027

Donkor ES, Newman MJ, Yeboah-manu D (2012) Epidemiological aspects of non-human antibiotic usage and resistance : implications for the control of antibiotic resistance in Ghana. Trop Med Int Heal 17:462–468. https://doi.org/10.1111/j.1365-3156.2012.02955.x

Edelson PJ, Harold R, Ackelsberg J, et al (2022) Climate change and the epidemiology of infectious diseases in the United States. Clin Infect Dis 02114:1–7. https://doi.org/10.1093/cid/ciac697

Edokpayi JN, Enitan-folami AM, Adeeyo AO (2020) Chapter 9—Recent trends and national policies for water provision and wastewater treatment in South Africa. Elsevier Inc.

Elfadaly HA, Hassanain NA, Hassanain MA et al (2018) Evaluation of primitive ground water supplies as a risk factor for the development of major waterborne zoonosis in Egyptian children living in rural areas. J Infect Public Health 11:203–208. https://doi.org/10.1016/j.jiph.2017.07.025

Elmonir W, Vetchapitak T, Amano T, et al (2022) Survival capability of *Campylobacter upsaliensis* under environmental stresses. BMC Res Notes 15. https://doi.org/10.1186/s13104-022-05919-2

Enany S, Piccirillo A, Elhadidy M, Tryjanowski P (2021) Editorial : The role of environmental reservoirs in Campylobacter—mediated infection 11:1–4. https://doi.org/10.3389/fcimb.2021.773436

Ferrari S, Frosth S, Svensson L et al (2019) Detection of *Campylobacter spp.* in water by dead-end ultrafiltration and application at farm level. J Appl Microbiol 127:1270–1279. https://doi.org/10.1111/jam.14379

Finley RL, Collignon P, Larsson DGJ et al (2013) The scourge of antibiotic resistance: The important role of the environment. Clin Infect Dis 57:704–710. https://doi.org/10.1093/cid/cit355

Fravalo P, Denis M, Chidaine B, Me F (2011) Description and sources of contamination by *Campylobacter spp.* of river water destined for human consumption in Brittany , France ' res destine ´ e Description et origines de contamination par *Campylobacter spp.* d ' eau de rivie. 59:256–263. https://doi.org/10.1016/j.patbio.2009.10.007

Garg AX, Pope JE, Clark WF, Ouimet J (2008) Arthritis risk after acute bacterial gastroenteritis. 200–204. https://doi.org/10.1093/rheumatology/kem339

Geissler AL, Carrillo B, Swanson K, et al (2017) Increasing *Campylobacter* infections, outbreaks , and antimicrobial resistance in the United States , 2004 – 2012. 65:. https://doi.org/10.1093/cid/cix624

Ghunaim H, Behnke JM, Aigha I et al (2015) Analysis of resistance to antimicrobials and presence of virulence/stress response genes in *Campylobacter* isolates from patients with severe diarrhoea. PLoS ONE 10:1–16. https://doi.org/10.1371/journal.pone.0119268

Gilpin BJ, Walker T, Paine S et al (2020) A large scale waterborne Campylobacteriosis outbreak, Havelock North, New Zealand. J Infect 81:390–395. https://doi.org/10.1016/j.jinf.2020.06.065

Gitter A, Mena KD, Wagner KL, et al (2020) Human health risks associated with recreational waters: Preliminary approach of integrating quantitative microbial risk assessment with microbial source tracking. Water (Switzerland) 12. https://doi.org/10.3390/w12020327

Goddard MR, Brien SO, Williams N, et al (2022) A restatement of the natural science evidence base regarding the source, spread and control of *Campylobacter* species causing human disease. R Soc Publ 289:. https://doi.org/10.1098/rspb.2022.0400

Hassen B, Abbassi MS, Benlabidi S, et al (2020) Genetic characterization of ESBL-producing *Escherichia coli* and *Klebsiella pneumoniae* isolated from wastewater and river water in Tunisia : predominance of CTX-M-15 and high genetic diversity. Environ Sci Pollut Res 27:44368–44377. https://doi.org/10.1007/s11356-020-10326-w

He S, Jia M, Xiang Y et al (2022) Biofilm on microplastics in aqueous environment : Physicochemical properties and environmental implications. J Hazard Mater 424:127286. https://doi.org/10.1016/j.jhazmat.2021.127286

Henry R, Schang C, Chandrasena GI, et al (2015) Environmental monitoring of waterborne *Campylobacter*: Evaluation of the Australian standard and a hybrid extraction-free MPN-PCR method. Front Microbiol 6:. https://doi.org/10.3389/fmicb.2015.00074

Hodgson K, Manus L (2006) A drinking water quality framework for South Africa. Water SA 32:673–678. https://doi.org/10.4314/wsa.v32i5.47853

Horman A, Rimhanen-Finne R, Maunula L et al (2004) Indicator organisms in surface water in South Western Finland, 2000–2001. Appl Environ Microbiol 70:87–95. https://doi.org/10.1128/AEM.70.1.87

Huang J, Yin H, Chapra SC, Zhou Q (2017) Modelling Dissolved Oxygen Depression in an Urban. Water 9:1–19. https://doi.org/10.3390/w9070520

Hubeny J, Harnisz M, Korzeniewska E, et al (2021) Industrialization as a source of heavy metals and antibiotics which can enhance the antibiotic resistance in wastewater , sewage sludge and river water. PLoS One 1–24. https://doi.org/10.1371/journal.pone.0252691

Huygens J, Daeseleire E, Mahillon J et al (2021) Presence of antibiotic residues and antibiotic resistant bacteria in cattle manure intended for fertilization of agricultural fields : A One Health Perspective. Antibiot 2021:10

Igwaran A, Okoh AI (2019) Human campylobacteriosis: A public health concern of global importance. Heliyon 5:e02814. https://doi.org/10.1016/j.heliyon.2019.e02814

Igwaran A, Okoh AI (2020) Occurrence, virulence and antimicrobial resistance-associated markers in *Campylobacter* species isolated from retail fresh milk and water samples in two district municipalities in the Eastern Cape Province, South Africa. Antibiotics 9:1–18. https://doi.org/10.3390/antibiotics9070426

Imran M, Das KR, Naik MM (2019) Co-selection of multi-antibiotic resistance in bacterial pathogens in metal and microplastic contaminated environments : An emerging health threat. Chemosphere 215:846–857. https://doi.org/10.1016/j.chemosphere.2018.10.114

Irfan M, Almotiri A (2022) Antimicrobial resistance and its drivers — A review. Antibiotics 11:. https://doi.org/10.3390/antibiotics11101362

Iwu CD, Korsten L, Okoh AI (2020) The incidence of antibiotic resistance within and beyond the agricultural ecosystem: A concern for public health. Microbiologyopen 9:1–28. https://doi.org/10.1002/mbo3.1035

Jokinen CC, Koot JM, Carrillo CD et al (2012) An enhanced technique combining pre-enrichment and passive filtration increases the isolation efficiency of *Campylobacter jejuni* and *Campylobacter coli* from water and animal fecal samples. J Microbiol Methods 91:506–513. https://doi.org/10.1016/j.mimet.2012.09.005

Jokinen CC, Schreier H, Mauro W, et al (2010) The occurrence and sources of *Campylobacter spp.*, *Salmonella enterica* and *Escherichia coli* O157: H7 in the Salmon River, British Columbia , Canada. 374–386. https://doi.org/10.2166/wh.2009.076

Kaakoush NO, Castaño-Rodríguez N, Mitchell HM, Man SM (2015) Global epidemiology of *Campylobacter* infection. Clin Microbiol Rev 28:687–720. https://doi.org/10.1128/CMR.00006-15

Karikari AB, Obiri-danso K, Frimpong EH, Krogfelt KA (2016) Occurrence and susceptibility patterns of *Campylobacter* isolated from environmental water sources. African J Microbiol Res 10:1576–1580. https://doi.org/10.5897/AJMR2016.8296

Kashoma IP, Kassem II, John J et al (2016) Prevalence and antimicrobial resistance of *Campylobacter* isolated from dressed beef carcasses and raw milk in Tanzania. Microb Drug Resist 22:40–52. https://doi.org/10.1089/mdr.2015.0079

Kelly JJ, London MG, Oforji N, et al (2020) Microplastic selects for convergent microbiomes from distinct riverine sources. 39:281–291. https://doi.org/10.1086/708934

Khan IUH, Gannon V, Jokinen CC et al (2014) A national investigation of the prevalence and diversity of thermophilic *Campylobacter* species in agricultural watersheds in Canada. Water Res 61:243–252. https://doi.org/10.1016/j.watres.2014.05.027

Kim J, Oh E, Kim J, Jeon B (2015) Regulation of oxidative stress resistance in *Campylobacter jejuni*, a microaerophilic foodborne pathogen. Front Microbiol 6:1–12. https://doi.org/10.3389/fmicb.2015.00751

Kobayashi M, Zhang Q, Segawa T et al (2022) Temporal dynamics of *Campylobacter* and *Arcobacter* in a freshwater lake that receives fecal inputs from migratory geese. Water Res 217:118397. https://doi.org/10.1016/j.watres.2022.118397

Koch N, Islam NF, Sonowal S et al (2021) Environmental antibiotics and resistance genes as emerging contaminants : Methods of detection and bioremediation. Curr Res Microb Sci 2:100027. https://doi.org/10.1016/j.crmicr.2021.100027

Kovanen S, Kivistö R, Llarena A-K et al (2016) Tracing isolates from domestic human *Campylobacter jejuni* infections to chicken slaughter batches and swimming water using whole-genome multilocus sequence typing. Int J Food Microbiol 226:53–60. https://doi.org/10.1016/j.ijfood micro.2016.03.009

Kreling V, Falcone FH, Kehrenberg C, Hensel A (2020) *Campylobacter sp*: Pathogenicity factors and prevention methods—new molecular targets for innovative antivirulence drugs? Appl Microbiol Biotechnol 104:10409–10436. https://doi.org/10.1007/s00253-020-10974-5

Kuhn KG, Nielsen EM, Mølbak K, Ethelberg S (2018) Determinants of sporadic *Campylobacter* infections in Denmark: a nationwide case-control study among children and young adults. Clin Epidemiol 10:1695–1707. https://doi.org/10.2147/CLEP.S177141

Kuusi M, Nuorti JP, Ha¨nninen M-L, et al (2005) A large outbreak of campylobacteriosis associated with a municipal water supply in Finland. Epidemiol Infect 133:593–601. https://doi.org/10.1017/S0950268805003808

Laffite A, Kilunga PI, Kayembe JM et al (2016) Hospital effluents are one of several sources of metal, antibiotic resistance genes, and bacterial markers disseminated in Sub-Saharan urban rivers. Front Microbiol 7:1–14. https://doi.org/10.3389/fmicb.2016.01128

Li J, Cheng W, Xu L et al (2015) Antibiotic-resistant genes and antibiotic-resistant bacteria in the effluent of urban residential areas, hospitals, and a municipal wastewater treatment plant system. Environ Sci Pollut Res 22:4587–4596. https://doi.org/10.1007/s11356-014-3665-2

Liu F, Lee SA, Xue J, Riordan SM (2022) Global epidemiology of campylobacteriosis and the impact of COVID-19. Front Cell Infect Microbiol 12:. https://doi.org/10.3389/fcimb.2022.979055

López-Gálvez F, Randazzo W, Vásquez A et al (2018) Irrigating Lettuce with Wastewater Effluent: Does Disinfection with Chlorine Dioxide Inactivate Viruses? J Environ Qual 47:1139–1145. https://doi.org/10.2134/jeq2017.12.0485

Luangtongkum T, Jeon B, Han J, et al (2009) Antibiotic resistance in *Campylobacter*: emergence, transmission and persistence 4:189–200. https://doi.org/10.2217/17460913.4.2.189

Magana-Arachchi DN, Wanigatunge RP (2020) Ubiquitous waterborne pathogens. Elsevier

Magnano San Lio R, Favara G, Maugeri A, et al (2023) How antimicrobial resistance is linked to climate change : An overview of two intertwined global challenges. Int J Environ Res Public Health 20:1681. https://doi.org/10.3390/ijerph20031681

Makaya E, Rohse M, Day R, et al (2020) Water governance challenges in rural South Africa : Exploring institutional coordination in drought management. 22:519–540. https://doi.org/10.2166/wp.2020.234

Mbanga J, Abia ALK, Amoako DG, Essack SY (2020) Quantitative microbial risk assessment for waterborne pathogens in a wastewater treatment plant and its receiving surface water body. BMC Microbiol 20:1–12. https://doi.org/10.1186/s12866-020-02036-7

Mboweni TJ, De Crom EP (2016) A narrative interpretation of the cultural impressions on water of the communities along the Vaal River, Parys, Free State. J Transdiscipl Res South Africa 12:1–7. https://doi.org/10.4102/td.v12i1.345

Meinersmann RJ, Snyder BJ, Berrang ME et al (2019) Recovery of thermophilic Campylobacter by three sampling methods from river sites in Northeast Georgia, USA, and their antimicrobial resistance genes. Lett Appl Microbiol 71:102–107. https://doi.org/10.1111/lam.13224

Meinersmann RJ, Berrang ME, Shariat NW, et al (2023) Despite shared geography, *Campylobacter* isolated from surface water are genetically distinct from *Campylobacter* isolated from chickens. Microbiol Spectr 11. https://doi.org/10.1128/spectrum.04147-22

Meng Z, Chandrasena G, Henry R et al (2018) Stormwater constructed wetlands : A source or a sink of *Campylobacter*. Water Res 131:218–227. https://doi.org/10.1016/j.watres.2017.12.045

Mortensen N, Jonasson SA, Lavesson IV et al (2021) Characteristics of hospitalized patients during a large waterborne outbreak of *Campylobacter jejuni* in Norway. PLoS ONE 16:1–11. https://doi.org/10.1371/journal.pone.0248464

Mpondo L, Ebomah KE, Okoh AI (2021) Multidrug-resistant *listeria* species shows abundance in environmental waters of a key district municipality in South Africa. Int J Environ Res Public Health 18:1–12. https://doi.org/10.3390/ijerph18020481

Mughini-gras L, Penny C, Ragimbeau C et al (2016) Quantifying potential sources of surface water contamination with *Campylobacter jejuni* and *Campylobacter coli*. Water Res 101:36–45. https://doi.org/10.1016/j.watres.2016.05.069

Mulder AC, Franz E, De RS et al (2021) Tracing the animal sources of surface water contamination with *Campylobacter jejuni* and *Campylobacter coli*. Water Res 187:116421. https://doi.org/10.1016/j.watres.2020.116421

Nilsson A, Johansson C, Skarp A, Ren E (2018) Survival of *Campylobacter jejuni* and *Campylobacter coli* water isolates in lake and well water. 762–770. https://doi.org/10.1111/apm.12879

Njage PMK, Buys EM (2017) Quantitative assessment of human exposure to extended spectrum and AmpC β-lactamases bearing *E. coli* in lettuce attributable to irrigation water and subsequent horizontal gene transfer. Int J Food Microbiol 240:141–151. https://doi.org/10.1016/j.ijfood micro.2016.10.011

Okada A, Tsuchida M, Rahman MM, Inoshima Y (2022) Two-round treatment with propidium monoazide completely inhibits the detection of dead *Campylobacter spp.* cells by quantitative PCR. Front Microbiol 13:1–7. https://doi.org/10.3389/fmicb.2022.801961

On SLW, Miller WG, Houf K, et al (2017) Minimal standards for describing new species belonging to the families Campylobacteraceae and Helicobacteraceae 5296–5311. https://doi.org/10.1099/ijsem.0.002255

Otigbu AC, Clarke AM, Fri J, et al (2018) Antibiotic sensitivity profiling and virulence potential of *Campylobacter Jejuni* isolates from estuarine water in the Eastern Cape Province, South Africa. Int J Environ Res Public Health 15. https://doi.org/10.3390/ijerph15050925

Paruch L, Paruch AM, Sørheim R (2020) DNA-based faecal source tracking of contaminated drinking water causing a large *Campylobacter* outbreak in Norway 2019. Int J Hyg Environ Health 224:113420. https://doi.org/10.1016/j.ijheh.2019.113420

Phungela TT, Maphanga T, Chidi BS, et al (2022) The impact of wastewater treatment effluent on Crocodile River quality in Ehlanzeni District, Mpumalanga Province, South Africa. S Afr J Sci 118:1–8. https://doi.org/10.17159/sajs.2022/12575

Pillay S, Amoako DG, Abia ALK, et al (2020) Characterisation of *Campylobacter spp.* isolated from poultry in Kwazulu-Natal, South Africa. Antibiotics 9:. https://doi.org/10.3390/antibiotics9020042

Pitkänen T (2013) Review of *Campylobacter spp.* in drinking and environmental waters. J Microbiol Methods 95:39–47. https://doi.org/10.1016/j.mimet.2013.06.008

Pitkanen T, Hanninen M-L (2017) Members of the family Campylobacteraceae : *Campylobacter jejuni, Campylobacter coli*. Glob Water Pathog Proj. http://hdl.handle.net/10138/232581

Pons W, Young I, Truong J, et al (2015) A Systematic review of waterborne disease outbreaks associated with small non- community drinking water systems in Canada and the United States. 1–17. PLoS ONE 10(10): e0141646. https://doi.org/10.1371/journal.pone.0141646

Profitós JMH, Mouhaman A, Lee S, Garabed R (2014) Muddying the waters : A new area of concern for drinking water contamination in Cameroon. Int J Environ Res Public Health 12454–12472. https://doi.org/10.3390/ijerph111212454

Radwan EK, Ghafar HHA, Ibrahim MBM, Moursy AS (2023) Recent trends in treatment technologies of emerging contaminants. Environ Qual Manag 32:7–25. https://doi.org/10.1002/tqem.21877

Raulo S, Llarena A, Kovanen S, Kivist R (2015) Antimicrobial resistance and multilocus sequence types of Finnish *Campylobacter jejuni* isolates from multiple sources. Zoonoses Public Health 63:10–19. https://doi.org/10.1111/zph.12198

Ravel A, Hurst M, Petrica N et al (2017) Source attribution of human campylobacteriosis at the point of exposure by combining comparative exposure assessment and subtype comparison based on comparative genomic fingerprinting. PLoS ONE 12:1–21. https://doi.org/10.1371/journal.pone.0183790

Rechenburg A, Kistemann T (2009) Sewage effluent as a source of *Campylobacter spp.* in a surface water catchment. Int J Environ Health Res 19:239–249. https://doi.org/10.1080/096031208024 60376

Reddy S, Zishiri OT (2018) Genetic characterisation of virulence genes associated with adherence, invasion and cytotoxicity in *Campylobacter spp.* Isolated from commercial chickens and human clinical cases. Onderstepoort J Vet Res 85:1–9. https://doi.org/10.4102/ojvr.v85i1.1507

Reddy S, Kaur K, Barathe P et al (2022) Antimicrobial resistance in urban river ecosystems. Microbiol Res 263:127135. https://doi.org/10.1016/j.micres.2022.127135

Revez J, Llarena A, Schott T et al (2014) Genome analysis of *Campylobacter jejuni* strains isolated from a waterborne outbreak. BMC Genomics 15:1–8

Rodgers K, Mclellan I, Peshkur T et al (2019) Can the legacy of industrial pollution influence antimicrobial resistance in estuarine sediments ? Environ Chem Lett 17:595–607. https://doi.org/10.1007/s10311-018-0791-y

Rodríguez-Martínez S, Cervero-Aragó S, Gil-Martin I, Araujo R (2013) Multilocus sequence typing of *Campylobacter jejuni* and *Campylobacter coli* strains isolated from environmental waters in the Mediterranean area. Environ Res 127:56–62. https://doi.org/10.1016/j.envres.2013.10.003

Rukambile E, Alders R, Rukambile E, Alders R (2019) Infection, colonization and shedding of Campylobacter and Salmonella in animals and their contribution to human disease : A review. Zoonoses Public Heal 66:562–578. https://doi.org/10.1111/zph.12611

Saifur S, Gardner CM (2021) Loading, transport, and treatment of emerging chemical and biological contaminants of concern in stormwater. Water Sci Technol 83(12):2863–2885. https://doi.org/10.2166/wst.2021.187

Sales-ortells H, Medema G (2015) Microbial health risks associated with exposure to stormwater in a water plaza. Water Res 74:34–46. https://doi.org/10.1016/j.watres.2015.01.044

Sales-ortells H, Agostini G, Medema G (2015) Quantification of waterborne pathogens and associated health risks in urban water. Environ Sci Technol. https://doi.org/10.1021/acs.est.5b0 0625

Schang C, Lintern A, Cook PLM et al (2016) Presence and survival of culturable *Campylobacter spp.* and *Escherichia coli* in a temperate urban estuary. Sci Total Environ 569–570:1201–1211. https://doi.org/10.1016/j.scitotenv.2016.06.195

Sekwadi PG, Ravhuhali KG, Mosam A, et al (2018) Waterborne outbreak of gastroenteritis on the KwaZulu-Natal Coast , South Africa , December 2016/January 2017. Epidemiol Infect 146:1318–1325. https://doi.org/10.1017/S095026881800122X

Shagieva E, Demnerova K, Michova H (2021) Waterborne isolates of *Campylobacter jejuni* are able to develop aerotolerance, survive exposure to low temperature, and interact with *Acanthamoeba polyphaga*. 12:1–10. https://doi.org/10.3389/fmicb.2021.730858

Shobo CO, Bester LA, Baijnath S et al (2016) Antibiotic resistance profiles of *Campylobacter* species in the South Africa private health care sector. J Infect Dev Cties 10:1214–1221. https://doi.org/10.3855/jidc.8165

Siddiqee MH, Henry R, Coleman RA et al (2019) *Campylobacter* in an urban estuary: Public health insights from occurrence, HeLa Cytotoxicity, and Caco-2 attachment cum invasion. Microbes Environ 34:436. https://doi.org/10.1264/JSME2.ME19088

Silva J, Leite D, Fernandes M et al (2011) *Campylobacter spp.* As a foodborne pathogen: A review. Front Microbiol 2:1–12. https://doi.org/10.3389/fmicb.2011.00200

Sithole V, Amoako DG, Abia ALK, et al (2021) Occurrence, antimicrobial resistance, and molecular characterization of *Campylobacter spp.* In intensive pig production in South Africa. Pathogens 10:. https://doi.org/10.3390/pathogens10040439

Smith AM, Tau NP, Smouse SL et al (2019) Outbreak of *Listeria monocytogenes* in South Africa, 2017–2018: Laboratory activities and experiences associated with whole-genome sequencing analysis of isolates. Foodborne Pathog Dis 16:524–530. https://doi.org/10.1089/fpd.2018.2586

Sproston EL, Wimalarathna HML, Sheppard SK (2018) Trends in fluoroquinolone resistance in *Campylobacter*. Microb Genomics 4:1–8. https://doi.org/10.1099/mgen.0.000198

Sterk A, Schijven J, Nijs T De, Maria AHDR (2013) Direct and indirect effects of climate change on the risk of infection by water-transmitted pathogens. Environ Sci Technol 47:12648–12660. https://doi.org/10.1021/es403549s

Strakova N, Shagieva E, Ovesna P et al (2022) The effect of environmental conditions on the occurrence of *Campylobacter jejuni* and *Campylobacter coli* in wastewater and surface waters. J Appl Microbiol 132:725–735. https://doi.org/10.1111/jam.15197

Strakova N, Korena K, Gelbicova T, et al (2021) A rapid culture method for the detection of *Campylobacter* from water environments. Int J Environ Res Public Health 18. https://doi.org/10.3390/ijerph18116098

Szczepanska B, Andrzejewska M, Spica D, Klawe JJ (2017) Prevalence and antimicrobial resistance of *Campylobacter jejuni* and *Campylobacter coli* isolated from children and environmental sources in urban and suburban areas. BMC Microbiol 17:85–94. https://doi.org/10.1186/s12866-017-0991-9

Tapela K, Rahube T (2019) Isolation and antibiotic resistance profiles of bacteria from influent , effluent and downstream : A study in Botswana. 13:279–289. https://doi.org/10.5897/AJMR2019.9065

Taviani E, Berg H Van Den, Nhassengo F, et al (2022) Occurrence of waterborne pathogens and antibiotic resistance in water supply systems in a small town in Mozambique. BMC Microbiol 22:1–11. https://doi.org/10.1186/s12866-022-02654-3

Taylor EV, Herman KM, Ailes EC et al (2013) Common source outbreaks of *Campylobacter* infection in the. Epidemics 141:987–996. https://doi.org/10.1017/S0950268812001744

Teh AHT, Lee SM, Dykes GA (2017) The influence of dissolved oxygen level and medium on biofilm formation by *Campylobacter jejuni*. Food Microbiol 61:120–125. https://doi.org/10.1016/j.fm.2016.09.008

Tzani M, Mellou K, Kyritsi M, et al (2020) ' Evidence for waterborne origin of an extended mixed gastroenteritis outbreak in a town in Northern Greece , 2019 .' Epidemiol Infect 149:1–8. https://doi.org/10.1017/S0950268820002976

Van HP, Zhang J, Hayashi M et al (2011) Genetic relatedness and identification of clinical strains of genus Campylobacter based on dnaJ, 16S rRNA, groEL, and rpoB gene sequences. Microbiol Cult Collect 27:1–12

Veltcheva D, Colles FM, Varga M et al (2022) Emerging patterns of fluoroquinolone resistance in *Campylobacter jejuni* in the UK [1998 – 2018]. Microb Genomics 8:1–9. https://doi.org/10.1099/mgen.0.000875

Vereen E, Lowrance RR, Jenkins MB et al (2013) Landscape and seasonal factors influence *Salmonella* and *Campylobacter* prevalence in a rural mixed use watershed. Water Res 47:6075–6085. https://doi.org/10.1016/j.watres.2013.07.028

Vouga M, Greub G (2016) Emerging bacterial pathogens : the past and beyond. Clin Microbiol Infect 22:12–21. https://doi.org/10.1016/j.cmi.2015.10.010

Wilkes G, Edge TA, Gannon VPJ et al (2011) Associations among pathogenic bacteria, parasites, and environmental and land use factors in multiple mixed-use watersheds. Water Res 45:5807–5825. https://doi.org/10.1016/j.watres.2011.06.021

Yang K, Lejeune J, Alsdorf D, et al (2012) Global Distribution of Outbreaks of Water-Associated Infectious Diseases 6. https://doi.org/10.1371/journal.pntd.0001483

Yitayew B, Woldeamanuel Y, Asrat D, et al (2022) Antimicrobial resistance genes in microbiota associated with sediments and water from the Akaki river in Ethiopia. Environ Sci Pollut Res 70040–70055. https://doi.org/10.1007/s11356-022-20684-2

Yuan Y, Zheng G, Lin M, Mustapha A (2018) Detection of viable *Escherichia coli* in environmental water using combined propidium monoazide staining and quantitative PCR. Water Res 145:398–407. https://doi.org/10.1016/j.watres.2018.08.044

Yuan T, Vadde KK, Tonkin JD, et al (2019) Urbanization impacts the physicochemical characteristics and abundance of fecal markers and bacterial pathogens in surface water. Int J Environ Res Public Health 16:1–19. https://doi.org/10.3390/ijerph16101739

Zaman MS, Sizemore RC (2017) Freshwater resources could become the most critical factor in the future of the earth. J Mississippi Acad Sci 62:348–397

Chapter 4
Odor-Causing Compounds in Drinking Water: Occurrence and Sources in Major Cities Across China

Chunmiao Wang, Jianwei Yu, and Min Yang

Abstract An investigation of odor distribution characteristics in source and drinking water of 98 drinking water treatment plants in 31 cities across China was carried out based on a high-throughput quantification method for 95 odorants using gas chromatography-triple quadrupole tandem mass spectrometry (GC–MS/MS) and Flavor Profile Analysis. The national investigation showed that about 90% of source water samples exhibited odor problems with earthy/musty (31.8%) and swampy/septic (45.4%) odors as the dominant ones. 77 odor-causing compounds were detected in source water samples, and about 40 compounds were associated with emerging chemical pollution, including ethers, cyclic acetals, phenols, and benzene-containing compounds, etc. 2-Methylisoborneol was identified as the main earthy/musty odor-causing compound and thioethers including dimethyl disulfide and dimethyl trisulfide were mainly associated with swampy/septic odor. In addition, some cyclic acetals were first detected in Huangpu River, which were the main cause of chemical odor episodes in Huangpu River source water. Cyclic acetals and bis(2-chloro-1-ethylethyl) ether could be associated with resin-related industrial pollution, and their potential health effects are also worthy of attention.

Keywords Drinking water · Taste and odor · Odorant quantification · Odor characteristics

4.1 Introduction

Clean and safe drinking water is a basic need for human beings. Taste and odor in drinking water can be directly perceived by human, thus are often associated with the perception of water safety for consumers (Lin et al. 2019). In recent years, despite our ability to provide safe drinking water was improved substantially, the frequently occurring emerging pollutants in source drinking water led

C. Wang · J. Yu (✉) · M. Yang
Research Center for Eco-Environmental Sciences, Chinese Academy of Sciences, Beijing, China
e-mail: jwyu@rcees.ac.cn

© UNESCO 2025

67

S. Zandaryaa et al. (eds.), *Emerging Pollutants*, Advances in Water Security,
https://doi.org/10.1007/978-3-031-71758-1_4

to poor taste and odor continues to be a culprit issue for water suppliers globally. As reported, a wide range of emerging pollutants including algal and/or bacterial metabolites and industrial pollutants can induce odor issues in drinking water. Two widely documented musty/earthy odor compounds, 2-methylisoborneol (2-MIB) and geosmin, are mainly produced by cyanobacteria and actinomycetes (Suffet et al. 1999; Suurnäkki et al. 2015). Thioethers (Wang et al. 2019a; Watson and Jüttner 2016), and indoles (Suffet et al. 1999) were reported to cause swampy/septic odor or similar odors. Chemicals generated from industrial activities, such as substituted benzenes, ethers, dioxanes, and dioxolanes, could cause chemical, solvent-like, sweet or aromatic odors (Quintana et al. 2015). These compounds with various structural features have different odor characteristics as well as diverse sources.

In China, public concern has been evoked recently due to the outbreak of a serious drinking water odor crisis in Wuxi City in 2007 (Yang et al. 2008), in Qinghuangdao City (2007), as well as in Lanzhou City (2014, 2015). An investigation of source waters among 34 major cities in China showed that 80% of samples exhibited odor issues (Sun et al. 2014), while the knowledge about the composition and sources of corresponding odorants is rather limited for most odor issues, except the most widely documented earthy/musty 2-MIB and geosmin. More systematic investigation is urgently needed to choose suitable treatment technologies and improve water quality standards for delivering much tasty water. This chapter presents the occurrence and potential sources of 95 odorants in source drinking water, based on a nationwide investigation of 98 drinking water treatment plants in China using a high-throughput quantification method of gas chromatography-triple quadrupole tandem mass spectrometry (GC–MS/MS) and Flavor Profile Analysis (FPA). The investigated odorants include aldehydes, compounds containing benzene-rings and no oxygen, cyclic acetals, ethers, indoles, terpenes and terpenoids, musks, phenols, pyrazines and thiazoles, thioethers. The results provide an informative reference for water quality management in the drinking water industry and would benefit the setting of related water quality standards for drinking water in China.

4.2 Simultaneous Quantification of 95 Odorants Using GC–MS/MS

A sensitive method combining liquid–liquid extraction with gas chromatography-triple quadrupole tandem mass spectrometry (GC–MS/MS) was established to simultaneously analyze 95 odor causing compounds in drinking water. Water samples underwent liquid–liquid extraction with dichloromethane by a factor of 1000. Three deuterated analogs of target analytes, dimethyl disulfide-d_6, benzaldehyde-d_6, o-cresol-3,4,5,6-d_4 and 1,4-dioxane-d_8 were used to correct the variations in recovery, five isotope-labeled internal standards (4-chlorotoluene-d_4, 1,4-dichlorobenzene-d_4, naphthalene-d_8, acenaphthene-d_{10}, phenanthrene-d_{10} respectively) were used prior analysis to correct the variations arising from instrument fluctuations and injection errors among different samples. Analyses were performed on a GCMS-TQ8040 (Shimadzu Co., Japan) equipped with a VF-624 ms column (length, 60 m; diameter, 0.32 mm; thickness, 1.8 μm; Agilent Technologies, USA). Multiple reaction monitoring (MRM) mode was used for the quantification of the target chemicals, as shown in Table 4.1. The procedure has been described in detail in a previous study (Wang et al. 2019b).

Of the 95 compounds, the detection limits of the developed quantitative method were in the range of 0.10–100 ng/L, most of which were well below the odor threshold concentration of each odorant. The detection limits of most odorants were much improved by GC–MS/MS compared to other methods. The average recoveries of the most analytes in tap water samples were between 65 and 119%, and the method was reproductive (RSD < 20%, n = 5).

4.3 Odor Characteristics of Source Water

Odor characteristics of water samples were evaluated by Flavor Profile Analysis based on Standard Methods for the Examination of Water and Wastewater (American Public Health Association 2012). Source water samples were collected from 98 different drinking water treatment plants (DWTPs) in 31 cities throughout China from September 2015 to December 2018, covering ten watersheds including Songhua River, Liaohe River, Haihe River, Huaihe River, Yellow River, Yangtze River, Pearl River, Taihu Lake, Chaohu Lake and Dianchi Lake (Wang et al. 2021). Results showed that more than 90% source water samples exhibited odor problems, and earthy/musty (31.8%) and swampy/septic (45.4%) odors were dominant odor descriptors (Fig. 4.1). Swampy/septic odor was the major odor type in Pearl River (FPA intensity ≤ 9), Taihu Lake (FPA intensity ≤ 6), and Yangtze River (FPA intensity ≤ 7.3), with FPA average intensity of 2–3. Except Chaohu Lake and Dianchi Lake, earthy/musty odors were detected widely with odor intensity of less than 6.7. The detection frequency of swampy/septic odors and earthy/musty odors are comparable in Yellow River, Haihe River, and Huaihe River.

4.4 Occurrence of the Investigated Odorants in Source Water

A total of 75 odorants were found in raw water with concentration ranges for individual odorants from not detected (n.d.) to hundreds of or thousands of ng/L (Table 4.2). About 30 odorants showed detection frequencies of above 30%, and 19 odorants were frequently detected (freq. > 50%). In lake/reservoir source water, 66 odorants were detected with 22 odorants, including two aldehydes, six non-oxygen benzene-containing compounds, one cyclic acetal, three terpenes and terpenoids, two

Table 4.1* Parameters of GC–MS/MS in multiple reaction monitoring mode

Odorants	CAS	Target ion m/z	Ch$_1$[a] CE[b]	Reference ion 1 m/z	Ch$_2$ CE	Reference ion 2 m/z	Ch$_3$ CE
Hexanal	66-25-1	82.00 > 67.10	6	56.00 > 41.10	12	72.00 > 57.10	12
Heptanal	111-71-7	70.00 > 55.10	9	70.00 > 42.00	6	81.00 > 41.10	18
Benzaldehyde	100-52-7	105.00 > 77.10	12	106.00 > 77.10	18	77.00 > 51.10	18
2,4-Heptadienal	4313-03-5	81.00 > 53.10	18	110.00 > 81.00	6	79.00 > 77.10	12
2-Octenal	2548-87-0	83.00 > 55.10	9	70.00 > 42.00	6	70.00 > 55.10	9
Nonanal	124-19-6	82.00 > 67.10	6	70.00 > 55.10	9	98.00 > 56.10	6
2,6-Nonadienal	557-48-2	69.00 > 41.00	6	70.00 > 42.10	9	69.00 > 39.00	21
Decanal	112-31-2	82.00 > 67.10	6	71.00 > 43.10	6	82.00 > 41.10	24
2,4-Decadienal	2363-88-4	81.00 > 53.10	18	67.00 > 41.10	12	95.00 > 67.10	9
Ethylbenzene	100-41-4	106.00 > 91.10	12	91.00 > 65.10	18	91.00 > 39.10	27
Para-xylene	106-42-3	106.00 > 91.10	12	91.00 > 65.10	18	91.00 > 39.10	27
1,4-Dichlorobenzene	106-46-7	146.00 > 111.00	18	146.00 > 75.10	24	111.00 > 75.00	12
Indan	496-11-7	118.00 > 115.10	24	115.00 > 89.10	18	118.00 > 91.10	24
Isopropylbenzene	98-82-8	105.00 > 77.10	18	120.00 > 105.10	9	105.00 > 79.10	9
Nitrobenzene	98-95-3	77.00 > 51.00	15	123.00 > 77.10	15	123.00 > 93.10	6
Biphenyl	92-52-4	154.00 > 152.10	27	153.00 > 151.10	30	154.00 > 115.10	27
Phenylethylene	100-42-5	104.00 > 78.10	15	78.00 > 52.10	21	104.00 > 52.10	27
3-Methylstyrene	100-80-1	117.00 > 115.10	15	117.00 > 91.10	21	118.00 > 91.10	27
1-Methylnaphthalene	90-12-0	141.00 > 115.10	21	142.00 > 115.10	27	115.00 > 89.10	18
bis(2-Chloro-1-methylethyl)ether	108-60-1	107.00 > 41.10	18	121.00 > 45.00	6	121.00 > 41.10	18
bis(2-Chloroethyl) ether	111-44-4	93.00 > 63.00	6	95.00 > 65.00	6	63.00 > 61.00	24
2,4,6-Trichloroanisole	87-40-1	195.00 > 166.90	18	197.00 > 169.00	18	210.00 > 194.90	12

(continued)

Table 4.1* (continued)

Odorants	CAS	Target ion m/z	Ch$_1$[a] CE[b]	Reference ion 1 m/z	Ch$_2$ CE	Reference ion 2 m/z	Ch$_3$ CE
2,3,4-Trichloroanisole	54,135-80-7	195.00 > 167.00	12	197.00 > 168.90	12	210.00 > 167.00	21
2,3,6-Trichloroanisole	50,375-10-5	210.00 > 166.90	21	212.00 > 168.90	21	210.00 > 194.90	12
Pentachloroanisole	1825-21-4	265.00 > 236.80	18	280.00 > 236.80	24	280.00 > 264.80	12
2,4,6-Tribromoanisole	607-99-8	344.00 > 328.50	21	329.00 > 300.50	24	344.00 > 300.40	33
Diphenyl ether	101-84-8	170.00 > 141.10	18	141.00 > 115.10	18	170.00 > 77.10	21
Ethyl-tert-butylether	637-92-3	87.00 > 59.10	9	59.00 > 43.00	21	59.00 > 41.00	9
Methyl-tert-amylether	994-05-8	73.00 > 43.00	18	43.00 > 41.10	6	73.00 > 45.10	9
1,4-Dioxane	123-91-1	88.00 > 58.00	9	88.00 > 44.10	6	88.00 > 43.00	18
1,3-Dioxane	505-22-6	87.00 > 59.10	9	87.00 > 41.10	18	59.00 > 41.00	6
2-Ethyl-2-methyl-1,3-dioxolane	126-39-6	87.00 > 43.00	18	57.00 > 42.00	27	43.00 > 41.00	6
1,3-Dioxolane	646-06-0	73.00 > 45.00	12	73.00 > 43.10	21	45.00 > 43.00	18
2,2-Dimethyl-1,3-dioxolane	2916-31-6	87.00 > 43.00	18	42.00 > 40.00	9	43.00 > 41.10	9
2-Methyl-1,3-dioxolane	497-26-7	73.00 > 45.00	12	73.00 > 43.00	24	45.00 > 43.00	12
2-Ethyl-4-methyl-1,3-dioxolane	4359-46-0	87.00 > 59.10	9	59.00 > 41.10	6	87.00 > 41.00	15
2-Ethyl-5,5-dimethyl-1,3-dioxane	768-58-1	56.00 > 41.00	9	69.00 > 41.00	9	56.00 > 39.00	21
Galaxolide	1222-05-5	243.00 > 213.20	9	258.00 > 243.20	9	243.00 > 143.20	21
Tonalid	21,145-77-7	258.00 > 243.20	9	243.00 > 187.20	9	243.00 > 57.20	21
Indole	120-72-9	117.00 > 90.10	18	89.00 > 63.00	18	90.00 > 63.10	24
3-Methylindole	83-34-1	130.00 > 77.10	24	130.00 > 103.10	15	103.00 > 77.10	12
Cyclohexene	110-83-8	67.00 > 65.10	12	67.00 > 51.00	24	54.00 > 51.10	24
1-Octene	111-66-0	43.00 > 41.10	6	41.00 > 39.00	9	55.00 > 39.00	18
3-Methyl-1-cyclohexene	591-48-0	81.00 > 79.10	12	81.00 > 53.10	15	67.00 > 65.00	12

(continued)

Table 4.1* (continued)

Odorants	CAS	Target ion m/z	Ch$_1$[a] CE[b]	Reference ion 1 m/z	Ch$_2$ CE	Reference ion 2 m/z	Ch$_3$ CE
Dicyclopentadiene	77-73-6	66.00 > 51.00	21	66.00 > 63.10	27	66.00 > 62.00	45
1-Penten-3-one	1629-58-9	84.00 > 55.00	6	83.00 > 55.10	15	84.00 > 41.10	12
Cyclohexanone	108-94-1	98.00 > 55.10	18	98.00 > 80.10	6	98.00 > 69.10	9
1-Octen-3-one	431-99-6	70.00 > 55.00	6	70.00 > 43.00	9	70.00 > 41.10	18
cis-3-Hexenol	928-96-1	82.00 > 67.10	6	67.00 > 41.10	12	41.00 > 39.00	9
n-Propyl butyrate	105-66-8	71.00 > 43.10	9	89.00 > 43.10	12	71.00 > 41.10	18
cis-3-Hexenyl acetate	3681-71-8	82.00 > 67.10	9	67.00 > 41.10	15	82.00 > 41.10	21
Butyl butyrate	109-21-7	71.00 > 43.10	6	89.00 > 43.10	12	71.00 > 41.00	18
o-Cresol	95-48-7	108.00 > 77.10	24	108.00 > 79.10	18	107.00 > 79.10	9
4-Bromophenol	106-41-2	172.00 > 65.10	21	174.00 > 65.10	24	93.00 > 65.10	9
p(m)-Cresol	106-44-5/ 108-39-4	108.00 > 77.10	27	108.00 > 79.10	18	107.00 > 51.10	27
o-Nitrophenol	88-75-5	139.00 > 81.10	18	139.00 > 109.10	9	139.00 > 65.10	21
2,6-Dimethylphenol	576-26-1	107.00 > 77.10	18	122.00 > 107.10	12	122.00 > 77.10	27
2-Chlorophenol	95-57-8	128.00 > 64.10	18	130.00 > 64.10	18	128.00 > 92.10	9
2-tert-Butylphenol	88-18-6	135.00 > 107.10	12	107.00 > 77.10	18	150.00 > 135.10	9
2,3,5,6-Tetramethylpyrazine	1124-11-4	136.00 > 54.10	12	136.00 > 95.10	9	136.00 > 121.10	12
Pyrazine	290-37-9	80.00 > 53.10	12	81.00 > 54.10	12	80.00 > 78.00	45
2-Isopropyl-3-methoxy pyrazine	25,773-40-4	152.00 > 137.10	9	137.00 > 109.10	9	152.00 > 124.10	6
2-Isobutyl-3-methoxy pyrazine	24,683-00-9	124.00 > 94.10	12	124.00 > 81.10	9	124.00 > 42.10	24
Dimethylpyrazine	108-50-9	108.00 > 42.10	15	108.00 > 40.10	21	108.00 > 81.10	9
2-Ethyl-5(6)-methylpyrazine	36,731-41-6	122.00 > 94.10	15	122.00 > 66.10	27	122.00 > 53.10	27

(continued)

Table 4.1* (continued)

Odorants	CAS	Target ion m/z	Ch$_1$ CE^b	Reference ion 1 m/z	Ch$_2$ CE	Reference ion 2 m/z	Ch$_3$ CE
2,3,5-Trimethylpyrazine	14,667-55-1	122.00 > 42.10	15	122.00 > 81.10	9	81.05 > 42.10	6
Diethyl sulfide	352-93-2	90.00 > 75.10	12	90.00 > 62.00	9	90.00 > 47.10	18
Dimethyl disulfide	624-92-0	94.00 > 79.00	15	94.00 > 61.00	9	94.00 > 64.00	27
Isopropyl Sulfide	625-80-9	103.00 > 61.00	6	118.00 > 103.10	9	118.00 > 43.10	18
Dipropyl sulfide	111-47-7	76.00 > 42.10	6	118.00 > 76.10	6	89.00 > 61.00	6
Diethyl disulfide	110-81-6	122.00 > 94.00	9	122.00 > 66.00	18	94.00 > 66.00	6
Dimethyl trisulfide	3658-80-8	126.00 > 79.00	18	79.00 > 64.00	18	126.00 > 61.10	6
Dibutyl sulfide	544-40-1	90.00 > 56.10	6	146.00 > 56.10	18	146.00 > 90.10	9
Propyl disulfide	629-19-6	150.00 > 43.10	18	108.00 > 43.10	12	150.00 > 108.10	6
Amyl sulfide	872-10-6	70.00 > 55.10	9	103.00 > 69.10	9	103.00 > 41.10	18
Butyl disulfide	629-45-8	178.00 > 57.20	18	122.00 > 57.10	6	178.00 > 122.10	6
Diamyl disulfide	112-51-6	206.00 > 43.10	21	136.00 > 43.10	18	103.00 > 69.10	12
Diphenyl sulfide	139-66-2	186.00 > 184.10	27	186.00 > 77.10	27	185.00 > 152.10	27
Isopropyl propyl sulfide	5008-73-1	118.00 > 76.10	9	103.00 > 61.00	6	118.00 > 103.10	9
Diisopropyl disulfide	4253-89-8	150.00 > 108.00	6	108.00 > 66.00	6	66.00 > 64.00	21
Diisopropyl trisulfide	5943-34-0	182.00 > 140.00	6	182.00 > 75.10	12	98.00 > 64.00	12
2-Methylisoborneol	2371-42-8	95.00 > 67.10	15	95.00 > 55.10	18	108.00 > 93.10	12
Geosmin	19,700-21-1	112.00 > 97.10	12	112.00 > 83.10	12	112.00 > 69.10	21
β-Cyclocitral	432-25-7	137.00 > 109.20	6	152.00 > 137.20	9	152.00 > 123.10	6
Linalool	78-70-6	71.00 > 43.00	9	93.00 > 77.10	12	93.00 > 91.10	9
(±)-Camphor	76-22-2	108.00 > 93.10	12	95.00 > 55.10	15	95.00 > 67.10	12
L-Menthol	2216-51-5	81.00 > 41.00	18	71.00 > 43.00	6	95.00 > 67.10	9

(continued)

Table 4.1* (continued)

Odorants	CAS	Target ion m/z	Ch₁ᵃ CEᵇ	Reference ion 1 m/z	Ch₂ CE	Reference ion 2 m/z	Ch₃ CE
Nerol	106-25-2	69.00 > 41.00	6	93.00 > 77.00	12	69.00 > 39.00	21
Geraniol	106-24-1	69.00 > 41.00	6	69.00 > 39.00	21	41.00 > 39.00	9
1,8-Cineole	470-82-6	108.00 > 93.10	9	93.00 > 77.10	18	81.00 > 41.10	18
Bornyl acetate	76-49-3	93.00 > 77.10	15	95.00 > 67.10	12	95.00 > 55.10	18
β-Ionone	79-77-6	177.00 > 147.10	24	177.00 > 162.20	18	91.00 > 65.10	18
Cinene	138-86-3	68.00 > 53.00	12	67.00 > 65.10	12	68.00 > 65.20	15
Thiazole	288-47-1	85.00 > 58.00	12	87.00 > 60.00	18	58.00 > 45.00	24
2-Acetylthiazole	24,295-03-2	99.00 > 59.00	21	99.00 > 54.10	15	99.00 > 72.10	9

ᵃ Ch: channel

ᵇ CE: collision energy at the corresponding channel, V

Source Wang et al. 2021, Wang et al. 2019b

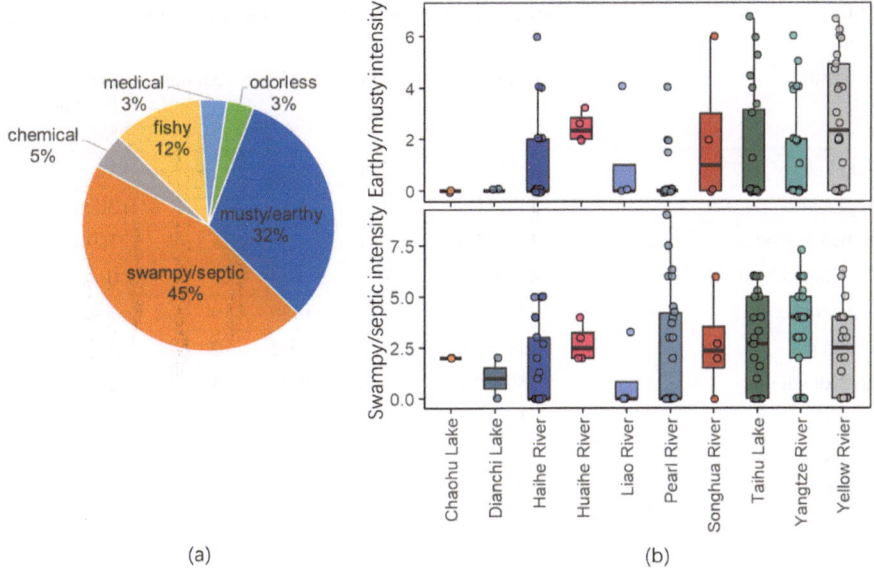

Fig. 4.1 Odor characteristics of water samples in raw water samples (**a**) and distribution of earthy/musty and swampy/septic odors in water samples from major watersheds (**b**). *Source* Authors

phenols, five pyrazines, two thiazoles, and two thioethers, being frequently detected. In contrast, 67 odorants were detected with 17 odorants, including one aldehyde, four non-oxygen benzene-containing compounds, one cyclic acetal, one musk, one phenol, five pyrazines, two thiazoles, and two thioethers, being frequently detected in river source water.

4.4.1   Odorants Related to Earthy/musty Odor

The occurrence of earthy/musty odors in drinking water has been attributed to the presence of several organic chemicals, including geosmin, 2-MIB, 2-isobutyl-3-methoxy pyrazine, 2-isopropyl-3-methoxy pyrazine and 2,4,6-trichloroanisole in literature (Pirbazari et al. 1992). These odorants could cause earthy/musty odor at low ng/L levels (Lin et al. 2019). In the current investigation, 2-isobutyl-3-methoxy pyrazine, 2-isopropyl-3-methoxy pyrazine, and 2,4,6-trichloroanisole were rarely detected during the investigation period (freq. < 5%), indicating their limited odor contribution.

2-MIB and geosmin have been reported to be the most emerging pollutants that contribute to the majority of earthy/musty odors in drinking water, and are produced by cyanobacteria and actinomycetes (Bruce et al. 2002). 2-MIB was detected in raw water at a concentration range of n.d.-250.97 ng/L with a detection frequency of

Table 4.2* Concentrations of odorants detected in raw water samples obtained for 140 sampling events

Odorants	CAS	Max	Mean	Median	Detection frequency, %
Hexanal	66-25-1	211	12.8	1.9	54
Heptanal	111-71-7	194	5.8	0	30
Benzaldehyde	100-52-7	351	12.6	5.1	70
2,4-Heptadienal	4313-03-5	0	0	0	0
2-Octenal	2548-87-0	22.1	0.28	0	1.4
Nonanal	124-19-6	150	11.2	0	41
2,6-Nonadienal	557-48-2	104	0.91	0	2.1
Decanal	112-31-2	135	13.3	0	30
2,4-Decadienal	2363-88-4	0	0	0	0
Ethylbenzene	100-41-4	140	9.1	2.0	56
Para-xylene	106-42-3	67.1	4.9	0.19	51
1,4-Dichlorobenzene	106-46-7	125	11.9	1.8	61
Indan	496-11-7	11.4	1.3	0.75	66
Isopropylbenzene	98-82-8	2.1	0.01	0	0.7
Nitrobenzene	98-95-3	88.7	8.1	0	21
Biphenyl	92-52-4	469	4.9	0	41
Phenylethylene	100-42-5	191	4.8	0	14
3-Methylstyrene	139-66-2	12.6	0.23	0	4.3
1-Methylnaphthalene	90-12-0	3644	36.4	0	41
bis(2-Chloro-1-methylethyl)ether	108-60-1	1280	35.8	0	42
bis(2-Chloroethyl) ether	111-44-4	379	15.4	0	23
2,4,6-Trichloroanisole	87-40-1	0.32	0	0	2.1
2,3,4-Trichloroanisole	54,135-80-7	0	0	0	0
2,3,6-Trichloroanisole	50,375-10-5	0	0	0	0
Pentachloroanisole	1825-21-4	1.3	0.01	0	0.7
2,4,6-Tribromoanisole	607-99-8	208	3.0	0	1.4
Diphenyl ether	101-84-8	1.0	0.01	0	0.7
Ethyl-tert-butylether	637-92-3	267	4.0	0	3.6
Methyl-tert-amylether	994-05-8	6722	82.0	0	19
1,4-Dioxane	123-91-1	7757	423	168	81
1,3-Dioxane	505-22-6	0	0	0	0
2-Ethyl-2-methyl-1,3-dioxolane	126-39-6	0	0	0	0
1,3-Dioxolane	646-06-0	309	5.9	0	5.0
2,2-Dimethyl-1,3-dioxolane	2916-31-6	0	0	0	0
2-Methyl-1,3-dioxolane	497-26-7	1644	42.4	0	22

(continued)

Table 4.2*   (continued)

Odorants	CAS	Max	Mean	Median	Detection frequency, %
2-Ethyl-4-methyl-1,3-dioxolane	4359-46-0	61.8	2.7	0	12
2-Ethyl-5,5-dimethyl-1,3-dioxane	768-58-1	12.6	0.18	0	2.1
Galaxolide	1222-05-5	84.1	12.7	0	47
Tonalid	21,145-77-7	0	0	0	0
Indole	120-72-9	1025	14.7	0	21
3-Methylindole	83-34-1	0	0	0	0
Cyclohexene	110-83-8	6.3	0.09	0	1.4
1-Octene	111-66-0	1816	34.5	0	7.1
3-Methyl-1-cyclohexene	591-48-0	111	3.5	0	10
Dicyclopentadiene	77-73-6	26.7	0.84	0	4.3
1-Penten-3-one	1629-58-9	10.5	0.08	0	0.7
Cyclohexanone	108-94-1	245	10.6	0	32
1-Octen-3-one	4312-99-6	144	11.2	0	9.3
cis-3-Hexenol	928-96-1	0	0	0	0
n-Propyl butyrate	105-66-8	8.1	0.07	0	2.9
cis-3-Hexenyl acetate	3681-71-8	0	0	0	0
Butyl butyrate	109-21-7	33.2	1.8	0	12
o-Cresol	95-48-7	11.1	1.4	0	47
4-Bromophenol	106-41-2	25.9	0.6	0	2.9
p(m)-Cresol	106-44-5/ 108-39-4	545	7.4	0.71	50
o-Nitrophenol	88-75-5	422	95.4	82.4	67
2,6-Dimethylphenol	576-26-1	7.9	0.49	0	23
2-Chlorophenol	95-57-8	5.4	0.66	0	21
2-tert-Butylphenol	88-18-6	528	3.9	0	1.4
2,3,5,6-Tetramethylpyrazine	1124-11-4	8.0	1.3	0.98	55
Pyrazine	290-37-9	32.7	9.6	8.2	94
2-Isopropyl-3-methoxy pyrazine	25,773-40-4	0	0	0	0
2-Isobutyl-3-methoxy pyrazine	24,683-00-9	0	0	0	0
Dimethylpyrazine	108-50-9	62.2	12.8	9.3	86
2-Ethyl-5(6)-methylpyrazine	36,731-41-6	19.7	4.6	5.1	64
2,3,5-Trimethylpyrazine	14,667-55-1	18.0	4.3	4.1	76
Diethyl sulfide	352-93-2	0.77	0.01	0	1.4
Dimethyl disulfide	624-92-0	714	15.0	4.5	86
Isopropyl sulfide	625-80-9	2.1	0.04	0	6.4

(continued)

Table 4.2* (continued)

Odorants	CAS	Max	Mean	Median	Detection frequency, %
Dipropyl sulfide	111-47-7	0.96	0.01	0	1.4
Diethyl disulfide	110-81-6	1.7	0.05	0	7.1
Dimethyl trisulfide	3658-80-8	84.4	2.1	0.39	60
Dibutyl sulfide	544-40-1	0	0	0	0
Propyl disulfide	629-19-6	0	0	0	0
Amyl sulfide	872-10-6	0	0	0	0
Butyl disulfide	629-45-8	6.3	0.04	0	0.7
Diamyl disulfide	112-51-6	0	0	0	0
Diphenyl sulfide	139-66-2	0	0	0	0
Isopropyl propyl sulfide	5008-73-1	0	0	0	0
Diisopropyl disulfide	4253-89-8	27.2	0.33	0	2.1
Diisopropyl trisulfide	5943-34-0	0	0	0	0
2-Methylisoborneol	2371-42-8	251	9.8	1.1	54
Geosmin	19,700-21-1	10.8	0.9	0.32	56
β-Cyclocitral	432-25-7	294	7.2	0	38
Linalool	78-70-6	130	2.8	0	3.6
(±)-Camphor	76-22-2	110	4.0	0	20
L-Menthol	2216-51-5	79.7	2.4	0	7.1
Nerol	106-25-2	110	0.79	0	0.7
Geraniol	106-24-1	803	31.6	0	16
1,8-Cineole	470-82-6	23.1	1.9	0	42
Bornyl acetate	76-49-3	9.6	0.07	0	0.7
β-Ionone	79-77-6	17.6	0.68	0	5.0
Cinene	138-86-3	198	11.9	0	14
Thiazole	288-47-1	13.8	5.9	5.9	89
2-Acetylthiazole	24,295-03-2	34.2	8.2	7.7	74

Source Wang et al. 2021

53.79%. Relatively higher concentrations (n.d. − 251 ng/L) of 2-MIB were observed with a detection frequency of 72% in lake/reservoir source water compared to river source water (n.d. − 148 ng/L). Geosmin, mainly exhibiting earthy odor, was detected at concentrations of < 11 ng/L and a detection frequency above 50%. By using Pearson correlation analysis, 2-MIB was identified as the major earthy/musty odor-causing compound in source water of China ($p < 0.05$). 2-MIB has been more prominent recently and has been commonly detected globally, especially in China. Some species of *Oscillatoria*, *Planktothricoides*, *Pseudanabaena*, and *Leptolyngbya bijugata* have been reported to be 2-MIB producers in China (Wang et al. 2015, 2011; Zhong et al. 2011).

In addition, concentrations of alkyl pyrazines were detected with a detection frequency of more than 50% (n.d. − 62.2 ng/L), these odorants typically possess earthy/musty/moldy/nutty odors, and might also contribute to earthy/musty odor by synthetic effects between multiple odorants (Wang et al. 2020).

4.4.2 Odorants Related to Swampy/septic Odor

Some small molecule organic compounds, such as thioethers, exhibit odors described as swampy/septic, rotten, rancid and stink ones, which could be perceived by human beings at low levels of ng/L or even less (Guo et al. 2016; Watson and Jüttner 2016). In China, concerns about swampy/septic odor problems caused by thioethers have greatly increased after the water crisis in Wuxi in 2007 (Yang et al. 2008).

Eight thioethers were detected in the major watersheds of China, among which dimethyl disulfide (DMDS) and dimethyl trisulfide (DMTS) were detected co-occurred widely in raw water ($r = 0.93$, $p < 0.05$), with concentrations of n.d.-714 ng/L (freq. 86%) and n.d.-84.4 ng/L (freq. 60%). DMDS and DMTS were reported to be commonly found in the decay of cyanobacteria, algae, or vegetation, wastewater discharges, and anoxic sediments (Watson and Jüttner 2016). Diisopropyl disulfide was also detected with a maximum concentration of 27.2 ng/L though with a very low detection frequency (2.1%). *Microcystis flos-aquae* was ever reported to be the causative agent for the generation of isopropyl-sulfur compounds (Hofbauer and Jüttner 1988; Jenkins et al. 1967; Xu et al. 2017). All other thioethers were less frequently detected (<10%).

The Pearson correlation analysis showed that thioethers were the major swampy/septic odor-causing compounds in source water of China ($r = 0.45$, $p < 0.05$). Thioethers were distributed widely in major watersheds, and higher concentrations were detected in the east and south parts of China. Based on Classification and regression tree analysis, TOC input and geographic location (watershed), were two major factors affecting the occurrence levels of thioethers (Fig. 4.2). Higher levels of DMDS in Taihu Lake, Yangtze River, and Yellow River at TOC of > 2.58 mg/L were observed. For DMTS, TOC was at the first level in the hierarchy, and samples taken from Taihu Lake, Pearl River, and Yellow River showed higher concentrations at TOC of ≥ 2.08 mg/L. Especially, DMTS at NH_4-N ≥ 0.456 mg/L showed higher concentration compared to lower NH_4-N in samples when TOC < 2.08 mg/L, illustrating the higher organic sulfides might originate from sulfur-containing amino acid (e.g., methionine or cysteine). The results showed that TOC input affected the distribution of thioethers in source waters predominantly. For example, dramatic increases in nutrient input of TOC and NH_4-N were observed in Taihu Lake (China) during the odor crisis in June 2007, which might be caused by urban and agricultural development in the watershed (Zhang et al. 2010).

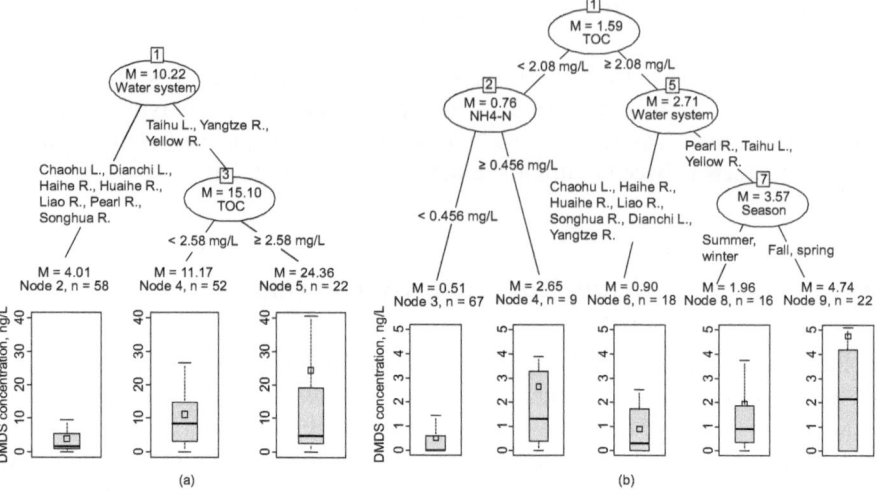

Fig. 4.2* Classification and regression tree of the explanatory factors regarding (**a**) DMDS concentrations, ng/L, and (**b**) DMTS concentrations, ng/L. Note: R., the abbreviation of "River"; L., the abbreviation of "Lake"; M, mean concentration in ng/L is shown at the top of each node and is marked with "□" in each box plot (outliers are not drawn). *Source* Wang et al. 2021

Besides, indoles are also reported to be strongly associated with septic odors in natural waters, with odor thresholds in the range of $0.1 - 300 \, \mu g/L$ (Lin et al. 2019). Indole was detected in raw water at concentrations of n.d. $- 1025 \, ng/L$ with a detection frequency of 21%. 3-Methylindole was not found in any of the water samples. Indoles are largely present in livestock and municipal wastewaters (Kim et al. 2016).

4.4.3 Odorants Related to Fishy Odor

Fishy odor problems often occur in cold water, even under ice-covered water with low nutrient concentrations. The occurrence of fishy odor has been associated with algal metabolites, mainly polyunsaturated aldehydes (PUAs), such as 2,4-heptadienal, 2,4-decadienal, and 2,4,7-dectridienal (Pohnert 2002; Watson 2010; Wendel and Jüttner 1996). Some saturated aldehydes, e.g., hexanal, heptanal, and benzaldehyde, were also assumed to contribute to fishy odor (Li et al. 2016; Venkateshwarlu et al. 2004). As sporadically reported, fishy odor issues often occur in low-temperature period with low nutrient concentrations. The potential fishy odor-causing algae include chrysophytes, diatoms, cryptophytes, and dinoflagellates (Jüttner 1981; Li et al. 2016; Shinfuku et al. 2022; Watson et al. 2001; Zhao et al. 2013).

A total of nine aldehydes was investigated, and seven kinds of aldehydes were detected in raw water. Hexanal (n.d.-211 ng/L), benzaldehyde (n.d.-351 ng/L), nonanal (n.d.-150 ng/L), heptanal (n.d.-158 ng/L), and decanal (n.d.-135 ng/L) were the dominant aldehydes detected (freq. > 40%). Relatively higher average concentrations in lake/reservoir source water samples were observed for hexanal, benzaldehyde, nonanal, and decanal compared to river source water samples. 2,4-Heptadienal, 2-octenal, 2,6-nonadienal and 2,4-decadienal were rarely detected.

The odor activity values (ratio of odorant concentration and odor threshold concentration) were in the range of 0–0.81, 0–0.13, 0–0.20, 0–0.19, and 0–0.19 for hexanal, benzaldehyde, nonanal, heptanal, and decanal, with average values of 0.10, 0.01, 0.02, < 0.01, and 0.02. Hexanal was described as grassy, herbal, or fatty odor, with odor thresholds of 0.3–14,000 µg/L (Ömür-Özbek and Dietrich 2008; Pripdeevech and Wongpornchai 2013). Benzaldehyde has cherry, almond odor, and odor threshold concentrations of 0.5–4.5 µg/L (Lin et al. 2019). Nonanal and decanal exhibit orange-fruity odor dominantly, with odor thresholds of 1 µg/L and 0.1–2.53 µg/L, respectively (Dietrich and Burlingame 2020; Lin et al. 2019). Heptanal exhibits oily-fatty odor and might has a fishy odor in the co-occurrence with (E, Z)-3,5-octadien-2one, another oxylipin common during and after algae bloom episodes (Malnic et al. 2004). Overall, hexanal and heptanal might contribute to fishy odor by synthetic effects in the simultaneous presence of multiple odorants.

4.4.4 Odorants Related to Chemical/hydrocarbon Odor

Chemical/hydrocarbon odor problems caused by emerging chemical pollution are common in some open water bodies like river source water. Certain emerging pollutants not only pose health risks but also contribute to unpleasant odors. Serious odor incidents induced by sudden chemical spills could disrupt consumers' normal lives for several days, bring about psychological fear and loss of confidence regarding the safety of drinking water, and even evolve into severe social public events (Gallagher et al. 2015). About 40 compounds were detected in the investigation that were associated with industrial pollution, including benzene-containing compounds, ethers, cyclic acetals, phenols, etc.

Several non-oxygen benzene-containing compounds, including ethylbenzene, para-xylene, indan, biphenyl, and 1-methylnaphthalene, were detected at concentrations of less than 3644 ng/L with detection frequencies of > 40% in raw water and correlated well. The co-occurrence of these substituted benzenes illustrated their similar source. Benzenes are important industrial chemicals, exhibiting sweet, varnish, gasoline, paint/putty, and solvent odors (Suffet et al. 2019), were reported mainly from discharging by petrochemical industries, like gasoline (Satoshi 2003).

Cyclohexanone was detected at concentrations of n.d.−351 ng/L with a detection frequency of 70%. The co-occurrence with benzaldehyde was observed in source water (r = 0.81, p < 0.05). Aldol condensation of cyclohexanone and benzaldehyde is an important industrial process manufacturing many fine chemicals of commercial interest (Tang et al. 2013). For example, α,α′–bisbenzylidene cyclohexanone is an important intermediate in the pharmaceutical industries, optics, rocket engineering, photolithography, and liquid crystalline polymers (Tabrizian et al. 2015; Vashishtha et al. 2015).

Bis (2-chloro-1-methylethyl) ether (2,2'-dichlorodiisopropyl ether, DCIP), is a b-haloether that is used as an extracting solvent in the chemical industry and as an organochlorine pesticide in agriculture (Iordache et al. 2009). It is also an undesirable by-product in the industrial production of propylene oxide and epichlorohydrin by the chlorohydrin process (Moreno Horn et al. 2003). Previous studies demonstrated the global detection of DCIP in the surface water of Ohio River of the U.S. (0.5–5 µg/L) (Kleopfer and Fairless 1972), the Rhine and Scheldt rivers of the Netherlands (Moreno Horn et al. 2003), the Elbe river of Germany (< 0.8 µg/L) (Franke et al. 1995), Odra River in Central Europe (0.8 µg/L) (Kuczyn´ska et al. 2004). DCIP was detected at concentrations of n.d.-1280 ng/L (freq. 42%) with a mean concentration of 35.8 ng/L in raw water of China (Wang et al. 2021). It is mainly distributed in the eastern and southern regions with higher concentrations detected in water samples from watersheds of Yellow River, Haihe River, Yangtze River, and Taihu Lake (Fig. 4.3). The occurrence of DCIP was in accordance with the distribution of industrial production of propylene oxide and epichlorohydrin, indicating DCIP might be a by-product of industrial activities related to epichlorohydrin/propylene oxide according to industrial distribution analysis.

Cyclic acetals, including dioxanes and dioxolanes, which are normally formed as byproducts during resin manufacturing, have been reported to be involved in several drinking water odor incidents (Carrera et al. 2019; Quintana et al. 2015; Schweitzer et al. 1999). Among this group of compounds, 2-alkyl-5,5-dimethyl-1,3-dioxanes and 2-alkyl-4-methyl-1,3-dioxolanes have been particularly focused because of their low odor thresholds (5 − 10 ng/L) with sickening sweet, olive, latex paint, varnish, solvent, green apple, marshy/sulfurous/decaying vegetation and fishy/

Fig. 4.3* Distribution of bis (2-chloro-1-methylethyl) ether in raw water samples of major watersheds and in different source water types (embedded plot) *Source* Wang et al. 2023. *Note* L, lake and reservoir source water; R, river source water; The extreme values were not plotted

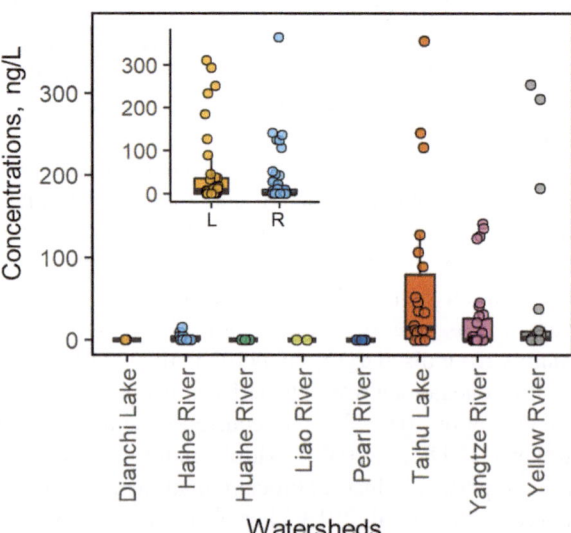

algal odors (Crump et al. 2014). The first documented odor incident with dioxanes and dioxolanes was in 1977. Since then, these chemicals intermittently reported as imparting odor issues in surface or ground waters globally (Bruchet et al. 2007; Quintana et al. 2015). Some recent odor incidents occurred in January-March, 2010 in North-East UK London and in Barcelona in the year of 2013–2014 (Quintana et al. 2015). In China, five cyclic acetals including 1,4-dioxane, 1,3-dioxolane, 2-methyl-1,3-dioxolane (2MDL), 2-ethyl-4-methyl-1,3-dioxolane (2E4MDL), and 2-ethyl-5,5-dimethyl-1,3-dioxane (2EDD) were detected according to the national investigation. 1,4-Dioxane (n.d.-7757 ng/L), 2MDL (n.d.-1644 ng/L), and 2E4MDL (n.d.-61.8 ng/L) were the predominant ones, with detection frequencies of 81%, 22% and 12%. 2EDD was rarely detected at concentrations of n.d.-12.6 ng/L, though it was the most identified malodor dioxane in literature.

In particular, relatively high occurrence levels and detection frequency were observed for dioxanes and dioxolanes in Huangpu River, China. Eight chemicals were detected, including 1,4-dioxane, 1,3-dioxane, 2,5,5-trimethyl-1,3-dioxane (TMD), 2EDD, 1,3-dioxolane, 2MDL, 2-ethyl-2-methyl-1,3-dioxolane (2E2MDL) and 2E4MDL. 1,4-Dioxane was the predominant dioxane (212 − 8310 ng/L, freq. = 100%) with a mean concentration of 1958 ng/L, followed by TMD (n.d. − 133 ng/L, freq. = 56.7%). 1,3-Dioxane (n.d. − 71.9 ng/L) and 2EDD (n.d. − 48.3 ng/L) were rarely detected in the river samples (< 5%). 2MDL (49.5 − 2278 ng/L), 2E4MDL (n.d. − 167 ng/L) and 1,3-dioxolane (n.d. − 225 ng/L) were the major dioxolanes as exhibited by their high detection frequencies (100%, 85.0% and 73.3%, respectively). 2E2MDL was less frequently detected (24.5%), with an average concentration of 2.6 ng/L. At the same time, DCIP was detected and co-occurred with the detected cyclic acetals at the concentration range of n.d. − 1094 ng/L with a detection frequency above 90%. The maximum odor activity values of 2E4MDL, TMD, 2EDD were 0.19 − 33.4, 13.3, and 4.8 − 76.7, respectively, should have contributed to the septic/chemical odor profiles in the river water.

4.4.5   Odorants Related to Medicinal/phenolic Odor

Phenols, especially chloro, and bromo substituted halophenols are reported to be associated with medicinal odors (Dietrich and Burlingame 2020; Suffet et al. 1999). Seven phenols were detected, and o-cresol (n.d.-11.4 ng/L), p(m)-cresol (n.d.-645 ng/L), o-nitrophenol (n.d.-422 ng/L), 2,6-dimethylphenol (n.d.-7.9 ng/L), 2-chlorophenol (n.d.-5.4 ng/L) were the dominated ones (> 20%) in raw water. Phenols are widely used in household products and as intermediates for the industrial synthesis of plastics, pesticides, insecticides, and petroleum (Murray et al. 2010), leading to their potential to cause odor issues in water sources. There was a serious plastic/chemical odor problem in the tap water of Hangzhou City, which was confirmed to be caused by the pollution of phenolic substances such as o-tert-butylphenol (Sun and Xiao 2014).

Specifically, p(m)-cresol showed that it co-occurred with geosmin in source waters ($r = 0.64$, $p < 0.05$). The two compounds were reported to co-occur in municipal wastewaters (Bylinski et al. 2019), industrial wastewaters (e.g. pulp and paper mill) (Cook and Hoy 2008), as well as in juices or wine when grapes containing fungal pathogens were processed (Welke 2019). Thus, the ability of microbes to produce geosmin and m-cresol suggests that their co-occurrence in raw water is due to natural causes.

Halophenols (chlorophenols, bromophenols) exhibit similar iodoform, phenolic, and medicinal odors, with odor thresholds in ng/L level (Young et al. 1996). The formation of halophenols in drinking water is usually related to the reaction of phenols and halogen ions during the chlorine process in water treatment plants as well as in water distribution system (Khiari et al. 1999; Whitfield et al. 1988).

4.5   Conclusion

The nationwide investigation of odor distribution characteristics across China showed widespread odor problems in source water samples. Earthy/musty, swampy/septic, fishy, chemical/hydrocarbon, and medicinal/phenolic odors were detected, and the potentially responsible odorants for each odor type were summarized. 2-Methylisoborneol and thioethers were identified as the main earthy/musty and swampy/septic odor-causing compounds, respectively. Some cyclic acetals were detected to contribute to chemical odor episodes in Huangpu River source water. Their co-occurrence with bis(2-chloro-1-ethylethyl) ether indicates the association with resin-related industrial pollution. The results would be helpful for the management of aesthetic quality of drinking water and provide references for setting water quality standards in China.

Acknowledgements   This work was supported by the National Key R&D Program of China (grant number 2023YFC3210100), Funds for the National Natural Science Foundation of China (No. 52100018), the Major Science and Technology Program for Water Pollution Control and Treatment (No. 2017ZX07207004).

References

American Public Health Association (2012) Standard methods for the examination of water and wastewater. In: EW Rice, RB Baird, AD Eaton, LS Clesceri (eds) American Public Health Association, 22nd ed., Washington, D.C
Bruce D, Westerhoff P, Brawley-Chesworth A (2002) Removal of 2-methylisoborneol and geosmin in surface water treatment plants in Arizona. J Water Supply: Res Technology—AQUA, 183–198
Bruchet A, Hochereau C, Campos C (2007) An acute taste and odour episode solved by olfactory GC-MS. Water Sci Technol 55:223–230

Bylinski H, Gebicki J, Namiesnik J (2019) Evaluation of health hazard due to emission of volatile organic compounds from various processing units of wastewater treatment plant. Int J Environ Res Public Health 16(10):1712

Carrera G, Vegue L, Ventura F, Hernandez-Valencia A, Devesa R, Boleda MR (2019) Dioxanes and dioxolanes in source waters: Occurrence, odor thresholds and behavior through upgraded conventional and advanced processes in a drinking water treatment plant. Water Res 156:404–413

Cook DL, Hoy D (2008) Wha's that Smell? analytical measurements for odorous compounds in pulp and paper mill wastewaters. Proc Water Environ Fed 2008:55–81

Crump D, Charlton A, Taylor J, Bevan R (2014) National assessment of the risks to water supplies posed by low taste and odour threshold compounds. In: The Institute of Environment and Health, Phil H (ed) Cranfield University Bedfordshire MK43 OAL UK: The Institute of Environment and Health

Dietrich AM, Burlingame GA (2020) A review: The challenge, consensus, and confusion of describing odors and tastes in drinking water. Sci Total Environ: 135061

Franke S, Hildebrandt S, Francke W, Reincke H (1995) The occurrence of chlorinated bis-(propyl)ethers in the Elbe River and tributaries. Naturwissenschaften 82:80–83

Gallagher DL, Phetxumphou K, Smiley E, Dietrich AM (2015) Tale of two isomers: complexities of human odor perception for cis- and trans-4-methylcyclohexane methanol from the chemical spill in West Virginia. Environ Sci Technol 49:1319–1327

Guo QY, Yu JW, Yang K, Wen XD, Zhang HF, Yu ZY, Li HY, Zhang D, Yang M (2016) Identification of complex septic odorants in Huangpu River source water by combining the data from gas chromatography-olfactometry and comprehensive two-dimensional gas chromatography using retention indices. Sci Total Environ 556:36–44

Hofbauer B, Jüttner F (1988) Occurrence of isopropylthio compounds in the aquatic ecosystem (Lake Neusiedl, Austria) as a chemical marker for Microcystis flos-aquae. FEMS Microbiol Lett 53:113–121

Iordache M, Pavel VL, Myunghee L, Iordache I (2009) Removal of bis (1-chloro-2-propyl) ether from wastewater using sonodegradation and biodegradation. Environ Eng Manag J 8:201–206

Jenkins D, Medsker LL, Thomas JF (1967) Odorous compounds in natural waters. Some sulfur compounds associated with blue-green algae. Environ Sci Technol 1:731–735

Jüttner F (1981) Detection of Lipid Degradation Products in the Water of a Reservoir During a Bloom of Synura uvella. Appl Environ Microbiol 41:100–106

Khiari D, Bruchet A, Gittelman T, Matia L, Barrett S, Suffet IM, Hund R (1999) Distribution-generated taste-and-odor phenomena. Water Sci Technol 40:129–133

Kim M, Lee JH, Kim E, Choi H, Kim Y, Lee J (2016) Isolation of Indole Utilizing Bacteria Arthrobacter sp. and Alcaligenes sp. From Livestock Waste. Indian Journal of Microbiology 56:158–166

Kleopfer RD, Fairless BJ (1972) Characterization of organic components in a municipal water supply. Environ Sci Technol 6:1036–1037

Kuczyn'ska A, Wolska L, Namienik J (2004) An Attempt to Identify Volatile and Semi-Volatile Organic Compounds Present in the Odra River Waters. Chromatographia 60: 279–289

Li X, Yu J, Guo Q, Su M, Liu T, Yang M, Zhao Y (2016) Source-water odor during winter in the Yellow River area of China: Occurrence and diagnosis. Environ Pollut 218:252–258

Lin T-F, Watson S, Suffet IM (2019) Taste and odour in source and drinking water: causes, controls, and consequences. IWA Publishing, London SW1H 0QS, UK

Malnic B, Godfrey PA, Buck LB (2004) The human olfactory receptor gene family. Proc Natl Acad Sci U S A 101:2584–2589

Moreno Horn M, Garbe L-A, Tressl R, Adrian L, Görisch H (2003) Biodegradation of bis(1-chloro-2-propyl) ether via initial ether scission and subsequent dehalogenation by Rhodococcus sp. strain DTB. Arch Microbiol 179:234–241

Murray KE, Thomas SM, Bodour AA (2010) Prioritizing research for trace pollutants and emerging contaminants in the freshwater environment. Environ Pollut 158:3462–3471

Ömür-Özbek P, Dietrich AM (2008) Developing hexanal as an odor reference standard for sensory analysis of drinking water. Water Res 42:2598–2604

Pirbazari M, Borow HS, Craig S, Ravindran V, McGuire MJ (1992) Physical Chemical Characterization of Five Earthy-Musty-Smelling Compounds. Water Sci Technol 25:81–88

Pohnert G (2002) Phospholipase A2 activity triggers the wound-activated chemical defense in the diatom Thalassiosira rotula. Plant Physiol 129:103–111

Pripdeevech P, Wongpornchai S (2013) Odor and Flavor Volatiles of Different Types of Tea. In: Preedy VR (ed) Tea in Health and Disease Prevention. Academic Press, London, UK, pp 307–120

Quintana J, Vegué L, Martín-Alonso J, Paraira M, Boleda MR, Ventura F (2015) Odor events in surface and treated water: the case of 1, 3-dioxane related compounds. Environ Sci Technol 50:62–69

Satoshi I (2003) Analysis of aromatic hydrocarbons in gasoline and naphtha with the Agilent 6820 series gas chromatograph and a single polar capillary column. Agilent Technologies: 1–16

Schweitzer L, Noblet J, Ye Q, Ruth E, Suffet IH (1999) The environmental fate and mechanism of formation of 2-ethyl-5,5′-dimethyl-1,3-dioxane (2EDD) — A malodorous contaminant in drinking water. Water Sci Technol 40:217–224

Shinfuku Y, Takanashi H, Nakajima T, Kasuga I, Akiba M (2022) The Status Quo of Causal Substance Exploration for Fishy Odor in Raw Water for Taps. J Water Environ Technol 20: 29–44

Suffet IM, Khiari D, Bruchet A (1999) The drinking water taste and odor wheel for the millennium: beyond geosmin and 2-methylisoborneol. Water Sci Technol 40:1–13

Suffet IM, Braithwaite S, Zhou Y, Bruchet A (2019) The drinking water taste-and-odour wheel after 30 years. In: Tsair-Fuh Lin SW, Andrea M. Dietrich, IH (Mel) Suffet (ed) Taste and Odour in Source and Drinking Water: Causes, Controls, and Consequences. IWA Publishing, London SW1H 0QS, UK

Sun DL, Yu JW, Yang M, An W, Zhao YY, Lu N, Yuan SG, Zhang DQ (2014) Occurrence of odor problems in drinking water of major cities across China. Front Environ Sci Eng 8:411–416

Sun Y, Xiao J (2014) Zhejiang Online. http://zjnews.zjol.com.cn/system/2014/01/17/019816103. shtml. In

Suurnäkki S, Gomez-Saez GV, Rantala-Ylinen A, Jokela J, Fewer DP, Sivonen K (2015) Identification of geosmin and 2-methylisoborneol in cyanobacteria and molecular detection methods for the producers of these compounds. Water Res 68:56–66

Tabrizian E, Amoozadeh A, Rahmani S, Salehi M, Kubicki M (2015) Synthesis, characterization, and crystal structures of α, α′-bis(substituted-benzylidene)cycloalkanone derivatives by nano-TiO2/HOAc. Res Chem Intermed 42:531–544

Tang Y, Xu J, Gu X (2013) Modified calcium oxide as stable solid base catalyst for Aldol condensation reaction. J Chem Sci (Bangalore, India) 125:313–320

Vashishtha M, Mishra M, Undre S, Singh M, Shah DO (2015) Molecular mechanism of micellar catalysis of cross aldol reaction: Effect of surfactant chain length and surfactant concentration. J Mol Catal a: Chem 396:143–154

Venkateshwarlu G, Let MB, Meyer AS, Jacobsen C (2004) Chemical and olfactometric characterization of volatile flavor compounds in a fish oil enriched milk emulsion. J Agric Food Chem 52:311–317

Wang Z, Xu Y, Shao J, Wang J, Li R (2011) Genes associated with 2-methylisoborneol biosynthesis in cyanobacteria: isolation, characterization, and expression in response to light. PLoS ONE 6(4):e18665

Wang Z, Xiao P, Song G, Li Y, Li R (2015) Isolation and characterization of a new reported cyanobacterium Leptolyngbya bijugata coproducing odorous geosmin and 2-methylisoborneol. Environ Sci Pollut Res Int 22:12133–12140

Wang C, Yu J, Guo Q, Sun D, Su M, An W, Zhang Y, Yang M (2019a) Occurrence of swampy/ septic odor and possible odorants in source and finished drinking water of major cities across China. Environ Pollut 249:305–310

Wang C, Yu J, Guo Q, Zhao Y, Cao N, Yu Z, Yang M (2019b) Simultaneous quantification of fifty-one odor-causing compounds in drinking water using gas chromatography-triple quadrupole tandem mass spectrometry. J Environ Sci-China 79:100–110

Wang C, Yu J, Gallagher DL, Byrd J, Yao W, Wang Q, Guo Q, Dietrich AM, Yang M (2020) Pyrazines: A diverse class of earthy-musty odorants impacting drinking water quality and consumer satisfaction. Water Res 182:115971

Wang C, Gallagher DL, Dietrich AM, Su M, Wang Q, Guo Q, Zhang J, An W, Yu J, Yang M (2021) Data analytics determines co-occurrence of odorants in raw water and evaluates drinking water treatment removal strategies. Environ Sci Technol 55:16770–16782

Wang C, Guo Q, Zhang B, An W, Wang Z, Zhang D, Yang M, Yu J (2023) Solvent-like bis (2-chloro-1-methylethyl) ether occurrence in drinking water: Multidimensional risk assessment integrated health and aesthetic aspects. J Hazard Mater 453:131446

Watson SB, Jüttner F (2016) Malodorous volatile organic sulfur compounds: Sources, sinks and significance in inland waters. Crit Rev Microbiol 43:210–237

Watson S, Satchwill T, Dixon E, McCauley E (2001) Under-ice blooms and source-water odour in a nutrient-poor reservoir: biological, ecological and applied perspectives. Freshw Biol 46:1553–1567

Watson SB (2010) Algal taste and odor. In: Algae: Source to Treatment. American Water Works Association, The United States of America, pp 329–376

Welke JE (2019) Fungal and mycotoxin problems in grape juice and wine industries. Curr Opin Food Sci 29:7–13

Wendel T, Jüttner F (1996) Lipoxygenase-mediated formation of hydrocarbons and unsaturated aldehydes in freshwater diatoms. Phytochemistry 41:1445–1449

Whitfield FB, Last JH, Shaw KJ, Tindale CR (1988) 2,6-Dibromophenol: the cause of an iodoform-like off flavour in some Australian crustacea. J Sci Food Agric 46:29–42

Xu Q, Yang L, Yang W, Bai Y, Hou P, Zhao J, Zhou L, Zuo Z (2017) Volatile organic compounds released from Microcystis flos-aquae under nitrogen sources and their toxic effects on Chlorella vulgaris. Ecotoxicol Environ Saf 135:191–200

Yang M, Yu J, Li Z, Guo Z, Burch M, Lin T-F (2008) Taihu Lake not to blame for Wuxi's woes. Science 319:158–158

Young W, Horth H, Crane R, Ogden T, Arnott M (1996) Taste and odour threshold concentrations of potential potable water contaminants. Water Res 30:331–340

Zhang XJ, Chao C, Ding JQ, Hou A, Yong L, Niu ZB, Su XY, Xu YJ, Laws EA (2010) The 2007 water crisis in Wuxi, China: Analysis of the origin. J Hazard Mater 182:130–135

Zhao Y, Yu J, Su M, An W, Yang M (2013) A fishy odor episode in a north China reservoir: Occurrence, origin, and possible odor causing compounds. J Environ Sci 25:2361–2366

Zhong F, Gao YN, Yu T, Zhang YY, Xu D, Xiao ER, He F, Zhou QH, Wu ZB (2011) The management of undesirable cyanobacteria blooms in channel catfish ponds using a constructed wetland: Contribution to the control of off-flavor occurrences. Water Res 45:6479–6488

The opinions expressed in this chapter are those of the author(s) and do not necessarily reflect the views of the UNESCO: United Nations Educational, Scientific and Cultural Organization, its Board of Directors, or the countries they represent.

Chapter 5
Assessment of Harmful Algae as an Emerging Pollutant in Domestic Water Supply from Rainwater Harvesting Facilities in Sudan

Wifag Hassan Mahmoud, Sabry Zagloul Wahba, and Muna Mohammed Musnad

Abstract The rural population in Sudan comprises approximately 67.1% of the total population, with 55% residing in areas outside the Nile system. A significant portion of the population relies on surface rainwater harvesting facilities, known as Hafirs, for their domestic and drinking water needs. These sources are susceptible to contamination due to the surrounding livelihood activities. Algae as a biological indicator has attracted little attention when the water quality is assessed, despite its potential harm if diagnosed as toxic species. The blooming of Cyanophyta (Blue-Green Algae), especially their toxins-producing genera and species, is the most harmful phenomenon triggered by nutrient contamination, and anthropogenic and environmental factors. As a result, this pioneering study aimed to classify the types of algae and identify the toxins producing algae genera and species. Water samples were collected during the wet and dry seasons from eleven sites in five states (Gazira, Khartoum, North Kordofan, Gadarif, and Sinnar) based on different practices, such as agriculture and rangeland. Laboratory results of this study indicated the presence of green and blue-green algae with a very remarkable percentage across different sites and different seasons. Blue-Green algae were dominated by the following very well-known toxic genera: *Anabaena, Cylindrospermum, Merismopedia,* and *Oscillatoria.* It is concluded that these water supply facilities become unsafe for domestic and

W. H. Mahmoud (✉)
Regional Center for Capacity Building and Research in Water Harvesting, under the auspices of UNESCO, Khartoum, Sudan
e-mail: wifagmahmoud@gmail.com

Water Harvesting Center, Faculty of Engineering Science, University of Nyala, Nyala, Sudan

S. Z. Wahba
Water Pollution Research Department, Institute of Environmental and Climate Change Research, National Research Center, Cairo, Egypt

M. M. Musnad
UNESCO Chair in Water Resources, Omdurman Islamic University, Khartoum, Sudan

© UNESCO 2025
S. Zandaryaa et al. (eds.), *Emerging Pollutants*, Advances in Water Security,
https://doi.org/10.1007/978-3-031-71758-1_5

drinking purposes due to the dominance of toxins-producing algae species, especially during the end of the dry season. It is recommended that attention be paid to and remedies be applied to mitigate the algae development. It is also of utmost necessity to develop a strategy plan and policies to control harmful algae to protect drinking water sources from pollution. Furthermore, comprehensive water quality monitoring and algae toxicity prevention measures are necessary to improve natural resource management, maintain sustainable ecosystems, and protect public health. This will eventually enhance achieving the Sustainable Development Goals (SDGs), and particularly the sixth goal by 2030, which urges universal access to safe drinking water.

Keywords Algae blooms · Cyanobacteria · Unconventional water supply · Water contamination · Sudan

5.1 Algae: An Emerging Harmful Substance

Algae are a natural and important part of a reservoir's ecosystem. They can play an important role in determining the quality of water in freshwater sources, e.g., lakes, reservoirs, and rivers. A change in phytoplankton diversity and redundancy seems to be the most reliable and applicable means for assessing water quality. The term algae refers to a wide variety of different organisms that use light to grow. They are the base of the aquatic food chain and the primary producers of organic matter and oxygen in the aquatic environment.

Certain environmental conditions, like warming and excess nitrogen and phosphorus in water bodies, can intensify algae growth, causing algal blooms. Some types of these blooms, when they become large and produce chemicals or toxins, are called harmful algal blooms (HABs). They are mainly the result of a type of algae called *cyanobacteria*, also known as blue-green algae (BGA). These blooms jeopardize freshwater resources globally and make water management more complicated (Bonilla, et. al., 2023). Hamilton et al. (2014) and Buratti et al. (2017) attributed the probable rise in rate and intensity of cyanoHABs to climate change factors represented by the rise in global temperature and extreme flooding and events, as well as the expansion of agricultural land use. Considering the deteriorating impact of climate change, harmful BGA blooms will badly affect the quality of water and pose threats to public health, safety, and economy (Vu et al. 2020).

Increases in cyanobacteria bloom occurrence were predicted globally in the early 2000s in systems experiencing more rapid eutrophication, changing nutrient inputs, warmer aquatic temperatures, and greater CO_2 concentrations (Harvell et al. 2000; Visser et al. 2016; and Beaver et al. 2018). Assessments of cyanobacteria bloom formation suggest that both nitrogen and phosphorus are major drivers of freshwater blooms (Bonilla, et. al., 2023). Nutrients exacerbate the decline of waterways by promoting the growth of harmful species and deteriorating others (Mackeigan et al. 2023).

Eutrophication (nutrient enrichment) is a much stronger driver of cyanobacteria than climatic gradients. Lake characteristics influence the relative role of phosphorus and nitrogen in predicting cyanobacterial biomass, with shallow lakes more susceptible to eutrophication than deeper ones (Mânica & de Lime 2023).

If HAB occurs in freshwater sources, e.g., lakes, reservoirs, rivers, ponds, bays, and coastal waters, it can pose harm to people, animals, and aquatic ecosystems, as some of the cyanoHABs can produce toxins. These toxins can damage the liver and neurological systems of both humans and animals, and in severe cases, they can cause death. Cyanotoxins can be produced by a wide variety of BGAs, i.e., cyanobacteria. Some of the most commonly occurring genera are *Microcystis, Anabaena, Fischerella, Gloeotrichia, Nodularia, Nostoc, Oscillatoria, Cylindrospermopsis,* and *Aphanizomenon.* The toxins can be found in cyanobacteria or water.

Buratti et al. (2017) have comprehensively reviewed the cyanotoxins producing BGA and their potential risk to human health. Furthermore, they have reported serious and acute health problems due to human exposure to cyanotoxin via contaminated water in Brazil, the USA, and other countries. Another case of the cyanotoxins producing BGA-contaminated drinking water reservoirs, mainly *microcystins* (MC), was studied by Habtemariam et al. (2021). It is found that the Ethiopian Legedadi Reservoir is contaminated with high concentrations of extracellular MCs, indicating the tremendously high susceptibility of the targeted population to health problems. Koreivienė et al. (2014) studied the potential risk to public health safety from exposure to cyanotoxin-contaminated water in recreational areas. Acute or chronic diseases may result from body exposure to toxins through mouth, nose, or skin contact. Furthermore, cyanotoxin algae seasonal and annual variation in 11 public water supply reservoirs in the Brazilian semiarid region was studied by Lorenzi et al (2018). A substantial concentration of MCs was found in all reservoirs, in addition to high levels of other toxins, mainly CYN, STXs/Neo-STX, and ATX-a, which were detected in four, ten, and two reservoirs, respectively, during the water scarce dry season. Mohamed and Al Shehri (2007) investigated the presence of eukaryotic algae and cyanobacteria in a semiarid city in Saudi Arabia. These toxins-producing algae were found in open treated-water storage and tap waters. Even though the concentration of toxins was below the WHO guidelines, they remain unchanged when heated, making the water unhealthy for consumption.

Box 1: Harmful Algal Blooms (HAB) Algal blooms are intensified algae growth as a result of warming and excess nitrogen and phosphorus in water bodies. Some large types of chemical or toxins producing algal blooms are called harmful algal blooms (HABs). They are mainly the result of a type of algae called cyanobacteria, also known as blue-green algae. These toxins can damage the liver and neurological system of both humans and animals.

In the rural areas of Sudan, storage reservoirs for rainwater harvesting are the main source of drinking water for a considerable population of humans and animals.

A consistent monitoring system is not in practice, only in cases of reported emerging outbreaks of a disease or animal death. Among other water quality parameters, the biological and microbiological parameters gain little to no attention. Despite its potential toxicity, algae as a biological indicator is not considered among the water quality testing parameters. This issue urged the investigation of the types and classes of the existing algae in connection with the achievement of the sixth development goal (SDG 6).

5.2   SDG 6 Achievement: Global Concern Versus National Action

People must have adequate access to water, both in quantity and quality, as it is a basic human right. The global concern is to make water sufficiently accessible without jeopardizing people's health. The target of the United Nations Millennium Development Goals (MDGs), to reduce the proportion of people without sustainable access to improved water sources to half, was achieved by the year 2015. It was globally achieved five years ahead of schedule. By the year 2015, 91% of the global population had access to improved water sources, compared to 76% in the 1990 baseline year (UN MDGs 2015). In 2015, the UN embraced the 17 post-MDGs' Sustainable Development Goals (SDGs), aiming at global eradication of poverty and a safe planet from climate change (CC) threats by the year 2030. Water is vital to the achievement of 11 out of the 17 goals. However, the sixth goal (SDG 6) targets having universal access to safe drinking water and 'leaving no one behind' by the year 2030 (UN SDGs 2019). Nevertheless, the challenge for achieving SDG 6 is the total targeted population, which is the proportion of people who remained without access to improved water sources in the year 2015 plus the projected population by the year 2030.

Nationally, as stated in the Sudan Millennium Development Goals Progress Report (SMDG, 2010), the proportion of the population having access to improved water sources was 64% in the year 1990. In 2009, it was 65% (rising by just1% in two decades). To achieve the MDG target by the year 2015, the challenge was to add only in 6 years—more than 18% of the population to those with access to improved water sources, i.e., 83% of the population with access to improved drinking water by the year 2015. Specifically, it is needed to fill the gap of 23% for the rural population to achieve up to 50% coverage.

Literally, the targeted population to achieve the 6th SDG by 2030, taking the year 2015 as the baseline, is 63%, including the 17% population without access to improved water sources and the 45% increase in the total population of Sudan by the year 2030 (Population Pyramids of the World 2019). However, according to MoIWR and UNICEF (2019), 73.7% of the total population in the year 2019 has access to basic drinking water supply, in comparison to 68% in the baseline year of 2014. The rural population's accessibility to basic drinking water supply was 63.5% and 68.8%

Fig. 5.1 Sudan 2030 population access to basic domestic water supply taking 2014 as baseline year. *Source* MIWR and UNICEF, 2019

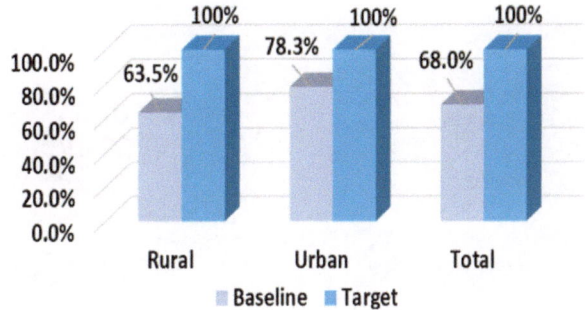

for the years 2014 and 2019, respectively (Fig. 5.1). Thus, for the rural population of Sudan, the achievement of the SDG6 will be reaching 100% coverage of the total population with basic accessibility by the year 2030, i.e., targeting 31.2% of the rural population.

> **Box 2: Basic vs safely managed drinking water supply** *A basic drinking water* is defined as drinking water from an improved source, provided collection time is not more than 30 minutes for a round trip. Improved water sources include piped water, bore-holes or tube wells, protected dug wells, protected springs, and packaged or delivered water.
>
> *A safely managed drinking water* is an improved water source that is accessible on premises, available when needed and free from faecal and priority chemical contamination.
>
> *Source* WHO/UNICEF Joint Monitoring Program (JMP)

5.2.1 Rural Sudan and the SDG6

The rural population in Sudan, according to CBS (2018), represents 67.1% of the total population. 55% live in areas apart from the Nile system and depend on non-Nilotic water resources, i.e., Wadis (ephemeral water courses) and groundwater, for drinking water supply. They access these water sources through wells and rainwater harvesting systems, such as Hafirs or earth dams. Hafir is a traditionally known natural depression or hand-dug or excavated pond to capture runoff water during the rainy season for later use (Fig. 5.2). Some long lasting Hafirs supply people and animals with drinking water during the dry season up to the onset of the second wet season.

However, the water supply is not adequately provided, neither in the desired quantity nor quality. ***Quantity-wise***, the Groundwater and Wadis Directorate (GWWD)

Fig. 5.2 Traditional rainwater harvesting technique known as Hafir (Left: after the rainy season; right: during the dry season) © Wifag Hassan Mahmoud

and the Drinking Water and Sanitation Unit (DWSU) of the Ministry of Irrigation and Water Resources (MIWR) are the authorized governmental institutions for the construction of groundwater and surface rainwater harvesting facilities, both at the national and state levels. The Dam Implementation Unit (DIU) of the MIWR has been authorized since 2010 to construct Hafirs and earth dams all over the country. *Quality-wise*, the Federal Ministry of Health (FMoH), through the Water, Environment, and Sanitation program (WES), authorized the monitoring of the water quality. According to MWRIE and FMoH (2017), the water quality monitoring microbiological (bacteriological) and chemical indicators in the country are limited to a high level of faecal contamination and specific chemicals, including fluoride, nitrate and nitrite.

> **Box 3: Definition of Hafir** A traditionally known natural depression or hand dug or excavated pond to capture runoff water during the rainy season for later use during the dry season for drinking water supply for people and animals.

The monitoring of the water quality of the surface rainwater harvesting facilities in rural areas of Sudan gains little attention, despite the high percentage of human and animal populations dependent on them as drinking water sources. RCWH and UNICEF (2019) assessed the Hafir systems in the rural areas of Sudan from technical, socio-economic, and environmental aspects. The study findings pertinent to the water quality of the Hafir systems in Sudan found that 88% of the Hafir systems in the rural areas of five representative states (Gadarif, Blue Nile, White Nile, North Darfur, and North Kordofan) are not equipped with any water treatment facilities.

The main trigger for contamination of these susceptible water sources is the practices surrounding livelihood activities and the extent of their protection. Figure 5.3 shows the dependency of people and animals on substantially deteriorated water quality from long-lasting Hafirs at the end of the dry season.

Fig. 5.3 Hafirs during the dry season serves as drinking water source for people and animals in rural area in Khartoum North © Wifag Hassan Mahmoud

With the consideration of the achievement of SDG6, the RCWH and UCWR recently conducted a study[1] to assess the water quality of these important water sources, i.e., physically, chemically, biologically, and micro-biologically. Five states were considered for the assessment, mainly Gazira, Khartoum, North Kordofan, Gadarif, and Sinnar states (Fig. 5.4). Water samples were collected from ten sites represented by Hafirs, dam reservoirs, and irrigation canals shortly after the rainy season and at the end of the dry season during the period 2020–2022. The surrounding livelihood in practice is agriculture and/or animal raising (Table 5.1). As an important but commonly ignored biological parameter, algae were taken for water contamination testing. Toxin- and chemical-producing algal blooms can be harmful to human and animal health. They result from algal intensified growth due to warming and excess nitrogen and phosphorus in the stagnant surface water of unprotected Hafirs and dam reservoirs.

5.2.1.1 Algae in Rural Rainwater Harvesting Reservoirs

As an important player in the determination of water quality in freshwater sources, algae were tested to classify the types of algae and identify the toxins' producing species as well as their relationship with some physiochemical characteristics in the rural areas of five states in Sudan. Agriculture and raising animals were the surrounding livelihood activities in the selected sampling sites. Figure 5.5 shows two sampling sites in Khartoum and Gezira States, where agriculture and raising animals' livelihood activities are practiced in one site and agricultural fields surround the irrigation canal in the other.

---

[1] "Assessment and Evaluation of the Water Quality of the Water Harvesting Facilities in Rural Areas of Sudan": conducted jointly by the UNESCO Category II Regional Center for Capacity Building and Research in Water Harvesting (***RCWH***) and the UNESCO Chair in Water Resources (***UCWR***) during the period (2020–2022).

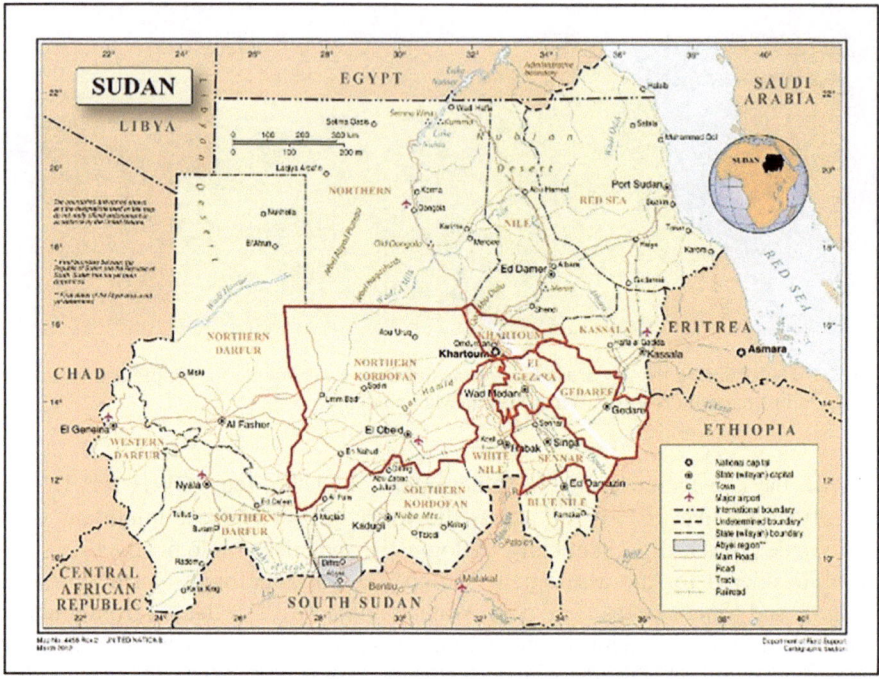

Fig. 5.4 States considered for algae investigation. *Source* Adapted on the basis of map No.4458 rev.2 (2012) from the United Nations Geospatial (https://www.un.org/geospatial/content/sudan)

Table 5.1 Location and respective surrounding livelihood activity at water sample sites

No	State	Purpose of use	Water source	Surrounding livelihood activity
1	Khartoum	Drinking water	Earth Dam	Agriculture
2	Gezira	Irrigation and drinking water	Irrigation canal	Agriculture
		Drinking water	Hafir	
3	North Kordofan	Drinking water	Hafir	Open grazing area
4	Gadarif	Drinking water	Hafir	Open grazing area
5	Sinnar	Drinking water	Hafir	–

Source Authors

5.3   Algae Classification and Distribution Pattern

This section summarizes the laboratory findings on the algae classes and their distribution with respect to the sample source. Laboratory tests were undertaken for algae and physiochemical parameters. The results were compared to the permissible limits

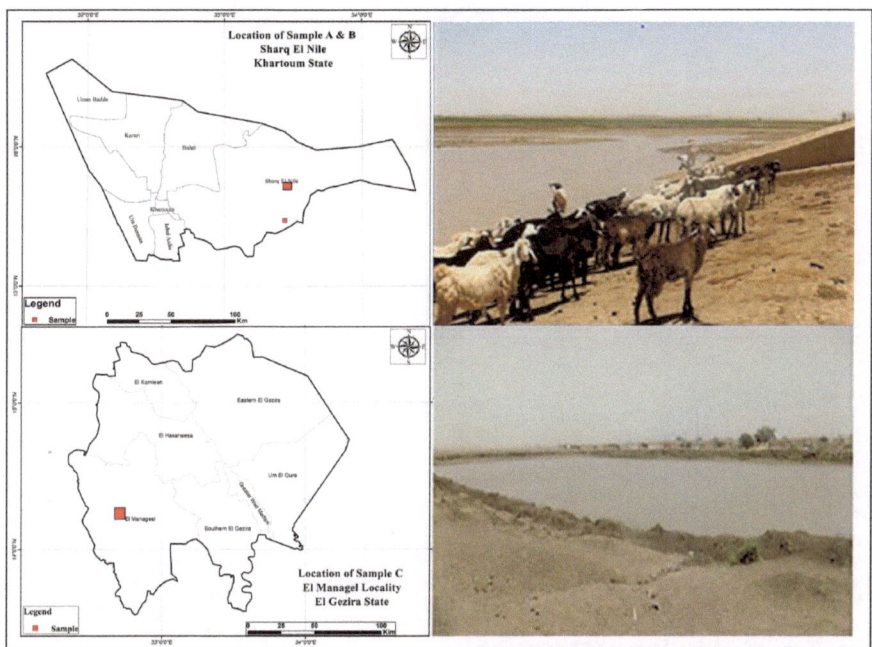

Fig. 5.5 Maps and photos of States location of the collected water samples: Hafir in Khartoum State (top) and irrigation canal in Gezira State (bottom). © Wifag Hassan Mahmoud *Source* Authors

according to Sudanese standards for the year 2016. Species of three algae groups/ communities, i.e., green algae, blue-green algae and diatoms, were detected, classified, and counted. Figure 5.6 shows a microscopic view of some algae species related to the green algae, blue-green algae, and diatom groups or communities.

The densities of the BGA counts in the sampling sites in the five states during the investigated dry and wet seasons were calculated using (Eq. 5.1)[2]. High turbidity values necessitated the use of 25 ml for the net volume. There was a great variance in the turbidity values within the 11 samples, especially during the wet season. Overall, the turbidity values ranged between 13.9–3068 NTU and 10.9–180.6 NTU for dry and wet seasons, respectively. Figure 5.7 shows a highly turbid water sample from Wadi Haseeb dam reservoir in Khartoum state. The focus of the investigation was mainly for the cyanobacteria species, as their high lighting intensity with the red color demonstrated their dominance within the total algal count, especially the organisms with toxins production ability.

$$\text{Count Density}\left(\frac{\text{Org}}{\text{ml}}\right) = \frac{\text{Actual field (1000 square)}}{\text{count field (100 square)}} * \frac{\text{Net volume (150 or 25 ml)}}{\text{Sample volume (1 L)* No. of organism}} \quad (5.1)$$

---

[2] Baird and Bridgewater (2017). Standard method for the examination of water and waste water - 23rd edition.

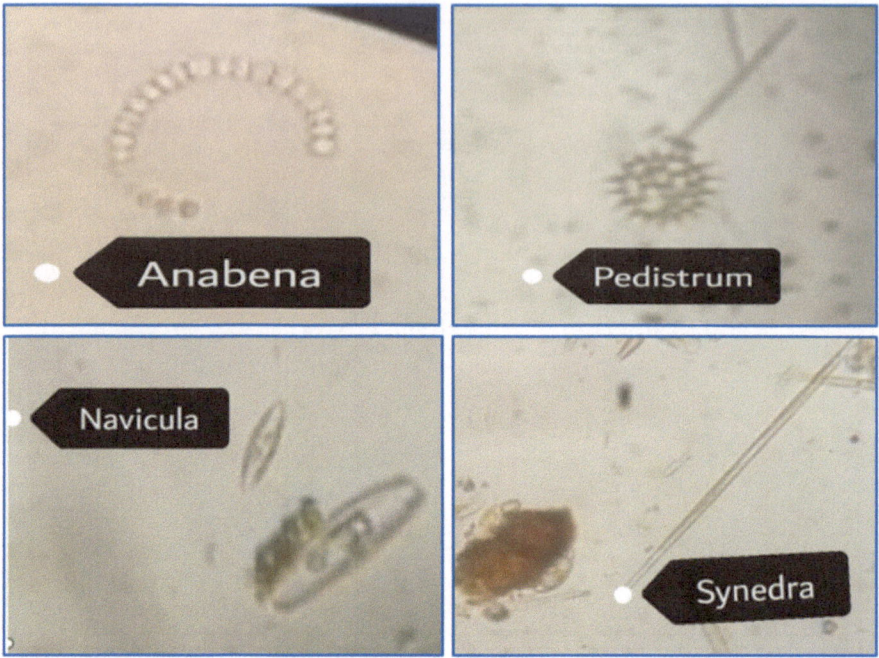

Fig. 5.6 Microscopic photographs of some species of cyanobacteria BGA (top left), green algae (top right) and diatoms (bottom left and right) © Muna Mohammed Musnad

Fig. 5.7 Microscopic view of Khartoum state water sample with high turbidity © Muna Mohammed Musnad

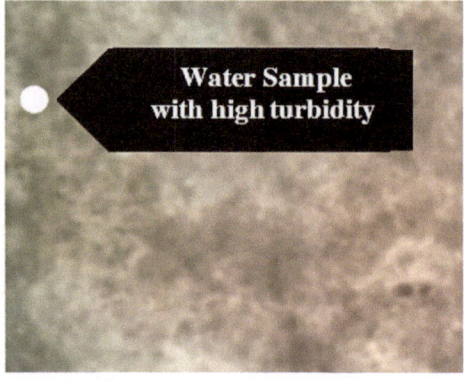

With regard to the temperature, the sampling sites are normally characterized by a moderate temperature but a hot summer. The water temperature ranged between 22.8–26.30 °C in all sites at the time of taking the samples. This temperature range could encourage some cyanotoxin species to develop outside the hot season. According to Salmaso et al. (2012), Dolichospermum, a previously considered Anabaena species, is commonly found in large blooms during the hot season. However, some of them have an optimal living temperature in the range of 19–26 °C.

5.3.1 Classification of Blue-Green Algae

Six species of cyanobactria BGA were identified: *Anabaena sp., Anacystis sp., Chroococcus sp., Cylindrospermum sp., Merismopedia sp.,* and *Oscillatoria sp.* In contrast to *Merismopedia*, the species *Anabaena, Cylindrospermum,* and *Oscillatoria* are toxic and nitrogen-fixing species. They are well-known species for the production of various toxins. On one hand, *Anabaena is* a producer of anatoxin and microcystin, which may cause gastrointestinal distress if present in water in small quantities. Yet, drinking water with a high concentration of anatoxin may cause serious neurological damage. On the other hand, *Cylindrospermum* is known for being harmful to the liver and kidney tissues, while *Oscillatoria* is a producer of liver, nerve, and skin irritant toxins.

5.3.2 Distribution Pattern of Blue-Green Algae

5.3.2.1 Gazira State

The results in Table 5.2 present the variation of algae count from three sampling points; El Mahata (Main Station) (Hafir), Galokah Village (Hafir), and Elhai Elahli (irrigation canal) in Gazira State. It was found that the water samples during the wet seasons were dominated by BGA groups, especially Anabaena sp., they represent 65%, 95.6%, and 40.6% of the total algae count for El Mahata, Galokah Village, and Elhai Elahli, respectively. It is important to notice that Anabaena is a toxin-producing species. So, the use of this water for drinking purposes without proper treatment is very risky to human health.

5.3.2.2 Khartoum State

Table 5.3 shows the results of the BAG count for Wad Haseeb Point A and Point B in Khartoum State. It is important to show that Anabaena sp. was found during only the wet season and *Cylindrospermum sp.* during the dry season. However, *Oscillatoria sp.* is present in both the dry and wet seasons. The percentage of BGA count to the total algal count ranged from 40.3% to 98.3%, which was dominated by three toxins-producing genera: *Anabaena sp., Cylindrospermum sp.,* and *Oscillatoria sp.*

Table 5.2 Algae Community Structure at Gazira State

Blue-Green algae species	Main Station (El Mahata) Hafir		Galokah village Hafir		Elhai Elahli irrigation canal	
	Dry season	Wet season	Dry season	Wet season	Dry season	Wet season
Anabaena	6096	80,106	3000	**414,690**	7500	21,798
Anacystis	45,720	ND	15,000	ND	45,000	ND
Chroococcus	ND	15,210	ND	25,242	ND	10,899
Cylindrospermum	ND	ND	ND	ND	6000	519
Merismopedia	ND	507	3000	ND	1500	519
Oscillatoria	ND	1014	7500	1202	12,000	4152
Total BGA Counts (Org/ml)	51,816	96,837	28,500	441,134	72,000	37,887
Total Algal Counts (Org/ml) *	248,412	149,058	114,000	461,564	333,000	93,420
(%) BGA of total Algal Count	20.9	65	25	95.6	21.6	40.6

Source Authors
*The summation of the count of Green Algae, BGA and Diatoms in the sample. NA: No water available; ND: not detected

Table 5.3 Algae Community Structure at Khartoum State

Blue-Green algae species	Wadi Haseeb Dam Reservoir			
	Point A		Point B	
	Dry season	Wet season	Dry season	Wet season
Anabaena	ND	**570,000**	ND	**907,500**
Anacystis	6176	ND	1050	ND
Chroococcus	ND	17,500	ND	37,500
Cylindrospermum	3288	ND	18,975	ND
Merismopedia	822	ND	ND	ND
Oscillatoria	26,304	7500	2250	2500
Total BGA Counts (Org/ml)	**36,590**	**595,000**	**22,275**	**947,500**
Total Algal Counts (Org/ml) *	**90,842**	**702,500**	**27,375**	**963,750**
(%) BGA of total Algal Count	40.3	84.7	81.4	98.3

Source Authors
*The summation of the count of Green Algae, BGA and Diatoms in the sample ND: not detected

5.3.2.3 North Kordofan State

The algal count of Bagara Point A and Alain Point B at North Kordofan (Shikan) in Table 5.4 shows the 78%–97.8% dominance of the BGA group in both the dry and

Table 5.4 Algae Community Structure at North Kordofan State

Blue-Green algae species	Bagara Hafir Point A		Alain Hafir Point B	
	Dry season	Wet season	Dry season	Wet season
Anabaena	ND	62,500	NA	60,528
Anacystis	10,120	ND	NA	ND
Chroococcus	ND	20,000	NA	70,616
Cylindrospermum	1518	**1,22000**	NA	**1,452,62**
Merismopedia	**73,370**	7500	NA	**35,308**
Oscillatoria	3289	37,500	NA	**5,044,00**
Total BGA Counts (Org/ml)	**88,297**	**1,34700**	**NA**	**6,663,14**
Total Algal Counts (Org/ml) *	**112,585**	**1,41700**	NA	**6,814,14**
(%) BGA of total Algal Count	78%	95.1%	NA	97.8%

Source Authors
*The summation of the count of Green Algae, BGA and Diatoms in the sampleNA: No water available; ND: not detected

wet seasons. At Bagara Point A, the algal count during the dry season was dominated by the nontoxic genus BGA Merismopedia spp. However, care and continuous monitoring of the algae community structure count are highly needed as a change to dominance by other toxic genus may happen in the next few seasons.

5.3.2.4   Gadarif State

Unfortunately, data for the dry seasons of the two sampling sites; Gadamblia and Azaza Hafirs in Gadarif, were not available. The results in Table 5.5 show that the algal count was absolutely dominated by BGA, which represents 95%–96.9% of the total algal count. Comparing the count of BGA in Gadarif State with the other states shows very low values. Yet, they are still dominated by toxin-producing BGA.

5.3.2.5   Sinnar State

The algae count of Galaat and Gebis Hafirs in Sinnar State is shown in Table 5.6. It is found that only the dry season of Galaat Hafir was dominated by the BGA toxin-producing Oscillatoria sp., i.e., 84.5% of the total algae count; however, it is very low in count. This means low algae density, which may cause no harm.

Table 5.5 Algae Community Structure at Gadarif State

Blue-Green algae species	Gadambalia Hafir		Azaza Hafir	
	Dry season	Wet season	Dry season	Wet season
Anabaena	NA	28,224	NA	100
Chroococcus	NA	2520	NA	300
Cylindrospermum	NA	1008	NA	ND
Merismopedia	NA	ND	NA	7100
Oscillatoria	NA	ND	NA	100
Total BGA Counts (Org/ml)	NA	**31,752**	NA	**7600**
Total Algal Counts (Org/ml) *	NA	**32,760**	NA	**8000**
(%) BGA of total Algal Count	NA	96.9	NA	95

Source Authors
*The summation of the count of Green Algae, BGA and Diatoms in the sample; NA: No water available; ND: not detected

Table 5.6 Algae Community Structure at Sinnar State

Blue-Green algae species	Galaat Hafir		Gebis Hafir	
	Dry season	Wet season	Dry season	Wet season
Anabaena	36	26	775	ND
Chroococcus	99	26	62	270
Cylindrospermum	36	ND	ND	30
Oscillatoria	4995	130	1116	90
Total BGA Counts (Org/ml)	**5166**	**182**	**1953**	**390**
Total Algal Counts (Org/ml) *	**6110**	**1300**	**5983**	**3600**
(%) BGA of total Algal Count	84.5	14	32.6	10.8

Source Authors
*The summation of the count of Green Algae, BGA and Diatoms in the sample, ND: not detected

5.4   Blue-Green Algae: An Emerging Pollutant in Rural Sudan?

Analysis of the laboratory results shows that the rural population that accesses drinking water from unimproved surface rainwater harvesting reservoirs is at risk. Toxin producing BGA cyanobacteria do exist in these reservoirs. The prevailing perception of algae being an unharmful contaminant and not being considered for testing has faded now. Table 5.7 and Fig. 5.8 summarize the presence and count percentage of the blue-green algae species in the inspected areas. In practice, the common livelihood is open grazing and cultivation, which encourage the growth of algal blooms in general and the BGA in particular. *Anabaena,* the producer of anatoxin and microcystin toxins, is found to be dominant in the wet season in four states. This species is responsible for producing anatoxin and microcystin toxins, among others. When digested in small amounts, they may cause gastrointestinal distress. Higher levels of anatoxin may result in neurological damage. However, in the states where agricultural activity is the main livelihood practice, the *Anabaena sp.* count was extremely high in comparison to the states with pasture activity for the same season, i.e., the wet season.

Oscillatoria and *Cylindrospermum* species are dominating the state with the greatest open natural pasture and dense animal population in the country (North Kordofan State). The extremely high count during the wet season is possibly attributed to the fact that animal waste is being leached by the runoff water during the rainy season into the water source. Both species are found in water rich in organic matter during the hot season. *Oscillatoria sp.* is known for being a producer of liver, nerve, and skin irritant toxins, e.g., microcystins, saxitoxin, and lipopolysaccharides, while *Cylindrospermum sp.* is known for being harmful to the liver and kidney tissues.

The non-toxic *Merismopedia sp.* is found mostly in all states with agricultural and pasture activities but varies in count from one state to another. Yet, the higher count seems to dominate the dry season, privileging the states with pasture livelihood activities. This coincides with the fact that *Merismopedia sp.,* is *found* in the sedimentation of freshwater sources. The other non-toxic species, i.e., *Chroococcus sp.* is found in all of the states without livelihood preference. It is found only in the wet season, as this species lives only in underwater environments.

It is worth mentioning that the Hafir sites in Sinnar state, where there is no presence of agricultural or open pastural activities, show minimal count values of cyanobacteria species. The count values in comparison to the other sites are negligible, with the exception of *Oscillatoria sp.;* which has a high value, especially during the dry season.

Table 5.7 Distribution of BGA species within the inspected areas

State	Livelihood Activity	Dominance of BGA sp. and the Season				
		Ana	Cyl	Chr	Osc	Mer
Gezira	Agriculture	W	-	W	-	D
Khartoum	Agriculture	W	-	W	-	D
North Kurdufan	Pasture	-	W	W	W	D/W
Gadarif	Pasture	-	-	W	-	W
Sinnar	-	-	-	-	-	-

Ana.: *Anabaena sp.;* Cyl.: *Cylindrospermum sp.;* Chr.: *Chroococcus sp.;* Ose.: *Oscillatoria sp.;* Mer.: *Merismopedia sp.;* W: Wet season (shortly after the rainy season); D: Dry season (end of the hot summer season)

Source Authors

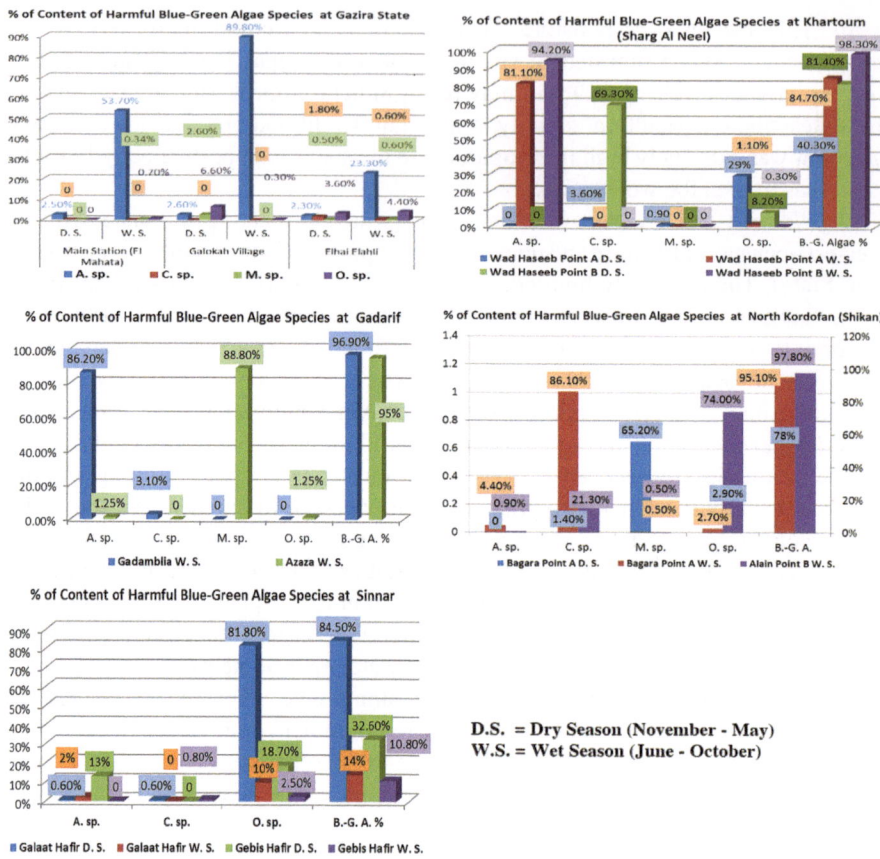

Fig. 5.8 Distribution and count percentage of the BGA species in the inspected areas. *Source* Authors

5.5 Conclusion and Way Forward

This is the pioneering study of its type that sheds light on a significant but rarely considered hazardous biological indicator of water quality. Algae investigation of remotely distributed rural rainwater harvesting facilities affirmed the presence of harmful species. The investigation was based only on the algae count. Koreivienė et al. (2014) recommended this methodology as a cost-effective monitoring method for the analysis of microcystin in reservoirs with the potential presence of toxic cyanobacteria. It is apparent that the rural population is at high risk for water-related diseases pertinent to cytotoxins producing blue-green algae species. These toxins are associated with severe liver and neurological diseases if they exist in water with high levels of concentration. This necessitates conducting further studies pertinent to the description of the toxins for the cyanotoxin species and their respective concentrations and compliance with the national and international standers. In addition, the favorable environment for the development of harmful BGA blooms should be eliminated via the rehabilitation or construction of a well-maintained, improved water source. Furthermore, comprehensive water quality monitoring and algae toxicity prevention measures are necessary to be in place for improved natural resources, sustainably maintained ecosystems, and well-preserved public health. Koreiviené et al. (2014) recommended cell count as a simple and cost-effective monitoring method for drinking water reservoirs. A threshold of 4000 cells/m is recommended for the analysis of Microcystin in reservoirs with potential presence of toxic cyanobacteria. From an economic point of view, Woodhouse et al. (2014) recommended investing in low-cost monitoring systems where toxic algae blooms can be early detected, and accordingly, the economic impact will be lessened by the timely prevention of accessibility to the contaminated water. Yet, rather expensive and sophisticated methods, e.g., surveillance by remote sensing, polymerase chain reaction (PCR), or fluorescence spectra, were recommended by Hamilton et al. (2014) for treating and monitoring water reservoirs and wastewater plants with the potential presence of cyanotoxin. Moreover, a grave need for sustainably managed and developed water resources is to be built on capacity development, citizen science, and political will. This way, the endeavor to achieve the SDGs by 2030 will eventually be enhanced, in particular the sixth goal that urges universal access to safe drinking water, leaving no one behind.

Acknowledgements The funds provided by the Ministry of Higher Education and Scientific Research and the Ministry of Finance and Economic Planning – Development Division—of Sudan are highly acknowledged. The authors would also like to acknowledge the technical team's provision for the whole study during the field visits and the laboratory data analysis and interpretations, as well as the provision of the shapefiles in Figures 5.4 and 5.5 by Mr. Safe Zarroug. The constructive feedback by two anonymous reviewers is highly appreciated and acknowledged by the authors.

References

Baird R, Bridgewater L (2017) Standard methods for the examination of water and wastewater, 23rd edn. D.C., American Public Health Association, Washington

Beaver JR, Tausz CE, Scotese KC, Pollard AI, Mitchell RM (2018) Environmental factors influencing the quantitative distribution of microcystin and common potentially toxigenic cyanobacteria in us lakes and reservoirs. Harmful Algae 78:118–128. https://doi.org/10.1016/j.hal.2018.08.004

Bonilla S, Aguilera A, Aubriot L, Huszar V, Almanza V, Haakonsson S, Antoniades D (2023) Nutrients and not temperature are the key drivers for cyanobacterial biomass in the Americas. Harmful Algae 121:102367

Buratti FM, Manganelli M, Vichi S, Stefanelli M, Scardala S, Testai E, Funari E (2017) Cyanotoxins: producing organisms, occurrence, toxicity, mechanism of action and human health toxicological risk evaluation. Arch Toxicol 91(3):1049–1130. https://doi.org/10.1007/s00204-016-1913-6. Epub 2017 PMID: 28110405

CBS (2018) Central Bureau of Statistics, Population Statistics, Population Indicators. http://cbs.gov.sd/index.php/ar/statistics/main/7

Habtemariam H, Kifle D, Leta S (2021) Cyanotoxins in drinking water supply reservoir (Legedadi, Central Ethiopia): implications for public health safety. SN Appl Sci 3:328. https://doi.org/10.1007/s42452-021-04313-0

Hamilton DP, Wood SA, Dietrich DR, Puddick J (2014) Costs of harmful blooms of freshwater cyanobacteria. In Cyanobacteria (eds N.K. Sharma, A.K. Rai and L.J. Stal).https://doi.org/10.1002/9781118402238.ch15

Harvell CD, Kim K, Burkholder JM, Colwell RR, Epstein PR, Grimes DJ, Hoffman EE, Lipp EK, Osterhaus ADME, Overstreet RM, Porter JW, Smith GW, Vasta GR (2000) Emerging marine diseases-climate links and anthropogenic factors. Science 285:1505–1510. https://doi.org/10.1126/science.285.5433.1505

Koreivienė J, Anne O, Kasperovičienė J, Burškytė V (2014) Cyanotoxin management and human health risk mitigation in recreational waters. Environ Monit Assess 186(7):4443–4459. https://doi.org/10.1007/s10661-014-3710-0. Epub 2014 Mar 25 PMID: 24664523

Lorenzi AS, Cordeiro-Araújo MK, Chia MA (2018) Cyanotoxin contamination of semiarid drinking water supply reservoirs. Environ Earth Sci 77:595. https://doi.org/10.1007/s12665-018-7774-y

Mackeigan PW, Taranu ZE, Pick FR, Beisner BE, Gregory-Eaves I (2023) Both biotic and abiotic predictors explain significant variation in cyanobacteria biomass across lakes from temperature to subarctic zones. Limnology and Oceanography

Mânica AN, de Lima Isaac R (2023) Seasonal dynamics and diversity of cyanobacteria in a eutrophied Urban River in Brazil. Water Supply

MIWR and UNICEF (2019) Sudan Plan to Achieve the SDG 6. Report developed by the Ministry of Irrigation and Water Resources and UNICEF WASH—Sudan and the Federal Ministry of Health

Mohamed ZA, Al Shehri AM (2007) Cyanobacteria and their toxins in treated-water storage reservoirs in Abha city. Saudi Arabia Toxicon 50(1):75–84. https://doi.org/10.1016/j.toxicon.2007.02.021

MWRIE and FMoH (2017). Contextual Analysis-Drinking Water Safety in Sudan. Report by Ministry of Irrigation and Water Resources and Federal Ministry of Health

Population Pyramids of the World (2019). Population Pyramids of the World from 1950 to 2100. https://www.populationpyramid.net/sudan/

RCWH and UNICEF (2019): Assessment of Hafir Systems' in Sudan: Technical, Socio-economic and Environmental Aspects. Report by UNESCO Category II Regional Center for Capacity Development and Research in Water Harvesting and UNICEF – WASH sector Sudan

Salmaso N, Buzzi F, Garibaldi L, Morabito G, Simona M (2012) Effects of nutrient availability and temperature on phytoplankton development: a case study from large lakes south of the Alps. Aquat Sci 74:555–570

SMDG (2010). Sudan Millennium Development Goals Progress Report www.undp.org (Accessed January 2020)

UN MDGs (2015). The United Nations Millennium Development Goals Report https://www.un.org/millenniumgoals/

UN SDGs (2019): UN sustainable development goals report https://sustainabledevelopment.un.org/sdg6

Visser PM, Verspagen JM, Sandrini G, Stal LJ, Matthijs HC, Davis TW, Paerl HW, Huisman J (2016) How rising co2 and global warming may stimulate harmful cyanobacterial blooms. Harmful Algae 54:145–159. https://doi.org/10.1016/j.hal.2015.12.006

Vu H, Luong N, Jakub Z, Tran N, Long N (2020) Blue-green algae in surface water: Problems and opportunities. Current Pollution Reports 6. https://doi.org/10.1007/s40726-020-00140-w

Woodhouse JN, Rapadas M, Neilan BA (2014) Cyanotoxins. In: Sharma NK, Rai AK, Stal LJ (eds) Cyanobacteria. https://doi.org/10.1002/9781118402238.ch16

The opinions expressed in this chapter are those of the author(s) and do not necessarily reflect the views of the UNESCO: United Nations Educational, Scientific and Cultural Organization, its Board of Directors, or the countries they represent.

Chapter 6
Prioritization of Chemicals in Aquatic Ecosystems in Türkiye

Esra Şıltu, Sibel Mine Güçver, Gülnur Ölmez, Burcu Cömert, and Aybala Koç Orhon

Abstract Chemicals have a variety of irreplaceable and beneficial uses in the modern life. However, their production, trade and use should be regulated due to the adverse effects they may have on human health, the ecosystems and the environment. This study was conducted with the aim of identifying the emerging aquatic pollutants in Türkiye and review and update the specific pollutants listed in the By-Law on Surface Water Quality. In scope of the study, 199 emerging pollutants from the list of specific pollutants of the By-Law on Surface Water Quality were undergone an ecological risk assessment by applying the NORMAN prioritization methodology. The data on physicochemical and toxicological properties and uses of the chemicals was collected from several publicly available databases. Moreover, ambient water monitoring data collected between 2011 and 2019 in Türkiye was used for exposure assessment. Exposure, hazard and risks scores and overall scores were calculated for 199 chemicals. The chemicals were ranked according to the overall score and 53 chemicals were proposed to have higher risks to aquatic environment and to be assessed more rigorously. The study will be elaborated with additional monitoring data to be obtained from ongoing and future studies and will inform the policy making procedure on updating the specific pollutants list to protect water resources and the human health. By this feature, the study demonstrates a good example of bridging the science and policy.

Keywords Emerging aquatic pollutants · Ecological risk assessments · Prioritization of chemicals

E. Şıltu (✉) · S. M. Güçver · G. Ölmez · A. Koç Orhon
Ministry of Agriculture and Forestry, General Directorate of Water Management, Ankara, Türkiye
e-mail: esra.siltu@tarimorman.gov.tr

B. Cömert
Ministry of Agriculture and Forestry, 21 Regional Directorate of State Hydraulic Works, Aydın, Türkiye

© UNESCO 2025
S. Zandaryaa et al. (eds.), *Emerging Pollutants*, Advances in Water Security,
https://doi.org/10.1007/978-3-031-71758-1_6

109

6.1 Introduction

Chemicals have a vast variety of beneficial and inevitable uses in our daily lives, industrial applications, agriculture and in any other activities of humankind. We owe significant economic and social developments to the use of xenobiotic chemicals. The chemical industry is still one of the largest industries in the world. The global production of chemicals has increased from 1 million tons in 1930 to several hundreds of million tons in 2010s (Koskinen 2017). Furthermore, it is estimated that there are 40,000 to 60,000 industrial chemicals in commerce and approximately 6,000 of these chemicals account for more than 99% of the total volume of industrial chemicals in commerce globally (UNEP and ICCA 2019). According to the data of Organization for Economic Cooperation and Development (OECD), the revenue of the global chemical industry is larger than EUR 4,000 billion annually and it is estimated to exceed EUR 20,000 billion by 2060 (OECD 2021).

Despite their irreplaceable beneficial uses in modern life, the synthetic chemicals may have various adverse effects on the ecosystems and human health. The World Health Organization estimated that 2 million people lost their life due to preventable exposures to chemicals which corresponds to the 3.6% of total deaths in 2019 (WHO 2019). Due to these adverse effects, the production, marketing, export, import, use, discharge, disposal, and residual levels of chemicals in various products, food and environmental compartments are regulated via national legislations and/or international agreements. As regulating all the chemicals existing in the market is not feasible and necessary, a risk assessment and prioritization study is the mandatory first step of any regulatory process. There are various approaches and methodologies currently adopted in the prioritization and risk assessment of chemicals depending on the target of the regulatory process. The focus of this study is the prioritization and risk assessment of chemicals which may create concern regarding the integrity of aquatic environments and pollution of water resources. In this context, this study is aiming to briefly introduce the background and evolvement of the regulatory action towards chemicals with the perspectives from the USA, European Union, and Türkiye as well as the international agreements, and to introduce a recent case study from Türkiye on identification of aquatic emerging chemicals. The presented case study is also supposed to assist the review and update of the specific pollutants list of the By-Law on Surface Water Quality.

In this context, the existing list of 250 specific pollutants which is composed of various types of chemicals of the By-Law were undergone an ecological risk assessment by applying the NORMAN (Network of Reference Laboratories and Related Organizations for Monitoring and Bio-monitoring of Emerging Environmental Substances) prioritization methodology. The methodology includes a comprehensive assessment of the persistency, bioaccumulation, and toxicological properties of the chemicals as well as the monitored levels and detection frequency. The details of the methodology are outlined in Sect.6.5.1. According to this methodology, hazard score, exposure score, risk score and a final prioritization score were calculated for each of the chemicals. However, the initial list of 250 chemicals contains a number of chemicals which have already been detected in aquatic environments for years or some that had already been banned and cannot be defined as "emerging chemicals". Hence, those chemicals were not included in the assessments in this study. The final list includes 199 chemical pollutants which are being used or were formerly used in several industrial applications, agriculture or for personal care and health applications or were detected in preliminary monitoring studies in Türkiye. These 199 chemicals were ranked comparatively based on their final prioritization score and 53 chemicals were claimed to pose higher risk to aquatic environments compared to the remaining of the list with having a final prioritization score of 1 or greater. Phthalates and pesticides were the major groups among the 53 chemicals. These 53 chemicals were proposed to be considered in the revision of the By-Law on Surface Water Quality. Finally, the results of the study were briefly discussed, and some suggestions were presented for the management of chemicals in Türkiye.

6.2 Background and Evolvement of Regulatory Action on Chemicals

In (1962), Rachel Carson discovered that pesticides have undesired effects on organisms and the environment. The discovery of Carson and the release of her book titled "Silent Spring" raised public awareness on the adverse effects of chemicals. Carson's discovery also attracted the attention of the scientific community towards the impacts of chemicals and many studies on detecting the levels of chemicals in the environment and investigating the adverse ecological and human health effects associated with their use have been conducted since then (Faber and Hickey 1973; Wiemeyer et al. 1984; Zhang et al. 1992; Taha and Gray 1993; Nurminen 1995; Safe 2000; Grandjean and Landrigan 2006). The studies revealed that exposure to synthetic chemicals may cause cancer (Oakley et al. 1996; Zahm and Ward 1998; Bassil et al. 2007; Van Maele-Fabry et al. 2011), reproductive (Saradha and Mathur 2006; Mendola et al. 2008) and neurodevelopmental disorders (Mendola et al. 2002; Needham et al. 2005; Rauh and Margolis 2016), cardiovascular diseases (Kristensen 1989; Cosselman et al. 2015; Huang et al. 2018; Fu et al. 2020) systemic-metabolic diseases and disruption of the endocrine system in humans and other biota (Colborn et al. 1993; Vos et al. 2000; Heindel et al. 2017; Le Magueresse-Battistoni et al. 2017). Moreover, aquatic, and terrestrial ecosystems are adversely affected by the chemical pollution as chemicals can disrupt the functions of pollinators, accelerates antimicrobial resistance, causing reproductive, immunological and neurological damage in animals and reducing biodiversity (UNEP 2019). For instance, chemicals in water resources may cause feminization of fish, amphibians, and reptiles and developmental delays, acceleration, and malformations in amphibians (UNEP 2013). In the European Union, it is reported that more than 34 million tons of carcinogenic, mutagenic and reprotoxic (CMR) chemicals were produced and consumed in 2020 (EUROSTAT 2021) alone. In addition to this, World Health Organization (2019) estimated the burden of disease from preventable exposure to chemicals as 1,550,334 deaths in 2019 alone which corresponds to 2.7% of total deaths.

The recognition of the adverse health effects of chemicals on human health and the environment entailed regulatory action on the production, trade, and use of chemicals and this eventually resulted in several regulations on chemicals. Besides the release of national regulations, global regulatory actions resulted in international treaties namely the Stockholm Convention and Rotterdam Convention. Furthermore, there is an ongoing process on drafting of an international treaty to control the plastic pollution led by United Nations Environment Program and it is aimed to be finalized by the end of 2024. The overall common target of these national regulations and international treaties is to minimize the exposure of human and other organisms to synthetic man-made chemicals and reduce the health and environmental burden of chemicals. Nevertheless, regulating all the chemicals in commerce is not feasible due to the cost that would be associated with the control measures and the indispensable uses of some chemicals. Hence, the foundation of all chemical-related regulations lies in the evaluation of risks associated with the use of chemicals, enabling the screening and prioritization of chemicals that warrant regulation. Since water is one of the main routes of exposure of humans and biota to chemicals, countries bring into effect several pieces of legislation on the control of chemicals in water resources. European Union took a particularly important step in 1976 and put the "Council Directive 76/464/EEC on Pollution Caused by Certain Dangerous Substances Discharged into the Aquatic Environment of the Community" into effect. The discharges of certain chemicals were controlled via the limits designated by the Directive. Afterwards, Directive 76/464/EEC was repealed by Water Framework Directive (2000/60/EC) and the terms "priority substances" and "specific pollutants" were introduced. The Directive identified 33 priority pollutants initially and obligated the European Union member states to identify their specific pollutants and environmental quality standards for these chemicals. Türkiye is also obliged to transpose the horizontal and framework environmental legislation of the European Union to the national legislation as one of the requirements of the European Union accession period. In this context, studies were initiated in 2011 to identify the river basin specific pollutants of Türkiye and to derive environmental quality standards for those chemicals. Consequently, 250 specific pollutants and their environmental quality standards were designated via By-Law on Surface Water Quality in 2016.

6.3   Regulations on Chemicals Management

The recognition of the adverse effects of chemicals on human health and the environment yielded bans or restrictions on the production, trade, and use of some of the widely used chemicals through regulatory action. Nevertheless, the regulations on management of chemicals are not confined to limitations on the production and use of chemicals. As response to scientific advances in the detection of chemicals in various environmental media as well as the bodies of various organisms from different trophic states and investigation of the health effects of chemicals, regulations also evolved, varied and became more target oriented. The legislative materials on management of chemicals enacted in countries can be categorized broadly based on their respective functions as:

- Regulations on registration and classification of chemicals
- Regulations defining the ambient environmental standards and emission limits for chemicals
- Regulations defining the standards for chemical residues in food
- Regulations defining the standards for chemicals in consumer products
- Regulations on biocides, pesticides and pharmaceuticals
- Regulations on the restriction and ban of manufacture and use of certain chemicals.

The framework of the legislation on chemicals management in the United States of America (USA), the European Union, and Türkiye are briefly outlined for the purpose of comparison in the subsequent sub-chapters.

6.3.1 Regulation of Chemicals in the USA

The regulatory actions on control of chemicals date back to 1950s in the USA. The release of Rachel Carson's book "Silent Spring" also proliferated widespread concern over the adverse effects of pesticides on wildlife and consequently, DDT was banned in the USA in 1972 by the Environmental Protection Agency (EPA) (EPA 2023a). Then, another group of famous chemicals, namely PCBs which were used widely in industrial and consumer products, were banned in the USA in 1979 (CDC 2017). The foundation of EPA and the issuing of bans or restrictions on several chemicals paved the way to enacting legislation on the management of chemicals. Currently, the main legislative documents on controlling the production, trade and use of chemicals are:

- Toxic Substances Control Act (TSCA)
- Pollution Prevention Act
- Federal Insecticide, Fungicide, and Rodenticide Act (FIFRA)
- Federal Food, Drug and Cosmetic Act (FFDCA).

The fundamental legislation on controlling the exposure to hazardous chemicals is the Toxic Substances Control Act. The Toxic Substances Control Act entered into force in 1976 and was amended by the Frank R. Lautenberg Chemical Safety Act of the 21st Century in 2016. The act regulates the production, use, trade, import, export and disposal of chemicals found to be hazardous to humans and the ecosystem. Drugs, pesticides, biocides, food additives and personal care products are not in the scope of the act. The manufacturers, importers, exporters or processors of the chemicals are obliged to test the chemicals to identify their risks to environment and human health and to declare the results of the test to EPA. The risk assessment conducted under the act is based on assessment of hazards and the exposure potential to humans. An inventory of chemicals produced or used in the USA has been compiled in accordance with the provisions of the act. As of 2024, there are more than 86,000 chemicals in the inventory. New chemicals produced commercially are placed on the list (EPA 2022).

The Pollution Prevention Act is another piece of legislation that aims to reduce pollution at the source by implementing certain practices to limit the emission of hazardous substances to the environment. The act entered into force in 1990 and designated "toxic substances" of which emissions must be reduced and the reduction reports must be submitted to EPA annually (EPA 2023b).

Although not solely developed to control chemical hazards, the Clean Water Act is a substantial legislation in the USA to protect water resources from the pollution caused by point sources. The foundation of the Clean Water Act (CWA) was established in 1948 under the name of the Federal Water Pollution Control Act. However, a substantial restructuring and expansion of the Act occurred in 1972, leading to its adoption of the title "Clean Water Act" following the 1972 amendments (EPA 2023c). The Clean Water Act introduced interrelated lists of chemicals. Initially, the "Toxic Substances List" was developed in 1976 and the discharge limits and water quality criteria recommendations were designated for the chemicals in the list. Then, the "Priority Pollutants List" was developed in 1977 based on the "Toxic Substances List" with the aim of facilitating the implementation of the "Toxic Substances List". The "Priority Pollutants List" currently contains 126 chemicals from the "Toxic Substances List", which were detected in water resources more frequently than 2.5% of the tested samples and were produced in substantial volumes (EPA 2023d). Testing standards were also developed for chemicals in the "Priority Pollutants List".

Federal Insecticide, Fungicide, and Rodenticide Act (FIFRA) and Federal Food, Drug and Cosmetic Act (FFDCA) introduce provisions related to the control of chemicals in cosmetic products and pharmaceuticals.

6.3.2   Regulation of Chemicals in the European Union

In the European Union, several pieces of legislation are enacted to control the production, authorization, trade and use of chemicals. The most prominent ones of these legislations can be listed as:

- Regulation (EC) No 1907/2006 of the European Parliament and of the Council of 18 December 2006 concerning the Registration, Evaluation, Authorisation and Restriction of Chemicals (REACH)
- Regulation (EC) No 1272/2008 of the European Parliament and of the Council of 16 December 2008 on Classification, Labelling and Packaging of Substances and Mixtures (CLP)
- The Biocidal Products Regulation (BPR, Regulation (EU) 528/2012)
- The Chemical Agents Directive (Directive 98/24/EC, CAD) and The Carcinogens, Mutagens or Reprotoxic Substances Directive (Directive 2004/37/EC, CMRD from 9 March 2022)
- Regulation (EU) No 2019/1021 of the European Parliament and of the Council of 20 June 2019 on Persistent Organic Pollutants (POPs Regulation)
- Prior Informed Consent Regulation (Regulation (EU) No 649/2012 of the European Parliament and of the Council of 4 July 2012 Concerning the Export and Import of Hazardous Chemicals; PIC Regulation).

Issues pertaining to the registration and authorization of chemicals are subject to regulation under the REACH (Registration, Evaluation, Authorization, and Restriction of Chemicals) framework. The fundamental objective of this regulatory framework is to enhance the protection of both human health and the environment by identifying and evaluating the intrinsic properties of chemicals. This objective is achieved through the rigorous processes of chemical registration, evaluation, authorization, and restriction. The REACH Regulation imposes a requirement upon the industrial sector to register chemicals that are either manufactured or imported in quantities exceeding one metric ton annually. Furthermore, it mandates industry stakeholders to effectively manage the risks associated with these chemicals. Information pertaining to the properties of chemicals, originating from manufacturers and importers, is compiled and stored within a database maintained and administered by the European Chemicals Agency (ECHA).

The CLP Regulation requires the proper classification, labeling, and packaging of hazardous chemicals which are used in industrial processes and consumer products by manufacturers, importers or downstream users prior to release to the market according to the physical, health, environmental and additional hazard criteria defined by the regulation.

Through the implementation of REACH and CLP Regulations, the manufacture, placing on the market and use of toxic chemicals are authorized and controlled and the exposure potential of humans and other biota to toxic chemicals is reduced.

On the other hand, the launch of biocidal products to the market which are designed for the protection of humans, animals, materials, or articles against harmful organisms, including pests and bacteria through the active substances contained are regulated via the Biocidal Products Regulation with a concurrent commitment to maintaining a high level of safeguarding for human health and the environment.

The Chemical Agents Directive and the Carcinogens, Mutagens or Reprotoxic Substances Directive designate the occupational limit values for carcinogens, mutagens or reprotoxic substances.

In addition to these, POPs and PICs regulations harmonize the provisions of Stockholm Convention on Persistent Organic Pollutants and Rotterdam Convention on Procedure for Certain Hazardous Chemicals and Pesticides in International Trade in the European Union legislation and defines rules and regulations for management of persistent organic pollutants designated by the Stockholm Convention and trade of certain hazardous chemicals and pesticides designated by the Rotterdam Convention.

On the other hand, the Water Framework Directive (2000/60/EC), Environmental Quality Standards Directives (Directive 2008/105/EC of the European Parliament and of the Council of 16 December 2008 on Environmental Quality Standards in the Field of Water Policy and Directive 2013/39/EU) are the fundamental regulations designating the European Union-wide priority substances and environmental quality standards which are the ambient water quality criteria developed based on the toxicity to aquatic organisms. The list of priority substances is analogous to priority pollutants list in the USA. In the Water Framework Directive, priority substances are defined as substances that identified to pose significant risks to or via aquatic environment by the risk assessment conducted in a precautionary manner. Some of the chemicals within the priority pollutants list are identified as priority hazardous substances due to higher risk posed to aquatic ecosystem and the human health. The

Directive requires the progressive reduction of pollution from priority substances and ceasing or phasing out emissions, discharges and losses of priority hazardous substances among the European Union. The identification of priority substances is designed as a dynamic process and the European Commission is required to update the list every six years following the first review. Initially, 33 priority substances were listed in the Water Framework Directive in 2000 when it was first released. Then the Directive was amended by the Environmental Quality Standards Directive (2008/105/EC) in 2008 and the list of 33 chemicals were published in the annexes of this directive together with the corresponding environmental quality standards. Later on, the list was reviewed and updated by the European Commission and the final list of priority substances comprised of 45 chemicals and environmental quality standards were released by the Directive 2013/39/EU. The levels of priority substances in surface waters are assessed under the chemical status of water bodies according to the provisions of the Water Framework Directive.

Moreover, these regulations impose on member states the requirement to designate their specific pollutants and the corresponding environmental quality standards. The specific pollutants are simply described by the Water Framework Directive as chemicals that are being discharged in significant quantities into the body of water. The levels of specific pollutants in surface water resources are assessed under the ecological status of water bodies according to the provisions of the Water Framework Directive. Therefore, each of the member states and accession countries are obliged to identify their specific pollutants and their environmental quality standards, monitor them properly in the surface water bodies, develop measures to comply with the environmental quality standards where necessary and report to the European Commission.

6.3.3 Regulation of Chemicals in Türkiye

Regulations on chemicals management in Türkiye are generally in line with the European Union legislation due to Türkiye's requirement pertaining to the European Union accession period. The list of basic legislation on the control of production, trade and use of chemicals are presented:

- By-Law on Registration, Evaluation, Authorization and Restriction of Chemicals
- By-Law on The Classification, Labeling, and Packaging of Substances and Mixtures
- By-Law on Health and Safety Measures in Working with Chemical Substances
- By-Law on Import and Export of Certain Hazardous Chemicals
- By-Law on Biocidal Products
- By-Law on Control of Plant Protection Products
- By-Law on Classification, Packaging, and Labeling of Plant Protection Products
- By-Law on Licensing and Placement on the Market of Plant Protection Products.

Although there are numerous regulations issuing provisions on production, use, and trade of chemicals, the fundamental legislation controlling the registration and uses of chemicals is the By-Law on Registration, Evaluation, Authorization and

Restriction of Chemicals. The by-law addresses the rules and procedures of registration, evaluation, placing on the market and use and trade restrictions for the chemicals whereas the rules and procedures for classification, labeling and packaging of chemicals are regulated by By-Law on The Classification, Labeling, and Packaging of Substances and Mixtures. An inventory of chemicals manufactured in or imported to Türkiye in quantities greater than one metric tons annually is compiled according to the provisions of the By-Law on Registration, Evaluation, Authorization and Restriction of Chemicals. The manufacturers or importers are obliged to present the required data on environmental and hazard properties of the chemicals. Both by-laws have been developed in compliance with the relevant European Union legislation, namely the REACH and CLP Regulations with the aim of protecting human health and the environment from the adverse effects of chemicals via reducing the potential of exposure.

The By-Law on Health and Safety Measures in Working with Chemical Substances on defines workplace limit values for certain hazardous substances to reduce occupational exposure to hazardous chemicals.

The By-Law on Import and Export of Certain Hazardous Chemicals was developed to harmonize the requirements of the Rotterdam Convention to national legislation and defines rules and procedures for the trade of certain chemicals to Türkiye and other countries.

The manufacture, use, labeling, packaging, placing on the market, authorization and use of biocidal and plant protection products are regulated by the relevant legislation presented in the list.

In addition to these regulations, By-Law on Surface Water Quality has been put into force in 2012 with the aim of controlling the pollution of surface water resources due to the hazardous chemicals. The by-law defines provisions on monitoring of the water quality and quantity, establishes the intended uses of these waters, and the procedures and principles necessary to achieve the "good water status" determined by the Water Framework Directive. The by-law also includes the priority substances and specific pollutants definitions in line with the Water Framework Directive. By-Law on Surface Water Quality was amended in 2016 and introduced 45 priority substances of the Water Framework Directive and 250 specific pollutants as well as their environmental quality standards.

6.3.4   International Treaties on Chemicals Management

The discovery of the substantial potential of some chemicals to transport to long ranges from the place of emission across the boundaries of nations prompted the world community to act globally to control the use of chemicals and these efforts yielded international treaties on the management of chemicals, eventually.

The first example and the substantial international treaty introducing global measures on the use of certain chemicals is the Vienna Convention in 1985 and its Montreal Protocol on Substances that Deplete the Ozone Layer in 1987. These international treaties succeeded in phasing out the emissions of chlorofluorocarbons (CFCs) which is a group of chemicals depleting the ozone layer. The effectiveness of these treaties encouraged countries to enact new and more comprehensive global measures to control exposure to hazardous chemicals and to protect the environment and human health. The Stockholm Convention, the Basel Convention and the Rotterdam Convention are adopted as the consequence of these needs.

6.3.4.1   The Stockholm Convention

The major global regulatory action towards the restriction/ban of synthetic chemicals is the Stockholm Convention on Persistent Organic Chemicals which is an international treaty signed by 179 countries in 2001 and entered into force in 2004. Persistent Organic Pollutants are chemicals that could be persistent in the environment for long durations following their emission to the environment, could be concentrated in the body of the organisms and accumulate through the food chain, could have toxic effects on humans and biota and long-range transport potential. Initially, 12 legacy persistent organic pollutants called "dirty dozen" including pesticides, industrial chemicals and unintentionally produced chemicals were listed and regulated globally by the convention. As of 2024, there are 34 persistent organic pollutants regulated under the convention. The convention bans or restricts the use and manufacture of industrial chemicals and pesticides and controls the emissions of unintentionally produced by-products with the aim of reducing the global burden of persistent organic pollutants.

6.3.4.2   The Basel Convention

The Basel Convention was adopted in 1989 with the aim of minimizing the generation of hazardous waste and encouraging the environmentally sound handling of hazardous wastes, limiting the cross-border transportation of hazardous wastes, except when it aligns with environmentally responsible management principles and implementing a regulatory framework globally for situations where cross-border waste movements are permissible.

6.3.4.3   The Rotterdam Convention

The Rotterdam Convention was adopted in 1998 in order to regulate international trade of certain hazardous substances and contribute to environmentally sound use of hazardous chemicals to protect human health and the environment. It defines rules and procedures of import and export of pesticides and industrial chemicals that have been banned or severely restricted for health or environmental reasons by the international treaties or nations.

6.4   Approaches Towards Identification of Emerging Pollutants and Prioritization of Chemicals in Aquatic Environments

The emerging pollutants may be emitted to the water resources for long periods of time, or they may be used and emitted to the environment recently. Due to their adverse effects on human health and the ecosystem and widespread detection in water resources, emerging pollutants are potential substances for which regulatory action is required and the control of emerging pollutants is one of the main requisites of ensuring access to safe drinking water. Therefore, the identification of emerging pollutants and assessing the risks associated with them can be viewed as the point of departure of constituting a list of chemicals to be regulated.

There are several definitions of emerging pollutants in the literature, and they are similar to each other in some sense. Tang et al. (2019) defines emerging pollutants as the organic substances that are released to the environment for which no regulations are currently established, while a broader definition exists as synthetic or naturally occurring substances and microorganisms that are typically not subject to regular environmental monitoring or regulation and are potentially associated with recognized or suspected adverse effects on ecological systems and human health (Kumar et al. 2020). The definition of NORMAN Association is a generally accepted one and defines emerging contaminants as currently not regulated (not submitted to a routine monitoring and/or emission control regime) but may be under scrutiny for future regulation (Dulio et al. 2018).

A study demonstrated that at least 700 emerging pollutants exist in the water resources of Europe (Geissen et al. 2015). Another study investigating the occurrence of emerging pollutants in Danube River reported that 22 out of the 235 emerging pollutants monitored at 55 sampling sites along the whole Danube River were present all sites in and 125 were found in at least 50% (Ginebreda et al. 2018). In Korea, 175 emerging pollutants were monitored at 8 sampling sites on Nakdong River between 2020 and 2021 and 130 of these pollutants were measured with a concentration above the LOQ in at least one of the 96 samples taken during the study. In addition to that, these chemicals were undergone a prioritization adopted from the risk-based approach of NORMAN Prioritization Framework applied in this study as well and the priority scores were calculated for 21 substances with the aim of identifying priority substances for nationwide monitoring in Korea (Lee et al. 2024). Peña-Guzmán et al. (2019) reviewed different monitoring studies performed in 11 countries in urban water cycle between 1999 and 2018 and reported that the highest concentrations in surface water were cholesterol (301,000 ng/L), caffeine (106,000 ng/L), stigmasterol (85,500 ng/L), and bisphenol (64,200 ng/L); those with lower concentrations in surface water were diclofenac sodium, naproxen, 4-octylphenol, and sulfamethoxazole among the 99 pollutants monitored. A reviewe conducted in 2010 assessed the relevance of 71 emerging pollutants in freshwater resources and concluded that the highest priority pollutants for regulation and treatment should include industrial chemicals (PFOA, PFOS and DEHP), pesticides (diazinon, methoxychlor, and dieldrin), and PPCPs (EE2, carbamazepine, βE2, DEET, triclosan, acetaminophen, and E1) because they occur frequently in the freshwater environment and pose human health hazards at environmental concentrations (Murray et al. 2010).

6.5 Materials and Methods

In order to identify the emerging pollutants with regulatory concern, countries commonly implement monitoring programs and record the chemicals with high production volumes. The information gathered from these studies is utilized to identify potential emerging pollutants. Afterwards, a risk assessment is applied to the emerging pollutants to designate the list of substances to be regulated. In Türkiye, the approaches adopted in the European Union was used for defining the chemicals to be regulated in water resources and the first list of specific pollutants were published in 2016 in the relevant legislation. Although the approaches of countries to identification and risk assessment of emerging pollutants may differ, all the approaches are subject to uncertainty arising from the availability of reliable data on both physicochemical and toxicological properties and exposure to chemicals. In case of lack of reliable data, the approach is generally to assume the worst-case scenario and use the default values which correspond to the highest risk. In this study, NORMAN Priority Setting Framework was applied to a list of 199 chemicals to identify which of the emerging pollutants needs to be regulated in Türkiye.

6.5.1 NORMAN Priority Setting Framework

NORMAN was established with the aim of facilitating the analysis of emerging environmental pollutants and sharing knowledge and experience regarding their environmental impacts. It is a research network that includes various research organizations primarily from European countries as well as from the USA, Canada, and Israel. In 2013, the NORMAN network developed a prioritization methodology that could also be applied to identify specific pollutants in EU countries. The aim of this method is to classify and prioritize chemicals based on the existing data before the evaluation process, by combining existing prioritization approaches. It also aims to provide an approach that allows the assessment of all chemicals, despite data gaps, during the review of the priority substances list for surface water quality management (Dulio and Ohe 2013). In this approach, a risk assessment is conducted based on the negative impacts on aquatic ecosystems and human health through aquatic ecosystems. This assessment is based on the exposure, effects and risk assessments and finally, an overall risk score is calculated for each of the chemicals.

Exposure assessment is conducted based on monitoring results and usage data of chemicals. The criteria considered in the assessment are as follows:

- Frequency of measurements above the LOQ (Limit of Quantification)
- Number of countries with positive measurement results
- Number of monitoring points with positive measurement results
- Concentration trend
- Presence in groundwater
- Production quantity/usage
- Use pattern.

Two different exposure scores can be calculated according to the method and both get a value between 0 and 1. If reliable monitoring data is available, it is better to account for those and engage the observed exposure score to the assessment. However, if the monitoring data is not available or does not satisfy the requirements of the method, predicted exposure score can be calculated based on the quantity and pattern of use.

Hazard assessment is conducted based on the physicochemical and toxicological properties of chemicals and related data. The criteria considered in the assessment are as follows:

- PBT (Persistent, Bioaccumulative, Toxic) properties
- Long-range transport potential
- Presence of non-standard toxicity measures
- Carcinogenic, mutagenic, and reproductive toxicity
- Endocrine-disrupting properties.

A risk score based on comparison of available ambient water monitoring data to the toxicity thresholds and frequency of detection is conducted. Two main indicators are considered in risk assessment:

- Frequency of exceeding the lowest PNEC (Predicted No-Effect Concentration) value
- Degree of exceeding the lowest PNEC value.

The frequency of exceeding the lowest PNEC value is calculated as n/N,

where:

n: Number of points where MEC (Maximum Environmental Concentration)/ Lowest PNEC > 1.

N: Number of points where measurements were taken for the relevant substance.

The degree of exceeding the lowest PNEC value is calculated as MEC_{95}/Lowest PNEC,

where:

MEC95: 95th percentile of Maximum Environmental Concentration.

The total risk score is calculated as half of the sum of these two indicator values:

$$\text{Risk score} = (n/N + MEC95/\text{Lowest PNEC})/2$$

The final prioritization score is calculated as sum of the exposure, hazard and risk scores as in the following formula.

$$PRIO = EXPO + HAZ + RISK$$

Exposure, hazard, and risk scores are normalized to a range of 0–1. The final prioritization score ranges from 0 to 3. Cut-off criteria applied to physicochemical and toxicological properties of chemicals and the score of each category calculated accordingly. The details of scoring are presented in Table 6.1.

In this study, an illustrative case study from Türkiye which was carried out with the aim of contributing to the review and update of the specific pollutant list of By-Law on Surface Water Quality was showcased. NORMAN Prioritization Framework was employed for the prioritization of emerging aquatic pollutants. There are 250 specific pollutants listed in the By-Law on Surface Water Quality in Türkiye as of 2024. The list of 250 substances was constituted as the outcome of an initial screening of a list of 3102 chemicals which were compiled from the chemicals produced or imported more than one metric ton annually and the chemicals that are declared to be used in the industry by the manufacturers (Şıltu 2015). This initial list was constituted with a precautionary approach, and it was also aimed to consider this list as a comprehensive watch list to obtain more monitoring data on chemical pollutants. However, the initial list of 250 chemicals contains a few chemicals which have already been detected in aquatic environments for years and some have already been banned. As it is not literally appropriate to classify them as "emerging pollutants", those chemicals were not included in the assessments in this study. The final list assessed in this study includes 199 chemical pollutants which are being used or were used in the past in several industrial and agricultural applications, ingredients of personal care products, pharmaceuticals and chemicals which were detected in preliminary monitoring studies in Türkiye.

The NORMAN Prioritization Framework was selected due to its comprehensive approach towards hazard and exposure assessment. The method also provides a comprehensive and detailed assessment of exposure based on the monitoring data and this feature addresses the main target of the case study.

The first step of the study is the compilation of physicochemical properties, toxicological properties, and environmental behavior data of the 250 chemicals. The physicochemical properties, environmental fate and transport data, and acute and chronic toxicity data are compiled from the US EPA's CompTox Chemicals Dashboard (https://comptox.epa.gov/dashboard/) and European Chemicals Agency (ECHA) database (https://echa.europa.eu/). Experimental data was selected to be used if available. QSAR/QSPR (Quantity Structure Activity Relationship/Quantity Structure Property Relationship) derived data was used in case the experimental data was not existing. CMR evaluation of the European Union under the CLP Regulation was used in the assessment of CMR properties. Similarly, the endocrine disruptors assessment list was considered for evaluating the endocrine disrupting properties of chemicals. The lowest PNEC values are collected from the NORMAN database (https://www.norman-network.com/nds/factsheets/).

Table 6.1 Calculation of scores for exposure, hazard and risk assessment

Category		Sub-category Indicator	Value	Score	Overall Category Score
Exposure	Observed Exposure (Monitoring)	A) Frequency of detection>LOQ	ratio of detection>LOQ	decimal value	EXPO_O = (A+B+C+D+E)/5
		B) Number of countries with detection>LOQ	Number of countries with detection>LOQ	0-1; 0 (or no data) = 0 ≥1; 0.10 ≥10; 0.20 ≥100; 0.50 ≥1000; 1	
		C) Number of monitoring sites with detection>LOQ	Number of monitoring sites with detection>LOQ	0-1; 0 (or no data) = 0 ≥1; 0.10 ≥10; 0.20 ≥100; 0.50 ≥1000; 1	
		D) Concentration trend	Trend analysis of long term monitoring data	Significant increase = 1 Increase = 0.5 No trend = 0.25 No data =0.1 Decrease = 0	
		E) Detection in groundwater	Yes = 1 No = 0	Yes = 1 No = 0	
	Predicted Exposure (Use data)	F) Amount of production/use	Annual production in tonnes	< 1; 0.1 1-10; 0.2 10-100; 0.5 > 100; 1	EXPO_P = (F+G)/2
		G) Use pattern	Used in the to environment =1 Wide dispersive use = 0.75 Non-dispersive use = 0.50 Unknown = 0.25 Controlled system = 0.1	value	
Hazard	Human Health and the Environment	H) PBT/vPvB	((P+B+T) + (PBT/vPvB))/4		HAZ = (H+ I+ J+ K+ L)/5
		I) LRTP (Long range transport potential)	Half-life air > 2 gün Vapor pressure < 1000 Pa	Half-life air > 2 days and Vapor pressure < 1000 Pa = 1 Half-life air ≤ 2 days or Vapor pressure > 1000 Pa = 0	
		J) Non-standard toxicity end-points	Immune system defects. number of offsprings etc.	exist =1 under investigation = 0.50 no study = 0.25 not exist = 0	
		K) CMR (Carcinogen. mutagen. reprotoxic)	CMR category	CMR Category 1 = 1 CMR Category 2 = 0.75 CMR Category 3 = 0.5 under investigation = 0.5 insufficient data = 0.25 no study = 0.25 not CMR = 0	
		L) ED(Endocrine disrupting property)	ED category	proved =1 suspected = 0.5 no study = 0.25 not ED = 0	
Risk	Number of Monitoring Sites where PNEC exceeded	N) Frequency of exceedance of PNEC	n/N	decimal value	RISK= (M + N)/2
	Extent of Exceedance of PNEC	M) Extent of exceedance of PNEC	MEC95/lowest PNEC	MEC95/lowest PNEC < 1; 0 1 < MEC95/lowest PNEC < 10; 0.1 10 < MEC95/lowest PNEC < 100; 0.25 100 < MEC95/lowest PNEC < 1000; 0.5 MEC95/lowest PNEC > 1000; 1	

Source Dulio and Ohe 2013

Moreover, freshwater monitoring studies conducted in Türkiye by the Ministry of Agriculture and Forestry between 2011 and 2020 were assessed in this study in order to calculate the "Risk" and "Exposure" scores of each of the chemicals according to NORMAN Prioritization Framework. For the 199 chemicals assessed in this study, the number of surface water monitoring stations from which samples collected ranges between 50 to 516 and the number of river basins in which monitoring studies conducted ranges between 2 to 12 for individual chemicals. In total, 61,246 surface water monitoring data was assessed in scope of this study.

6.6   Results and Discussion

The exposure, hazard and the risk scores and final prioritization scores were calculated for each of the 199 chemicals in accordance with the methodology presented in Sect. 6.5.1. The final score was calculated as a value between 0 and 3. Among the 199 chemicals, benzylbutyl phtalate got the maximum final score as 2.22 and 2-methyl-4,6-dinitrophenol and isodrin got the minimum score as 0. Both of the arithmetic mean and the median of the final scores calculated for 199 chemicals is 0.82 with a standard deviation of 0.39, indicating that most of the chemicals get scores close to the central tendency estimate. The distribution of chemicals according to the final scores is demonstrated in Fig. 6.1.

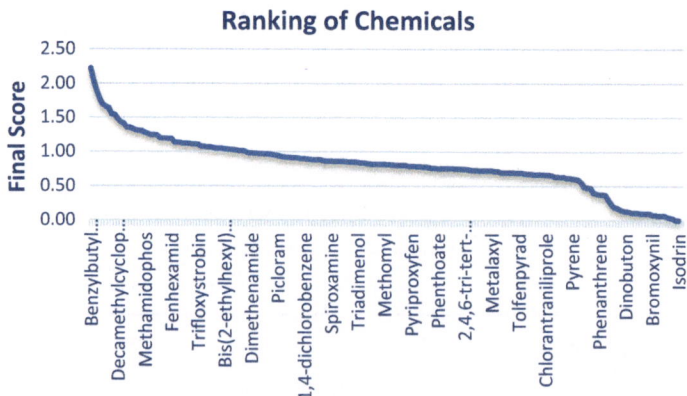

Fig. 6.1 Prioritization of emerging aquatic pollutants in Türkiye according to final score of NORMAN prioritization methodology (0–3). *Source* Authors

Although there is no significant break point in the distribution of final scores, distribution represents a more stable linear path below the value of 1. Therefore, chemicals having a score greater than 1 considered as posing higher potential risk and of regulatory relevance. This corresponds to a list of 53 chemicals which should be assessed more rigorously in terms of risks posed to aquatic environment to identify the potential specific pollutants of Türkiye. Nevertheless, the uncertainties in the input data should be considered while taking regulatory action. The list of 53 chemicals and their assessment scores are presented in Table 6.2.

6.7   Conclusion

The identification of emerging aquatic pollutants which need to be regulated is a dynamic process. It requires the prioritization and risk assessment of the potential chemicals and should be handled on a regular basis. The pollutant list must be updated in response to scientific developments, national and international requirements.

There are a variety of methods and approaches to risk assessment and prioritization of chemicals for identifying the chemicals to be regulated. In this study, a case study conducted in Türkiye for the prioritization of emerging aquatic pollutants with the aim of supporting the efforts on identification of specific pollutants was presented. A cut-off criteria based prioritization methodology was applied and 53 pollutants were proposed to have a higher regulatory priority due to the risks they pose to aquatic environments in Türkiye. However, due to the limitations on the spatial and temporal extent of the monitoring data used in the study, it is not appropriate to use the outcomes of this study directly in the regulatory action. Although the approach adopted in the study is quite comprehensive and provides a solid risk assessment enabling the assessment of exposure and hazard jointly, the study still needs to be extended with a more comprehensive and representative set of monitoring data. Nevertheless, the monitoring data used in this study is still capable of representing a snapshot of the impact of a diverse range of industrial activities and agricultural practices being carried out in Türkiye. The extent and reliability of the monitoring data is of critical importance in identifying the pollutants that actually pose threat to the aquatic environments.

On the other hand, experimental data is not available for toxicological, physico-chemical, and environmental health properties of majority of the emerging chemicals since derivation of these data is a resource and time consuming process. To fill the data gap and address the data requirements, further development of QSAR/QSPRs and pp-LFERs (Polyparameter Linear Free Energy Relationships) is essential. In relation to that, New Approach Methodologies (NAMs) that are described as any in silico, in chemico, in vitro, and ex vivo technology, methodology, approach, or combination thereof that can furnish insights into chemical hazard and risk assessment without resorting to animal experimentation are becoming more commonly used and proposed by several regulatory agencies for the risk assessment of chemicals

Table 6.2 The list of 53 chemicals suspected to pose higher risks to aquatic ecosystems and human health

Chemical	CAS No	Exposure score	Hazard score	Risk score	Final score
Benzylbutyl phtalate (BBP)	85-68-7	0.55	0.69	0.98	2.22
Bifenyl	92-52-4	0.55	0.50	0.99	2.04
4,5-dichloro-2-octyl-2H-isothiazol-3-one	64359-81-5	0.57	0.32	1.00	1.88
4-Chloro-3-methylphenol	59-50-7	0.57	0.38	0.81	1.76
Flutolanil	66332-96-5	0.67	0.88	0.15	1.69
Metam potassium	137-41-7	0.40	1.00	0.26	1.66
Linuron	330-55-2	0.57	0.81	0.26	1.64
Bentazon	25057-89-0	0.57	0.69	0.28	1.55
Benzyl benzoate	120-51-4	0.56	0.32	0.66	1.54
Decamethylcyclopentasiloxane	541-02-6	0.47	0.88	0.14	1.48
Penconazole	66246-88-6	0.50	0.88	0.05	1.43
Carbon tetrachloride	56-23-5	0.40	0.84	0.17	1.41
4-Aminoazobenzene	60-09-3	0.55	0.19	0.61	1.36
Diethyl phthalate	84-66-2	0.57	0.73	0.06	1.36
Bromophos-ethyl	4824-78-6	0.44	0.84	0.05	1.33
Prochloraz	67747-09-5	0.50	0.81	0.00	1.31
Hexythiazox	78587-05-0	0.43	0.88	0.00	1.31
Ethofumesate	26225-79-6	0.57	0.60	0.14	1.30
Methamidophos	10265-92-6	0.40	0.88	0.00	1.28
N, N-dimethyl-N'-phenylsulphamide	4710-17-2	0.40	0.81	0.05	1.26
Buprofezin	69327-76-0	0.44	0.19	0.62	1.25
Triasulfuron	82097-50-5	0.23	0.51	0.51	1.25
Tebuconazol	107534-96-3	0.40	0.84	0.00	1.24
Diisobutyl adipate	141-04-8	0.67	0.13	0.40	1.20
Fenthion	55–38-9	0.10	0.59	0.50	1.20
Fluazifop-P-butyl	79241-46-6	0.23	0.50	0.46	1.19
Azinphos-methyl	86-50-0	0.46	0.57	0.17	1.19
Fenhexamid	126833-17-8	0.57	0.45	0.17	1.19
Difenoconazole	119446-68-3	0.10	0.78	0.26	1.14
Fluorene	86-73-7	0.57	0.57	0.00	1.14
Imidachloprid	138261-41-3	0.33	0.63	0.16	1.13
Propham	122-42-9	0.43	0.69	0.00	1.12
Molinate	2212-67-1	0.23	0.75	0.14	1.12
Bisphenol-A	80-05-7	0.44	0.50	0.18	1.12
Propetamphos	31218-83-4	0.23	0.88	0.00	1.11

(continued)

Table 6.2 (continued)

Chemical	CAS No	Exposure score	Hazard score	Risk score	Final score
Sulfamethoxazole	723-46-6	0.23	0.81	0.06	1.11
Trifloxystrobin	141517-21-7	0.10	0.75	0.25	1.10
tert-butyl-4-methoxyphenol	25013-16-5	0.20	0.88	0.00	1.08
N-butyltin trichloride	1118-46-3	0.10	0.72	0.25	1.07
Diethofencarb	87130-20-9	0.43	0.63	0.00	1.06
Metazachlor	67129-08-2	0.50	0.44	0.13	1.06
Lenacil	2164-08-1	0.40	0.66	0.00	1.06
Dibutyl phthalate	84-74-2	0.67	0.25	0.13	1.05
Clopiyalid	1702-17-6	0.23	0.81	0.00	1.05
Propazine	139-40-2	0.23	0.81	0.00	1.05
Bis(2-ethylhexyl) terephthalate	6422-86-2	0.30	0.38	0.36	1.04
Cyclanilide	113136-77-9	0.10	0.88	0.06	1.03
Thiometon	640-15-3	0.40	0.57	0.06	1.03
Cyromazin	66215-27-8	0.40	0.63	0.00	1.03
Pyraclostrobin	175013-18-0	0.83	0.19	0.00	1.02
Acetamiprid	135410-20-7	0.11	0.63	0.28	1.02
Imidazolidine-2-thione	96-45-7	0.23	0.51	0.27	1.01
Propyzamide	23950-58-5	0.20	0.75	0.05	1.00

Source Authors

(Stucki et al. 2022). In this context, use of aggregated exposure and fate estimation modeling frameworks such as PROTEX-HT (Li et al. 2021) in the risk assessment of chemicals can be a promising approach in case there is no monitoring data for the chemicals. Nevertheless, these models are required to be further developed to yield more reliable estimates of exposure and fate. For the case of Türkiye, the case study presented here should be extended by using the latest available monitoring data. Moreover, the use of NAMs and exposure and fate estimation models should be considered, and the findings of NORMAN methodology and the modeling-based exercise can be compared in order to get better insights on the risks associated with the use of chemicals already registered in the country. Moreover, the capacity of the laboratories of the government institutions in terms monitoring of emerging pollutants should be improved in Türkiye and the production, use, import and export data of the chemicals should be made available. In the mid-term, a comprehensive monitoring campaign to detect the levels of chemicals in all the environmental compartments and should be initiated to have a snapshot of distribution of chemicals among the environmental compartments and the biota. This may aid science and evidence-based decision making on the management of chemicals.

References

Bassil KL, Vakil C, Sanborn M, Cole DC, Kaur JS, Kerr KJ (2007) Cancer health effects of pesticides: systematic review. Can Fam Phys 53(10):1704–1711

Carson R (1962) Silent spring. Houghton Mifflin

Centers for Disease Control and Protection (CDC) (2017) Dioxins, furans and dioxin-like polychlorinated biphenyls factsheet. https://www.cdc.gov/biomonitoring/DioxinLikeChemicals_FactSheet.html

Colborn T, Vom Saal FS, Soto AM (1993) Developmental effects of endocrine-disrupting chemicals in wildlife and humans. Environ Health Perspect 101(5):378–384

Cosselman KE, Navas-Acien A, Kaufman JD (2015) Environmental factors in cardiovascular disease. Nat Rev Cardiol 12(11):627–642

Dulio V, Ohe PC (2013) NORMAN prioritisation framework for emerging substances. NORMAN Association, France

Dulio V, van Bavel B, Brorstrom-Lunden E, Harmsen J, Hollender J, Schlabach M, Slobodnik J, Thomas K, Koschorreck J (2018) Emerging pollutants in the EU: 10 years of NORMAN in support of environmental policies and regulations. Environ Sci Eur 30:5. https://doi.org/10.1186/s12302-018-0135-3

EPA (2022) Summary of the toxic substances control act. https://www.epa.gov/laws-regulations/summary-toxic-substances-control-act

EPA (2023a) DDT - a brief history and status. https://www.epa.gov/ingredients-used-pesticide-products/ddt-brief-history-and-status

EPA (2023b) Pollution prevention act of 1990. https://www.epa.gov/p2/pollution-prevention-act-1990#:~:text=The%20Congress%20hereby%20declares%20it,be%20prevented%20or%20recycled%20should

EPA (2023c) Summary of the clean water act. https://www.epa.gov/laws-regulations/summary-clean-water-act

EPA (2023d) Toxic and priority pollutants under the clean water act. https://www.epa.gov/eg/toxic-and-priority-pollutants-under-clean-water-act

EUROSTAT (2021) Production and consumption of chemicals by hazard class. https://ec.europa.eu/eurostat/databrowser/view/env_chmhaz/default/table?lang=en

Faber RA, Hickey JJ (1973) Eggshell thinning, chlorinated hydrocarbons, and mercury in inland aquatic bird eggs, 1969 and 1970. Pestic Monit J 7(1):27–36

Fu X, Xu J, Zhang R, Yu J (2020) The association between environmental endocrine disruptors and cardiovascular diseases: a systematic review and meta-analysis. Environ Res 187:109464

Geissen V, Mol H, Klumpp E, Umlauf G, Nadal M, van der Ploeg M, van de Zee SEATM, Ritsema CJ (2015) Emerging pollutants in the environment: a challenge for water resource management. Int Soil Water Conserv Res 3(1):57–65

Ginebreda A, Sabater-Liesa L, Rico A, Focks A, Barceló D (2018) Reconciling monitoring and modeling: an appraisal of river monitoring networks based on a spatial autocorrelation approach-emerging pollutants in the Danube River as a case study. Sci Total Environ 618:323–335

Grandjean P, Landrigan PJ (2006) Developmental neurotoxicity of industrial chemicals. The Lancet 368(9553):2167–2178

Peña-Guzmán C, Ulloa-Sánchez S, Mora K, Helena-Bustos R, Lopez-Barrera E, Alvarez J, Rodriguez-Pinzón M (2019) Emerging pollutants in the urban water cycle in Latin America: a review of the current literature. J Environ Manag 237:408–423

Heindel JJ, Blumberg B, Cave M, Machtinger R, Mantovani A, Mendez MA, Nadal A, Palanza P, Panzica G, Sargis R, Vandenberg LV, Vom Saal F (2017) Metabolism disrupting chemicals and metabolic disorders. Reprod Toxicol 68:3–33

Huang M, Jiao J, Zhuang P, Chen X, Wang J, Zhang Y (2018) Serum polyfluoroalkyl chemicals are associated with risk of cardiovascular diseases in national US population. Environ Int 119:37–46

Koskinen M (2017) Chemicals in our life - why are chemicals important. https://echa.europa.eu/-/chemicals-in-our-life-why-are-chemicals-important

Kristensen TS (1989) Cardiovascular diseases and the work environment: a critical review of the epidemiologic literature on chemical factors. Scand J Work, Environ Health 245–264

Kumar M, Borah P, Devi P (2020) Priority and emerging pollutants in water. In: Inorganic pollutants in water, pp 33–49. Elsevier

Le Magueresse-Battistoni B, Labaronne E, Vidal H, Naville D (2017) Endocrine disrupting chemicals in mixture and obesity, diabetes and related metabolic disorders. World J Biol Chem 8(2):108

Lee S, Choi Y, Kang D, Jeon J (2024) Proposal for priority emerging pollutants in the Nakdong river, Korea: application of EU watch list mechanisms. Environ Pollut 341:122838

Li L, Sangion A, Wania F, Armitage JM, Toose L, Hughes L, Arnot JA (2021) Development and evaluation of a holistic and mechanistic modeling framework for chemical emissions, fate, exposure, and risk. Environ Health Perspect 129(12):127006-2-16

Mendola P, Selevan SG, Gutter S, Rice D (2002) Environmental factors associated with a spectrum of neurodevelopmental deficits. Ment Retard Dev Disab Res Rev 8(3):188–197

Mendola P, Messer LC, Rappazzo K (2008) Science linking environmental contaminant exposures with fertility and reproductive health impacts in the adult female. Fertil Steril 89(2):e81–e94

Murray KE, Thomas SM, Bodour AA (2010) Prioritizing research for trace pollutants and emerging contaminants in the freshwater environment. Environ Pollut 158(12):3462–3471

Needham LL, Barr DB, Caudill SP, Pirkle JL, Turner WE, Osterloh J, Jones RL, Sampson EJ (2005) Concentrations of environmental chemicals associated with neurodevelopmental effects in US population. Neurotoxicology 26(4):531–545

Nurminen T (1995) Maternal pesticide exposure and pregnancy outcome. J Occup Environ Med 37:935–940

Oakley GG, Devanaboyina US, Robertson LW, Gupta RC (1996) Oxidative DNA damage induced by activation of polychlorinated biphenyls (PCBs): implications for PCB-induced oxidative stress in breast cancer. Chem Res Toxicol 9(8):1285–1292

OECD (2021) OECD work on chemical safety and biosafety. France, Paris

Rauh VA, Margolis AE (2016) Research review: environmental exposures, neurodevelopment, and child mental health–new paradigms for the study of brain and behavioral effects. J Child Psychol Psychiatr 57(7):775–793

Safe SH (2000) Endocrine disruptors and human health-is there a problem? An update. Environ Health Perspect 108(6):487–493

Stucki AO, Barton-Maclaren TS, Bhuller Y, Henriquez JE, Henry TR, Hirn C, Miller-Holt J, Nagy EG, Perron MM, Ratzlaff DE, Stedeford TJ, Clippinger AJ (2022) Use of new approach methodologies (NAMs) to meet regulatory requirements for the assessment of industrial chemicals and pesticides for effects on human health. Front Toxicol 4:964553

Şıltu E (2015) Su ortaminda bulunabilecek tehlikeli maddelerin önceliklendirilmesi açisindan Türkiye'de uygulanabilecek metodolojinin belirlenmesi. Expertise thesis, TR Ministry of Forestry and Water Affairs. Ankara, Türkiye. https://www.tarimoman.gov.tr/SYGM/Belgeler/TEZLER/uzmanl%C4%B1ktezi_EsraSiltu_22.11.15.pdf

Taha TE, Gray RH (1993) Agricultural pesticide exposure and perinatal mortality in central Sudan. Bull World Health Organ 71:317–321

Tang Y, Yin M, Yang W, Li H, Zhong Y, Mo L, Liang Y, Ma X, Sun X (2019) Emerging pollutants in water environment: occurrence, monitoring, fate, and risk assessment. Water Environ Res 91(10):984–991

UNEP (2013) Global chemicals outlook - towards sound management of chemicals

UNEP (2019) global chemicals outlook II, from legacies to innovative solutions: implementing the 2030 agenda for sustainable development

UNEP and ICCA (2019) Knowledge management and information sharing for the sound management of industrial chemicals

Van Maele-Fabry G, Lantin AC, Hoet P, Lison D (2011) Residential exposure to pesticides and childhood leukaemia: a systematic review and meta-analysis. Environ Int 37(1):280–291

Vos JG, Dybing E, Greim HA, Ladefoged O, Lambré C, Tarazona JV, Brandt I, Vethaak AD (2000) Health effects of endocrine-disrupting chemicals on wildlife, with special reference to the European situation. Crit Rev Toxicol 30(1):71–133

Wiemeyer SN, Lamont TG, Bunck CM, Sindelar CR, Gramlich FJ, Fraser JD, Byrd MA (1984) Organochlorine pesticide, polychlorobiphenyl, and mercury residues in bald eagle eggs—1969–79—and their relationships to shell thinning and reproduction. Arch Environ Contam Toxicol 13:529–549

World Health Organization (2019) The public health impact of chemicals: knowns and unknowns-2019 Adendum. World Health Organization

Zahm SH, Ward MH (1998) Pesticides and childhood cancer. Environ Health Perspect 106(suppl 3):893–908

Zhang J, Cai WW, Lee DJ (1992) Occupational hazards and pregnancy outcomes. Am J Ind Med 21:397–408

Chapter 7
Evolution of Water Research in South Africa from Legacy Pollutants to Contaminants of Emerging Concern: Successes and Opportunities

Nonhlanhla Kalebaila and Samkelisiwe Hlophe-Ginindza

Abstract South Africa, like many other countries, faces ongoing chemical water pollution challenges, intricately linked to its industrialization journey. This challenge is exacerbated by constrained water resource availability and the impacts of climate change, limiting the dilution capacity. The extent of water resources pollution by contaminants of emerging concern (CECs) and its impacts on humans, and ecosystems health is a complex and evolving area of high research interest, not only in South Africa but worldwide. This chapter presents a historical reflection of the evolving landscape of South Africa's chemical water pollution research, navigating the shift from legacy chemical pollutants to contaminants of emerging concern (CECs). Additionally, this chapter presents a comprehensive account of the state of research and knowledge on the sources, distribution and potentials risks associated with exposure to CECs in South African water systems. This is followed by an examination of the critical issues surrounding the management of CECs in water in South Africa by highlighting the successes achieved and challenges that remain in addressing water pollution issues related to CECs. This analysis is based on the nation's 20 years proactive and progressive research on CECs. The chapter concludes by outlining future directions for research and the need for leveraging opportunities, interdisciplinary collaboration, effective communication, being adaptable to the changing environment, and a commitment to evidence-based decision-making.

Keywords Water pollution · Emerging pollutants · Persistent organic pollutants · Endocrine disrupting compounds · Environmental chemicals

N. Kalebaila (✉) · S. Hlophe-Ginindza
Water Research Commission, Pretoria, South Africa
e-mail: nonhlanhlak@wrc.org.za

© UNESCO 2025
S. Zandaryaa et al. (eds.), *Emerging Pollutants*, Advances in Water Security,
https://doi.org/10.1007/978-3-031-71758-1_7

7.1 Introduction and Historical Context

Chemicals are part of our everyday life, and to date, a considerable number of chemicals have been produced and used to the benefit of humankind, including pharmaceuticals, household and personal care products, agricultural chemicals, as well as for industrial-scale applications (UNEP 2006). According to Persson et al. (2022), the annual production of chemicals and releases into the environment are increasing at a pace that outstrips the global capacity for assessment, monitoring, and management. For this reason, environmental chemical pollution is considered as one of the planetary challenges, as it directly impacts the biosphere integrity (Rockstrom et al. 2009). The long-lasting environmental and health impacts resulting from the historical use, release, and persistence of chemical pollutants has been on the global agenda since the latter half of the twentieth century (WHO 2017).

Similar to other countries, South Africa faces ongoing chemical water pollution challenges, intricately linked to its path to industrialization (Department of Environmental Affairs and Tourism 2005; UNEP 2020). The late nineteenth century was marked by an exponential growth of industries, including mining, manufacturing, and other extractive activities. However, the legacy of these historical industrial activities, improper waste disposal and the lack of environmental regulations during earlier stages of industrialization, has left a lasting negative impact on water resources. The water pollution challenge in South Africa is further exacerbated by the limited availability of water resources (Department of Water and Sanitation 2018). The dichotomy between ensuring the sustainable use of limited water resources for economic development while simultaneously safeguarding human and ecosystem health, by preventing pollution that could further degrade water quality represents one of the long-standing complex, and yet pressing challenge in South Africa.

The pivotal role of research and innovation in addressing this challenge was recognised as early as in 1945, through the establishment of the Council for Scientific and Industrial Research (CSIR 2022). The CSIR housed the first ever dedicated National Water Research Institute set up to address pollution and promote water quality management. The research conducted at this institute aligned with global interests, primarily focusing on monitoring the release of persistent chemical pollutants into the environment with little awareness on their long-term impacts on the ecosystem and public health. One of the major discoveries of that time was the prevalence of a group of compounds that later became known as persistent chemical pollutants, or "legacy pollutants". Persistent chemical pollutants, are substances that resist degradation and persist in the environment for extended periods of time, sometimes spanning decades or even longer (Scheringer et al. 2012; Sheriff et al. 2022). Persistent organic pollutants (POPs) are a class of highly hazardous synthetic organic chemicals, either intentionally or unintentionally produced/released, that remain intact in the environment for extended periods of time; and possess the ability to accumulate in the living organisms including humans. POPs were discovered to occur abundantly in the environment after World War II, when thousands of synthetic chemicals were introduced into commercial use (Ashraf 2017).

In 1971, South Africa took a strategic decision to establish a dedicated national water research entity, the Water Research Commission (WRC) (Water Research Commission 2022). The WRC was mandated to ensure the generation of credible scientific knowledge and the development of management tools to prevent contamination, minimize waste, and ensure the sustainable use of water. By the end of the twentieth century, South Africa's research on POPs was on par with global standards, positioning the nation as a significant contributor to the understanding and management of these environmental challenges. The establishment of the Water Research Commission (WRC) in South Africa elevated the nation's research endeavours to a global standard. Research work by Colborn and colleagues played an instrumental role in this elevation, as their findings on the presence, persistence, and potential long-term impacts of POPs on ecosystems and human health contributed significantly to a deeper global understanding (Myers 2016). This work led to the unanimous global adoption of the term endocrine disrupting chemicals (EDCs), further emphasizing their influence on international discourse and action regarding environmental contaminants of concern (CECs).

Endocrine disruptors are a large group of non-natural chemicals or mixtures of chemicals that possess the ability to mimic, block, or interfere with the endocrine system, which regulates hormonal functions in humans and animals, and can lead to various adverse health effects (Endocrine Society 2024). According to Arman et al. (2021), the terms "emerging pollutants", "emerging contaminants", "contaminants of emerging" or "chemicals of emerging concern" can be used interchangeably catch all terms to refer to pollutants that fall into one or more of the following criteria: (i) not necessarily a new compound, (ii) a compound that has long existed in the environment but whose presence has only recently been detected and whose significance is beginning to be recognized, and (iii) a long-known compound whose potential negative impact on humans and the environment has only recently been realized. Examples of compounds that are regarded as contaminants of emerging concern (CECs) include pharmaceuticals and personal care products (PPCPs), steroids, flame retardants, illicit drugs, hormones, microplastics, and engineered nanomaterials (Geissen et al. 2015; Fontes et al. 2020). In most developing countries, including South Africa, the term CECs may encompass pollutants regarded as 'known' or well established in developed countries, such as heavy metals, EDCs, POPs, and pesticides, since research in these areas only reached its peak around 2005 (Visanji et al. 2019).

South Africa's national research plan on CECs was formally launched by the Water Research Commission (WRC) in 2001 (Burger 2008). The first phase of the research programme focused on (EDCs) as contaminants of emerging concern. One of the objectives of this programme was to establish baseline knowledge on EDCs and build humans resources capacity and analytical capability for conducting research on CECs in the local water and sciences sector. Recognizing the country's constraints in resources, including expertise and laboratory infrastructure, the WRC made a strategic decision to collaborate with the Global Water Research Coalition (GWRC) on implementation of the research programme on EDCs. GWRC is a non-profit organization that has been in operation since 2002, that serves as a collaborative research platform for member organisation that deal with water supply, wastewater issues and renewable water resources. Under this partnership, South Arica's research landscape has evolved to include a range of other emerging pollutants of concern suspected to be endocrine disruptors or to cause other adverse health effects.

Research conducted to date has generated valuable knowledge and insights on various aspects related to CECs in water, including their sources, their prevalence in aquatic systems, analytical methods capable of detecting these contaminants, even at low concentrations, potential adverse effects, and pinpointing contamination hotspots. Owing to the reported presence, persistent nature, and mobility of a variety of CECs in water sources in South Africa, current efforts are aimed at investigating the adverse effects of individual CECs and their mixtures, as means of understanding the environmental and human health risks. In this context, the evolving journey of South African's water quality research has been vital in laying a solid foundation for establishing the current state of knowledge and capacity for conducting CEC research and monitoring in water systems. The sections that follow provides insights into the current state of knowledge, and gaps based on available research on the sources, occurrence, levels, and impacts. The chapter also highlights the complex and multifaceted journey from research to policy and practice, bringing to the fore some of the successes and challenges, which may well serve as lessons for other developing countries. The chapter concludes by outlining future directions for research and the need for leveraging opportunities, interdisciplinary collaboration, effective communication, being adaptable to the changing environment, and a commitment to evidence-based decision-making.

7.2   State of Knowledge on CECs in South African Water Systems

7.2.1   Sources, Entry, and Fate of CECs in Water Environments in South Africa

Water resources in South Africa are impacted by pollution due to various activities linked to both land and water uses. Such activities have been identified to significantly contribute to the entry and presence of industrial chemicals, heavy metals, acids, and CECs in water resources in South Africa (Department of Water and Sanitation 2017). Amongst these, pesticide use for agricultural purposes has been identified as the leading contributor to the pollution of South Africa's water resources (Tang et al. 2021). To put this into context, South Africa has over 700 different active ingredients registered for agricultural use, making the country one of largest consumers of pesticides in Africa, and globally (Dabrowski 2015a; FAO 2021). The role of runoff in transporting pesticide residues from the point of application into water resources in South Africa has been studied.

According to Schulz (2001), water pollution via the use of pesticides in adjacent areas occurs as a function of the time interval between the application of pesticides and the first heavy rainfall event, the slope and soil types of the catchment, the pesticide application method, and the size and characteristics of buffer strips. Based on this Schulz (2001) was able to attribute the presence of pesticides such as endosulfan, chlorpyrifos and azinphos-methyl in the Lourens River and its tributaries to a rainstorm event, which facilitated their transport from an adjacent orchard farm. In a much comprehensive and national-scale study, Dabrowski (2015a) investigated the extent and level of contamination of South African water resources by agricultural chemicals (pesticides, herbicides, and plant growth regulants), including those exhibiting endocrine disruptive (ED) properties. Dabrowski (2015a) developed crop production spatial distributions maps and overlaid these with application rate data of 206 pesticide active ingredients per magisterial district. Using pesticide use data and surface water mobility index, Dabrowski (2015b) showed that almost two-thirds of the applied volume potentially reaches water resources) and can be used as direct indication of the extent of water resources contamination per magisterial district.

South Africa is yet to implement a risk-based and holistic plan for addressing agricultural water pesticide pollution, considering a combination of stricter regulatory measures, sustainable agricultural practices, public education, and ongoing water resources monitoring. The detection of persistent organic pollutants (POPs) and legacy pesticides, such as atrazine, DDT and dieldrin, continue to be of concern in South Africa due to their long-lasting presence in water environments (Dalvie et al. 2014; Dabrowski 2015a; Curchod et al. 2020; Ojemaye et al. 2020). As such, these persistent pesticides and other agricultural chemicals of concern continue to be treated as contaminants of emerging concern in South Africa.

The discharge of inadequately treated municipal wastewater into water bodies is also identified as another significant contributor to presence of CECs in freshwater, marine water and treated drinking in South Arica. South Africa has a vast network of wastewater treatment plants (WWTPs), consisting of about 850 plants managed by 144 municipalities, and with a total installed treatment capacity of 6971 ML/day. The wastewater network in South Africa collects and treats both domestic and industrial wastewater. The performance of municipalities for wastewater quality management is monitored under the Green Drop Regulation programme. The Green Drop Regulation Programme, a national incentive-based regulation programme, was introduced by the national Department of Water and Sanitation (DWS) in South Africa in 2008 (Department of Water and Sanitation 2022). Green Drop audits are conducted every year and municipalities performance is assessed against the following five (5) main key performance areas (KPAs); A: capacity management, B: environmental management, C: financial management, D: technical management, and E: effluent and sludge compliance. Each KPA and sub-criteria carry a different weighting and are based on the relative regulatory priorities. As means of managing environmental pollution, effluent and sludge compliance carries the highest weighting (30%). A wastewater system that achieves a $\geq 90\%$ Green Drop score, is regarded as excellent, while a score of $< 31\%$ suggests that it is dysfunctional (Fig. 7.1).

According to the 2022 Green Drop report, only about 3% (22/850) of the WWTPs were regarded as excellent and 39% (334/850) are dysfunctional, while the remainder (about 58%) fall in between (Fig. 7.2). The current data indicates that

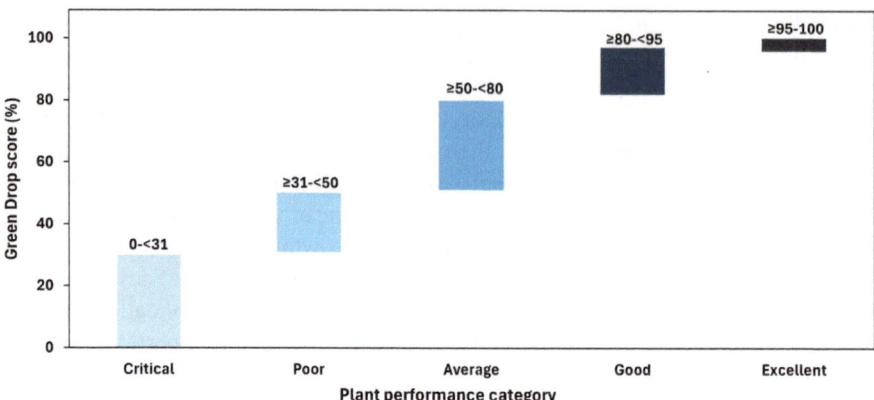

Fig. 7.1 Green Drop scoring based on the performance of a wastewater treatment system. *Source* Drawn by Authors using 2023 data from the Department of Water and Sanitation of South Africa

over 90% of the municipal wastewater plants release inadequately treated effluents into receiving water bodies. Consequently, a variety of CECs, including, pharmaceuticals, hormones, and personal care products, such as antibiotics, birth control hormones, fragrances, microbeads and microplastics, as well as a variety of industrial chemicals, have been detected in the raw and final treated effluents from WWTPs. Most wastewater treatment plants in South Africa consist of conventional wastewater treatment process trains, and as such are not designed for the effective removal of CECs (Olujini et al. 2013; Archer et al. 2017a, b; Coetzee et al. 2017; Swartz et al. 2018; Oluwalana et al. 2022).

Furthermore, release of untreated or partially treated wastewater into rivers, streams, and coastal areas, leading to the introduction of CECs, has been reported

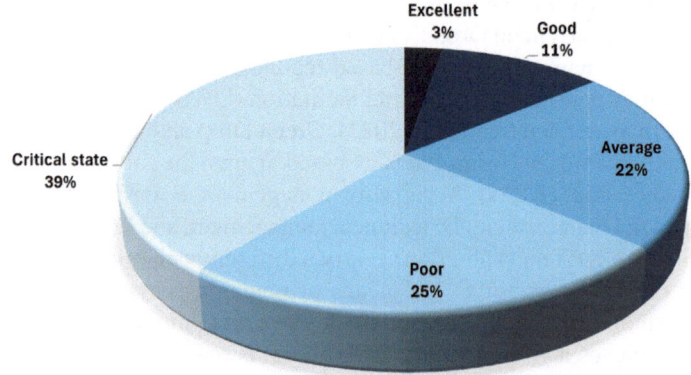

Fig. 7.2 National Green Drop performance of municipalities in South Africa in 2021. *Source* Drawn by Authors using 2022 data from the Department of Water and Sanitation of South Africa

to occur via incidental municipal sewer system overflows or system malfunctions (Masindi et al. 2022). The ingress of stormwater from urban and unserved informal settlements into both municipal wastewater collection systems and directly into water resources, has also been reported to be a contributor to the presence of a wide range of CECs, including, petrochemicals, heavy metals, pesticides, and microplastics, etc. and these are washed off from building materials, and other surfaces during stormwater events (Ojemaye 2020).

Once within the environment, CECs can be partitioned into different compartments and/or reach the marine environment. The extent of the contaminant transport is determined by their respective chemical properties, which dictate their persistence and availability (speciation) for accumulation by humans and aquatic life or complexation into soil and sediment particles. Given the risks and impacts related to exposure to CECs, there is need to strengthen research aimed at developing and testing modelling and/or forensic approaches for predicting the sources, fate, and behaviour of CECs within the aquatic environment. The adoption of forensic analytical techniques into policy can aid tracking of CECs sources, enabling regulatory agencies to monitor compliance to water quality standards and discharge permits and improve the enforcement of environmental pollution prevention strategies (Kruger et al. 2020). Although in its infancy, to date several research initiatives have been commissioned by the WRC to test and pilot the application of environmental analytical and forensic techniques for tracking the sources of different CECs in water.

Therefore, despite having wastewater quality legislation and regulatory programmes in place, water resources pollution due to non-compliance is still a concern in South Africa. Efforts to address the contribution of municipal wastewater to the entry of CECs in water resources are underway. The National Water and Sanitation Masterplan document released in 2019 proposes the development and implementation of a national water and wastewater treatment performance turnaround plan by 2030 (Department of Water and Sanitation 2018).

7.2.2 Distribution of CECs in Water Sources in South Africa

Generally, the socio-economic conditions in developing countries prior to 2005 have hampered efforts in the identification and monitoring of CECs in aquatic environments. Development and implementation of the EDC research programme in South Africa as early as 2001 has been one of the most progressive achievements of the WRC (Water Research Commission 2022). Research conducted under this programme has revealed trends and patterns of a wide variety of CECs, including presence, levels, pollution sources, and their impact on ecosystem health over time. EDCs as contaminants of emerging concern have been in the research spotlight in South Africa since the late 1990s (Burger 2008). Over the years, the national EDC research programme has broadened to investigate the presence and levels of a wide variety of chemical classes, including POPs, natural and synthetic hormones, plant constituents, compounds used in the plastics industry and in consumer products, as well as industrial by-products and pollutants.

The first project on establishing baseline knowledge on the detection and quantification of EDCs as emerging pollutants in water sources in South Africa was commissioned in 2002 (Burger 2008). This project was undertaken in partnership with collaborators from the Global Water Research Coalition (GWRC), involving the active participation of 7 countries and 14 laboratories, as follows: Al control Laboratories (UK), Anglian Water (UK), Berliner Wasser Betrieb (Germany), CAE—Veolia Environment (France), CRC WQT (Australia), CIRSEE – Suez Environment (France), EAWAG (Switzerland), Institut for Environmental Studies (Netherlands), Kiwa Water Research (Netherlands), National Laboratory Service (UK), TZW (Germany), USEPA (USA), USGS (USA), and WRC (South Africa). The compounds targeted for analysis in water samples in this test were three (3) natural hormones; 17β-Estradiol, 17α-Estradiol, Estrone, and one synthetic hormone; Ethinylestradiol. synthetic hormones: estradiol (alpha and beta), estrone and ethinylestradiol, as well as other EDCs compounds, where the methods were available. The presence and levels of these hormones were determined in the following water sources; raw wastewater influent, treated wastewater, wastewater sludge, surface water, groundwater and drinking water.

Based on the findings, no positive results were reported for natural or synthetic hormones, industrial chemicals, organic compounds and metals, pesticides in drinking water and groundwater for South Africa. In contrast, compared to other GWRC members, very high values for pesticide concentrations, such as aldrin (1–1,000,000 μg/L), dieldrin (1–1,000,000 μg/L), DDE (0.1–0.28 μg/L), and lindane (6.4 μg/L), were reported in surface waters. Similarly, phthalate compounds, such as nonylphenol (up to 5 μg/L), and di-n-butyl phthalate (in the range 0.04–76 μg/L), were detected in some of the surface water samples collected in South Africa. In terms of organic compounds and metals (such as, lead, and cadmium) were also detected in the ranges, 10–1110 μg/L and 10–260 μg/L, respectively, in surface water. Due to unavailability of resources (skills and methods) in South Africa, no samples were collected and analysed for wastewater and sludge (GWRC 2003). Though not conclusive, these findings served as a baseline and highlighted the need for strengthening the capacity for EDCs monitoring in water resources.

This initial investigation was followed up by a series of investigations on the persistence of some of the EDCs. Such studies are worthwhile as the findings can be reliably used to draw cause-and-effect relationships as they are biologically and ecologically persistent (Ashraf 2017). South Africa's drive to investigate the distribution of POPs in water resources was also fuelled by both this initial study, as well as the country's ambitions to implement the Stockholm Convention, after ratification in 2002. At that time, the convention was aimed at eliminating or reducing the production or releases of the initial group of 12 POPs, which includes a number of pesticides (aldrin, dieldrin, DDT, endrin, heptachlor, chlordane, hexachlorobenzene, mirex and toxaphene) and three chemicals belonging to the classes of dioxins (PCDD), dibenzofurans (PCDF) and polychlorinated biphenyls (PCB) (UNEP 2009).

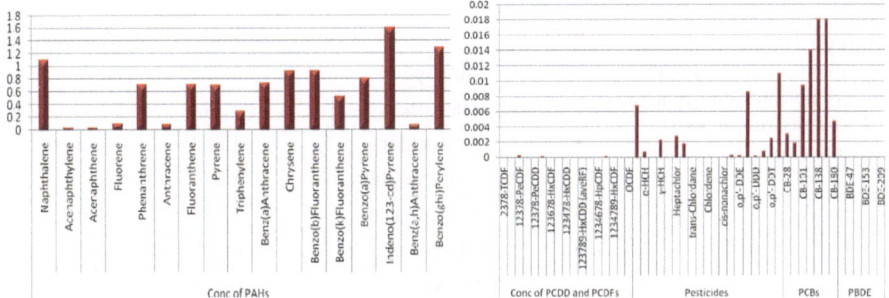

Fig. 7.3 Maximum detected concentrations (m/kg) of selected POPs in water environments in South Africa. *Source* Roos et al. 2011, reproduced with permission from the Water Research Commission (WRC) of South Africa

Extensive initial studies on monitoring POPs were conducted in the period between 2002–2005, with the first study focusing on establishing the levels and distribution of the original 12 POPs of the Stockholm Convention, namely; aldrin, chlordane, DDT, dieldrin, endrin, heptachlor, hexachlorobenzen, mirex, polychlorinated biphenyls, polychlorinated dibenzo-p-dioxins, polychlorinated dibenzofurans, and toxaphene) was assessed (Roos et al. 2011). Results from this initial assessment indicated that polycyclic aromatic hydrocarbons (PAHs) were the most abundant of all the groups of compounds investigated (Fig. 7.3). Polychlorinated biphenyls (PCBs) were present in intermediate concentrations, while polybrominated diphenyl ethers (PBDEs) were the least abundant and present in the lowest concentrations. The organochlorine pesticides (OCPs), aldrin and chlordane were not detected at any of the sites, whereas nonachlor, chlordane and oxychlordane were present at only a few of the sites in minor concentrations. HCB, HCH and DDT were the predominant OCPs, while heptachlor and mirex were present in lower concentrations (Fig. 7.3).

In 2015, the Water Research Commission commissioned a national-level risk mapping study aimed at identifying specific catchments within South Africa at risk from contamination using selected pesticides. Outputs from this study included maps of pesticide use, which provide a more refined spatial overview of hotspot areas in South Africa. This information is important as it shows the high priority pesticide contamination hotspots, where interventions are urgently required (Dabrowski 2015a). One other output from this study is a list of 63 high priority pesticides, based on their quantity of use, environmental mobility and risk to human health, whose use need to be tightly regulated or managed per application area. Pesticide prioritisation was performed by first listing all the toxic active ingredients of all pesticides used in enormous quantities in agricultural crop production within South Africa. The total toxicity score for each pesticide (TP) was then calculated, and then multiplied by a mobility score to derive the hazard potential (HP). Lastly, the score weighted HP (WHP), the potential hazard of the chemical was determined and expressed as a function of its total use in relation to the total use of all active ingredients applied in

Table 7.1* Top 10 high priority 10 pesticides/active ingredients of applied in the country posing significant risk to water resources contamination

Rank	Active ingredient	WHP	Mobility
1	Atrazine	3.6260	High
2	Mancozeb	3.4445	Low
3	Acetochlor	1.5875	Medium
4	Ethylene-dibromide	1.2210	High
5	Terbuthylazine	1.1344	Medium
6	Glyphosate	0.9779	Low
7	Sulphur	0.8600	Low
8	Copper oxychloride	0.8377	Low
9	Imidacloprid	0.7953	High
10	Metolachlor	0.6531	Medium

Source Dabrowski 2015b, reproduced with permission from the Water Research Commission (WRC) of South Africa

the country. Table 7.1 shows the top 10 priority pesticide / active ingredients based on the WHP index (Dabrowski 2015a).

To provide context on the persistence, commercial pesticide uses in South Africa dates back more than 100 years ago, to the passing of the Fertilisers, Farm Foods, Seeds, and Pest Remedies Act in 1917. Currently, there are over 700 different active ingredients registered for agricultural use, making the country one of largest consumers of pesticides in Africa (Dabrowski 2015a, FAO 2021). Amongst these are several organochlorine pesticides, including DDT (dichlorodiphenyltrichloroethane) and Lindane (a component of BHC—benzene hexachloride), which are listed under the Stockholm convention, but still in use under special permutations in South Africa (Department of Environmental Affairs and Tourism 2005). Furthermore, some pesticides (e.g., HCB) are still produced for industrial applications. The persistence of POPs and legacy pesticides like atrazine, DDT, and dieldrin continues to raise concerns in South Africa, primarily attributable to their use and enduring presence within aquatic ecosystems (Dalvie et al. 2014; Dabrowski 2015a; Curchod et al. 2020; Ojemaye et al. 2020; FAO 2021). Table 7.2 shows some of the recent maximum concentrations reported for some of the pesticides detected in water sources in South Africa.

Apart from studies focussed on the distribution of persistent and endocrine disrupting compounds, over 100 reports have been published in South Africa to date, each one reporting on the occurrence of a wide range of known and other "newer" CECs and their potential health effects (including endocrine disruption) in aquatic environments (Horak et al. 2021). The list of compounds that have been found to exhibit endocrine disrupting properties is continuously growing, with more potential EDCs introduced in an effort to replace currently existing ones (Gani et al. 2021).

Amongst some of the CECs that have been reported to occur in abundance are pharmaceuticals and personal care products (PPCPs). Particularly, environmentally persistent pharmaceutical products (EPPPs) have recently been recognized in the

Table 7.2 Distribution of pesticides in water sources in South Africa

Compound	Maximum reported concentration in water (μg/L)
Aldrin	0.12
Atrazine	18
Azinphos-methyl	0.87
Carbaryl	2.5
Carbofuran	1.6
γ-Chlordane	0.12
Chlorpyrifos	19.1
Cyfluthrin	0.0006
Cypermethrin	40.7
DDT and metabolites	3086
Deltamethrin	1.43
Dieldrin	2852
Endosulfan	6344
Endrin	65.9
Heptachlor	1285
Simazine	3.2
Terbuthylazine	4.0

Source Gani et al. 2021

Strategic Approach to International Chemicals Management (SAICM) policy framework as emerging pollutants of concern due to their health risks (UNEP 2006). Pharmaceutical use in South Africa remains one of the highest in Sub-Saharan Africa due to a high burden of diseases, as well as trauma-related ailments, such as interpersonal violence, such as gunshot- and stab wounds. In 2023, the South African pharmaceutical market was estimated to be worth a total of US$6.3 billion (United States of America International Trade Administration 2023). Due to the inadequacy of wastewater management infrastructure, residual concentrations of pharmaceuticals enter water environments.

Examples of pharmaceutical and personal care compounds that have been detected at increased levels in South Africa from 2003 to date in wastewater, drinking water and surface water include antibiotics, antiretroviral drugs, analgesics, anti-inflammatory drugs, lipid regulators, hormones, steroids, etc. (Patterton 2013; Osunmakinde et al. 2013; Odendaal et al. 2015; Swanepoel et al. 2015; Archer et al. 2017a,b; Schoeman et al. 2017; Kanama et al. 2018; Swartz et al. 2018; Mhuka et al. 2020; Gani et al. 2021; Horn et al. 2022). Personal care products—Examples detected in water sources in South Africa include parabens, triclocarban, and triclosan (Archer et al. 2017a, b; Kanama et al. 2018; Mhuka et al. 2020; Gumbi et al. 2022)). Regarding illicit drugs, available data shows that cocaine is the most predominant illicit drug in wastewater and surface water sources, followed by methamphetamine, and opioids. (Archer et al. 2017a, b; Lawrence et al. 2023). Table 7.3

shows maximum concentrations and classes of selected pharmaceuticals recently detected in wastewater treatment plants and surface water bodies in South African water resources.

Table 7.3 Reported maximum concentrations of selected pharmaceutical compounds in water

Pharmaceutical class	WWTP (final effluent)	Surface water
Analgesic and anti-inflammatory drugs		
Acetaminophen	0.2	0.2
Ibuprofen	1.2	8.5
Naproxen	2.9	1.9
Diclofenac	2.5	2.2
Ketoprofen	4.3	8.2
Antibiotics		
Ampicillin	8.9	14.5
Chloramphenicol	-	10.7
Ciprofloxacin	20.5	16.9
Erythromycin	0.2	22.6
Norfloxacin	0.4	0.001
Nalidixic acid	25.2	30.8
Ofloxacin	0.09	0.07
Streptomycin	-	8.4
Sulfamethoxazole	1.6	5.3
Tetracycline	3.2	5.7
Trimethoprim	1.6	1.1
Antiepileptic drugs		
Carbamazepine		
Tenofovir	0.3	0.25
Lamivudine	20.9	0.24
Zidovudine	53.0	0.6
Nevirapine	2.8	1.5
Personal care products		
Triclosan	22.9	0.9
Triclocarbanilide	0.5	-
Steroid hormones		
Oestrone (E_1)	0.08	0.03
Oestriol (E_3)	0.05	-
Oestradiol (E_2)	0.11	0.07
Ethynyl-oestradiol (EE_2)	4.6	0.06
Testosterone	0.03	0.02

Source Archer et al. 2020

The data presented in Table 7.3, confirms that WWTPs in South Africa are the major contributors to the presence of pharmaceuticals in water bodies and the environment. The data also shows that, some of the drinking water treatment process trains are not capable of removing these residuals resulting in their presence in the final water reaching the consumer. Knowledge on the presence, levels, and sources of pharmaceutical compounds in water sources that serve as the source for drinking water production is imperative for safeguarding environmental and public health. The reported presence of PPCPs in water resources in South Africa has emerged as a significant concern, prompting a critical examination of their potential role of the aquatic environment as a medium for the development and dissemination of antimicrobial resistance.

Other newer classes of CECs where baseline studies on their presence and levels in South African water resources have been undertaken include microplastics (MPs), engineered nanomaterials (ENMs) and per- and polyfluorinated substances (PFAS). The occurrence and abundance of MPs in South Africa has been investigated in different aquatic environments including marine, freshwater, wastewater, drinking water and sediments (Bouwman et al. 2018; Naidoo et al. 2020; Mgaba et al. 2022; Ramaremisa et al. 2022; Apetogbor et al. 2023; Swanepol et al. 2023). According to Malematja et al (2023), most of the studies done to date in South Africa have been on their abundance and bioaccumulation in marine organisms. Very few studies have investigated the occurrence of MPs in water bodies, wastewater treatment plant effluents and drinking water. Based on the number of reports available, it can be estimated that up 10 investigations have been done to date on the abundances of MPs in freshwater systems in South Africa (Malematja et al. 2023). Despite this, comparing the results from these studies is challenging due to the differences in the methods and units used to express MP abundances. MP abundances in freshwater systems range from 0.13 to 2.5 particles/m^3 (Ramaremisa et al. 2022), and studies on their characterisation are still at an early stage.

Regarding engineered nanomaterials (ENMs), although their presence in the environment has been flagged as an emerging threat, conclusive evidence on their presence, fate and impacts on human and environmental health is very scarce. This is because their detection and availability in water matrices is dependent on factors, such as, surface reactions, stability, mobility, and dissolution (Wagner et al. 2014). Consequently, studies done to date have focussed on development and application of modelling approaches to predict their presence, fate and behaviour in the aqueous environment. Based on available South African literature, ENMs in water have been detected in the order TiO2 > SiO2 > ZnO > Al2O3 > Ag > CNT (Nota et al. 2010; Musee et al. 2015). It is estimated that these concentrations, particularly for nAg, nAl2O3, carbon nano tubes (CNTs), nSiO2, nTiO2, and ZnO observed in 2015 will more than double by 2025 (Musee et al. 2015).

The presence and abundance of per- and polyfluoroalkyl substances (PFAS) in water environments has gained considerable global attention due to their persistence and bioaccumulative nature. Though, their risks are not very well understood, they have rapidly emerged as chemicals of concern in recent years due to their widespread presence in the environment. To date, studies done in South Africa have reported on

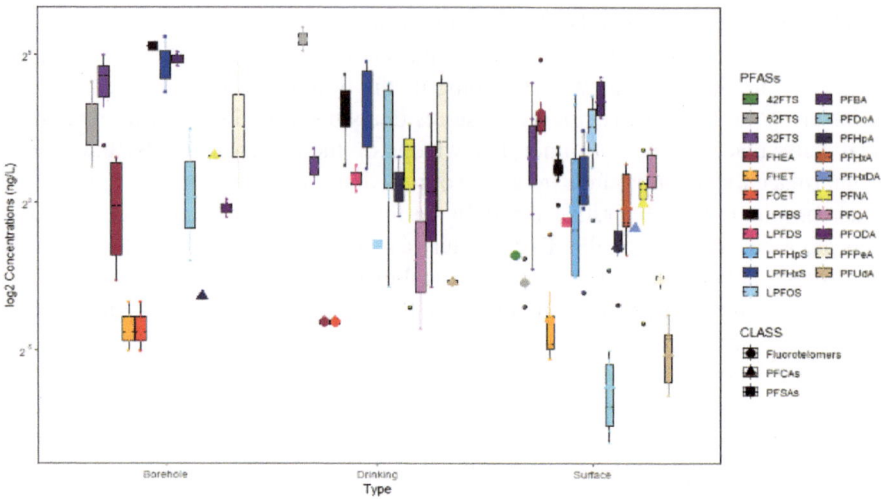

Fig. 7.4 Diversity and prevalence of PFAS in water sources in South Africa. *Source* Okonkwo et al. 2023, reproduced with permission from the Water Research Commission (WRC) of South Africa

their presence and levels in environmental water, wastewater, and drinking water (Mudumbi et al. 2014; Batayi et al. 2021). Recently, the Water Research Commission initiated a national monitoring baseline study to establish the presence, levels and classes of PFAS in different water sources in South Africa. For example, analysis of water samples collected from South Africa's most populous and industrialised province, Gauteng, have shown the presence of diverse classes of PFAS, with concentrations of up to 5 ng/L reported in both untreated and treated water (Fig. 7.4).

Generally, compared to other water sources, fewer studies have been done on the presence and levels of CECs in groundwater and the marine environment in South Africa (Selwe et al. 2022). Examples of studies conducted in groundwater include those reporting on the presence of industrial chemicals (such as flame retardants, bisphenol A), and pharmaceuticals (such as carbamazepine, cyclopentane, efavirenz, nevirapine, and tonalide) (Okonkwo et al. 2015; Wanda et al. 2017; Rimayi et al. 2018). Studies that report on the prevalence of CECs in the marine include those by Petrik et al. (2017) and Ojemaye et al. (2022). Examples of CECs that have been reported to occur in marine environments in these studies include perfluorodecanoic acid, perfluorononanoic acid, and perfluoroheptanoic acid, ranging from: (20.13–179.2 ng/g), (21.22–114.0 ng/g) and (40.06–138.3 ng/g), respectively. This aspect highlights a clear need for a source-to-sea approach for CECs monitoring as part of environment risk management. With regards to drinking water, a variety of CECs including, agricultural chemicals and pharmaceutical compounds have been reported to occur in drinking water (Patterton 2013). Currently, the focus has shifted from establishing prevalence to assessing and managing the risks associated with the presence of individual CECs and their mixtures.

In summary, extensive research work has been done on the distribution of CECs in water sources in South Africa, particularly those with known and suspected endocrine disruption properties, however, their monitoring, regulation and management still lags. A recent article by Gani et al. (2021) provides a comprehensive analysis of the presence and levels of CECs that have been detected in various geographical regions and water sources in South Africa to date. The authors of this article suggest that the levels and diversity of CECs detected in water sources in South Africa has been on the increase. Maximum CEC concentrations of up to of up to 7 mg/L in wastewater, followed by surface water, with maximum concentrations of up to 0.04 mg/L and lastly in drinking water, where up to 330 ng/L has been reported so far (Gani et al. 2021). The observed increase in CEC levels in water sources reflects the numerous challenges related to wastewater management, urbanization, water scarcity, and the need to improve wastewater infrastructure and treatment processes that the country continually faces. Other insights from the paper by Gani et al. (2021) concerns the types, occurrence, and levels of CECs, which have been shown to vary significantly in different geographical regions due to a variety of factors related to land use/cover activities, point sources, consumer behaviours, level of compliance to water quality regulations, as well as monitoring activities.

Lastly, the current level of understanding of the distribution of CECs in water sources in South Africa is a direct reflection of the maturity of the country's research landscape, in terms of human and analytical capabilities. Particularly, availability and access to various advanced analytical techniques and instrumentation has undoubtedly also played a pivotal role in the detection and quantification of CECs, especially at trace levels (Swartz et al. 2018). According to Mhuka et al. (2020), the use of hyphenated techniques, such as, gas chromatography-mass spectrometry (GC–MS) or liquid chromatography-tandem mass spectrometry (LC–MS/MS), provides enhanced capabilities for identifying and quantifying CECs. Furthermore, their use in non-targeted screening of CECs, have enabled the discovery of previously unknown contaminants by identifying compounds not targeted in initial analyses.

Indeed, the application of advanced analytical techniques for the detection and quantification has been invaluable in supporting research aimed at establishing baseline data on the occurrence and levels of CECs in South Africa. However, application of the same technologies for routine and regulatory monitoring of CECs in South Africa is lacking as there are no water reference or accredited analytical laboratories for CECs in south Africa. Establishment of such laboratories has been hindered by the excessive costs related to the instruments, software, and training. Furthermore, maintaining accreditation of laboratories and software updates can be expensive, and these costs can strain already limited budgets (Swartz et al. 2018; Mhuka et al. 2020).

7.2.3   Current Understanding on the Risks and Long-Term Impacts of CECs

Owing to the confirmed presence, persistent nature, and mobility of a variety of CECs with endocrine disrupting properties in water sources in South Africa, there has been dedicated research effort in establishing the potential environmental risks. Such studies have been conducted based on the health risk assessment protocol for EDCs developed by Genthe and Steyn (2008), which involves using a battery of in vitro and in vivo tests to screen for endocrine activity. Consequently, there is now adequate local evidence showing that exposure to EDCs impacts the regulation and expression of important genes associated with normal hormone function in fish species (Truter et al. 2015). Furthermore, exposure to residual concentrations of EDCs, such as 17-β oestradiol, p-Nonyl phenol, PCBs, estrone, estriol, organochlorine pesticides, and heavy metals in water results in the disruption of hormonal and reproductive systems resulting in intersex in aquatic and wildlife organisms (Horak et al. 2021).

With regards to human health risks, in vitro *tests* conducted to date have shown that exposure to EDCs through drinking water can elicit a wide range of endocrine disruption effects ranging from disruption of the reproductive system and other hormone signalling systems like those driven by progestogen, glucocorticoid, retinoid, and thyroid hormones (Aneck-Hahn et al. 2009, 2017; Archer and Van Wyk 2015). Results from these studies are concerning and highlight the need to strengthen the drinking water safety planning practices and water quality standards to also include the use of effect-based methods. Research on the assessment of toxicity and risks associated to exposure to some of the 'newer' CECs such as PFAS (Batayi et al. 2021), ENMs (Musee et al. 2015) and microplastics (Naidoo et al. 2020; Mgaba et al. 2022) is still in its infancy in South Africa. The results obtained in studies done to date have been deemed inconclusive, as there are still uncertainties on the methods.

To improve the assessment and prediction of toxicities, risks and long-term impacts related to exposure to EPs, the Adverse Outcome Pathway (AOP) framework has proposed a plausible approach (Ankley and Edwards 2018). Currently, there are no published studies demonstrating the use of the AOP framework for CECs screening in South Africa. Application of this framework has been hindered by the limited availability and ability to conduct the relevant bioassay tests in South Africa.

7.3   Advancements in Managing Environmental and Public Health Implications of CECs in Water

Since the formal launch of a research program in 2001, South Africa has remained at the global forefront of CECs research, with the WRC playing a vital role in funding and facilitating interdisciplinary collaborations among scientists, policymakers, and stakeholders to reduce risks and other complex challenges associated with the presence of CECs in water. Research on the occurrence, sources, behaviour, and potential risks due to exposure to CECs in water environments has been critical in establishing a firm knowledge base that can empower policymakers, practitioners, and communities to take informed decisions on their fitness for use and proactive measures toward sustainable water management. However, the journey from scientific research to practical application and policy development is complex and multifaceted. It is imperative that efforts to manage CECs in water encompass a range of strategies and initiatives. Such initiatives should be aimed at improving understanding, monitoring, and mitigating the impacts of these contaminants. The sub-sections below provide a brief description of some of the approaches adopted in South Africa, each with its own successes and opportunities.

7.3.1 Strengthening CECs Research, Monitoring and Innovations Development

Continual research is conducted to identify new emerging pollutants, understand their behaviour in aquatic environments, and assess their potential risks to ecosystems and human health. In South Africa's case, international collaboration and knowledge sharing has been very crucial in fostering knowledge exchange and capacity building, sharing of best practices, and deepening the understanding on emerging pollutants in water resources. Amongst these, collaborations on monitoring research programs to track the presence and levels of CECs in water sources has helped guide management strategies. For example, as informed by research, a National (water) Toxicity Monitoring Programme was promulgated under the National Water Act (1998) (Department of Water Affairs and Forestry 2005). This programme was specifically set up to report on the status of dichloro-diphenyl-trichloroethane and other persistent organic pollutants in water resources, as required under the Stockholm Convention. Maintaining all this programme, however, has been challenging due to the lack of financial resources, resulting in paucity of national water quality related data on emerging pollutants.

Nonetheless, given the risks and impacts related to exposure to these chemicals, the need to strengthen research aimed at developing and testing modelling and/or forensic approaches for predicting the sources, fate, and behaviour of chemical pollutants within the aquatic environment still stands. Recommendations from recent research state that the adoption of forensic analytical techniques into policy can aid tracking of their sources, enabling regulatory agencies to monitor compliance to water quality standards and discharge permits and improve the enforcement of environmental pollution prevention strategies (Kruger et al. 2020). Although in its infancy, to date several research initiatives have been commissioned by the WRC to assess and pilot the application of environmental analytical and forensic techniques for tracking the sources of different chemicals in water.

Extensive research aimed at the development, optimisation, and testing of innovative water treatment technologies targeted at removing a broader range of contaminants, including CECs such as ozonation, activated carbon filtration, membrane filtration, as well as other nanotechnologies has been undertaken in South Africa as part of a journey to transition into more effective CECs removal technologies (Amis and Lugogo 2018; Water Research Commission 2022). To promote the commercialisation and adoption of water treatment innovations, the Department of Science and Innovation (DSI) and the Water Research Commission (WRC) established the Water Technologies Demonstration Programme (WADER) in 2015. WADER acts as the country's innovation intermediary for facilitating high-level, collaborative technology demonstrations from public and private sectors to maximise the potential of the water innovation value chain. Under this platform, development, testing and implementation of advanced water treatment technologies for the removal of CECs such microplastics, pharmaceuticals, and other emerging contaminants is ongoing. The demonstration of innovative technologies is expected to pave way for the practical application of different advanced water treatment technologies for CECs removal in water and wastewater. Viewed differently, this period allows sufficient time for re-purposing current implementation of innovation and commercialisation programmes and developing new ones focused on fast tracking the deployment of appropriate advanced water treatment technologies in the sector.

As an initial step towards generating evidence for strengthening the monitoring and regulation of CECs in water, the South African government, through the national water research agency—the Water Research Commission, is currently involved in a global project coordinated by the GWRC on the harmonisation and use of effect-based methods for statutory water quality assessment programmes and regulation (Brack et al. 2019). This current project is a culmination of long-term research studies over the past two decades conducted by GWRC members, which have shown that effect-based methods can be used as valuable regulatory tools for screening the presence and monitoring long term impacts of CECs in the water environment (Kruger et al. 2022).

7.3.2 Promulgation of Stricter Drinking Water Quality Standards

Scientific findings related to the presence and levels of CECs in source water for drinking water production have led to the inclusion of representative/indicator CECs, including a variety of disinfection by-products, phenols, and atrazine, in a 2015 version of the national standards for drinking water quality (SANS 241). Recently, Swartz et al (2018) compiled a list of priority CECs in water treated for potable purposes and proposed their inclusion as part of the list of parameters in the next revision of SANS 241. In the absence of the revised standard, the guidelines by Swartz et al (2018) are invaluable in influencing municipal practices on the monitoring and communication of risks related to CECs in source and drinking water. For example, due to the high-risk potential linked to the use agricultural chemicals, a health-based limit of 100 μg/L has been proposed for adoption in the regulated national standards for drinking water quality. With regards to water resource regulations, there are efforts underway to update and establish regulations and guidelines specifically targeting CECs. These frameworks set standards for permissible levels of contaminants and guide industries in reducing their release into water sources.

7.3.3 Development of a National Implementation Plan for CECs

The government of South Africa is a signatory to several conventions on the sound management of chemicals, including those that are currently classified as CECs (Department of Environmental Affairs and Tourism 2005). The country's commitment to addressing emerging pollutants dates as far back as the early 2000s, and this was initially demonstrated by the ratification of the Basel, Stockholm, and Rotterdam conventions, as well as key measures like banning liquid waste from being dumped in landfills (UNEP 2020). In 2006, the South African government became party to the Strategic Approach to International Chemicals Management (SAICM), a policy framework with a commitment to achieving the sound management of chemicals by 2020 (UNEP 2020). The SAICM initiative identifies eight classes/types of EPs, namely, lead in paint; chemicals in products; hazardous substances within the life cycle of electrical and electronic products; nanotechnology and manufactured nanomaterials; EDCs; environmentally persistent pharmaceutical pollutants (EPPPs); and perfluorinated chemicals; and highly hazardous pesticides (HHPs) for prioritisation (UNEP 2006). Although South Africa failed to meet the 2020 target (UNEP 2019), the country has renewed its commitment under a new phase known as "Beyond 2020".

Under this new initiative, the South African government, via the Department of Forestry, Fisheries and the Environment, has embarked on an ambitious two-year project as part of the implementation of the Strategic Approach to International Chemicals Management (SAICM) policy framework to strengthen the country's institutional approach towards sound management of chemicals and wastes as means of achieving the sustainable development and safeguarding the survival of future generations (UNEP 2020). Using available scientific evidence on the presence, levels and potential impacts of CECs, South Africa has developed a national implementation plan (NIP) for addressing the current institutional capacity challenges impeding the management of CECs in South Africa. The NIP also includes action plans for reducing the identified risks and impacts related to exposure to EDCs; environmentally persistent pharmaceutical pollutants (EPPPs); and highly hazardous pesticides (HHPs) (UNEP 2020).

Efforts focused on promoting responsible industry practices, consumer behaviour and implementation of better waste management practices are already underway. Scientific studies on the risks and potential long-term impacts of CECs have led to the voluntary implementation of global sustainability policies and interventions on CECs within the South African chemicals industry. Industry self-regulation has played a significant role in complementing government regulations (CAIA 2022). The Chemical and Allied Industries' Association (CAIA) was established in 1993, and is a signatory to Responsible Care, which is the global chemical industry's unifying commitment to the safe management of chemicals throughout their life cycle, while promoting their role in improving quality of life and contributing to sustainable development. In the 2022 annual report, CAIA acknowledged the importance of collaborating in research to strengthen risk assessment as a means of supporting the attainment of the country targets for chemicals management. Furthermore, CAIA has committed to fast track the implementation of the Globally Harmonized System of Classification and Labelling of Chemicals (GHS) to aid public education on EPs and the communication related risks (CAIA 2022).

To further strengthen knowledge dissemination and communication of risks associated with CECs as a means of influencing consumer behaviour, the WRC and partners have since developed a knowledge hub for CECs (Botha et al. 2023). This knowledge hub is a GIS and web-based platform that serves as a repository for all data on CECs in South African water resources.

Through continued research and collaboration, the safety and integrity of water resources can be ensured for current and future generations. Implementation for this plan opens new opportunities to strengthen policy development and implementation, as well as promoting multistakeholder collaboration in research, innovation, and capacity building for managing CECs.

7.4 Summary, Conclusion, and Future Research Direction

Since the dawn of the twenty-first century, South Africa has witnessed a dynamic evolution in its understanding of emerging pollutants (EPs) within aquatic environments. Undoubtedly, the initial research focus on legacy, endocrine disrupters and POPs has played a pivotal role in shaping the chemical water pollution research paradigm in South Africa. The initial focus on EDCs has provided a robust foundation for understanding the intricate interplay between environmental contaminants and their far-reaching impacts on aquatic ecosystems and human health. Furthermore, the EDC research programme has helped build institutional and human resources capacity for research on emerging contaminants. Furthermore, the initial focus on EDCs has stimulated new research interests on tracking the emergence of newer contaminants of concern driven by changes in chemical consumption patterns, advancements in detection technologies, as well as the recognition of their potential impacts on ecosystems and human health. To date, the emerged research interest has addressed emerging pollutants, such as pharmaceuticals and personal care products, microplastics, engineered nanomaterials, newer pesticides, and other emerging substances affecting water quality.

While it is apparent that the South African research landscape on CECs has matured, it is equally important to note that CECs remain largely unregulated and continue to be released into natural water bodies. The disparities in management of CECs between developed and developing countries arise from several interconnected factors related to economic development, human resources and laboratory infrastructure, waste management infrastructure, regulatory frameworks, and societal priorities or chemical consumption patterns in the different worlds. For example, developed countries have more sophisticated waste and water treatment systems, reducing the likelihood of CECs entering the aquatic environment. On the contrary, poor waste management in developing nations can contribute to the widespread occurrence of CECs in the environment.

Nonetheless, the gradual transition from legacy pollutants towards emerging pollutants represents a significant milestone in environmental research in South Africa. In all this, the WRC has been instrumental in fostering capacity building, nurturing expertise and raising awareness about water quality issues. As a result, South Africa is currently at a much better position in terms of the state of knowledge on emerging pollutants, facilitating informed decision-making and proactive measures to safeguard water resources. It is essential to recognise that South Africa's successes in emerging pollutants research were not achieved without overcoming significant challenges, such as limited resources, technological constraints, and gaps in knowledge. Establishment of the relevant institutions and the country's commitment to advancing knowledge on emerging pollutants has led to noteworthy successes in the field of water quality research.

It is evident that access to advanced analytical techniques and availability of human capacity to undertake specialised water quality analysis for CECs is essential in accurately characterizing, monitoring, and regulating the presence of CECs in water, even at trace levels. Accredited laboratories play a crucial role in enabling the enforcement of water quality regulations and litigation matters on CECs due to their importance in ensuring the accuracy, reliability, and credibility of water quality data. Like other developing countries, South Africa has fewer accredited facilities for CECs, which is an aspect that has hindered the incorporation of CECs in water and other environmental regulations.

Considering that South Africa is a developing and water scarce country, it is important to understand that addressing water quality issues on CECs is a complex and ongoing process that is characterised by both remarkable successes and challenges. Over the years, research has served as the cornerstone for establishing baseline knowledge on CECs in different water environments. The knowledge, data, and insights generated have been crucial for improving understanding on the occurrence, levels, sources, transport, and fate of CECs within the aquatic environment. Such research has also played a crucial role in informing public awareness campaigns and catalysing discussions on new policies and regulatory measures on CECs in South Africa.

Although South Africa has made considerable efforts to advance research work on CECs, there remain challenges in environmental chemicals pollution, particularly in economic hubs and underserved communities. Water quality surveillance studies have confirmed that municipal, industrial, and domestic effluents, as well as runoff from unserved settlements and agricultural lands, serve as primary sources and pathways for the entry and wide dissemination of EPs into the aquatic environment.

In conclusion, the challenge of persistent chemicals and CECs in water resources is an area of growing concern, not only in South Africa, but globally. As the nation continues to navigate the intricate intersections of environmental sustainability and public health, a deeper exploration into the realm of CECs has become increasingly urgent. Given the wide-ranging risks and potential long-term impacts of CECs, as well as the close connections between human, animal, and environmental health, managing CECs in the environment demands the implementation of innovative and transformative research approaches. A transformative research approach represents a dynamic and revolutionary approach that transcends conventional boundaries, reshaping our understanding of complex challenges and opening new avenues for innovation and progress.

A transformative water quality and One Health research approach is being proposed by the Water Research Commission as the next research framework for South Africa for addressing knowledge gaps on CECs. One Health is an interdisciplinary and unifying approach that that aims to sustainably balance and optimize the health of people, animals, and ecosystems. It emphasises the need for a collaborative effort to address health issues that affect humans, animals, and the environment. This approach involves a holistic and interdisciplinary perspective that addresses the complex interactions between environmental contaminants, human health, animal health, and ecosystem integrity. This approach also promotes the implementation of integrated surveillance systems to monitor the presence and levels, as well the impacts of CECs in humans, animals, and the environment. The One health approach also advocates for evidence-based formulation of policies and regulations to address CECs at local, national, and international levels.

References

Amis MA and Lugogo S (2018) The South African water innovation story. Water Research Commission Report No. SP 126/18. Pretoria South Africa

Aneck-Hahn NH, Bornman MS, De Jager C (2009) Oestrogenic activity in drinking waters from a rural area in the Waterberg District. Limpopo Province, South Africa WaterSA 35(3):245–251

Aneck-Hahn NH, Van Zijl MC, de Jager C, Simba H and Ngcobo S (2017) Extending the EDC Toolbox 1 to include thyroid and androgenic bioassays. WRC Report No 2303/01/17. ISBN 978-1-4312-0924-8. Pretoria South Africa

Ankley GT, Edwards SW (2018) The adverse outcome pathway: A multifaceted framework supporting 21st century toxicology. Curr Opin Toxicol 1(9):1–7. https://doi.org/10.1016/j.cotox.2018.03.004.PMID:29682628;PMCID:PMC5906804

Apetogbor K, Pereao O, Sparks C, Opeolu B (2023) Spatio-temporal distribution of microplastics in water and sediment samples of the Plankenburg river, Western Cape. South Africa Environ Pollut 323(15):121303

Archer E, Wolfaardt GM, van Wyk JH (2017a) Pharmaceutical and personal care products (PPCPs) as endocrine disrupting contaminants (EDCs) in South African surface waters. Water SA 43(4):684–706. https://doi.org/10.4314/wsa.v43i4.16

Archer E, Petrie B, Kasprzyk-Hordern B, Wolfaardt GM (2017b) The fate of pharmaceuticals and personal care products (PPCPs), endocrine disrupting contaminants (EDCs), metabolites and illicit drugs in a WWTW and environmental waters. Chemosphere 174:437–446

Archer E and Van Wyk JH (2015) The potential anti-androgenic effect of agricultural pesticides used in the Western Cape: In vitro investigation of mixture effects WaterSA 41(1):129–137. https://doi.org/10.4314/wsa.v41i1.16

Archer E, Wolfaardt GM and Tucker KS (2020) Substances of emerging concern in South African aquatic ecosystems. Volume 1: Fate, environmental health risk characterisation and substance use epidemiology in surrounding communities. Water Research Commission Report No. 2733/1/20. ISBN 978–0–6392–0189–4

Arman NZ, Salmiati S, Aris A, Salim MR, Nazifa TH, Muhamad MS, Marpongahtun M (2021) A review on emerging pollutants in the water environment: existences Health Effects and Treatment Processes. Water 13:3258. https://doi.org/10.3390/w13223258

Ashraf MA (2017) Persistent organic pollutants (POPs): a global issue, a global challenge. Environ Sci Pollut Res 24:4223–4227. https://doi.org/10.1007/s11356-015-5225-9

Batayi B, Okonkwo O and Daso AP (2021) Poly- and perfluorinated substances in environmental water from the Hartbeespoort and Roodeplaat Dams, South Africa. Water SA. 47. https://doi.org/10.17159/wsa/2021.v47.i1.9445

Botha TL, Bamuza-Pemu E, Roopnarain A, Ncube Z, De Nysschen G, Ndaba B, Mokgalaka N, Bello-Akinosho M, Adeleke R, Mushwana A, van der Laan M, Mphahlele P, Vilakazi F, Jaca P, Ubomba-Jaswa E (2023) Development of a GIS-based knowledge hub for contaminants of emerging concern in South African water resources using open-source software: Lessons learnt, Heliyon, 9(1):e13007, ISSN 2405–8440. https://doi.org/10.1016/j.heliyon.2023.e13007

Bouwman H, Minnaar K, Bezuidenhout C and Verster C (2018) Microplastics in freshwater environments. WRC Report No.2610/1/18. ISBN 978–0–6392–0005–7. Pretoria South Africa

Brack W, Ait-Aissa S, Backhaus T, Dulio V, Escher B, Faust M, Hilscherová K, Hollender J, Hollert H, Müller C, Munthe J, Posthuma L, Seiler T, Slobodník J, Teodorović I, Tindall A, Umbuzeiro G, Altenburger R (2019) Effect-based methods are key. The European Collaborative Project SOLUTIONS recommends integrating effect-based methods for diagnosis and monitoring of water quality. Environ Sci Eur 31. https://doi.org/10.1186/s12302-019-0192-2

Burger AEC (2008) WRC research programme on Endocrine disrupting compounds (EDCs) volume 2: Implementation of a Research Programme for Investigating Endocrine Disrupting Contaminants in South African Water Systems, 2008, WRC Report No 1402/1/08, ISBN 1–77005–402–2. Pretoria. South Africa

CAIA (Chemical and Allied Industries' Association) (2022) Responsible Care performance report

Coetzee MAA, Momba MNB, Kibambe GM, Thobela KT, Kgositatu T and Mahlangu P (2017) The removal of endocrine disrupting compounds by wastewater treatment plants. WRC Report No. 2474/1/18. ISBN 978–0–6392–0117–7. Pretoria, South Africa

CSIR (2022) A story of our times—the CSIR's 75 remarkable years. Council for Scientific and Industrial Research. ISBN 978–0–7988–5666–9. Pretoria South Africa

Curchod L, Oltramare C, Junghans M, Stamm C (2020) Temporal variation of pesticide mixtures in rivers of three agricultural watersheds during a major drought in the Western Cape, South Africa. Water Research X 6:1–12. https://doi.org/10.1016/j.wroa.2019.100039

Dabrowski JM (2015a) Development of pesticide use maps for South Africa. S Afr J Sci 111(1–2) 07. https://doi.org/10.17159/sajs.2015/20140091

Dabrowski JM (2015b) Investigation of the contamination of water resources by agricultural chemicals and the impact on environmental health. Volume 1: Risk assessment of agricultural chemicals to human and animal health. WRC Report No. 1956/1/15

Dalvie MA, Sosan MB, Africa A, Cairncross E, London L (2014) Environmental monitoring of pesticide residues from farms at a neighbouring primary and pre-school in the Western cape in South Africa. Sci Total Environ 466–467:1078–1084. https://doi.org/10.1016/j.scitotenv.2013. 07.099

Department of Environmental Affairs and Tourism (2005) South African National Profile 2002–2005. A comprehensive assessment of the national infrastructure relating to the legal, administrative, and technical aspects of chemicals management in South Africa. Chemical and Hazardous Waste Management, Department of Environmental Affairs and Tourism. Pretoria South Africa. ISBN No 0–620–35993–5

Department of Water and Sanitation (2017) Water quality management and strategies for South Africa. Integrated Water Quality Strategy, DWS Report Number 3.2. Pretoria South Africa

Department of Water and Sanitation (2018) National Water and Sanitation Master Plan Volume 1: Call to Action. https://www.gov.za/sites/default/files/gcis_document/201911/national-water-andsanitation-master-plandf.pdf. Accessed on 10 August 2023

Department of Water and Sanitation (2022) Green Drop National report. Pretoria South Africa. https://ws.dws.gov.za/IRIS/releases/Report_NATIONAL%20_FINAL_30March22_MNEdit_web.pdf Accessed on 13 September 2023

Endocrine Society (2024) Endocrine-Disrupting Chemicals (EDCs) | Endocrine Society. Endocrine.org, Endocrine Society, 11 April 2024. https://www.endocrine.org/patient-engagement/endocrine-library/edcs

Fontes MK, Maranho LA, Pereira CDS (2020) Review on the occurrence and biological effects of illicit drugs in aquatic ecosystems. Environ Sci Pollut Res 27:30998–31034. https://doi.org/10.1007/s11356-020-08375-2

Food and Agriculture Organization of the United Nations (FAO). Statistical database of the food and agricultural organization of the United Nations. https://www.fao.org/statistics/data-dissemination/agrifood-systems/en. Accessed 26 January 2024

Gani KM, Hlongwa N, Abunama T, Kumari S, Bux F (2021) Emerging contaminants in South African water environment—A critical review of their occurrence, sources and ecotoxicological risks. Chemosphere 269:128737 Https://doi.org/10.1016/j.chemosphere.2020.128737

Geissen V, Mol H, Klumpp E, Umlauf G, Nadal M, van der Ploeg M, van de Zee S, Ritsema CJ (2020) Emerging pollutants in the environment: A challenge for water resource management, International Soil and Water Conservation Research, 3(1), 2015. ISSN 57–65:2095–6339. https://doi.org/10.1016/j.iswcr.2015.03.002

Genthe B and Steyn M. (2008) Health risk assessment protocol for endocrine disrupting chemicals. WRC Report No. KV 206/08. ISBN 978–1–77005–686–2. Pretoria. South Africa

Gumbi B, Moodley B, Birungi G, Ndungu P (2022) Risk assessment of personal care products, pharmaceuticals, and stimulants in Mgeni and Msunduzi Rivers, KwaZulu-Natal South Africa. Frontiers in Water 4:867201. https://doi.org/10.3389/frwa.2022.867201

GWRC (2003) Endocrine Disrupting Compounds. Occurrence of EDC in Water Systems. Prepared by: Water Research Commission (South Africa). In cooperation with: Kiwa Water Research (Netherlands) and TWZ – Water Technology Center (Germany). Global Water Research Coalition. United Kingdom

Horak I, Horn S, Pieters R. (2021) Agrochemicals in freshwater systems and their potential as endocrine disrupting chemicals: A South African context. Environ Pollut 268(Pt A):115718. https://doi.org/10.1016/j.envpol.2020.115718. Epub 2020 Sep 24. PMID : 33035912 ; PMCID : PMC7513804

Horn S, Vogt T, Gerber E, Vogt B, Bouwman H, Pieters R. (2022) HIV-antiretrovirals in river water from Gauteng, South Africa: Mixed messages of wastewater inflows as source. Sci Total Environ 806(Pt 2):150346. https://doi.org/10.1016/j.scitotenv.2021.150346. Epub 2021 Sep 21. PMID: 34601177

Kanama KM, Daso AP, Mpenyana-Monyatsi L and Coetzee MA. (2018) Assessment of pharmaceuticals, personal care products, and hormones in wastewater treatment plants receiving inflows from health facilities in Northwest province, South Africa. J Toxicol

Kruger A, Pieters R, Horn S, van Zijl C and Aneck-Hahn N (2022) The role of effect-based methods to address water quality monitoring in South Africa: a developing country's struggle. Environ Sci Pollut Res 29:84049–84055 (2022). https://doi.org/10.1007/s11356-022-23534-3

Lawrence TI, Sims N, Kasprzyk-Hordern B, Jonnalagadda SB and Martincigh BS. (2023) Wastewater profiling of illicit drugs, an estimation of community consumption: A case study of eThekwini Metropolitan Municipality, South Africa. Environmental Pollution 335. 122270, ISSN 0269-7491. https://doi.org/10.1016/j.envpol.2023.122270.

Malematja KC, Melato FA, Mokgalaka-Fleischmann NS (2023) The occurrence and fate of microplastics in wastewater treatment plants in south africa and the degradation of microplastics in Aquatic environments—A critical review. Sustainability 15(24):16865. https://doi.org/10.3390/su152416865

Masindi TK, Gyedu-Ababio T, Mpenyana-Monyatsi L (2022) Pollution of sand river by wastewater treatment works in the bushbuckridge local municipality South Africa. Pollutants 2:510–530. https://doi.org/10.3390/pollutants2040033

Mgaba N, Griffin N, Mensah PK, Tumwesigye E, Owowenu E, Mtintsilana Z, Nnadozie C and Odume ON. (2022) Microplastics as emerging contaminants: method development, ecotoxicity testing and biomonitoring in South African water resources. WRC Report No. 2919/1/22. ISBN 978–0–6392–0470–3. Pretoria South Africa

Mhuka V, Dube S, Selvarajan R and Nindi MM (2020) Emerging and persistent contaminants/pathogens: development of early warning system and monitoring tools. WRC Report No. 2516/1/20. ISBN 978–0–6392–0140–5. Pretoria. South Africa

Mudumbi JBN, Ntwampe SKO, Muganza FM and Okonkwo JO (2014) Perfluorooctanoate and perfluorooctane sulfonate in South African river water. Water Sci. Technol 69(1):185e194

Department of Water Affairs and Forestry (2005) National Toxicity Monitoring Programme (NTMP) for Surface Waters Phase 2: Prototype Implementation Manual. In: K Murray, RGM Heath, JL Slabbert, B Haasbroek, C Strydom and PM Matji (eds). Report No.: N0000REQ0605 ISBN No.: 0–621–36403–7. Resource Quality Services, Department of Water Affairs and Forestry, Pretoria, South Africa.

Musee N, Ondiaka M, Chimphango A and Aldrich C (2015) Modelling the fate, behaviour, and toxicity of engineered nanomaterials in aquatic systems. WRC Report No. 2107/1/14. ISBN 978–1–4312–0608–7. Pretoria. South Africa

Myers P. (2016) Our stolen future. Twent years later. Environmental Health. San Francisco Medicine. December 2016. https://www.healthandenvironment.org/docs/2016SFMMyersOurStolen.pdf. Accessed on 23 January 2024

Naidoo T, Naidoo S, Thompson RC, Rajkaran A (2020) Quantification and characterisation of microplastics ingested by selected juvenile fish species associated with mangroves in KwaZulu-Natal South Africa. Environ Pollut 257:113635. https://doi.org/10.1016/j.envpol.2019.113635. Epub 2019 Nov 18 PMID: 31767237

Nota N, Musee N, Aldrich C (2010) Estimation of Titanium dioxide and Silver engineered nanoparticles environmental exposure risks in water: a case of Gauteng Province, South Africa. Nanosci Young Res Symp, UWC, Belville Campus, Cape Town 17:17

Odendaal C, Seaman MT, Kemp G, Patterton HE, Patterton HG. (2015) An LCMS/MS based survey of contaminants of emerging concern in drinking water in South Africa. South Afr J Sci 111(9e10):1e6

Ojemaye CY, Pampanin DM, Sydnes MO, Green L, Petrik L (2022) (2020) The burden of emerging contaminants upon an Atlantic Ocean marine protected reserve adjacent to Camps Bay, Cape Town, South Africa. Heliyon 8(12):e12625. https://doi.org/10.1016/helicon.2022.e12625.PMID:36619409;PMCID:PMC9816787

Okonkwo OJ, Forbes PCB, Odusanya DOA, and Mnisi M. (2015) Screening study to determine the distribution of common brominated flame retardants in water. WRC Report No. 2153/1/15. ISBN 978–1–4312–0741–1. Pretoria South Africa

Okonkwo OJ, Batayi B, Morethe MF, Mashiloane K, Maliga Z, Rapoo S, Daso AP, Zwane BN, Monyatsi L, Jordaan E, Thaoge M, Chokwe T, Schoeman C and Rimayi CC (2023) Nationwide monitoring of per- and polyfluoroalkyl substances in water in South Africa. Volume III: Summary report on distribution, sources and health effects of per- and polyfluoroalkyl substances in water. Water Research Commission Report No. TT 931/3/23. ISBN 978–0–6392–0585–4. February 2023

Olujini OO, Fatoki OS, Daso AP, Akinsoji, OS, Oputu OU, Oluwafemi OS and Songca SP (2013) Levels of Nonylphenol and Bisphenol A in Wastewater Treatment Plant Effluent, Sewage Sludge and Leachates around Cape Town, South Africa. Handbook of Wastewater Treatment

Oluwalana AE, Musvuugwa T, Sikwila ST, Sefadi JS, Whata A, Nindi MM, Chaukura N (2022) The screening of emerging micropollutants in wastewater in Sol Plaatje Municipality, Northern Cape, South Africa. Environ Pollut. 2022 Dec 1; 314:120275. https://doi.org/10.1016/j.envpol.2022.120275. Epub 2022 Sep 24. PMID: 36167166

Osunmakinde CS, Tshabalala OS, Dube S and Nindi MM (2013) Verification and validation of analytical methods for testing the levels of PPHCPs (pharmaceutical & Personal Health Care Products) in treated drinking water and sewage. WRC Report No. 2094/1/13. ISBN 978–1–4312–0441–0. Pretoria South Africa

Patterton HG (2013) Scoping study and research strategy development on currently known and emerging contaminants influencing drinking water quality. WRC Report No. 2093/1/13. Pretoria South Africa

Persson L, Carney Almroth BM, Almroth CD, Collins CD, Cornell S, de Wit CA, Diamond ML, Fantke P, Hassellöv M, MacLeod M, Ryberg MW, Søgaard Jørgensen P, Villarrubia-Gómez P, Wang Z, Hauschild MZ (2022) Outside the safe operating space of the planetary boundary for novel entities Environ. Sci Technol 56(2022):1510–1521. https://doi.org/10.1021/acs.est.1c04158

Petrik L, Green L, Abegunde AP, Zackon M, Sanusi CY and Barnes J (2017) Desalination and seawater quality at Green Point, Cape Town: A study on the effects of marine sewage outfalls (with corrigendum). South Afr J Sci 113(11/12):10. https://doi.org/10.17159/sajs.2017/a0244

Ramaremisa G, Ndlovu M, Saad D (2022) Comparative assessment of microplastics in surface waters and sediments of the vaal river, south africa: abundance, composition, and sources. Environ Toxicol Chem 41:3029–3040. https://doi.org/10.1002/etc.5482

Rimayi C, Odusanya D, Weiss JM, de Boer J, Chimuka L (2018) Contaminants of emerging concern in the Hartbeespoort Dam catchment and the uMngeni River estuary 2016 pollution incident, South Africa. Sci Total Environ 627:1008–1017. https://doi.org/10.1016/j.scitotenv.2018.01.263

Rockström J, Steffen W, Noone K, Persson Å, Chapin III FS, Lambin E, Lenton TM, Scheffer M, Folke C, Schellnhuber HJ, Nykvist B (2009) Planetary boundaries: exploring the safe operating space for humanity. Ecol Soc 14(2)

Roos C, Pieters R., Genthe B, Bouwman H (2011) Persistent Organic Pollutants (POPs) in the Water Environment. Water Research Commission Report No. 1561/1/11, Water Research Commission, Pretoria

Scheringer M, Strempel S, Hukari S, Ng CA, Blepp M, Hungerbuhler K (2012) How many persistent organic pollutants should we expect? Atmos Pollut Res 3(4):383–391. ISSN 1309-1042. https://doi.org/10.5094/APR.2012.044

Schoeman C, Dlamini M, Okonkwo O (2017) The impact of a wastewater treatment works in Southern Gauteng, South Africa on efavirenz and nevirapine discharges into the aquatic environment. Emerg Contam 3(2):95e106

Schulz R (2001) (2001) Rainfall-induced sediment and pesticide input from orchards into the Lourens River, Western Cape, South Africa: importance of a single event. Water Res 35(8):1869–1876. https://doi.org/10.1016/s0043-1354(00)00458-9. PMID: 11337831

Selwe K, Thorn J, Desrousseaux A, Dessent C and Sallach J. (2022). Emerging Contaminant Exposure to Aquatic Systems in The Southern African Developmental Community. Environ Toxicol Chem 41. https://doi.org/10.1002/etc.5284

Sheriff I, Debela SA, Mans-Davies A (2022) The listing of new persistent organic pollutants in the stockholm convention: Its burden on developing countries. Environ Sci & Policy 130:9–15. ISSN 1462-9011. https://doi.org/10.1016/j.envsci.2022.01.005

Swanepoel C, Bouwman H, Pieters R and Bezuidenhout C (2015) Presence, concentrations, and potential implications of HIV-anti-retrovirals in selected water resources in South Africa. WRC Report No 2144/1. ISBN 978-1-4312-0637-7. Pretoria. South Africa

Swanepoel A, du Preez H and Bouwman H (2023) A baseline study on the prevalence of microplastics in South African drinking water: from source to distribution. Water SA, 49(4 October)

Swartz CD, Genthe B, Chanier J, Petrik LF, Tijani JG, Adeleye A, Coomans CJ, Ohlin A, Falk D, Menge JG (2018) Emerging contaminants in wastewater treated for direct potable reuse: The human health risk priorities in South Africa. WRC Report No. TT 742/2/17. Pretoria South Africa

Tang FHM, Lenzen M, McBratney A, Maggi F (2021) Risk of pesticide pollution at the global scale. Nat Geosci 14:206–210. https://doi.org/10.1038/s41561-021-00712-5

Truter JC, Van Wyk JH, Newman BK (2015) In vitro screening for endocrine disruptive activity in selected South African harbours and river mouths. Afr J Mar Sci 37(4):567–574, https://doi.org/10.2989/1814232X.2015.1105296

UNEP (United Nations Environment Programme) (2006) Strategic Approach to International Chemicals Management: SAICM texts and resolutions of the International Conference on Chemicals management. ISBN: 978-92-807-2751-7. Geneva

UNEP (United Nations Environment Programme) (2009). Stockholm Convention on Persistent Organic Pollutants (POPs) as amended in 2009: Text and Annexes. https://wedocs.unep.org/20.500.11822/27568

UNEP (United Nations Environment Programme) (2019) Global Chemicals Outlook II from Legacies to Innovative Solutions: Implementing the 2030 Agenda for Sustainable Development. ISBN No: 978-92-807-3745-5. Job No: DTI/2230/GE. Geneva

UNEP (United Nations Environment Programme) (2020) South Africa set to tackle emerging contaminants. Chemicals & Pollution Action Story. 19 March 2022.https://www.unep.org/news-and-stories/story/south-africa-set-tackle-emerging-contaminants#:~:text=Through%20this%20project%2C%20South%20Africa,to%20chemicals%20and%20waste%20management. Accessed on 21 August 2023

United States of America International Trade Administration (2023) South Africa—Healthcare: Medical Devices and Pharmaceuticals. https://www.trade.gov/country-commercial-guides/south-africa-healthcare-medical-devices-and-pharmaceuticals. Accessed on 25 January 2024

Visanji Z, Sadr SMK, Johns M, Savic D and Memon FK (2019) Optimising wastewater treatment solutions for the removal of contaminants of emerging concern (CECs): a case study for application in India. J Hydroinformatics 22. https://doi.org/10.2166/hydro.2019.031

Wagner S, Gondikas A, Neubauer E, Hofmann T, von der Kammer F (2014) Spot the difference: engineered and natural nanoparticles in the environment–release, behavior, and fate. Angew Chem Int Ed Engl 53(46):12398–12419. https://doi.org/10.1002/anie.201405050. Epub 2014 Oct 27 PMID: 25348500

Wanda EMM, Nyoni H, Mamba BB, Msagati TAM (2017) Occurrence of emerging micropollutants in water systems in Gauteng, Mpumalanga, and Northwest Provinces, South Africa. Environ Res Public Health 14(79):8–20. https://doi.org/10.3390/ijerph14010079

Water Research Commission (2022). WRC@50. Celebrating a half-century of excellence. WRC Report No SP 148/21. Pretoria South Africa

WHO (World Health Organization) (2017). Roadmap to Enhance Health Sector Engagement in the Strategic Approach to International Chemicals Management Towards the 2020 Goal and Beyond

Secretariat report A70/36 for the 70th World Health Assembly. Geneva, Switzerland: World Health Organization, 2017. http://apps.who.int/gb/ebwha/pdf_files/WHA70-REC1/A70_2017_REC1-en.pdf. (Accessed: 18 August 2023)

Part II
Emerging Pollutants and Groundwater

Chapter 8
Emerging Pollutants in Groundwater: The Origin, Transport Pathways, Remediation, and Challenges

Zehao Chen, Yinuo Wang, Junyuan Zhang, and Hongbin Zhan

Abstract Emerging pollutants (EPs) constitute a diverse group of novel or evolving chemicals and compounds, which usually have concentrations ranging from nanograms (ng) to micrograms (μg) per liter in aquatic environments. As the human population and activities continue to escalate, EPs are increasingly released into the groundwater. Due to their physical characteristics, the EPs would be transferred into groundwater through natural water cycling, leaching, and direct disposal, potentially threatening living organisms. The lack of comprehensive understanding of EPs makes it challenging to establish relevant legislation or effectively remediate EPs in groundwater. The absence of policies and regulations addressing EPs can lead to widespread repercussions, impacting groundwater quality and the health, economy, and social fabric of communities reliant on this vital resource. This review specifically identifies and analyzes four EPs: per- and polyfluoroalkyl substances (PFAS), nano- and micro-plastics (MNPs), pharmaceuticals and personal care products (PPCPs), and microorganisms. The transport pathways, global distribution, and remediation strategies of such four EPs are discussed. Additionally, we examine the challenges encountered and propose avenues for future research. This review article aims to address the substantial knowledge gap faced worldwide, and it will help establish and enforce effective regulations crucial to safeguarding groundwater quality and ensuring a sustainable and healthy environment for present and future generations.

Keywords Emerging pollutants · Per- and polyfluoroalkyl substances · Nano- and micro-plastics · Pharmaceuticals and personal care products · Microorganisms

Z. Chen · H. Zhan (✉)
Water Management and Hydrological Science, Texas A&M University, College Station, TX, USA
e-mail: zhan@tamu.edu

Y. Wang · J. Zhang · H. Zhan
Department of Geology and Geophysics, Texas A&M University, College Station, TX, USA

S. Zandaryaa et al. (eds.), *Emerging Pollutants*, Advances in Water Security,
https://doi.org/10.1007/978-3-031-71758-1_8

8.1 Introduction

Emerging pollutants (EPs) encompass a diverse category of novel or evolving environmental contaminants that may pose risks to human health and ecosystems. The list of all abbreviations used in this article can be seen in the Appendix. The group of EPs includes three categories of substances that (1) have recently been introduced into the environment, exemplified by newly developed industrial compounds; (2) have been introduced over a period and are only recently detected, owing to the diversity in chemical properties, the complexity of matrices, and generally low concentrations (such as micro- and nano-plastics (MNPs)); (3) have been known and measured for a long time and were only recently recognized as potential sources of adverse effects on living organisms and ecosystems (such as hormones, part of pharmaceuticals and personal care products (PPCPs) contaminants) (Houtman 2010). Currently, over 700 emerging pollutants, including their metabolites and transformation products, are documented as present in the European aquatic environment. Despite often occurring at concentrations ranging from nanograms to micrograms per liter (ng/L to µg/L) (or ng/kg to µg/kg in solid matrices), exposure to EPs frequently induces toxicological effects in biota (Sanganyado and Kajau 2022).

The complexity in detecting and quantifying EPs arises from their widespread discharging sources, making the analysis, regulation, and remediation of EPs challenging. In terms of their origins, EPs can be classified into artificial products (e.g., per- and polyfluoroalkyl substances (PFAS), MNPs, PPCPs) and natural substances (microorganisms). If not previously present in the environment, they can be released from point pollution sources, such as wastewater treatment plants. In comparison, nonpoint source pollution occurs through diffusion, atmospheric deposition, agricultural activities, or human activities. Regulating and monitoring point source pollution is comparatively more manageable than addressing nonpoint source pollution (Malkoske et al. 2016).

After entering the environment, potential sources of EPs to groundwater primarily include leaks in sewer lines, landfills, animal feeding operations, agricultural practices, wastewater irrigation, and leachates (Kapelewska et al. 2018). The presence of EPs in groundwater is also intricately linked to the geological characteristics of aquifers, which are predominantly detected in porous aquifers, facilitating the transport of contaminants. EPs can also be found in clay and sediments in the unsaturated vadose zone as well as in groundwater (Stefano et al. 2022).

The geographic patterns of their usage notably influence the distribution of EPs. For instance, certain pharmaceuticals intended for diseases prevention in Africa exhibit higher groundwater concentrations than those used in Western countries. Conversely, artificial sweeteners are less frequently used in third-world nations, resulting in significantly lower groundwater concentrations than in developed countries. In non-agricultural regions, the presence of artificial sweeteners serves as an indicator of groundwater contamination due to their higher concentrations than that in agricultural land. In contrast, other EPs, such as pesticides and antibiotics, are more likely to be found in agricultural settings (Lee et al. 2019; Li et al. 2021).

The pollution of groundwater by EPs is concerning due to groundwater's crucial role as a source of fresh water for human consumption, irrigation, and ecosystem requirements. Adverse effects of EPs in both humans and animals have been observed at concentrations as low as μg/L in groundwater, but long-term effects remain unknown. Water quality legislation addressing emerging pollutants is absent due to a lack of understanding concerning contaminant sources, pathways, properties, and analytical detection techniques (Egbuna et al. 2021). The absence of policies and regulations regarding EPs can have significant consequences. For example, the lack of specific regulations may allow harmful substances from EPs to enter groundwater uncontrollably, posing risks to both individual health and the balance of ecosystems. Once groundwater is contaminated, the recovery process can be lengthy, leading to persistent environmental degradation. In summary, the absence of policies and regulations addressing EPs can lead to widespread repercussions, impacting groundwater quality and the health, economy, and social fabric of communities reliant on this vital resource. Establishing and enforcing effective regulations is crucial to safeguarding groundwater quality and ensuring a sustainable and healthy environment for present and future generations.

This paper identifies four of the most concerning EPs in groundwater in recent years, which are PFAS, MNPs, PPCPs, and microorganisms. Figure 8.1 depicts the significant timeline for these four EPs, which will be discussed in detail in the subsequent sections. Our research illustrates their origins, distribution, transport pathways and mechanisms, potential remediation strategies, long-term effects, and regulations. Additionally, we examine the challenges encountered and propose paths for future research. This paper aims to address the substantial knowledge gap faced worldwide.

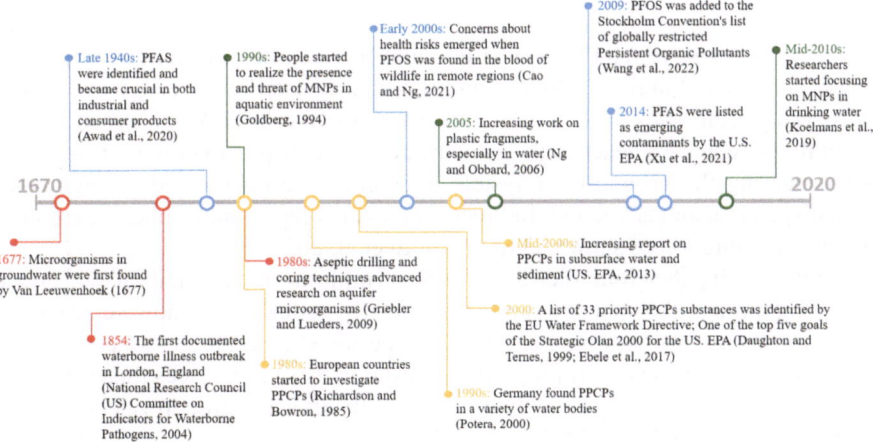

Fig. 8.1 Timeline of the development of four emerging pollutants (PFAS, MNPs, PPCPs, and microorganisms). *Source* Authors

8.2   Per- and Polyfluoroalkyl Substances (PFAS)

8.2.1   Background

Per- and polyfluoroalkyl substances (PFAS), a group of over 4,000 synthetic chemicals, are characterized by aliphatic structures containing carbon atoms fully (or partially) substituted with fluorine atoms. They were first identified in the late 1940s and have been extensively utilized in various industrial and commercial applications due to their unique properties (Awad et al. 2020). PFAS possess remarkable biological, chemical, and physical stability owing to their stable C–F bond, making them valuable for various industrial applications, including automotive parts.

In the United States, there are now 620 identified sites contaminated with PFAS, including drinking water systems, spread across 43 different states (Li et al. 2023). The United States Environmental Protection Agency (U.S. EPA) officially recognized PFAS as an emerging contaminants in 2014 (Xu et al. 2021). According to Saawarn et al. (2022), the concentration range of PFAS is from 0.025 to 1.2×10^8 ng/L in wastewater, 0.01 to 8.9×10^5 ng/L in surface water, and < 0.01 to 1.3×10^4 ng/L in groundwater globally.

This section aims to critically discuss the presence, transport mechanism, and remediation strategies of PFAS in the groundwater, which are crucial for effective management.

8.2.2   Transport Pathways

The routes through which PFAS enter groundwater are categorized into two primary sources: direct and indirect. Direct sources primarily entail the seepage of contaminated water from sites such as firefighting training areas. PFAS absorbed in agricultural soil can also leach into groundwater through precipitation or irrigation. After entering the groundwater system, they could be transported through diffusion, dispersion, and advection. Indirect sources encompass atmospheric deposition and the precipitation of snow and rainfall (Xu et al. 2021).

Upon reaching the ground surface, PFAS can infiltrate into the shallow subsurface, migrating through the vadose zone and potentially polluting groundwater. As many PFAS are surfactants, they have a tendency to be adsorbed at the air–water and solid–water interfaces in soils. This property can result in these substances lingering in the vadose zone for extended periods before eventually affecting groundwater (Brusseau et al. 2020; Zeng and Guo 2021).

Among all the potential origins of PFAS mentioned above, leakage of firefighting foams is one of the major sources of PFAS in groundwater. In the study by Schaefer et al. (2018), groundwater samples were examined in two distinct areas: one exposed to firefighting foams and the other without any such exposure. The results suggest that different PFAS, such as perfluorooctanoic acid (PFOA), perfluorobutanoic acid

(PFBA), and perfluorobutane sulfonate (PFBS), significantly increased from 1.3 to 3.8 times when firefighting foams were used.

Additionally, the disposal of pertinent industrial products in landfills represents a significant source of contamination since waste in landfill leachate contains PFAS. In their analysis, Hepburn et al. (2019) found 17 types of PFAS in the groundwater around a legacy landfill in an urban redevelopment area in Australia. Moreover, landfills are known to be a source of PFAS emissions into the atmosphere (Ahrens 2011).

Wastewater treatment plants (WWTPs), which are unable to remove all contaminants effectively, also contribute to PFAS in the groundwater (Schultz et al. 2006). After treating sewage, effluents from WWTPs have two end members: treated wastewater (TWW) and solid sewage sludge. TWW would be disposed of in surface water, used for irrigation, or injected into the subsurface. Solid sewage sludge rich in organic compounds (also called *biosolids*) is used in agriculture if it meets the criteria and is disposed of in landfills if it does not meet the criteria (Kinney and Heuvel 2020). The biosolids used in land applications allow PFAS to enter the soil, permeate groundwater, or runoff to surface water. Therefore, with numerous discharge routes such as effluent release, air emissions, biosolids disposal, and ineffective PFAS removal systems, WWTPs have become one of the most concerning point sources (Lindstrom et al. 2011) and collection points for PFAS compounds. Most WWTPs are not equipped with proven methods of PFAS removal (e.g., granular activated carbon or reverse osmosis) because they are cost prohibitive. As such, effluent from WWTPs often have concentrations of individual PFAS compounds higher than the respective compound detected in the influent. For example, in 2020, the testing result from Winslow WWTP showed that PFOA increased by 7.9 ng/L in March and 17.8 ng/L in July from the influents to the effluents (Murraysmith 2021).

Furthermore, the hydrologic cycle plays an important role in transporting PFAS. First, the contaminated surface water can seep into the underlying groundwater when there is a hydraulic connection between the surface water and the aquifer. Second, PFAS can be washed off from contaminated surfaces during rainfall and transported into stormwater systems. If these systems are connected to groundwater recharge areas, PFAS can infiltrate into the groundwater. Third, industries that use or produce PFAS-containing products may release PFAS into wastewater. If this wastewater is not properly treated, it can enter local water bodies or be discharged into the groundwater, potentially contaminating groundwater (Saawarn et al. 2022).

8.2.3 Worldwide Distribution of PFAS

PFAS are frequently detected in groundwater globally, with variations in the types of PFAS contamination observed among different countries. Numerous examples highlight the occurrence of PFAS in groundwater across diverse nations. In China, PFOA and perfluorooctane sulfonic acid (PFOS), which have been detected in seven major river systems, are the predominant pollutants. For example, a high concentration of

PFAS was detected immediately beneath the industrial park in Fuxin, China. The groundwater immediately beneath the park had the highest PFOA concentration of 524 ng/L (Bao et al. 2011).

High levels of PFAS contamination have been detected in the groundwater along the Ganges, the largest river in India, serving as a crucial drinking water source of India. Research by Sharma et al. (2016) revealed that PFBA and PFBS exhibited the highest concentrations in groundwater samples. The pollution of perfluoroalkyl carboxylates (PFCAs) in this region is likely attributed to discharge from urban residents residing along the banks of the Ganges River. Additionally, PFAS contamination in the groundwater is likely a result of infiltration, accidental spills, and the utilization of river water for irrigation purposes in this area.

Duong et al. (2015) conducted a study on the presence of PFAS in Vietnam, finding that the highest recorded concentrations of PFOA and PFOS in groundwater were 4.5 and 8.2 ng/L, respectively. In a separate study, Lam et al. (2017) examined PFAS levels in surface, tap, and groundwater samples from eight regions in Vietnam. They observed an average PFAS concentration of 0.32 ng/L in groundwater samples, notably lower than those in other industrialized countries.

An illustrative case can be found in Australian military sites where firefighting foams were employed. At the military base in Williamtown, New South Wales, levels of PFOA and PFOS were identified, reaching peak groundwater concentrations of 1,800 ng/L and 5,560 ng/L, respectively. Notably, the levels of contaminants in the groundwater declined with increasing distance from the source areas. There is another study at an agricultural site in Werribee South, near Melbourne, Australia, wastewater was used for irrigation, and measurements from groundwater samples reported a maximum PFOS concentration of 34 ng/L and a mean PFOA concentration of 2.2 ng/L (Szabo et al. 2018). The high PFAS concentration underscores the potential for recycled wastewater used in irrigation to serve as a diffuse source of PFAS, impacting groundwater quality.

In North America, surface water and groundwater serve as the primary drinking water sources, and they have been found to contain detectable levels of PFAS. According to the New Jersey Department of Environmental Protection (2014), PFOS was identified in 33% of groundwater samples between 2009 and 2010, with concentrations ranging from 9 to 57 ng/L. Oliaei et al. (2013) also reported that the Washington County Landfill, a closed and unlined site that historically accepted waste from the 3 M company, a primary manufacturer of PFOS in the United States. Groundwater at this location exhibited notably high concentrations of PFOA (42,000 ng/L) and PFOS (2,700 ng/L). Presently, the site is undergoing remediation through groundwater extraction and activated carbon treatment.

8.2.4 Health Concern and Remediation

PFAS are characterized by their high stability and resistance to natural breakdown processes, leading to their persistence in the environment. The stability of PFAS is also responsible for their potential to bioaccumulate in organisms, which raises health concerns. For example, with regard to perfluorooctanoic acid (PFOA), a common PFAS, the half-life of PFOA in human blood is approximately 3.5 years, which can lead to adverse effects such as genotoxicity, immunotoxicity, neurotoxicity, and hepatotoxicity (Rodea-Palomares et al. 2015).

Conventional treatment methods like coagulation, flocculation, sedimentation, oxidation, and disinfection have shown limited effectiveness in removing PFAS from water. While some promising strategies have been explored, further research is required to develop more effective and reliable approaches for PFAS remediation.

Boonya-atichart et al. (2016) utilized a nanofiltration membrane to extract PFOA from groundwater and noted an exceptional removal efficiency of over 99%. However, it is worth noting that membrane processes may diminish performance over time when issues like fouling and reduced water flux are present. The study also indicated that higher pressure and PFOA concentration could enhance removal efficiency, albeit at an increased operating cost. Consequently, it is imperative to conduct further research on the pivotal factors influencing membrane performance in PFOA removal from groundwater. Possible methods involve the development of innovative, high-permeability, and antifouling membranes, along with rigorous, long-term performance testing using authentic groundwater samples.

Sorption has proven to be a valuable method for eliminating PFOA from water. Various materials, such as activated carbon, carbon nanotubes, polymeric resins, and biomaterials, have been employed as adsorbents. However, these adsorbents have presented certain limitations, including low adsorption capacity, extended equilibrium time, high cost, and the potential for secondary pollution. These limitations have spurred research efforts toward developing new nanomaterials with enhanced capabilities for PFOA removal. For example, Gong et al. (2016) successfully employed starch-stabilized magnetite nanoparticles to effectively eliminate aqueous PFOA, demonstrating their potential for groundwater remediation. Nevertheless, it is important to acknowledge that materials utilized for PFAS adsorption may also pose a potential risk of secondary contamination.

Recently, there has been significant interest in utilizing zeolites and zeolite-based materials for the adsorption and catalytic degradation of PFAS. Zeolites are crystalline, hydrated aluminosilicates with a porous structure. Their unique properties make them valuable for various applications, including gas separation, adsorption processes, and catalytic conversion of biomass. In cases where hydrophobic interactions dominate PFAS adsorption, PFAS with longer chain lengths exhibit higher affinity due to their increased hydrophobicity.

Plasma represents the fourth state of matter, consisting of high-energy atoms, atom fragments, ions, free electrons, etc. Typically, the non-thermal plasma treatment processes are employed to degrade organic compounds, including PFAS, in aqueous environments (Saawarn et al. 2022). Several studies indicate plasma treatment systems can oxidize and reduce PFAS molecules through plasma-induced electron release. This method proves more effective in degrading PFAS chemicals than other advanced chemical treatment systems (Saleem et al. 2020). However, generating numerous gaseous and non-fluoride by-products after PFAS compound disintegration via the plasma process necessitates further research. Moreover, it is important to acknowledge that plasma technology has limitations, including potential interference from organic and inorganic co-ions, high voltage requirements, and safety concerns.

8.3 Micro- and Nano-Plastics

The invention of plastic has greatly changed human life in the past many decades. Plastics are durable, versatile, and cost-effective. These characters allow them to play an irreplicable role in a wide range of sectors (Okoffo et al. 2021). However, not until recent years did people realize that fragmented plastic particles have become an emerging pollutant to the environment and even the human body, and their impact on groundwater is not well-studied (Gündoğdu et al. 2023; Huang et al. 2021; Kumar et al. 2023). This section briefly reviews the definition of micro- and nano-plastics, the sources and effects of plastics in groundwater, and current knowledge gaps.

There is currently no consensus on the plastic particle categorization system, making it challenging to establish standardized experimental protocols, compare data, and formulate policies (Cerasa et al. 2021). Scientists generally distinguish plastic particles based on size and prefixes like mega, meso, micro, and nano (Hartmann et al. 2019). The most common in research are microplastics (MPs) and nanoplastics (NPs), especially in the subsurface environment.

8.3.1 Definition of Micro- and Nano-Plastics

Though it is commonly agreed that MPs and NPs are synthetic particles degraded from industrial plastics, the main disagreement focuses on how to set the upper and lower size limits for MPs and NPs. Frias and Nash (2019) defined MPs as particles with sizes ranging from 1 μm to 5 mm. Some other authors used 1 to 20 μm as the lower boundary for MNPs and an upper boundary for centimeter-scale. Gigault et al. (2018) proposed that NPs range between 1 nm and 1 μm, and Bleeker et al. (2013) used a 1 nm to 100 nm range for NPs. Further, no clear scientific evidence supports categorizing MPs and NPs through size.

Moreover, some concerns go beyond size. Plastics refers to a wide range of engineered materials. Merely distinguishing plastic particles based on size may overlook their distinctions in degradation rate, toxicity, and physical characteristics. Other potential critical characteristics of plastic particles include chemical composition, solubility, and color (Hartmann et al. 2019).

In some contexts, we can also refer to the origin of the particles to distinguish them. Particles are considered primary if they were produced in the microscopic range and secondary if they resulted from fragmented or degraded processes in the environment. This method is also a widely used and commonly agreed criterion and is a convenient concept in certain contexts (Cerasa et al. 2021; Cole et al. 2011).

The term micro-and nano-plastics (MNPs) is used for simplification in the following paragraphs. The term MNPs would be a generalized concept that refers to particles ranging from μm to mm in size, with no distinction in their material and origin unless otherwise specified.

8.3.2 Distribution, Origin, and Pathways

MNPs are found in surface water, marine, soil, groundwater, and drinking water. Compared to other systems, MNPs occurrence in groundwater or subsurface environments generally receives less attention (Huang et al. 2021). For example, in the literature review by Koelmans et al. (2019), only 1 out of 49 records of reported MNPs in water sources is from groundwater. Huang et al. (2021) collected and analyzed 42 studies, and 2 of them tested MNPs in groundwater. Among those limited studies, most reported data come from China and Europe.

Existing studies mainly conducted site-specific testing. Alvarado-Zambrano et al. (2023) found 330 MPs particles in 18 groundwater samples collected in an agricultural-intensive area in northwestern Mexico. The analysis result gives MPs concentration range between 10 to 34 particles/L, with isotactic polypropylene being the most abundant in samples. For a karst aquifer in Illinois, U.S., 16 out of 17 groundwater samples contained MPs, and the median concentration is 6.4 particles/ L (Panno et al. 2019). In contrast, a study by Mintenig et al. (2019) only detected MPs of 0.0007 particles/L in groundwater, which is almost neglectable. These measurements were taken under different criteria and could only give us an insight into the MNPs contamination status of limited spatial extent.

The sources of MNPs come from various anthropogenic activities. Industrial and municipal wastewater discharge, leaching from landfills, and septic tank outflows could release MNPs into the environment (Gündoğdu et al. 2023). Agricultural activities would directly release MNPs into groundwater or indirectly affect aquifers by elevating MNPs concentration in soil. Practices include sewage sludge, irrigation, using fertilizers and plastic mulch films (Chia et al. 2021). In some studies of MPs occurrence in China and Europe, scientists found that compared to non-agricultural soils, agricultural soils contain a much higher MPs concentration (Huang et al. 2021).

After entering the soil, MNPs could be vertically transported from soil to groundwater aquifers (Gündoğdu et al. 2023). Though insufficient scientific data reveals the detailed mechanisms, scientists have proposed potential pathways for MNPs. Direct pathways include the movement of MNPs through pore spaces larger than the particle diameter, e.g., in coarse sand. The vertical and horizontal movement would be possible if the water table level is high (Bläsing and Amelung 2018). Other factors that may affect MNPs transport include porosity, diffusivity, flow velocity, and properties of particles such as ionic strength, shape, and chemical composition (Gündoğdu et al. 2023).

Several researchers have tried to quantify the mobility of MNPs in the subsurface. O'Connor et al. (2019) showed that the number of wet-dry cycles would affect the vertical migration of MPs in soil. They forecasted a migration depth of 5.24 m in the next century for 347 cities in China, indicating soil as a feasible pathway for MNPs to groundwater aquifer. An experiment conducted by Goeppert and Goldscheider (2021) showed that MNPs ranging from 1 to 5 μm can be easily transported in sand-and-gravel aquifers. During the 171-day and 200 m scale experiment, researchers found that MNPs can travel a long distance horizontally while not being greatly affected by retardation and filtration as expected.

MNPs could also be transported through bioturbation. For example, the particles could adhere to plant roots and migrate to deeper soil through root growth. Besides, living organisms like earthworms and mites could also carry and spread MNPs (Guo et al. 2020; Huang et al. 2021). Freeze–thaw cycles, stormwater runoff, and agricultural irrigation could also accelerate MNPs' transport (Gündoğdu et al. 2023).

However, as of now, we still lack a thorough understanding of the transport mechanism of MNPs, especially for nano-size particles. More research is desired to test the hypothesized origin and pathways of MNPs mentioned above.

8.3.3 Detection and Quantification

Detection and quantification of MNPs are challenging due to their small size and large variety in morphology (Adhikari et al. 2022; Huang et al. 2021). In general, procedures for analyzing MNPs include (1) Sample collection: Choose a sampling site and approach based on the research target; (2) Extraction of MNPs: Techniques such as separation, digestion, and manual extraction could be applied; (3) Identification and quantification: Visual identification or using chromatography and vibration spectroscopy (Adhikari et al. 2022; Huang et al. 2021).

However, lacking standardized sampling and analyzing protocol, difficulties arise from every step, and data quality control is the most concerned (Koelmans et al. 2019). Meteorological conditions could affect particle migration in soil and groundwater, which should be carefully considered and controlled for better data quality (Adhikari et al. 2022). Background contamination of samples could happen during the collection step, for example, if plastic equipment is used or the sampling holes are covered with plastic film. Even in the laboratory, airborne plastic particles might contaminate samples, affecting the testing result (Torre et al. 2016).

In their research about the quality of MNPs data, Koelmans et al. (2019) rated existing studies according to sampling method, sample processing and storage, clean air conditions, etc. Only four of the 50 studies received non-zero scores for all nine criteria, indicating a need to improve data quality in MNPs studies. Creating blank groups, providing clean air and lab conditions, and changing sampling equipment would improve research quality (Adhikari et al. 2022; Koelmans et al. 2019; Tamminga et al. 2018).

8.3.4   Impact and Remediation Strategy

In recent years, human exposure to plastic particles has raised public concerns, but risks to the environment, biota, and human body by MNPs require further investigation. Since MNPs have been detected in groundwater, health-related concerns have risen from the consumption of raw groundwater, especially in developing countries (Gündoğdu et al. 2023). MNPs could possibly enter the human system via ingestion, so the consumption of raw groundwater may induce health-related issues (Gopinath et al. 2022). Nowadays, MNPs have already been detected in human blood and breast milk (Gündoğdu et al. 2023), posing an immediate need for in-depth research.

The effectiveness of the current water treatment system in removing MNPs is also a question. In a study conducted by Bäuerlein et al. (2022), scientists found that their investigated treatment plants successfully reduced MPs to lower than 2 particles/L. However, the targeted particles have a diameter larger than 20 μm. Thus, the question remains whether NPs still exist in treated drinking water.

For now, not enough remediation strategies have been proposed. Rain gardens may act as an MNPs barrier and are worth further investigation, but carefully treating MNPs before discharge or intake might be more efficient (Gündoğdu et al. 2023).

8.3.5   Knowledge Gap

Currently, knowledge about MNPs is still lacking. The following section provides a summary of current knowledge gaps:

(1) Unified classification system: lacking a clear definition and classification system complicated the research process from every aspect, such as the size of filters needed and the range of data to be compared.
(2) Standardized research protocols: designing universal research protocols could improve data quality.
(3) Systematic and large-scale research: more data are desired to thoroughly understand groundwater contamination by MNPs. In addition to site-specific studies, regional-scale research would provide valuable insight into the non-point source pollution of MNPs.
(4) Fundamental research: understanding the detailed mechanisms of MNPs transport and how it interacts with the surrounding environment still needs improvement. Based on this knowledge, better models could be developed to help assess MNPs.

(5) Impact on human health and environment: The long-term toxicity of MNPs and their endurance in the human body is unclear and should be studied as one of the highest priorities.

8.4 Pharmaceuticals and Personal Care Products (PPCPs)

8.4.1 Definitions, Origins, and Pathways

Pharmaceuticals and personal care products (PPCPs) have been considered an emerging pollutant in recent decades. PPCPs include diverse organic compounds from drugs and personal care products, such as soap, toothpaste, and sunscreens (Ebele et al. 2017; Liu and Wong 2013). PPCPs are widely used in the daily lives of individuals. Still, not until the last two or three decades have people realized that PPCPs could enter the aquatic environment and, subsequentially, bring detrimental effects to aquatic organisms and eventually harm the human body.

PPCPs are comprised of two major groups of products. Generally, pharmaceuticals, or medicines, could be divided into active pharmaceutical ingredients (APIs) and finished pharmaceutical products (FPPs), where APIs are raw materials, and FPPs are drugs used for clinical treatment. For example, caffeine is API, and aminopurine caffeine tablet is FPP (Wang et al. 2022). This paper only considers PPCPs and does not distinguish between API and FPP unless otherwise specified.

PPCPs are heavily used in daily life. Common pharmaceuticals include antibiotics (such as anti-infection drugs), hormones (such as steroid estrogens), analgesics and anti-inflammatory drugs (such as ibuprofen), β-blockers (such as metoprolol used to treat high blood pressure), and caffeine. Personal products include disinfectants (used in hand sanitizers and toothpaste.), fragrances, insect repellants, preservatives, and sunscreen UV filters (Liu and Wong 2013). Currently, more than 4,000 different kinds of pharmaceutical substances are available for use, and the consumption volume is over 100,000 tons/year worldwide (Hawash et al. 2023). In China, more than one thousand kinds of APIs were produced in 2018, with a total production of 18.48 million tons (Liu and Wong 2013; Wang et al. 2022).

PPCPs mainly find their way to the environment through municipal/industrial sewage discharge (Daughton and Ternes 1999). For example, medicines would be excreted, and toothpaste would be washed into municipal sewage systems. Though they are expected to be treated by WWTPs, studies have reported the occurrence of PPCPs in WWTPs effluents and biosolids worldwide, such as Greece, U.S., and China (Kosma et al. 2014; Sui et al. 2011; Yang et al. 2011).

PPCPs would eventually enter the environment after exiting WWTPs. Discharging of TWW, such as irrigation and wastewater injection, would directly introduce unremoved PPCPs to surface water, soil, and subsurface (Drewes et al. 2003a). PPCPs would leach from biosolids used as fertilizers or from landfills. Over time, these contaminants could potentially migrate to groundwater aquifers through infiltration, leaching, or surface water-groundwater interaction (Ebele et al. 2017; Hawash et al.

2023). Improper disposal or discharge of untreated wastes and sewage exfiltrate further complicated and contributed to the issue (Krishnan et al. 2021; Kuroda et al. 2012).

Besides human usage, pharmaceuticals (specifically, APIs) are also widely applied in livestock farming, such as for veterinary purposes (Topp et al. 2008). Osorio et al. (2016) found positive relationships between pharmaceutical level and live-stock density in their study sites in Spain. Further, surface runoff would transport PPCPs already released to the environment in agricultural and livestock-raising sites. According to Topp et al. (2008), some PPCPs concentration could be high enough in surface runoff to hurt the environment and organisms at a distance.

8.4.2 Distribution

PPCPs have been reported in various subsurface environments around the world, such as in karst aquifers (Einsiedl et al. 2010), artificial recharge sites (Drewes et al. 2003b), and aquifers in the vicinity of landfill sites (Peng et al. 2014). PPCPs in groundwater are usually at low ng/L to μg/L levels. However, abnormally high PPCPs concentration have been detected in some areas. For example, a study in Delhi, India, found a diclofenac concentration of 1.3 mg/L in the aquifer. In the U.S., an API concentration of around 0.098 to 0.12 mg/L was detected in groundwater (Kibuye et al. 2019). Highly concentrated PPCPs are usually found near TWW/biosolid disposal sites (Silori et al. 2022).

PPCPs are usually considered more concentrated in urban areas and developed countries, where the usage of related products is higher (Sun et al. 2016). Accordingly, groundwater in developed countries such as the USA and Spain was found to be heavily polluted by PPCPs. However, new studies showed that developing countries such as China, India, and Nigeria also have abundant PPCPs in groundwater (Peng et al. 2014; Silori et al. 2022).

The spatial variation of PPCPs concentration also depends on geological settings and the properties of the compounds. Karst aquifers are usually more vulnerable to PPCPs pollution due to their high hydraulic conductivity and connected pathways. Einsiedl et al. (2010) studied two common pharmaceuticals, ibuprofen and diclofenac, and found both entered and stored in the fractured system. Oppel et al. (2004) proposed that aquifers covered by sufficiently thick soil could be protected against PPCPs leaching. Among the studied PPCPs, clofibric acid and iopromide were very mobile, while ibuprofen and diazepam were less likely to leach. Besides, soil properties (e.g., pH and grain size) would affect the leaching behavior of PPCPs (Chen et al. 2011).

In contrast, Ma et al. (2018) found a vertical distribution of PPCPs in the vadose zone soil with a depth of 16 m, indicating a potential threat to the local ground-water aquifer. In addition, the geochemical properties of soil (i.e., potassium and silt content) serve as an important impact factor. These discrepancies in research results point to the importance of carefully characterizing the properties of study sites and target compounds.

PPCPs also show a seasonal variation in concentration and distribution. Sui et al. (2011) found peak PPCPs concentration in winter in the WWTP effluent, possibly due to the consumption of different products during the year or removal efficiency affected by temperature. Peng et al. (2014) found higher PPCPs concentration in reservoirs during spring, which might be affected by rainfall, hydrological processes, and temperature that affect the degradation rate (Karnjanapiboonwong et al. 2011). However, this temporal variation of PPCPs is also highly site-specific.

8.4.3 Risks

The risks of having PPCPs in groundwater come from several aspects (Ebele et al. 2017).

Firstly, some PPCPs are relatively hard to degrade and can present in subsurface over a long time (Wang et al. 2022). PPCPs are also considered *pseudopersistent*. Since they are continuously released into the environment, the degradation/transformation of compounds is compensated by the high replenishment rate, making some PPCPs behave more persistent than they actually are (Barceló and Petrovic 2007). Generally, the PPCPs risks in the groundwater are related to their residence times (Silori et al. 2022).

Secondly, bioaccumulation of PPCPs and their metabolites could potentially impact the food chain. Bioaccumulation refers to PPCPs entering and accumulating in the food chain through all possible routes, such as water, sediment, and diet, and accumulating PPCPs to a high concentration level may trigger more severe issues afterward (Srain et al. 2021).

Thirdly, exposure and accumulation to the active ingredients of PPCPs may trigger unwanted changes in organisms. PPCPs have been found in human blood, milk, and urine. Consumption of groundwater and aquatic organisms are potential pathways for PPCPs to enter human body. Though acute toxicity is not likely to happen, chronic effects, for example, on metabolism, cannot be neglected (Liu and Wong 2013; Silori et al. 2022).

8.4.4 Identification, Removal, and Remediation

Since PPCPs are usually present at a low level, detecting them is not easy. Traditional techniques include the use of gas chromatography. The latest technologies, such as liquid chromatography coupled with mass spectrometry, are more sensitive (Galindo-Miranda et al. 2019). The development of equipment allows more accurate detection of PPCPs.

Once entered the groundwater system, remediation becomes a difficult task. Current technology includes adsorption, treatment, filtration, etc. However, more advanced technologies are still desired to improve the effectiveness (Georgin et al. 2023). The large variety of PPCPs also makes it hard to target all contaminants using a single remediation approach.

Removing as many PPCPs as possible in the WWTPs effluent is crucial and more practical to reduce its environmental concentration. However, the removal efficiency of conventional WWTPs is low. Kosma et al. (2014) took samples from eight conventional WWTPs in Greece and found the occurrence of all target PPCPs in the wastewater samples, including paracetamol and caffeine. In contrast, the advanced wastewater reclamation plants studied by Yang et al. (2011) could remove major target contaminants through various treatment steps. It is important to note that PPCPs concentration in effluent from WWTPs that deal with pharmaceutical facilities is much higher than that from other plants, pointing to the importance of using better removal technology to treat pharmaceutical wastewater (Krishnan et al. 2021).

8.4.5 Knowledge Gap

Despite its ubiquitous presence in groundwater systems, current knowledge about PPCPs is insufficient. New PPCPs are continuously added to the already large family, but the fundamental mechanisms of major compounds are still lacking (Daughton and Ternes 1999). Compared to other environmental settings, there is still much room for research on PPCPs contamination of groundwater aquifers. More regulations should be implemented to reduce PPCPs discharge from sources such as WWTPs or landfill sites. Overall, researchers from different research areas and policymakers are expected to work together to reduce the PPCPs in the environment.

8.5 Microorganisms

Unlike the artificial pollutants discussed above, microorganisms are natural creatures, including prokaryotes (bacteria, archaea, etc.) and eukaryotes (animals, plants, fungi, etc.). However, these natural creatures, such as *Escherichia coli* (*E. coli*), sometimes harm humans. They may directly influence the human body or generate unwanted biomass, which may cause bioclogging in groundwater.

Before introducing the details of microorganisms in groundwater, knowing the development of the subject is essential, which can indicate how the background knowledge is established and what area has already been studied. Using a well water sample, Van Leeuwenhoek (1677) first found the microorganisms in groundwater, described as the first bacteria-like particles. After 200 years, a breakthrough was made: the first iron- and manganese-oxidizing bacteria were identified in groundwater wells. In 1854, modern society first recorded an instance of a waterborne illness outbreak that occurred in London, England (National Research Council (US) Committee on Indicators for Waterborne Pathogens, 2004). A century later, the indigenous microbial community in groundwater was found to be different from surface water based on the first systematic taxonomic assessment of bacteria in shallow aquifers. In the 1980s, the development of aseptic sampling techniques using drilling and coring supported research on microorganisms in aquifers (Griebler and Lueders 2009). After that, systematic research on microorganisms in groundwater began.

8.5.1   Origin and Distribution

Microorganisms in groundwater have various origins. On the one hand, some of them are indigenous species. They may enter the shallow groundwater early and gradually evolve into local species (Griebler and Lueders 2009). On the other hand, many other species may enter the groundwater through water seepage/leakage from their original habitat.

Compared to soil/surface water, groundwater is a relatively closed system. However, nutrients, organisms, and energy can still be transferred through the transition zone (soil, vadose zone, and hyporheic zone), allowing abundant diversity of microorganisms in groundwater (Gibert et al. 1990). For example, up to 40% of prokaryotic biomass resides in the groundwater, making subsurface aquifers the second largest habitat for prokaryotic (Griebler and Lueders 2009). Microorganisms are widely distributed in groundwater also because their habitat is only limited by extreme temperatures and pH conditions (Hendry 2006). In contrast, groundwater systems generally do not pose such challenges to microorganisms. For instance, Rampelotto (2013) mentioned that the archaeal Methanopyrus kandleri strain grows at 122 °C, and the genus Picrophilus grows at a pH of 0.06.

Most of the known microorganisms can be found in the groundwater, including Bacteria (Gounot 1994), Archaea (Detmers et al. 2004), Flagellata (Novarino et al. 1997), Amoebae (Novarino et al. 1997), etc. Generally, Prokaryotes (Bacteria and Archaea) and Flagellatam are attached to the sediment particles, rock surface, and other physical or biological structures. Respectively, the Amoebae are mainly distributed near-surface groundwater (Griebler and Lueders 2009). The different distribution patterns may be due to the variation in the energy sources and mobility of the microorganisms.

8.5.2   Environmental and Health Impact

Although microorganisms in groundwater are natural, they still negatively impact human life, especially when people use groundwater as drinking water. Generally, there are three ways that microorganisms could cause contamination (Egboka et al. 1989).

Firstly, microorganism is pathogenic (Egboka et al. 1989). In groundwater, a large amount of microorganism species, most of which are from outside sources, could directly cause diseases. Human activity can make more microorganisms leak into groundwater, for example, leak from septic tanks. Enteropathogenic E. coli, a species that belongs to bacteria, can cause enteritis. Besides, protozoans, such as Entamoeba histolytica, can cause amoebic dysentery and cause about 55,000 deaths per year (Shirley et al. 2018). Thus, paying attention to microorganisms in groundwater used as drinking water sources is especially important.

Secondly, microorganisms can produce superabundant biomass, which may not directly influence the human body but can cause hydrology issues such as bioclogging (Egboka et al. 1989). Bioclogging occurs because microbial biomass could occupy pores inside porous media (Xian et al. 2019). For example, research on geothermal has earned wide attention in recent years. But at the same time, the thermal injection may cause bioclogging in the aquifer by increasing the aquifer temperature and introducing oxygen, promoting the growth of microorganisms and contaminating the aquifer. Ni et al. (2018) showed that the aquifer thermal energy storage system would not cause biological clogging due to the high groundwater flow near the injection/pumping well. However, Kim and Lee (2019) found eight types of bacteria near the groundwater heat pump. Until now, it is still unclear under which circumstance the microorganism will cause bioclogging in geothermal systems, but the possibility is worth attention and further investigation.

Thirdly, microorganisms can generate toxic metabolites and adversely impact human health (Egboka et al. 1989). For instance, Cyanotoxin, produced by Cyanobacteria, is believed to pose a danger to humans in recent years. The production rate of Cyanotoxin is greatly dependent on the environment (water temperature, pH, trace element, etc.), which means it is not a controllable process. According to the U.S. EPA (2023), Microcystins, a type of Cyanobacteria, have some relationship with liver and colorectal cancers in humans. Although Cyanotoxin can only be produced in light-available conditions (e.g., lake, river, and surface water), it can be transported to groundwater through the transition zone. However, the toxicology of Cyanobacteria has not been thoroughly studied, and there is no direct evidence showing Cyanobacteria can cause cancer (U.S. EPA, 2023). Similar to Cyanobacteria and its metabolites, a large portion of microorganisms are still lacking in study nowadays.

8.5.3 Application of Microorganisms in Groundwater

Unlike other EPs, microorganisms have some positive effects in groundwater. One of the applications of microorganisms is to be used as a tracer. Harvey (1997) indicated that in the nineteenth century, bacteria forming red or yellow pigments, were first injected into groundwater to delineate flow paths in karst/fracture rock aquifers. Researchers still use microorganisms such as Patescibacteria, Actinobacteria, and

Bacteroidetes to investigate groundwater hydraulic properties (Rizzo et al. 2020). Microorganisms can also remove heavy metals or artificial pollutants in groundwater. Katsoyiannis and Zouboulis (2004) found that Gallionella ferruginea and Leptothrix ochracea can help increase arsenic removal rate through biotic oxidation of iron when arsenic is adsorbed onto iron oxides. Kane et al. (2001) discovered that microorganism species relative to β-Rubrivivax could directly link to promoting methyl tert-butyl ether degradation in aquifers.

Moreover, microorganisms can be used as ecological indicators for assessing groundwater quality. Stein et al. (2010) linked the composition of fauna in groundwater with nitrate concentrations, reflecting the impact of agricultural activities. Therefore, when considering microorganism remediation in groundwater, species with positive effects could be retained in the subsurface while only being removed when the groundwater is extracted for consumption.

8.5.4 Detection, Remediation, and Future Work

Groundwater microorganisms can be controlled with several factors: temperature, native groundwater organisms, dissolved oxygen level, and solid media attachment (John and Rose 2005). Besides, previous research (John and Rose 2005) shows that the inactivation rate of bacteria in groundwater varies with different species. Specifically, coliform bacteria in that research, for example, has the highest inactivation rate within 15°C-20°C in the research range (0–37°C).

Generally, microorganisms can be detected using membrane filtration, and different methods have varying detection efficiency for different species. For bacteria, a combined membrane filter-most probable number procedure is applied to increase the detection efficiency. For protozoa, the traditional method includes passing large volumes of water (up to 1,000 L) through polypropylene yarn-wound filter cartridges with a nominal porosity of 1 μm or through 1-μm absolute porosity pleated membrane capsules (National Research Council (US) Committee on Indicators for Waterborne Pathogens, 2004).

Many methods can be used to remediate the unwanted pathogens (bacteria and protozoa) in groundwater, including sand filtration, disinfection UFF, ozone, and chlorine), adsorption, and activated sludge systems (Da'ana et al. 2021). Among these techniques, sand filtration can be directly applied in the house, while others need a large-scale water treatment station.

Sand filtration set in the house is called a household slow sand filter. Cooperating with sodium hypochlorite, household slow sand filters can simultaneously remove *Escherichia coli*, *Giardia muris cysts*, and *Cryptosporidium parvum oocysts* from groundwater. Disinfection can couple with sand filtration, killing *Escherichia coli* and *Klebsiella pneumoniae* from groundwater (Alvear-Daza et al. 2018).

Among these large-scale remediation methods, adsorption is a cost-effective technology to remove pathogenic microorganisms. Besides, new materials such as nanocomposites can also be applied to remove microorganisms from groundwater.

Experiment (Sivaselvam et al. 2020) shows that different shapes of MgO nanostructures can inhibit bacterial growth by inducing intracellular production of reactive oxygen species and damaging the bacterial cell's surface physically.

Though microorganisms have been present in groundwater for a long time, we still do not understand them thoroughly. Knowledge gaps exist in the fundamental study of microorganisms because of their large variety and complicated biological impact. Regulations may be needed for some microorganisms after fully investigating their environmental impact.

8.6   Summary and Conclusion Remarks

Although the studies of PFAS, MNPs, PPCPs, and microorganisms began decades ago, much of their understanding is still lacking. Description of the unknown knowledge and required future research are discussed as follows:

(1) PFAS: there are more than 9,000 different PFAS classes, but analytical chemists have reference standards for fewer than 200 compounds (Hagstrom et al. 2021). Besides, some types of PFAS contain unknown chemical substances. Their composition, by-products, and biological materials could also be too complex to characterize fully. Moreover, a large number of PFAS compounds have different molecular structures. The lack of understanding of the physical–chemical properties of each constituent compound and biogeochemical interactions precludes the development of reliable models. Thus, it would be difficult to predict the transport and fate of PFAS and the change of PFAS composition with transport distances and contaminant ages. Therefore, future studies on the physical–chemical properties of each constituent compound and their biogeochemical interaction are needed.

(2) MNPs: fundamental study is not sufficient for MNPs. The classification of MNPs is unclear. A unified and standardized classification system based on the size and category of MNPs is needed to make it more convenient and efficient for researchers to analyze and compare data. Migration and aggregation mechanisms also need attention. Future works should include developing and validating models to assess the occurrence, transport mechanism and pathways, and fate of the plastic particles in groundwater.

(3) PPCPs: A thorough understanding of the physical and chemical properties of PPCPs is lacking, especially their behaviors in the subsurface environment. Processes such as sorption on geological material, migration in porous media, and degradation of PPCPs in groundwater aquifers should be further studied to develop more accurate predictive models. Specific subsurface zones, such as confined, unconfined, or karstic aquifers, should be distinguished in research. More importantly, current studies mainly focus on the site-specific groundwater contamination status and point-source pollution. In the future, regional-scale investigation should be considered to better characterize the temporal and spatial

behavior of PPCPs under the impact of large-scale hydrologic processes. More data is needed from developing countries, and long-term monitoring data are critical for all regions.

(4) Microorganisms: although microorganisms in groundwater have been studied for more than three centuries, the toxicology of most microorganisms is still unknown. Basic statistical data cannot indicate the biological impact of microorganisms on the human body. Lab experiments are needed to assess the toxicological mechanisms of microorganisms in humans. The secondary effect of human activity on microorganisms is another significant aspect that requires consideration. Geothermal practice is one example that dramatically changes the environmental conditions of aquifers. It could trigger the growth of microorganisms and lead to bioclogging, but the mechanism and long-term impact are still unknown. The mathematical model that couples the transport/distribution of microorganisms and the geothermal environment is needed to evaluate the potential efficiency and cost of geothermal energy.

Common challenges exist for different kinds of EPs mentioned in this article. Pollutant detection is especially difficult for PFAS and MNPs. For example, the actual PFAS concentrations in groundwater are lower than in previous removal experiments. Thus, detecting low PFAS concentrations requires sensitive equipment, which is expensive (Li et al. 2023; Verma et al. 2021). Improving the technique and equipment to detect and quantify PFAS and MNPs in groundwater is essential for future research.

Moreover, policymaking is also a significant factor in controlling these emerging pollutants. For example, stronger regulations are required to protect groundwater from PFAS and PPCPs pollution. Feasible measures include enforcing proper sewage disposal, upgrading wastewater treatment facilities, and setting threshold concentration values for contaminants in drinking water and sewage.

References

Adhikari S, Kelkar V, Kumar R, Halden RU (2022) Methods and challenges in the detection of microplastics and nanoplastics: A mini-review. Polym Int 71(5):543–551. https://doi.org/10.1002/pi.6348

Ahrens L (2011) Polyfluoroalkyl compounds in the aquatic environment: A review of their occurrence and fate. J Environ Monit 13(1):20–31. https://doi.org/10.1039/C0EM00373E

Alvarado-Zambrano D, Rivera-Hernández JR, Green-Ruiz C (2023) First insight into microplastic groundwater pollution in Latin America: The case of a coastal aquifer in northwest mexico. Environ Sci Pollut Res 30(29):73600–73611. https://doi.org/10.1007/s11356-023-27461-9

Alvear-Daza JJ, Sanabria J, Gutiérrez-Zapata HM, Rengifo-Herrera JA (2018) An Integrated drinking water production system to remove chemical and microbiological pollution from natural groundwater by a coupled prototype Helio-photochemical/h2o2/rapid sand filtration/chlorination powered by photovoltaic cell. Sol Energy 176(December):581–588. https://doi.org/10.1016/j.solener.2018.10.070

Awad R, Zhou Y, Nyberg E, Shahla Namazkar Wu, Yongning QX, Sun Y, Zhu Z, Bergman Å, Benskin JP (2020) Emerging Per- and polyfluoroalkyl substances (PFAS) in human milk from

Sweden and China. Environ Sci Process Impacts 22(10):2023–2030. https://doi.org/10.1039/D0EM00077A

Bao J, Liu W, Liu Li, Jin Y, Dai J, Ran X, Zhang Z, Tsuda S (2011) Perfluorinated compounds in the environment and the blood of residents living near fluorochemical plants in fuxin, China. Environ Sci Technol 45(19):8075–8080. https://doi.org/10.1021/es102610x

Barceló D, Petrovic M (2007) Pharmaceuticals and Personal Care Products (PPCPs) in the environment. Anal Bioanal Chem 387(4):1141–1142. https://doi.org/10.1007/s00216-006-1012-2

Bäuerlein PS, Hofman-Caris RCHM, Pieke EN, ter Laak TL (2022) Fate of microplastics in the drinking water production. Water Res 221(August):118790. https://doi.org/10.1016/j.watres.2022.118790

Bläsing M, Amelung W (2018) Plastics in soil: Analytical methods and possible sources. Sci Total Environ 612(January):422–435. https://doi.org/10.1016/j.scitotenv.2017.08.086

Bleeker EAJ, de Jong WH, Geertsma RE, Groenewold M, Heugens EHW, Koers-Jacquemijns M, van de Meent D et al (2013) Considerations on the EU definition of a nanomaterial: science to support policy making. Regul Toxicol Pharmacol 65(1):119–125. https://doi.org/10.1016/j.yrtph.2012.11.007

Boonya-atichart A, Boontanon SK, Boontanon N (2016) Removal of Perfluorooctanoic Acid (PFOA) in groundwater by Nanofiltration membrane. Water Sci Technol 74(11):2627–2633. https://doi.org/10.2166/wst.2016.434

Brusseau ML, Hunter Anderson R, Guo Bo (2020) PFAS Concentrations in Soils: Background Levels versus Contaminated Sites. Sci Total Environ 740(October):140017. https://doi.org/10.1016/j.scitotenv.2020.140017

Cerasa M, Teodori S, Pietrelli L (2021) Searching Nanoplastics: From sampling to sample processing. Polymers 13(21):3658. https://doi.org/10.3390/polym13213658

Chen H, Gao B, Li H, Lena QM (2011) Effects of pH and Ionic strength on sulfamethoxazole and ciprofloxacin transport in saturated porous media. J Contam Hydrol 126(1):29–36. https://doi.org/10.1016/j.jconhyd.2011.06.002

Chia RW, Lee J-Y, Kim H, Jang J (2021) Microplastic pollution in soil and groundwater: A review. Environ Chem Lett 19(6):4211–4224. https://doi.org/10.1007/s10311-021-01297-6

Christian S, Ternes T (1999) Pharmaceuticals and personal care products in the environment: Agents of subtle change? Environ Health Perspect 107(Supplement 6). https://doi.org/10.1289/ehp.99107s6907

Cole M, Lindeque P, Halsband C, Galloway TS (2011) Microplastics as contaminants in the marine environment: A review. Mar Pollut Bull 62(12):2588–2597. https://doi.org/10.1016/j.marpolbul.2011.09.025

Da'ana DA, Nabil Z, Ashfaq MY, Abu-Dieyeh M, Khraisheh M, Hijji YH, Al-Ghouti MA (2021) Removal of toxic elements and microbial contaminants from groundwater using low-cost treatment options. Curr Pollut Rep 7(3):300–324. https://doi.org/10.1007/s40726-021-00187-3

Detmers J, Strauss H, Schulte U, Bergmann A, Knittel K, Kuever J (2004) FISH Shows That Desulfotomaculum Spp. Are the Dominating Sulfate-Reducing Bacteria in a Pristine Aquifer. Microb Ecol 47(3). https://doi.org/10.1007/s00248-004-9952-6

Drewes JE, Heberer T, Rauch T, Reddersen K (2003) Fate of pharmaceuticals during ground water recharge. Groundw Monit & Remediat 23(3):64–72. https://doi.org/10.1111/j.1745-6592.2003.tb00684.x

Duong HT, Kadokami K, Shirasaka H, Hidaka R, Chau HTC, Kong L, Nguyen TQ, Nguyen TT (2015) Occurrence of Perfluoroalkyl acids in environmental waters in Vietnam. Chemosphere 122(March):115–124. https://doi.org/10.1016/j.chemosphere.2014.11.023

Ebele AJ, Abdallah M-E, Harrad S (2017) Pharmaceuticals and Personal Care Products (PPCPs) in the freshwater aquatic environment. Emerg Contam 3(1):1–16. https://doi.org/10.1016/j.emcon.2016.12.004

Egboka BC, Nwankwor GI, Orajaka IP, Ejiofor AO (1989) Principles and problems of environmental pollution of groundwater resources with case examples from developing countries. Environ Health Perspect 83(November):39–68. https://doi.org/10.1289/ehp.898339

Egbuna C, Amadi CN, Patrick-Iwuanyanwu KC, Ezzat SM, Awuchi CG, Ugonwa PO, Orisakwe OE (2021) Emerging pollutants in Nigeria: A systematic review. Environ Toxicol Pharmacol 85(July):103638. https://doi.org/10.1016/j.etap.2021.103638

Einsiedl F, Radke M, Maloszewski P (2010) Occurrence and transport of pharmaceuticals in a karst groundwater system affected by domestic wastewater treatment plants. J Contam Hydrol 117(1–4):26–36. https://doi.org/10.1016/j.jconhyd.2010.05.008

Frias JPGL, Nash R (2019) Microplastics: Finding a consensus on the definition. Mar Pollut Bull 138(January):145–147. https://doi.org/10.1016/j.marpolbul.2018.11.022

Galindo-Miranda JM, Guízar-González C, Becerril-Bravo EJ, Moeller-Chávez G, León-Becerril E, Vallejo-Rodríguez R (2019) Occurrence of emerging contaminants in environmental surface waters and their analytical methodology—a review. Water Supply 19(7):1871–1884. https://doi.org/10.2166/ws.2019.087

Georgin J, Franco DSP, Meili L, Dehmani Y, dos Reis GS, Lima EC (2023) Main advances and future prospects in the remediation of the antibiotic amoxicillin with a focus on adsorption technology: A critical review. J Water Process Eng 56(December):104407. https://doi.org/10.1016/j.jwpe.2023.104407

Gibert J, Dole-Olivier MJ, Marmonier P, Vervier P (1990) The ecology and management of aquatic-terrestrial ecotones. Naiman RJ, Décamps H (ed) Man and the Biosphere Series 4. Paris: Unescou.a

Gigault J, Ter Halle A, Baudrimont M, Pascal P-Y, Gauffre F, Phi T-L, El Hadri H, Grassl B, Reynaud S (2018) Current opinion: What Is a Nanoplastic? Environ Pollut 235(April):1030–1034. https://doi.org/10.1016/j.envpol.2018.01.024

Goeppert N, Goldscheider N (2021) Experimental field evidence for transport of microplastic tracers over large distances in an alluvial aquifer. J Hazard Mater 408(April):124844. https://doi.org/10.1016/j.jhazmat.2020.124844

Gong Y, Wang L, Liu J, Tang J, Zhao D (2016) Removal of aqueous Perfluorooctanoic Acid (PFOA) using starch-stabilized magnetite nanoparticles. Sci Total Environ 562(August):191–200. https://doi.org/10.1016/j.scitotenv.2016.03.100

Gopinath PM, Parvathi VD, Yoghalakshmi N, Kumar SM, Athulya PA, Mukherjee A, Chandrasekaran N (2022) Plastic particles in medicine: A systematic review of exposure and effects to human health. Chemosphere 303(September):135227. https://doi.org/10.1016/j.chemosphere.2022.135227

Gounot AM (1994) Groundwater ecology. Gibert J, Danielopol D, Arthur Stanford J (eds). Aquatic Biology [Sic] Series. San Diego: Academic Press

Griebler C, Lueders T (2009) Microbial biodiversity in groundwater ecosystems. Freshw Biol 54(4):649–677. https://doi.org/10.1111/j.1365-2427.2008.02013.x

Gündoğdu S, Mihai F-C, Fischer EK, Blettler MCM, Turgay OC, Akça MO, Aydoğan B, Ayat B (2023) Micro and Nano plastics in groundwater systems: A review of current knowledge and future perspectives. TrAC, Trends Anal Chem 165(August):117119. https://doi.org/10.1016/j.trac.2023.117119

Guo H, Chen Yi, Huiying Hu, Zhao K, Li H, Yan S, Xiu W, Coyte RM, Vengosh A (2020) High hexavalent chromium concentration in groundwater from a deep aquifer in the Baiyangdian Basin of the North China plain. Environ Sci Technol 54(16):10068–10077. https://doi.org/10.1021/acs.est.0c02357

Hagstrom AL, Anastas P, Boissevain A, Borrel A, Deziel NC, Fenton SE, Fields C et al (2021) Yale school of public health symposium: An overview of the challenges and opportunities associated with per- and Polyfluoroalkyl Substances (PFAS). Sci Total Environ 778(July):146192. https://doi.org/10.1016/j.scitotenv.2021.146192

Hartmann NB, Hüffer T, Thompson RC, Hassellöv M, Verschoor A, Daugaard AE, Rist S et al (2019) Are we speaking the same language? Recommendations for a definition and categorization framework for plastic debris. Environ Sci Technol 53(3):1039–1047. https://doi.org/10.1021/acs.est.8b05297

Harvey RW (1997) Microorganisms as tracers in groundwater injection and recovery experiments: A review. FEMS Microbiol Rev 20(3–4):461–472. https://doi.org/10.1111/j.1574-6976.1997.tb00330.x

Hawash HB, Moneer AA, Galhoum AA, Elgarahy AM, Mohamed WAA, Samy M, El-Seedi HR, Gaballah MS, Mubarak MF, Attia NF (2023) Occurrence and spatial distribution of Pharmaceuticals and Personal Care Products (PPCPs) in the aquatic environment, their characteristics, and adopted legislations. J Water Process Eng 52(April):103490. https://doi.org/10.1016/j.jwpe.2023.103490

Hendry P (2006) Extremophiles: There's more to life. Environ Chem 3(2):75. https://doi.org/10.1071/ENv3n2_ES

Hepburn E, Madden C, Szabo D, Coggan TL, Clarke B, Currell M (2019) Contamination of groundwater with Per- and Polyfluoroalkyl Substances (PFAS) from legacy landfills in an urban re-development precinct. Environ Pollut 248(May):101–113. https://doi.org/10.1016/j.envpol.2019.02.018

Houtman CJ (2010) Emerging contaminants in surface waters and their relevance for the production of drinking Water in Europe. J Integr Environ Sci 7(4):271–295. https://doi.org/10.1080/1943815X.2010.511648

Huang J, Chen H, Zheng Y, Yang Y, Zhang Y, Gao B (2021) Microplastic pollution in soils and groundwater: Characteristics, analytical methods and impacts. Chem Eng J 425(December):131870. https://doi.org/10.1016/j.cej.2021.131870

John DE, Rose JB (2005) Review of factors affecting microbial survival in groundwater. Environ Sci Technol 39(19):7345–7356. https://doi.org/10.1021/es047995w

Kane SR, Beller HR, Legler TC, Koester CJ, Pinkart HC, Halden RU, Happel AM (2001) Aerobic biodegradation of Methyl *Tert* -Butyl Ether by aquifer bacteria from leaking underground storage tank sites. Appl Environ Microbiol 67(12):5824–5829. https://doi.org/10.1128/AEM.67.12.5824-5829.2001

Kapelewska J, Kotowska U, Karpińska J, Kowalczuk D, Arciszewska A, Świrydo A (2018) Occurrence, removal, mass loading and environmental risk assessment of emerging organic contaminants in leachates, groundwaters and wastewaters. Microchem J 137:292–301

Karnjanapiboonwong A, Suski JG, Shah AA, Cai Q, Morse AN, Anderson TA (2011) Occurrence of PPCPs at a wastewater treatment plant and in soil and groundwater at a land application site. Water Air Soil Pollut 216(1):257–273. https://doi.org/10.1007/s11270-010-0532-8

Katsoyiannis IA, Zouboulis AI (2004) Application of biological processes for the removal of arsenic from Groundwaters. Water Res 38(1):17–26. https://doi.org/10.1016/j.watres.2003.09.011

Kibuye FA, Gall HE, Elkin KR, Ayers B, Veith TL, Miller M, Jacob S, Hayden KR, Watson JE, Elliott HA (2019) Fate of pharmaceuticals in a spray-irrigation system: From wastewater to Groundwater. Sci Total Environ 654(March):197–208. https://doi.org/10.1016/j.scitotenv.2018.10.442

Kim H, Lee J-Y (2019) Effects of a groundwater heat pump on thermophilic bacteria activity. Water 11(10):2084. https://doi.org/10.3390/w11102084

Kinney CA, Heuvel BV (2020) Translocation of pharmaceuticals and personal care products after land application of biosolids Current Opinion in Environmental Science & Health. Environ Pollut: Biosolids 14(April):23–30. https://doi.org/10.1016/j.coesh.2019.11.004

Koelmans AA, Nor NHM, Hermsen E, Kooi M, Mintenig SM, De France J (2019) Microplastics in freshwaters and drinking Water: Critical review and assessment of data quality. Water Res 155(May):410–422. https://doi.org/10.1016/j.watres.2019.02.054

Kosma CI, Lambropoulou DA, Albanis TA (2014) Investigation of PPCPs in wastewater treatment plants in Greece: Occurrence, removal and environmental risk assessment. Sci Total Environ 466–467(January):421–438. https://doi.org/10.1016/j.scitotenv.2013.07.044

Krishnan RY, Manikandan S, Subbaiya R, Biruntha M, Govarthanan M, Karmegam N (2021) Removal of emerging micropollutants originating from Pharmaceuticals and Personal Care

Products (PPCPs) in water and wastewater by advanced oxidation processes: A review. Environ Technol Innov 23(August):101757. https://doi.org/10.1016/j.eti.2021.101757

Kumar V, Singh E, Singh S, Pandey A, Bhargava PC (2023) Micro- and Nano-Plastics (MNPs) as emerging pollutant in ground water: environmental impact, potential risks, limitations and way forward towards sustainable management. Chem Eng J 459(March):141568. https://doi.org/10.1016/j.cej.2023.141568

Kuroda K, Murakami M, Oguma K, Muramatsu Y, Takada H, Takizawa S (2012) Assessment of groundwater pollution in Tokyo Using PPCPs as Sewage markers. Environ Sci Technol 46(3):1455–1464. https://doi.org/10.1021/es202059g

Lam NH, Cho C-R, Kannan K, Cho H-S (2017) A nationwide survey of perfluorinated alkyl substances in waters, sediment and biota collected from aquatic environment in Vietnam: Distributions and bioconcentration profiles. J Hazard Mater 323(February):116–127. https://doi.org/10.1016/j.jhazmat.2016.04.010

Lee H-J, Kim KY, Hamm S-Y, Kim MoonSu, Kim HK, Jeong-Eun Oh (2019) Occurrence and distribution of pharmaceutical and personal care products, artificial sweeteners, and pesticides in groundwater from an agricultural area in Korea. Sci Total Environ 659(April):168–176. https://doi.org/10.1016/j.scitotenv.2018.12.258

Li Z, Xiaopeng Yu, Furong Yu, Huang X (2021) Occurrence, sources and fate of pharmaceuticals and personal care products and artificial sweeteners in groundwater. Environ Sci Pollut Res 28(17):20903–20920. https://doi.org/10.1007/s11356-021-12721-3

Li H, Junker AL, Wen J, Ahrens L, Sillanpää M, Tian J, Cui F, Vergeynst L, Wei Z (2023) A recent overview of Per- and Polyfluoroalkyl substances (PFAS) removal by functional framework materials. Chem Eng J 452(January):139202. https://doi.org/10.1016/j.cej.2022.139202

Lindstrom AB, Strynar MJ, Delinsky AD, Nakayama SF, Larry McMillan E, Libelo L, Neill M, Thomas L (2011) Application of WWTP biosolids and resulting perfluorinated compound contamination of surface and well water in decatur, Alabama, USA. Environ Sci Technol 45(19):8015–8021. https://doi.org/10.1021/es1039425

Liu J-L, Wong M-H (2013) Pharmaceuticals and Personal Care Products (PPCPs): A review on environmental contamination in China. Environ Int 59(September):208–224. https://doi.org/10.1016/j.envint.2013.06.012

Ma L, Liu Y, Zhang J, Yang Q, Li G, Zhang D (2018) Impacts of irrigation water sources and geochemical conditions on vertical distribution of Pharmaceutical and Personal Care Products (PPCPs) in the Vadose zone soils. Sci Total Environ 626(June):1148–1156. https://doi.org/10.1016/j.scitotenv.2018.01.168

Malkoske T, Tang Y, Xu W, Yu S, Wang H (2016) A review of the environmental distribution, fate, and control of tetrabromobisphenol A released from sources. Sci Total Environ 569:1608–1617

Mintenig SM, Löder MGJ, Primpke S, Gerdts G (2019) Low numbers of microplastics detected in drinking water from ground water sources. Sci Total Environ 648(January):631–635. https://doi.org/10.1016/j.scitotenv.2018.08.178

Murraysmith (2021) COBI CECs Testing Result Summary January 2021

National Research Council (US) committee on indicators for waterborne pathogens (2004) Indicators for Waterborne Pathogens. In: Indicators for Waterborne Pathogens. National Academies Press (US). https://www.ncbi.nlm.nih.gov/books/NBK215666/.

Ni Z, Van Gaans P, Rijnaarts H, Grotenhuis T (2018) Combination of aquifer thermal energy storage and enhanced bioremediation: Biological and chemical clogging. Sci Total Environ 613–614(February):707–713. https://doi.org/10.1016/j.scitotenv.2017.09.087

New Jersey Department of Environmental Protection (2014) Occurrence of perfluorinated chemicals in untreated New Jersey drinking water sources: final report. https://doi.org/10.7282/T3VD7159

Novarino G, Warren A, Butler H, Lambourne G, Boxshall A, Bateman J, Kinner NE, Harvey RW, Mosse RA, Teltsch B (1997) Protistan communities in aquifers: A review. FEMS Microbiol Rev 20(3–4):261–275. https://doi.org/10.1111/j.1574-6976.1997.tb00313.x

Okoffo ED, Donner E, McGrath SP, Tscharke BJ, O'Brien JW, O'Brien S, Ribeiro F et al (2021) Plastics in Biosolids from 1950 to 2016: A function of global plastic production and consumption. Water Res 201(August):117367. https://doi.org/10.1016/j.watres.2021.117367

Oliaei F, Kriens D, Weber R, Watson A (2013) PFOS and PFC releases and associated pollution from a PFC production plant in minnesota (USA). Environ Sci Pollut Res 20(4):1977–1992. https://doi.org/10.1007/s11356-012-1275-4

Oppel J, Broll G, Löffler D, Meller M, Römbke J, Ternes Th (2004) Leaching behaviour of pharmaceuticals in soil-testing-systems: A part of an environmental risk assessment for groundwater protection. Sci Total Environ 328(1):265–273. https://doi.org/10.1016/j.scitotenv.2004.02.004

Osorio V, Larrañaga A, Aceña J, Pérez S, Barceló D (2016) Concentration and risk of pharmaceuticals in freshwater systems are related to the population density and the livestock units in Iberian rivers. Sci Total Environ 540(January):267–277. https://doi.org/10.1016/j.scitotenv.2015.06.143

Panno SV, Kelly WR, Scott J, Zheng W, McNeish RE, Holm N, Hoellein TJ, Baranski EL (2019) Microplastic contamination in karst groundwater Systems. Groundwater 57(2):189–196. https://doi.org/10.1111/gwat.12862

Peng X, Weihui Ou, Wang C, Wang Z, Huang Q, Jin J, Tan J (2014) Occurrence and ecological potential of pharmaceuticals and personal care products in groundwater and reservoirs in the vicinity of municipal landfills in China. Sci Total Environ 490(August):889–898. https://doi.org/10.1016/j.scitotenv.2014.05.068

Rampelotto P (2013) Extremophiles and extreme environments. Life 3(3):482–485. https://doi.org/10.3390/life3030482

Rizzo P, Petrella E, Bucci A, Salvioli-Mariani E, Chelli A, Sanangelantoni AM, Raimondo M, Quagliarini A, Celico F (2020) Studying hydraulic interconnections in low-permeability media by using bacterial communities as natural tracers. Water 12(6):1795. https://doi.org/10.3390/w12061795

Rodea-Palomares I, Makowski M, Gonzalo S, González-Pleiter M, Leganés F, Fernández-Piñas F (2015) Effect of PFOA/PFOS pre-exposure on the toxicity of the herbicides 2,4-D, atrazine, diuron and paraquat to a model aquatic photosynthetic microorganism. Chemosphere 139(November):65–72. https://doi.org/10.1016/j.chemosphere.2015.05.078

Saawarn B, Mahanty B, Hait S, Hussain S (2022) Sources, occurrence, and treatment techniques of per- and polyfluoroalkyl substances in aqueous matrices: A comprehensive review. Environ Res 214(November):114004. https://doi.org/10.1016/j.envres.2022.114004

Saleem M, Biondo O, Sretenović G, Tomei G, Magarotto M, Pavarin D, Marotta E, Paradisi C (2020) Comparative performance assessment of plasma reactors for the treatment of PFOA; Reactor design, kinetics, mineralization and energy yield. Chem Eng J 382(February):123031. https://doi.org/10.1016/j.cej.2019.123031

Edmond S, Kajau TA (2022) The fate of emerging pollutants in aquatic systems: An overview. In: Emerging Freshwater Pollutants, 119–35. Elsevier. https://doi.org/10.1016/B978-0-12-822850-0.00002-8

Schaefer CE, Sarah Choyke P, Ferguson L, Andaya C, Burant A, Maizel A, Strathmann TJ, Higgins CP (2018) Electrochemical transformations of Perfluoroalkyl Acid (PFAA) precursors and pfaas in groundwater impacted with aqueous film forming foams. Environ Sci Technol 52(18):10689–10697. https://doi.org/10.1021/acs.est.8b02726

Schultz MM, Higgins CP, Huset CA, Luthy RG, Barofsky DF, Field JA (2006) Fluorochemical mass flows in a municipal wastewater treatment facility. Environ Sci Technol 40(23):7350–7357. https://doi.org/10.1021/es061025m

Sharma BM, Bharat GK, Tayal S, Larssen T, Bečanová J, Karásková P, Whitehead PG, Futter MN, Butterfield D, Nizzetto L (2016) Perfluoroalkyl Substances (PFAS) in river and ground/drinking water of the Ganges River Basin: Emissions and implications for human exposure. Environ Pollut 208(January):704–713. https://doi.org/10.1016/j.envpol.2015.10.050

Shirley, Debbie-Ann T, Farr L, Watanabe K, Moonah S (2018) A review of the global Burden, new diagnostics, and current therapeutics for Amebiasis. Open Forum Infect Dis 5(7):ofy161. https://doi.org/10.1093/ofid/ofy161

Silori R, Shrivastava V, Singh A, Sharma P, Aouad M, Mahlknecht J, Kumar M (2022) Global groundwater vulnerability for Pharmaceutical and Personal Care Products (PPCPs): The scenario of second decade of 21st century. J Environ Manage 320(October):115703. https://doi.org/10.1016/j.jenvman.2022.115703

Sivaselvam S, Premasudha P, Viswanathan C, Ponpandian N (2020) Enhanced removal of emerging pharmaceutical contaminant ciprofloxacin and pathogen inactivation using morphologically tuned MgO nanostructures. J Environ Chem Eng 8(5):104256. https://doi.org/10.1016/j.jece.2020.104256

Srain HS, Beazley KF, Walker TR (2021) Pharmaceuticals and personal care products and their sublethal and lethal effects in aquatic organisms. Environ Rev 29(2):142–181. https://doi.org/10.1139/er-2020-0054

Stefano PHP, Roisenberg A, Santos MR, Dias MA, Montagner CC (2022) Unraveling the occurrence of contaminants of emerging concern in groundwater from urban setting: a combined multidisciplinary approach and self-organizing maps. Chemosphere 299134395. https://doi.org/10.1016/j.chemosphere.2022.134395

Stein H, Kellermann C, Schmidt SI, Brielmann H, Steube C, Berkhoff SE, Fuchs A, Hahn HJ, Thulin B, Griebler C (2010) The potential use of fauna and bacteria as ecological indicators for the assessment of groundwater quality. J Environ Monit 12(1):242–254. https://doi.org/10.1039/B913484K

Sui Q, Huang J, Deng S, Chen W, Gang Yu (2011) Seasonal variation in the occurrence and removal of pharmaceuticals and personal care products in different biological wastewater treatment processes. Environ Sci Technol 45(8):3341–3348. https://doi.org/10.1021/es200248d

Sun Q, Li M, Ma C, Chen X, Xie X, Chang-Ping Yu (2016) Seasonal and spatial variations of PPCP occurrence, removal and mass loading in three wastewater treatment plants located in different urbanization areas in Xiamen, China. Environ Pollut 208(January):371–381. https://doi.org/10.1016/j.envpol.2015.10.003

Szabo D, Coggan TL, Robson TC, Currell M, Clarke BO (2018) Investigating recycled water use as a diffuse source of Per- and Polyfluoroalkyl Substances (PFASs) to groundwater in Melbourne, Australia. Sci Total Environ 644(December):1409–1417. https://doi.org/10.1016/j.scitotenv.2018.07.048

Tamminga M, Hengstmann E, Fischer EK (2018) Microplastic analysis in the South Funen archipelago, Baltic Sea, implementing Manta trawling and bulk sampling. Mar Pollut Bull 128(March):601–608. https://doi.org/10.1016/j.marpolbul.2018.01.066

Topp E, Monteiro SC, Beck A, Coelho BB, Boxall ABA, Duenk PW, Kleywegt S et al (2008) Runoff of pharmaceuticals and personal care products following application of biosolids to an agricultural field. Sci Total Environ 396(1):52–59. https://doi.org/10.1016/j.scitotenv.2008.02.011

Torre M, Digka N, Anastasopoulou A, Tsangaris C, Mytilineou C (2016) Anthropogenic microfibres pollution in Marine Biota. A new and simple methodology to minimize airborne contamination. Mar Pollut Bull 113(1–2):55–61. https://doi.org/10.1016/j.marpolbul.2016.07.050

Van Leeuwenhoek A (1677) About little animals observed in rain-well-sea- and snow-water; as also in water wherein pepper had lain infused. Philosophical Transactions of the Royal Society of London 12

Verma S, Varma RS, Nadagouda MN (2021) Remediation and mineralization processes for Per- and Polyfluoroalkyl Substances (PFAS) in water: A review. Sci Total Environ 794(November):148987. https://doi.org/10.1016/j.scitotenv.2021.148987

Wang R, Yonglong Lu, Song S, Yang S, Yanqi Wu, Cui H (2022) Industrial source discharge estimation for pharmaceutical and personal care products in China. J Clean Prod 381(December):135129. https://doi.org/10.1016/j.jclepro.2022.135129

Xian Y, Jin M, Zhan H, Liu Y (2019) Reactive transport of nutrients and bioclogging during dynamic disconnection process of stream and groundwater. Water Resour Res 55(5):3882–3903. https://doi.org/10.1029/2019WR024826

Xu B, Liu S, Zhou JL, Zheng C, Weifeng J, Chen B, Zhang T, Qiu W (2021) PFAS and their substitutes in groundwater: Occurrence, transformation and remediation. J Hazard Mater 412(June):125159. https://doi.org/10.1016/j.jhazmat.2021.125159

Yang X, Flowers RC, Weinberg HS, Singer PC (2011) Occurrence and removal of pharmaceuticals and personal care products (PPCPs) in an advanced wastewater reclamation plant. Water Res 45(16):5218–5228. https://doi.org/10.1016/j.watres.2011.07.026

Zeng J, Guo Bo (2021) Multidimensional simulation of PFAS transport and leaching in the vadose zone: Impact of surfactant-induced flow and subsurface heterogeneities. Adv Water Resour 155(September):104015. https://doi.org/10.1016/j.advwatres.2021.104015

Chapter 9
Emerging Contaminants in Groundwater: Challenges, Management, and Policy Perspectives

Roya Narimani, Ioana Murgulet, and Dorina Murgulet

Abstract This chapter examines emerging contaminants (ECs) in groundwater, underscoring the regional variability of their prevalence and outlining some of the major challenges in remediation, treatment, and exposure risks in the environment. It addresses the importance of understanding specific contaminants and their risks to groundwater quality in different geographic contexts. A review of major laws and policies in regions such as the United States, European Union, and India is provided, revealing different approaches to managing these contaminants. Discussion includes the sources, transport pathways, and fate of ECs in soils and aquifers. Particular attention is given to per- and poly-fluoroalkyl substances, pharmaceuticals and personal care products, and nanoparticles (NPs) due to their widespread prevalence and associated health risks. The socioeconomic impacts of groundwater contamination, especially on low-income communities, and the technological limitations in current wastewater treatment infrastructure are explored. Regulatory and policy frameworks are examined, highlighting the complexity of managing contaminants due to ambiguous jurisdictional boundaries and the balance between their perceived benefits and harmful effects. The chapter concludes by emphasizing the importance of integrated management, cutting-edge technologies, and community participation in ensuring sustainable groundwater quality while considering the benefits and risks of using ECs, particularly NPs, in industrial and commercial applications.

Keywords Emerging contaminants · Groundwater pollution · Environmental policy · Remediation technologies

R. Narimani · D. Murgulet (✉)
Department of Physical and Environmental Sciences, Center for Water Supply Studies, Texas A&M University-Corpus Christi, Corpus Christi, TX, USA
e-mail: dorina.murgulet@tamucc.edu

I. Murgulet
Department of Biosciences, Rice University, Houston, TX, USA

© UNESCO 2025

189

S. Zandaryaa et al. (eds.), *Emerging Pollutants*, Advances in Water Security,
https://doi.org/10.1007/978-3-031-71758-1_9

9.1 Introduction

Groundwater is a crucial component of the Earth's hydrological cycle and is increasingly recognized as a critical resource for various human uses, including drinking, irrigation, and domestic and industrial applications (Gude and Maganti 2021). The issue of groundwater contamination has become increasingly apparent recently, with emerging contaminants (ECs) posing a serious threat to the environment and public health. However, ECs are increasingly receiving greater attention in countries with high population density and inadequate sanitation infrastructure/wastewater treatment. ECs primarily include micropollutants, endocrine disruptors (EDs), pesticides, pharmaceuticals, hormones, toxins, industrially related synthetic dyes, and hazardous pollutants containing dyes (Khan et al. 2022). The diverse nature of these contaminants has resulted in significant impacts that require immediate attention. However, most ECs do not have established regulatory safeguards or water quality standards/limits, leading to scarce monitoring of their presence in groundwater (Lapworth et al. 2019). The presence of contaminants in groundwater is an especially concerning issue, particularly in regions with high population density and insufficient sanitation infrastructure. Several studies have shown that the problem is widespread and persistent, as highlighted by Ashfaq et al. (2019), Luo et al. (2014), and Mukherjee et al. (2021). The implications of this issue are wide-ranging, including potential health risks (Fig. 9.1), environmental damage, and economic consequences. Chronic exposure to ECs, even at very low concentrations such as parts per trillion (ppt) or less, can cause several negative effects on organisms (Francisco et al. 2019; Montagner et al. 2017). A number of studies have shown that the introduction of these contaminants through different exposure routes can lead to human health problems such as skin irritation, headaches, and the development of chronic diseases like arthritis, heart disease, neurological effects, asthma, diabetes, and cancer (Koutros et al. 2016; Raanan et al. 2015; Starling et al. 2014).

A diverse range of ECs in groundwater is derived from anthropogenic activities, such as pharmaceuticals and personal care products (PPCPs), cyanotoxins, nanoparticles, flame retardants (Gavrilescu et al. 2015; Sauvé and Desrosiers 2014), pesticides, and per and polyfluoroalkyl substances. The Environmental Protection Agency (EPA) has identified twelve ECs at federal facilities, highlighting the variety of sources, health implications, and exposure pathways (US EPA 2014). Of these, PPCPs represent a significant group, including antibiotics, analgesics, beta-blockers, and hormones, and are frequently found in aquatic environments (Golet et al. 2001; Yang et al. 2017). In 2006, Schwarzenbach et al. (2006) reported about 30,000 to 70,000 registered chemicals in household products, and Dulio et al. (2018) suggested that 4,000 new chemicals were registered daily. As such, many newly emerging substances are being introduced into the environment, which may harm human health and ecosystems (Brion et al. 2019; Freeling et al. 2019). In addition, micro- and nanoplastics (NMPs) have emerged as a potentially significant organic ECs group in groundwater, as reviewed by Re (2019), and their potential impact on human health and the environment has gained attention in recent years (Lukač Reberski et al. 2022;

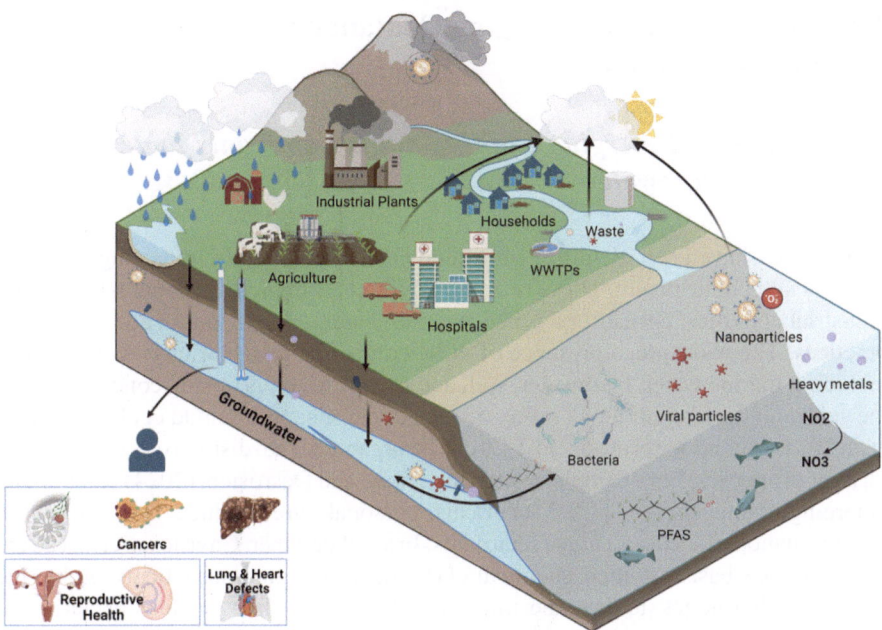

Fig. 9.1 Surface and groundwater interactions and the resulting distribution of ECs from various sources. *Source* Authors

Stock et al. 2022; Wong et al. 2020). The transport pathways of these contaminants to aquifers are diverse (Fig. 9.1), involving seepage from soils or the ground surface and direct infiltration from wells (Barnes et al. 2008; Kolpin et al. 2002; Krall et al. 2018; Pradhan et al. 2023).

This chapter is a valuable addition to the book, offering an in-depth examination of emerging contaminants in groundwater and effectively bridging the gap between broader aquatic ecosystem studies and the specific challenges of groundwater contamination. It provides a detailed and diverse perspective, complementing and enriching the book's other chapters. A quick search on Google Scholar yielded 232,000 results on emerging contaminants in groundwater and 44,400 on their risks and remediation, underscoring this topic's vast scope and significance. While recognizing the need for more research to fully comprehend these contaminants' sources, pathways, and fate, this chapter draws attention to their regional variability. It underscores the critical need to understand specific contaminants and assess their impact on groundwater quality across different geographic contexts.

9.2 Current State of Emerging Contaminants in Groundwater

9.2.1 Types of Emerging Contaminants Commonly Found in Groundwater

Various ECs have been identified in groundwater, significantly impacting the environment and public health. These ECs can originate from multiple sources, including industrial activities, agricultural practices, municipal waste disposal, and household practices. The widespread presence of these contaminants in various environments, including marine water, freshwater, soil, and groundwater, is raising concerns due to the substantial and continuous influx of such contaminants (Ashfaq et al. 2019; Luo et al. 2014; Mukherjee et al. 2021). ECs can be organized into distinct categories based on their properties. For example, a study by Sauvé and Desrosiers (2014) categorized emerging chemicals into pharmaceuticals, personal care products (PPCPs), cyanotoxins, nanoparticles, and flame retardants. In another paper, Gavrilescu et al. (2015) analyzed ECs based on their chemical class (including pharmaceuticals, endocrine disruptors, hormones, toxins, and more) and biological class (such as bacteria and viruses). The EPA has identified twelve ECs, which vary in their sources, health impacts, and exposure pathways (Richardson 2010). According to Lee et al. (2021), various contaminants have been categorized based on their usage in different industries. These categories include pharmaceuticals (amoxicillin, diclofenac, ibuprofen, acetaminophen (ACE), sulfamethoxazole (SMX), carbamazepine), food additives (acesulfame potassium (ACE-K), sucralose), pesticides (atrazine, glyphosate), industrial chemicals (1,4-dioxane, caffeine, perfluorooctane sulfonate (PFOS) and perfluorooctanoic acid (PFOA), N-Nitroso-dimethylamine (NDMA), bisphenol A (BPA)), personal care products (PCPs) (diethyltoluamide (DEET), galaxolide), disinfection by-products (DBPs) (2,6-dichloro-1,4-benzoquinone, dibromoiodomethane (DBIM), chlorodiiodomethane (CDIM), bromodiiodomethane (BDIM), iodoform (TIM)).

Building upon the established categorization of ECs, previous research has identified various classes that underscore the diverse nature of these pollutants in groundwater. Among these categories, PPCPs constitute a prominent group, encompassing substances like antibiotics, analgesics, beta-blockers, anti-inflammatories, hormones, and lifestyle drugs. Studies, such as those conducted by Golet et al. (2001) and Yang et al. (2017), have documented the presence of these pharmaceuticals in aquatic environments, revealing concentrations that range from nanograms ($ng{\cdot}L^{-1}$) to milligrams per liter ($mg{\cdot}L^{-1}$). In addition to PPCPs, the ECs extend to include bacteria and viruses, cyanotoxins, nanoparticles, flame retardants, per- and polyfluoroalkyl substances (PFAS), nano- and microplastics, industrial chemicals (e.g., bisphenols (BPs)), pesticides, illicit drugs, algal toxins, pesticides, and herbicides. While heavy metals and radionuclides are well-established pollutants, certain aspects of their presence, behavior, or impact on the environment can evolve in a way that places them in the category of ECs (Naidu and Biswas, 2024). However, they are not covered in this chapter.

9.2.2 Major Sources of Contamination, Transport Pathways, and Fate in Aquifers

ECs can enter the environment through various pathways, such as municipal, industrial, and agricultural. The primary origins of pollution include chemicals employed in agriculture and industrial effluent releases (Tomiyama et al. 2019; Xue et al. 2023), polluted waterways, household and municipal waste discharges, and disposal of laboratory waste (Pradhan et al. 2023). The main pathways through which contaminants infiltrate groundwater (Fig. 9.1) include seepage from soils or surface water, direct entry from wells, cross-contamination or "short-circuiting" from wells screened in multiple hydro-stratigraphic units (for aging or improperly installed wells), and pumping-mediated flow of contaminated water into freshwater aquifers (Pradhan et al. 2023). Additionally, septic systems contribute to these contamination pathways by leaching and percolating untreated wastewater. Untreated effluents contain many contaminants, including pathogens and PPCPs, which can seep through the soils to the saturated zone. Furthermore, micro-pollutants have been identified in groundwater nationwide (Barnes et al. 2008) and find their way into aquatic ecosystems by discharging wastewater effluent (Kolpin et al. 2002), agricultural runoff, industrial discharge, septic systems, and leaching from landfills.

The groundwater chemistry in an aquifer is primarily influenced by a series of hydrogeochemical reactions and interactions between water and sediment, including hydrolysis, oxidation–reduction, adsorption, cation exchange, and other related processes (Chakraborty et al. 2023). The transport of pollutants to groundwater is impacted by various factors, including soil type, hydrological conditions, and environmental characteristics, as noted by Krall et al (2018). Additionally, natural or artificial barriers such as impermeable clay lenses or dams can play a role. As contaminants pass through the soil and aquifer materials, they may undergo filtration, degradation, dilution, and/or other reactions, ultimately affecting their groundwater concentration. Assessing the transport, complexation, and degradation of ECs is most challenging in coupled terrestrial-aquatic systems such as linked groundwater-surface water systems due to the dynamics of the hydrogeochemical characteristics of porewaters. Detections in groundwater are indicative of compounds that are both persistent and mobile (Selwe et al. 2022).

Anti-inflammatory drugs, blood lipid-lowering agents, antibiotics, and hormones are among the PPCPs often found in soils and groundwater. The primary sources of these types of ECs in soil and groundwater are often wastewater treatment plants, septic tanks, landfills, hospitals, animal farms, aquatic farms, and agricultural lands treated with manure and biosolids (Awad et al. 2014; Fick et al. 2009). The improper disposal of household effluents and the release of treated industrial effluents and medical waste can lead to the emergence of partially degraded and refractory PPCPs. These contaminants can infiltrate shallow groundwater as they pass through the soil (Mompelat et al. 2009). DEET, a pesticide commonly found in insect repellents, has a moderate soil partitioning coefficient (relatively high mobility) and high stability, making it a concern for groundwater contamination. Studies have found levels of DEET in urine as high as 5,690 $\mu g \cdot L^{-1}$ in US park employees (Smallwood et al. 1992), indicating that it can travel through the vadose zone and pollute groundwater. In fact, DEET has been detected in 80% of groundwater samples from 23 European countries (Loos et al. 2010), highlighting the need for monitoring and prevention measures.

The fate and transport of PPCPs in soils and groundwater are significantly influenced by the interaction capacity of the compounds with soils, determined mainly by the physicochemical characteristics of PPCPs (e.g., molecular structure, hydrophobicity, polarity, etc.) and, to some extent, soil types and their physicochemical characteristics and organic matter content (Schwarzenbach et al. 2006; Zhi et al. 2019). For instance, most PPCPs are organic compounds with certain degrees of hydrophobicity, which means they have a high affinity for soils (e.g., tend to be absorbed by soil solids) and have variable retention in soils and aquifer matrices (Gao 2022). However, the mobility of PPCPs in groundwater can be enhanced by colloidal transport or dissolved organic matter. The properties of PPCPs vary greatly between groups, making their transport and fate in soils and aquifers more complex, particularly due to the heterogeneity of the medium. For example, finer-textured grains have greater sorption capacities owing to their larger surface area. Some silt loam soils have higher sorption capacity than sandy loam for many PPCPs (Karnjanapiboonwong et al. 2010). Furthermore, metal oxides (e.g., iron or aluminum oxides) can provide additional positively charged attraction sites or complexation sites (Chen et al. 2013a, b). Clay and colloids are highly reactive due to their large surface area and high cationic exchange capacity (CEC). Some forms of PPCPs have been found to positively correlate with the soil's clay content (Rabølle and Spliid 2000). The functional groups on the surface of soil organic matter and its large surface area can increase the sorption of PPCPs by providing additional sorption sites on the soil or sediment grains where the organic matter is bound to the soil matrix. However, suppose the sediment or soil has a high sorption capacity for PPCPs. In that case, the organic matter coating the mineral surfaces can act as a shielding agent and reduce the degree of PPCP sorption (Yao et al. 2016). Conversely, dissolved organic matter may compete with PPCPs for surface sorption sites or complex with PPCPs, enhancing their mobility (Oh et al. 2016).

The mobility of PPCPs, much like other pollutants, is also affected by pH, which has been shown to influence the speciation and ionization of PPCPs, as well as the predominant surface charge of the soil or sediment matrix. pH levels are critical in determining PPCP's attraction to soil or sediment (Gao 2022). The surface charge of soil minerals is determined by the pH of the soil solution at the point of zero charge (PZC). In many studies of the sorption of PPCPs, it is observed that as the pH of the soil solution increases, the sorption decreases, especially for soil particles with relatively low PZC. However, if the soil particles have high PZC, such as those containing certain iron or aluminum minerals, the sorption may increase with increased pH (Gu et al. 2015). The sorption of PPCPs is a complex matter, particularly when considering the presence of coexisting cations, such as Na^+ and K^+. This becomes especially pertinent if PPCPs are absorbed onto soil or sediment surfaces through cation exchange, as the cations present in the soil pore space solution may compete with the PPCPs and diminish their sorption. However, it should be noted that once the concentration of these cations in the solution reaches a certain threshold, they can also impede the sorption of PPCPs altogether (Chen et al. 2013a, 2013b).

Following sorption and desorption, most PPCPs undergo degradation. In addition to the sorption processes, many factors influence PPCP degradation, including their intrinsic properties and initial concentrations, microbial composition, redox conditions, and soil texture. For instance, Sassman and Lee (2005) and Xu et al. (2021b) showed that soil organic matter and texture influenced the degradation of six selected PPCPs from US agricultural soils associated with reclaimed wastewater reuse. Other studies found that the degradation of four other types of PPCPs from three different soils resulted from the association with soil organic carbon, microbial content, and soil type (Yu et al. 2013). In addition, the degradation of PPCPs in soils was found to follow first-order exponential decay kinetics, yielding half-lives ranging between 0.97 and 10.21 days and mainly associated with microbial activities (Chen et al. 2013a, b; Zhang et al. 2019). Conkle et al. (2012) found that aerobic versus anaerobic soils enhanced the degradation of ibuprofen, DEET, and gemfibrozil.

PFAS are a significant group of emerging persistent organic pollutants found globally in the environment, which are very stable and difficult to decompose (Li et al. 2019). Although there are approximately 10,900 scholarly studies on PFAS in groundwater, much work is still required to understand their environmental behavior, particularly in soil and groundwater. To assess environmental contamination, it is vital to understand the sources, transport pathways, and fate of PFAS in aquifers. PFAS contamination in groundwater often stems from direct sources like surface water and soils. Soils are susceptible to contamination from numerous sources, including landfills, biosolid treatments, septic systems, and sewers, as well as specific point sources. These different contributors can lead to the introduction of PFAS into the environment. The presence of PFAS in soils poses a significant risk as they can easily infiltrate water systems through surface runoff and downward leaching, thus affecting both environmental health and water quality. For example, in rural areas of eastern China near Yuqiao Reservoir and the Ganges River Basin of India, surface water was the primary contributor to groundwater PFAS levels (Cao et al. 2019; Sharma et al. 2016). The contamination often occurs through seepage, diffusion, dispersion, and advection (Liu et al. 2016; Xiao et al. 2015). Soil is also a significant source of PFAS contamination in groundwater, mainly due to leaching caused by its high solubility in water and low log Koc (organic carbon–water partition coefficient) values (Gellrich et al. 2012; Xiao et al. 2015). Studies show a strong correlation between PFAS concentrations in soil and groundwater, indicating a common origin (Cao et al. 2019). Using aqueous film-forming foams (AFFFs) in firefighting activities has become a significant concern for PFAS contamination in groundwater as PFAS concentrations in groundwater impacted by AFFF use have increased (Schaefer et al. 2018).

Indirect sources of PFAS in groundwater include atmospheric deposition and precipitation, with studies reporting PFAS concentrations in air and surface snow (Brusseau et al. 2020; Bryant et al. 2022; Johnson et al. 2022; Pepper et al. 2021; Wang et al. 2019; Zhao et al. 2020). Other sources include non-stick cookware, pesticides, fast-food packaging, and discharge from PFAS production facilities (Xu et al. 2021a). Consequently, PFAS in groundwater can be transported to tap/drinking water, contributing to human exposure and potential health risks. The potential risks of PFAS in groundwater, especially as a source of drinking water, are significant. Several studies have highlighted that conventional water treatment processes do not effectively remove PFAS, posing health risks to local communities (Qi et al. 2016; Zhu et al. 2017). High concentrations of PFOA and PFOS in groundwater have been reported in various locations, exceeding regulatory guideline values and necessitating heightened attention to prevent further contamination and associated health risks (Bao et al. 2019; Hepburn et al. 2019; Liu et al. 2019, 2020; Sammut et al. 2019; Schaefer et al. 2018; Sims et al. 2022; Tang et al. 2023).

Ongoing investigations are being conducted to determine the effects of substitutes such as GenX (trade name for 2,3,3,3-tetrafluoro-2-[heptafluoropropoxy]propanoic acid), F-53B (trade name for Chlorinated polyfluoroalkyl ether sulfonic acids [Cl-PFESAs]), and OBS (sodium p-perfluorous nonenoxybenzene sulfonate), which have been introduced as replacements for PFOA and PFOS. These substitutes have been found in various environmental media and human tissues, indicating their widespread presence (Gebbink and Van Leeuwen, 2020; Li et al. 2020). Despite this, a lack of comprehensive studies on OBS has made identifying sources and impacts difficult. Recent research into GenX, F-53B, and OBS suggests that these substitutes may not be safer alternatives to PFOA and PFOS. As a result, it is imperative that careful risk assessments are conducted and further research is carried out to evaluate their potential impacts on both environmental and human health (Xu et al. 2021a).

The behavior of PFAS in soils and groundwater, particularly regarding their fate, transport, and accumulation in aquifers, is intricately linked to their chemical and physical properties. Initially valued for their resistance to biodegradation and water-proof capabilities, PFAS compounds, due to these same qualities, pose significant health risks through accumulation in soil and potential bioaccumulation. Studies, such as the one by East et al. (2021), have analyzed soils from various US Air Force bases, finding that soil characteristics like total organic carbon (TOC), pH, and clay content significantly influence PFAS sorption. This research indicates that TOC and clay content correlate with the soil's capacity to transport PFAS, with hydrophobic properties of organic carbon playing a pivotal role in PFAS partitioning in the soil. Furthermore, the length of the perfluorinated chain in PFAS compounds, such as PFOS and PFOA, affects their sorption rates, with longer chains exhibiting increased hydrophobicity and, thus, a higher likelihood of partitioning. In addition, PFAS are preferentially retained at interfaces between solid/air and air/water, enhancing retention and accumulation in soils and groundwater (Brusseau 2018; Brusseau and Guo 2022; Bryant et al. 2022).

Milinovic et al. (2015) further confirmed these findings, demonstrating through liquid chromatography and mass spectrometry that soil sorption levels depend on the perfluorinated chain length and the soil's organic carbon content. Their study revealed that longer-chain compounds like PFOS are retained in higher concentrations and are less reversible than shorter-chain compounds. This suggests that longer-chain PFAS compounds have lower soil degradation rates and are more likely to mobilize into saturated zones (Brusseau 2018; Brusseau and Guo 2022). Additionally, the anionic properties of certain PFAS compounds, such as the carboxylic acid groups in PFOAs and sulfonic acid groups in PFOS, contribute to their partitioning behavior. These amphiphilic properties, combining hydrophilic and hydrophobic characteristics, allow PFAS to interact with different charged clay minerals in soils. East et al. (2021) noted that acidic soils with high clay content could retard anionic PFAS compounds, leading to their accumulation at the air–water interface and potential leaching into saturated zones. This complex interplay of chemical properties, soil composition, and environmental factors underscores the persistent and complicated nature of PFAS in the environment and their potential impact on human health and ecosystems.

Like other types of ECs, the development and expanded use of various nanoparticles (NPs) are occurring without standardized environmental laws or guidelines, potentially leading to unfavorable environmental changes and impacting human health. Engineered nanoparticles and nanoparticle-enabled products enter the environment through a variety of pathways. Natural processes like forest fires, volcanic eruptions, and soil erosion contribute to this release, as do human activities, including vehicle exhaust, mining, and the use and disposal of nanoparticle-containing products. Once in the environment, these nanoparticles can find their way into soil and groundwater, posing risks to human and animal health through direct contact, inhalation, or ingestion (Mishra and Sundaram 2023). Each type of engineered NP has unique properties, making them suitable for specific applications, ranging from industrial uses to groundbreaking medical treatments. However, their small size and novel properties also necessitate careful consideration of potential health and environmental impacts.

A very important topic is the accidental release of NPs into the environment; this subject is not considered in NP inventories, but it is expected to become an important threat in the coming years (Ramirez et al. 2023). Most of these accidental releases are spills; for this reason, the soil is considered an important destination for different types of NPs. NPs spilled onto soil can later reach groundwater, and, in specific cases, they may later impact other environmental compartments where aquatic ecosystems are present, such as lakes, rivers, and wetlands. Due to these accidental spills, the acute exposure of marine ecosystems to NPs is expected. However, there are currently no defined standards for determining nanoparticle effects, and few investigations have been conducted on nanoparticles' environmental and hazardous repercussions for explicit and implicit exposure. A recent review by Mishra and Sundaram (2023) highlights the detrimental effects of nanoparticles on humans, plants, animals, and the environment, along with various sources, exposure routes, and transportation of nanoparticles. Nonetheless, studies on the migration of nanoparticles from solid waste to liquid phases, such as in landfill leachates (Reinhart et al. 2010), or their release into the air during recycling processes or incineration operations, have been somewhat limited in scope (Bouillard et al. 2013; Rajput et al. 2020). Therefore, a large gap remains in understanding how nanoparticles interact with soils and aquifer matrixes and what processes may drive their fate and accumulation in groundwater using field studies (Ramirez et al. 2023).

9.2.3 Global and Regional Trends in Groundwater Contamination through Case Studies

Groundwater is an essential resource supporting agriculture and supplying drinking water to billions of people worldwide. However, increasing anthropogenic activities and industrialization have led to groundwater contamination, posing serious environmental and public health concerns. Fertilizers and pesticides from agricultural practices, pharmaceuticals and pathogens from human waste, and industrial plasticizers are the main sources of ECs, leading to chronic contamination of shallow and deep groundwater worldwide. The long-term cumulative effects of these contaminants in groundwater exacerbate the environmental and public health issues caused by groundwater pollution. Degradation of groundwater quality due to EC contamination diminishes the available 'safe' quantity for specific needs, thus affecting water resource sustainability. Diminished groundwater quality disproportionately impacts unincorporated communities, often underserved and impoverished, relying solely on groundwater for daily use.

In developed and developing countries, PPCPs are used to improve living standards. The most important scientific and medical breakthrough of the 20th century is thought to have been the discovery of antibiotics that are extensively used in human and animal medicine (Chakraborti et al. 2015). According to a review paper by Lapworth et al. (2012), five PPCPs, including diclofenac, ibuprofen, caffeine, carbamazepine, and sulfamethoxazole, have been found in groundwater across 14 countries in North America, Europe, the Middle East, and Asia (Lapworth et al. 2012). Carbamazepine with a concentration of 390 ng·L^{-1} was reported in 42% of groundwater samples collected from 164 locations across 23 European countries (Loos et al. 2010). K'oreje et al. (2016) report that Kenya's surface and groundwater contain 24 different pharmaceuticals, including antiretroviral therapy, psychiatric medications, antibiotics, and analgesic/anti-inflammatory drugs. A study out of France reports that carbamazepine was detected at around 20% of sites in confined and semi-confined aquifers (Faille 2010). According to Bunting et al. (2021), the most frequently seen ECs in Europe were carbamazepine and caffeine, with maximum concentrations of 2.3 and 14.8 μg·L^{-1}, respectively.

Endocrine disruptors (EDs) include various chemicals that can leak into the environment from multiple sources, such as industrial discharge and runoff from agriculture. According to a study by Saha et al. (2022), Di-n-butyl phthalate, benzyl butyl phthalate, di (2-ethylhexyl) phthalate, triclosan, triclocarban, 4-tert octyl phenol, 4-nonylphenol, propylparaben, and butylparaben were all found in 5 out of 26 collected groundwater samples in different agro-climatic zones of India during 2019–2020. In another study conducted by Nazifa et al. (2020), eight EDs were detected in five rivers of Malaysia, including the Klang River, Pahang River, Kuantan River, Kelantan River, and Terengganu River. Figure 9.2 is constructed from 298 previous studies for three types of contamination: pharmaceuticals, plasticizers, and pesticides. Based on a review of 153 articles in Google Scholar related to pharmaceutical contamination in groundwater, 45 studies mainly focused on European countries, with 7 out of 45 studies based in Spain. Furthermore, 25 studies were conducted in the western parts of the United States, primarily focusing on pharmaceutical contamination. Some studies have been undertaken in Australia, India, China, Iran, South Korea, Africa, Canada, and South America. Most Canadian studies were primarily concentrated in Ontario. Notably, out of 65 studies on plasticizers, more than 70% focused on countries in Europe and the United States. More research is being conducted throughout Europe, particularly in Germany. Of the 65 studies, a few were found for Mexico, Singapore, Malaysia, China, Brazil, Australia, and India.

Out of the 80 studies that primarily address pesticide contamination in groundwater, 35 were focused on the east and west coasts of the United States, 6 in South America, 3 in Australia, more than 11 in India, and more than 23 in Europe (Fig. 9.2). To better understand global contamination distribution and related impacts, the number of studies focusing solely on groundwater pollution needs to be increased, particularly in the middle of Africa and the north of Canada. It can be concluded that most EC research is conducted in developed regions. This emphasizes the importance of a more inclusive global approach to environmental research, as this concentration in developed regions raises questions about our understanding of ECs in less studied or economically developing areas where large human populations, often impoverished and with little or no political power, exist.

9.3   PFAS, PPCPs, and NPs in Groundwater: Health Risks and Socio-Economic Impacts

As new contaminants emerge, their adverse effects disproportionately exacerbate low-income communities and countries' challenges. Without the necessary infrastructure to maintain freshwater resources, contaminated groundwater delays socioeconomic development and worsens health conditions. Water shortages induced by contamination can result in food insecurity, increased poverty, and conflicts within and between communities and countries (Li et al. 2021). There is a growing concern over the effects of exposure to PFAS, PPCPs, and NPs on human health. Moreover, many wastewater treatment plants, particularly in low-income regions, are not equipped to remove these contaminants, which are typically stable and difficult to break down. These contaminants can infiltrate groundwater, potentially increasing the risk of various cancers, hormone disruptions, and other toxic effects.

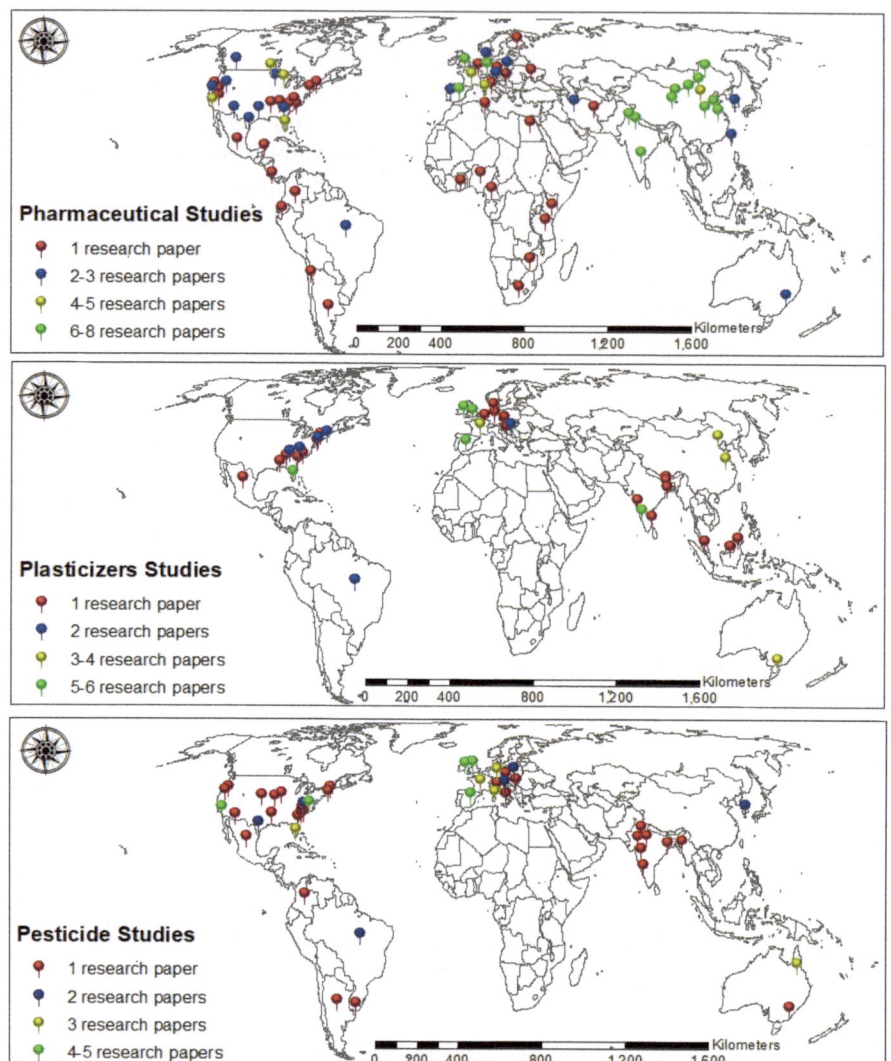

Fig. 9.2 Global studies on pharmaceuticals, plasticizers, and pesticides in groundwater. *Source* Adapted by the authors based on the world map from the United Nations Geospatial (https://www.un.org/geospatial)

PFOA is one of the most detected synthetic byproducts of fluoropolymers in groundwater (EPA 2017). PFOA has been primarily used in producing polytetrafluoroethylene (PTFE) for food and beverage industries. Once in the environment, PFOA is unable to break down due to its strong carbon–fluorine bonds. PFOA has hydrophobic and lipophobic properties, making it capable of bioaccumulation and sorption in the blood and tissue (Calafat et al. 2019; Wee and Aris 2023). Thus, it can persist in the body for extended periods. Studies have detected a broad class of PFAS in the blood and urine in widespread populations, raising concerns about the extent of exposure (Lewis et al. 2015). PFOA persists in the environment despite being phased out in the U.S. and EU and is a common source of water contamination. Most concerning remains the impact on low-income countries that often lack the mechanisms/resources to regulate and remove PFAS.

PFOA has been classified as a possible carcinogen and endocrine (hormone) disruptor ("PFAS Forever Chemicals (also PFOA, PFOS)—Breast Cancer Prevention Partners (BCPP),"). Well-established studies suggest that liver, testicular, and pancreatic cancer may be related to PFOA exposure (Nicole 2013). Recent in-vitro experiments have demonstrated that PFOA can elicit tumor cell proliferation, migration, and invasion in breast epithelial cells (Pierozan et al. 2018). Around two-thirds of breast cancers are hormone-dependent or positive for estrogen-type receptors. Estrogen is a group of female sex hormones responsible for maintaining secondary sex characteristics (Hewitt et al. 2000). Secretion of estrogen by the ovaries allows for the thickening of the endometrium, a process essential for regulating the menstrual cycle. However, altered levels or exposure to estrogen influenced by substances like PFOA may increase the risk of breast cancer ("Risk Factors" 2015). In vitro studies suggest that PFOA can potentially interfere with estrogen homeostasis. Sonthithai et al. (2016) demonstrated that in T47D breast cancer cells, there is an amplification in the estrogenic effects of 17b-estradiol. Additionally, serum levels of PFOA were positively correlated with breast cancer risk in a study of females of Greenland Inuit descent. Notably, the Greenlandic Inuit population is heavily exposed to PFAS by consuming seafood and marine mammals (Wielsøe et al. 2017), where PFAS is bioaccumulated in the marine life that forms a significant part of their diet.

Through disruption of sex hormone production and their receptor functions, PFOA may also harm reproductive health. While most research has focused on measuring levels of PFOA in serum and urine, there is an increasing interest in its presence in follicular fluid (Shen et al. 2023). Follicular fluid plays a crucial role in ovarian function, particularly in the growth and maturation of oocytes (Revelli et al. 2009). During development, PFOA can cross the blood follicle barrier and accumulate in follicular fluid, exposing oocytes to PFOA. Studies have shown that women with infertility have elevated levels of PFOA, suggesting its link to poor fertility (Heffernan et al. 2018). In animal studies, PFOA exposure decreased the number of follicles in different stages of maturation. During pre-puberty, exposure to PFOA resulted in more frequent irregular estrous cycles (Du et al. 2019). In adult mice, it accelerated the development of preantral and antral follicles, altering the normal follicle cycle (Huang et al. 2022). Additionally, PFOA exposure lowered the levels of estrone (E1) and estradiol (E2), primary forms of estrogen. This interruption of steroid hormone production can induce oxidative stress on oocyte meiosis and preimplantation development (Huang et al. 2022). Furthermore, growing evidence indicates an association between exposure to PFOA and the development of reproductive diseases such as endometriosis, a chronic condition in which an endometrial lining grows outside the uterus instead of inside, causing severe pain and infertility. In hospital-based case–control studies, increased serum PFOA concentrations raised the odds of developing endometriosis (Szczęsna et al. 2022).

PPCPs have gained increasing attention as ECs. PPCPs enter groundwater through human excretion and improper disposal. As wastewater treatment plants were not originally designed to remove such contaminants, PPCPs heavily persist in numerous bodies of water. The release of hormones and antibiotics in the water is particularly concerning. There is well-established evidence of the adverse effects of pharmaceuticals on aquatic life, including the detection of numerous pharmaceuticals in fish tissues ("Drugs in the water" 2011). In areas with high concentrations of contaminants, such as estrogen, fish populations have exhibited altered sex ratios, with a higher female-to-male ratio. However, the effects of exposure to pharmaceuticals on humans are uncertain. Antibiotic resistance is a major public health concern, contributing to an increase in severe illness and spreading of disease. The release of resistant bacteria may spread throughout environmental compartments and reach humans, potentially resulting in asymptomatic colonization of the bacteria or treatment-resistant infections. The characterization of ecological antibiotic resistance has been difficult to quantify. Thus, low concentrations of antibiotics cannot be correlated with the load of antibiotic resistance. However, antibiotics are often present with other contaminants, such as personal care products (PCPs). Non-antibiotic compounds have been observed to disrupt microbial communities through oxidative stress, stimulating the acquisition of antibiotic-resistance genes. Beyond just fewer antibiotics, expanding wastewater treatment should be a global goal in reducing resistant bacteria (Manaia et al. 2023).

Engineered NPs are utilized in the manufacturing of certain PCPs as well as NP-based drug delivery systems. While NPs have significantly improved imaging diagnostics and treatments such as chemotherapy and radiotherapy, they have also emerged as contaminants in groundwater (Westerhoff et al. 2018). Their small size and high reactivity contribute to their effectiveness and can also induce adverse environmental effects. There has been a rapid growth of studies investigating the impact of NPs on human health, as their effects still need to be fully understood (Kumah et al. 2023). In mammalian cells, exposure to NPs was shown to induce high levels of cytotoxicity, such as cell membrane damage, oxidative stress, and glutathione depletion (Mukherjee et al. 2012). Exposure of NPs to human epithelial cells was also associated with cell membrane damage, along with reduction of cell viability and changes in cell structure and cell cycle (Eldawud et al. 2018).

9.4 Challenges in Monitoring, Data Gaps, and Research Needs

Pollution is a complex issue with multiple sources, and it poses several challenges that require practical solutions. These challenges span various factors, from technological limitations to dynamic environmental factors. Addressing them effectively requires reliable monitoring protocols considering the changing nature of pollutants and data interpretation and resource allocation issues. The interrelated nature of these issues highlights how crucial it is to develop thorough and adaptable monitoring strategies to efficiently monitor and control ECs in various environmental matrices and compartments. The problems associated with new pollutants encompass the requirement for sophisticated detection techniques, understanding the long-term impacts on ecosystems, evaluating human health risks, and developing effective remediation strategies. These challenges must be carefully considered when creating efficient monitoring plans. Monitoring efforts are continuously challenged by the dynamic nature of emerging pollutants, defined by the introduction of new contaminants and the evolution of existing ones, as well as the various unknown interactions between chemicals, environmental factors (e.g., pH, ionic strength, among others), and multiple matrices (e.g., soils, water types, among others). However, there's a chance that the most sophisticated detection methods will always have limits regarding sensitivity, selectivity, and the capacity to find new pollutants at trace levels.

One of the most important parts of monitoring programs is resource allocation since comprehensive monitoring initiatives can take time to implement due to conflicting priorities and scarce resources. Furthermore, adequate funding and strategic resource allocation are essential to guarantee that monitoring programs successfully identify and track emerging pollutants. However, because of the various origins and routes of newly emerging pollutants, interpreting monitoring data is difficult and necessitates a thorough comprehension of how contaminants behave in different environmental matrices. Monitoring efforts are further complicated by the changing landscape of emerging pollutants, influenced by modifications in industrial processes, environmental factors, and human activities. To effectively track and manage ECs in groundwater, surface water, and other ecological matrices/compartments, it is imperative to have a flexible and adaptive monitoring approach that considers the interlinkages among these challenges. Therefore, a comprehensive strategy combining enhanced detection technologies, astute resource management, and a thorough comprehension of evolving contaminant patterns is necessary to monitor emerging pollutants effectively. By addressing these problems, we can develop monitoring plans that provide precise and timely data, supporting well-informed environmental management decision-making.

Despite improvements in monitoring techniques and technologies, filling in data gaps and determining additional research are essential steps to manage emerging pollutants comprehensively. Because environmental systems are dynamic and ECs are complex, research constantly needs to improve monitoring procedures to fill knowledge gaps. Ongoing research is necessary to find additional emerging pollutants that might not be properly understood or monitored. Monitoring programs must be vigilant and flexible, considering the potential introduction of new contaminants by the constantly changing industrial and consumer product landscape. The research goal should be to develop and improve detection technologies to get around current constraints. Comprehensive monitoring requires improvements in sensitivity, selectivity, and the capacity to identify contaminants at trace levels. Cooperative efforts among scientists, engineers, and technology developers can foster innovation in this field. More investigation is needed to comprehend how emerging pollutants behave in diverse environmental matrices. For instance, it is necessary to investigate their transformation, movement, and fate, especially in complex connected systems like surface and groundwater.

According to Hu et al. (2022), expanding coverage to specific population segments, such as those reliant on private wells, is crucial. However, challenges arise when promptly addressing newly detected contaminants due to sporadic testing schedules and delayed reporting of monitoring outcomes. To address these gaps, an improved monitoring framework is necessary. With new contaminants emerging in various water sources, especially in areas heavily reliant on private wells or prone to contamination, a vigilant approach is crucial for quickly identifying these pollutants. This requires an improved monitoring system that prioritizes increasing coverage and establishes frequent and timely test protocols. Additionally, it necessitates the development of analytical technologies that are cost-effective, easy to acquire, and provide rapid turnaround times. However, the costs, time frames for method development (especially for new contaminants), and availability of such technologies often make it not feasible, particularly for low-income regions. Nevertheless, such advancements are necessary to identify emerging pollutants that pose environmental and public health risks. Detecting these contaminants in groundwater is complex. Thus, a sophisticated and feasible monitoring strategy is needed for contaminant prevention and efficient remediation of groundwater resources.

Gaps in monitoring efforts can be filled by applying big data analytics and modeling techniques. A more comprehensive understanding of contaminants' distribution, sources, and possible effects can be obtained by combining data from multiple sources and using sophisticated modeling techniques, including machine learning (ML) and artificial intelligence (AI). It can be helpful to prioritize testing efforts in areas with a higher chance of elevated contamination levels by using forecasting models specifically designed for water quality. Furthermore, according to Hu et al (2022), these models can advance our understanding of the variables affecting the patterns of water quality's spatial distribution. To forecast PFAS concentrations in private wells in the states of North Carolina and New Hampshire, for example, Bayesian networks and random forest models have been developed (Hu et al. 2021; Roostaei et al. 2021). However, states and individual investigators generally house large or vital datasets that can be used to train and inform models. It is crucial to encourage collaboration and sharing of these datasets to improve the training capabilities for developing more robust models for improved characterization and prediction. Standardizing monitoring protocols is also essential to ensure data collection and reporting consistency. The ultimate aim of research should be to establish uniform data collection, processing, and reporting procedures, thereby facilitating data comparisons from diverse studies and regions. Long-term monitoring programs must be set up to screen changes in contaminant levels over time. Over long periods, effective strategies for maintaining monitoring programs should be investigated, taking into account the associated financial limitations and changing environmental circumstances. Citizen science projects that involve local communities in monitoring can improve environmental stewardship and provide useful data (Conrad and Hilchey 2011). Effective ways to involve communities, raise awareness, and provide citizens with the ability to take part in monitoring operations should be investigated through research. More research is needed to develop reliable risk assessment techniques for emerging pollutants. It is imperative to comprehend the possible hazards linked to the exposure of these contaminants to develop efficient strategies for mitigation and management. Communities, legislators, and researchers must collaboratively address data gaps and research needs. The scientific community can aid in creating more sustainable and successful methods for tracking and controlling ECs in a variety of environmental contexts by giving priority to these research areas.

9.5 Regulatory and Policy Framework

Major laws and regulations governing groundwater in the United States include the Ground Water Rule (GWR), Underground Injection Control (UIC) program, and Source Water Protection (SWP) program. GWR was enacted in 2006 to reduce microbial contamination in drinking water sourced from groundwater. The GWR targets groundwater systems susceptible to fecal contamination. The Safe Drinking Water Act (SDWA) established the UIC program in 1974. It aims to prevent groundwater contamination from injecting fluids into the subsurface. SWP is a voluntary program that manages potential sources of contamination with a wide variety of actions, such as regular testing and improvements in infrastructure (US EPA 2015, 2013). Other laws, such as the Resource Conservation and Recovery Act (RCRA) and Toxic Substances Control Act (TSCA), regulate the use, storage, and disposal of solid and hazardous chemicals and wastes that can reach groundwater ("Federal Laws on Groundwater Protection | Environmental Health & Safety | Michigan State University").

Similarly, the European Parliament and the Council of the European Union passed the Groundwater Directive 2006/118/EC in 2006 (Directive 2006) to protect groundwater against contaminants and water quality deterioration. This directive enacted regular chemical assessments and changes in farming and forestry practices. In India, the Central Ground Water Authority (CGWA) was constituted under the Environment Protection Act of 1986 and 1997. CGWA plays a crucial role in managing groundwater development in the country by issuing No Objection Certificates, which require industries to gain permission to extract groundwater (Chakraborti et al. 2011).

Governments have engaged in cooperative efforts to manage transboundary groundwater resources. An example is the 2020 Guarani Aquifer Agreement, a formal accord between Argentina, Brazil, Paraguay, and Uruguay that promotes the monitoring, management, and sustainable utilization of the Guarani Aquifer system. The agreement also outlines protocols for settling potential disputes among these countries (Manganelli 2020).

Since the early 2000s, policies have been implemented to mitigate the prevalence of per- and polyfluoroalkyl substances (PFAS) in the environment. The production of long-chain PFAS, such as perfluorooctane sulfonic acid (PFOS) and perfluorooctanoic acid (PFOA), was phased out in numerous countries in the early 2000s. However, in 2018, the International Pollutants Elimination Network found that most PFAS substances were not regulated in several countries despite being a part of the Stockholm Convention, which enforced the elimination of its production and use. Moreover, it was not until 2018 that the adverse health effects of many short-chain PFAS were recognized. Since then, U.S. federal funding of EPA initiatives that drive state-level monitoring and remediation of PFAS has risen. Despite this, its national and international regulation remains complicated due to unclear regulatory jurisdictions and the ongoing deliberation over the benefits of developing appropriate regulatory guidance to combat the deleterious effects of PFAS (Brennan et al. 2021).

Current infrastructure in most countries lacks the technologies to remove emerging, nondegradable contaminants. For example, most wastewater treatments are not designed to remove PPCPs (Nebot et al. 2015). Given the need for national and international policies regulating pharmaceutical use, efforts to combat this issue tend to be sector-specific. An example of localized initiatives includes the 'National Take Back Day' conducted by the U.S. Drug Enforcement Administration (DEA) and other public safety officials. Today, the DEA offers free disposal services of unwanted medications to prevent environmental contamination in certain areas ("Take Back Day" 2023). While the World Health Organization (WHO) has proposed guidelines, they lack detailed provisions for managing PPCP waste. Instead, the guidelines focus on regulating the manufacturing of pharmaceuticals (Miettinen and Khan 2022).

9.6 Remediation and Management Strategies

9.6.1 Current Remediation Technologies and Their Effectiveness

Remediation technologies for ECs in groundwater are crucial for safeguarding the environment and public health by focusing on reducing or eliminating these pollutants. Several critical factors, such as the contaminant characteristics (e.g., concentration, cycling, behavior, degradation potential), the specific conditions of the site, and regulatory compliance, all significantly impact the effectiveness of any given remediation method. Pharmaceuticals, personal care products, industrial chemicals, and other new pollutants are just a few of the wide range of ECs that require a flexible and adaptive set of remediation techniques. Traditional methods like activated carbon adsorption (Sellaoui et al. 2023; Torrellas et al. 2015), advanced oxidation processes (Dewil et al. 2017), and bioremediation (Agrawal et al. 2020; Sutherland and Ralph 2019) remain fundamental, demonstrating high effectiveness across a spectrum of contaminants. Site-specific conditions, such as soil and aquifer characteristics, dictate the selection of remediation technologies. Thus, a suitable, case-by-case strategy is necessary.

As regulations expand to include more ECs, remediation has adopted innovative technologies such as nanotech, microbial fuel cells, and advanced biological treatment systems. The continuous search for effective and sustainable remediation techniques highlights the value of research and development, driving breakthroughs that guarantee a thorough and flexible response to the challenges presented by ECs in soil and groundwater. Innovative techniques that combine conventional methods with state-of-the-art technologies have become more prevalent in groundwater remediation in recent years, aiming to improve efficiency and fully address ECs. Integrating multiple remediation strategies and exploring synergistic effects highlights these efforts' dynamic and evolving nature.

One notable advancement involves the integration of electrokinetic-enhanced methods with traditional techniques. Electrokinetic processes, such as electrokinetic remediation (EKR), leverage electric fields to mobilize and transport contaminants (Sun et al. 2023; Vocciante et al. 2021). A synergistic effect can be achieved by combining EKR with advanced oxidation techniques or activated carbon adsorption to remove ECs specifically and efficiently. Combining traditional methods with phytoremediation is another promising approach. Using living green plants, phytoremediation is cost-effective and environmentally friendly for addressing soil or groundwater pollution. Phytoremediation, which uses plants to absorb and accumulate contaminants, can be integrated with activated carbon adsorption or in situ chemical oxidation treatment techniques (Yin et al. 2022). This combination takes advantage of plants' natural ability to absorb contaminants and improves overall remediation efficiency.

Furthermore, conventional bioremediation techniques can be successfully paired with cutting-edge technologies like in situ chemical reduction (ISCR). ISCR utilizes chemical amendments to reduce (i.e., redox, reduction reactions) contaminants (Dolfing et al. 2007; Tratnyek et al. 2014), and when paired with biostimulation (Herrero et al. 2019) or bioaugmentation, it creates a dual-action remediation system targeting both chemical reduction and microbial degradation. Combined with traditional sorption methods (Guerra et al. 2018), engineered NPs (Khan et al. 2021; Zhang et al. 2019) have been among the more recent remediation technologies. For instance, nanoparticles can provide a high surface area for adsorbing contaminants. When paired with materials like activated carbon or sorbent resins (Mane and Bhandari 2022), it results in a multifaceted remediation approach for diverse ECs. These hybrid approaches are prime examples of the dynamic evolution of groundwater remediation practices, where a customized and effective response to the challenges presented by ECs is achieved through the integration of diverse approaches. Higher remediation efficiencies may be possible with the continued development of such integrated technologies, especially when handling the complex and varied nature of ECs in groundwater.

Recent advancements in remediation technologies for PFAS, particularly PFOA, in groundwater have been explored between 2016 and 2020, focusing on various novel methods and their efficacy (see Xu et al. 2021a and references herein for the information in this paragraph). These include membrane technologies like nanofiltration, which show high removal rates but face challenges like membrane fouling and high costs. Chemical redox reactions, such as activated persulfate and starch-stabilized magnetite nanoparticles, offer lower costs but vary in effectiveness and potential for secondary pollution. Electrochemical treatments, including electrocoagulation and photoelectrochemical systems, have been noted for their convenience and efficiency, though they may face limitations in real-world applications. Photocatalysis, particularly with materials like BOHP (Benzoyl peroxide [BPO] and hydrogen peroxide [H_2O_2], which are used as photocatalysts in advanced oxidation processes [AOPs]) microparticles and TiO_2/Potassium monopersulfate (PMS), has been noted to be a cost-effective option with high degradation efficacy but it requires external light sources. Hybrid treatments combining membrane filtration and photocatalysis have also been proposed for enhanced effectiveness. These novel approaches show promise in addressing PFAS contamination in groundwater, each with advantages, limitations, and considerations for practical application.

9.6.2 Innovative and Emerging Solutions: Circular Economy and Advanced Treatments

According to the EPA, the term "circular economy" describes a system that uses recycling, composting, remanufacturing, reuse, maintenance, and refurbishment to keep materials and products in circulation for as long as possible. Circular economy strategies emphasize the recycling and reuse of water resources to close the water use loop in the context of water management. This covers techniques like harvesting water, repurposing wastewater for non-potable purposes (e.g., agriculture, irrigation, land), and incorporating green infrastructure to improve the purification processes of naturally occurring water. Additionally, there are water purification or blending techniques for potable water uses, including converting sewage-treated water into potable sources and its use for recharging groundwater, whether for indirect or direct potable purposes (Radini et al. 2021). The elimination of pollutants from water sources is made possible by technological developments in water treatment. Modern technologies and procedures are used in advanced treatment methods to increase the effectiveness of water purification. So far, methods like nanotechnology, membrane filtration, and sophisticated oxidation processes have helped completely and successfully remove contaminants from water. These techniques are especially important for dealing with ECs that might not respond well to traditional treatment methods.

One of the significant ECs in the aquatic environment is pharmaceutical residues, and their impacts on human health and the environment are a growing concern. Comparing membrane processes, reverse osmosis, and advanced oxidation processes, PPCP removal through adsorption by activated carbon is thought to be straightforward, inexpensive, safe, effective, and not produce any harmful byproducts (Reyes et al. 2021; Shahid et al. 2021). While activated carbon and advanced oxidation processes can assist in removing PPCPs, the resulting waste products, such as fully saturated activated carbon and reverse osmosis concentrates, may present challenges for the circular economy. Proper disposal of these wastes could entail new landfills or other storage or destruction methods, which may impede the sustainability of the process and hinder the circular economy's goal of waste upcycling and reuse. The first successful synthesis of Fenton catalytic material from solid wastes was achieved by Dat et al. (2023). This demonstrated a possible path toward a circular economy, wherein environmental pollution may be reduced, and economic benefits could rise (Dat et al. 2023). On the other hand, the creation of an effective adsorbent sludge-based activated carbon (SBAC) from sewage sludge, as suggested by Mohamed et al. (2023), is a sustainable integrated circular economy solution for stabilizing ECs, such as including polycyclic aromatic hydrocarbons, polybrominated diphenyl ethers and per- and polyfluoroalkyl substances, and enhancing the quality of wastewater effluents in a closed-loop system without creating waste. Adopting different strategies makes it possible to build more robust and effective techniques that support the long-term well-being of human communities and the environment. These strategies focus on maximizing resource efficiency and minimizing waste, which aligns with the circular economy concept.

9.6.3 Integrated Management Approaches for Sustainable Groundwater Quality

Comprehensive plans to protect and improve the quality of groundwater resources are part of integrated management approaches for sustainable groundwater quality. While several in situ remediation technologies are available for reducing contaminants in groundwater, future considerations will center on how best to integrate these techniques with environmental sustainability objectives. Conjunctive water management is recommended to set the path for an effective integrated management strategy that guarantees sustainable groundwater usage. This strategy involves managing surface and groundwater resources in concert to maintain environmental sustainability and strengthen the security of the water supply. Accordingly, it is unlikely that a universal concept of sustainability that would work in every country or area can be established because of limitations imposed by social and demographic variables, current technological advancements, and environmental concerns. However, a unified strategy with appropriate technology interventions, legislative frameworks, and community involvement is necessary for workable, long-lasting solutions.

The main elements of sustainable groundwater quality are monitoring and evaluating groundwater quality and developing and implementing appropriate lawful restrictions, source protection and land management, groundwater recharge and storage plans, community involvement and education programs, technological advancements, teamwork, and collaborative partnerships. Good monitoring and assessment set the stage by supplying the necessary information for well-informed decision-making. Regulatory measures provide a legal framework to safeguard groundwater resources and depend on strong policies to support them. Groundwater resources have been protected from contaminants through proactive source protection and land management programs. In addition, encouraging education and community involvement fosters a sense of shared accountability, thereby enhancing the participation of the local populace and the community to engage in conservation initiatives actively.

9.7 Case Studies of Successful Management

9.7.1 Examples of Successful Management and Remediation of Emerging Contaminants in Groundwater

Proficiently managing and removing ECs in groundwater demonstrates noteworthy accomplishments that offer invaluable perspectives and optimal methodologies. Innovative technologies, extensive monitoring strategies, and community engagement are often combined in successful cases. Remediation techniques that have demonstrated effectiveness in minimizing the concentrations of ECs include advanced oxidation processes, bioremediation, adsorption processes, and phytoremediation (Masud et al. 2023; Sarker et al. 2023; Shweta et al. 2021). Advanced oxidation processes (AOPs) are highly effective in breaking down persistent antibiotics and personal care products in the environment (Calenciuc et al. 2022). Activated carbon (AC) catalysts have been successfully used in wastewater treatment owing to their stability, lack of toxicity, and substantial surface area, facilitating the prompt activation of the oxidant (Van Tran et al. 2017; Wong et al. 2018). Electrochemical advanced oxidation methods eliminate toxic pollutants from wastewater before it is discharged into rivers (Moreira et al. 2017). Lee et al. (2023) presented an effective preparation method for activated carbon ball-milled (ACBM) media from granular activated carbon (GAC) for groundwater treatment. ACBM promotes effective pollutant degradation and the activation of peroxymonosulfate (PMS) due to its improved colloidal stability. Adsorption studies show that ACBM achieves 92% ibuprofen degradation in the groundwater system, outperforming GAC in active sites and maximum adsorption capacity. The study illustrates the suitability and efficacy of the ACBM/PMS system for actual groundwater treatment by clarifying the degradation pathways, identifying byproducts, and evaluating toxicity (Lee et al. 2023).

Pharmaceutical residues in wastewater are becoming more prevalent due to increased pharmaceutical use, causing environmental concern. These substances pose a serious problem in aquatic environments because of their stability, complexity, and persistence. By producing extremely reactive and non-selective oxidizing radicals, Advanced Oxidation Processes (AOPs) like UV, ozone, Fenton-based AOPs, and heterogeneous photocatalysis are effective in removing a variety of drugs from wastewater effluents that are prepared in laboratories as well as those that are generated in real wastewater treatment facilities (Umair et al. 2023). Under such conditions, AOPs can be cost-effective, adaptable, and exceptionally efficient techniques for degrading persistent organic compounds (Taoufik et al. 2021; Wu et al. 2021). Additionally, nano remediation for pesticides is a novel approach employing nanomaterials to break down and eliminate pesticides from polluted sites. It is a technology widely expanding in its use and application (Sarker et al. 2023).

9.7.2 Lessons Learned and Best Practices

The best practices and lessons learned from the successful management and remediation of ECs in groundwater emphasize the value of proactive source protection, early detection, comprehensive monitoring strategies, innovative technologies, and cooperative efforts between residents, government organizations, and researchers. Putting into practice sustainable techniques like natural attenuation and green infrastructure (Liu et al. 2020) further assists with management and remediation results. The instances mentioned above highlight the importance of adopting a comprehensive and flexible strategy for dealing with the difficulties of newly discovered groundwater pollutants. Certain remediation and management plans might have unintended, undesired side effects. Therefore, to guarantee the long-term success of groundwater quality initiatives, it is imperative that comprehensive risk assessments, ongoing monitoring, and adaptive management strategies be implemented.

Although EC concentrations are effectively reduced by remediation techniques such as AOPs, bioremediation, adsorption processes, and phytoremediation, it is crucial to be aware of any potential negative side effects. For example, AOPs effectively remove a wide variety of water pollutants. However, the high operating costs prevent them from being widely adopted, and several studies have indicated that byproducts may be produced in the process (Giwa et al. 2021; Sharma et al. 2018). These by-products are present in various water matrices and can be more toxic than the original contaminants (Lloret et al. 2012; Sein et al. 2008). The chlorination process of water purification systems has the potential to produce disinfection by-products (DBPs), raising concerns due to their possible involvement in carcinogenesis, teratogenesis, and mutagenesis. Dissolved organic nitrogen is the main source of DBPs. It can produce nitrogenous disinfection by-products (N-DBPs) when chlorinated and chloraminated (Dotson et al. 2009; Sharma et al. 2018). As such, N-DBPs are more toxic than disinfection products made from carbonaceous sources (Wagner et al. 2012).

Sustainable groundwater management greatly depends on community involvement, education, and technology advancements. Increasing public knowledge of the possible dangers of ECs fosters a sense of responsibility and promotes ecologically friendly behavior. Educational programs and development motivation allow communities to participate actively in groundwater protection efforts. Moreover, cooperation and affiliation are important pillars in achieving sustainable groundwater quality. Knowledge, resources, and expertise can be shared more easily when academic institutions, industry stakeholders, and regulatory bodies work together effectively. In the end, cutting-edge technologies, legal frameworks, community involvement, and cooperative partnerships are needed to address the complex issues of sustainable groundwater quality management. Stakeholders can navigate the complexities of ECs by adopting a comprehensive and flexible approach, further promoting and helping to ensure the preservation and protection of groundwater resources for present and future generations.

9.8 Future Directions and Research Opportunities

9.8.1 Emerging Trends and Future Challenges in Managing Groundwater Contamination

One enduring problem in groundwater protection is detecting and handling new pollutants. Nevertheless, several substances, such as pharmaceuticals, personal care items, and industrial chemicals, threaten groundwater quality. Effective mitigation strategies require ongoing research into the fate and movement of these contaminants to guarantee that management techniques remain adaptive to the changing nature of groundwater pollution. Groundwater contamination management is transforming due to recent developments in sensing technologies, data analytics, and remote monitoring. These technologies permit real-time groundwater quality monitoring, enabling prompt and efficient responses to emerging pollution issues. Improved monitoring systems make it easier to implement remediation strategies that are specifically targeted and contribute to a more thorough understanding of aquifer dynamics and related transport and fate of contaminants in these systems. Even with advancements in remote sensing /monitoring technologies, the effects of climate change on groundwater management pose a complex array of challenges. They will be considered for appropriate management actions and solutions in the future. Variations in temperature, precipitation patterns, and the frequency of extreme weather events can significantly impact aquifer susceptibility, pollutant transport and fate channels/processes, and groundwater recharge rates. Adapting management strategies to these climatic alterations ensures groundwater resources remain resilient.

However, groundwater quality is directly impacted by land-use changes brought about by urbanization, agricultural practices, and shifting social demands. Pollutant introduction and increased runoff are two ways urbanization and intensified agriculture can lead to contamination. Groundwater quality maintenance necessitates balancing development demands and sustainable land use practices. Reducing the detrimental effects of changing land use on groundwater quality requires effective rules and sustainable land management practices. An integrated and multidisciplinary approach is crucial as we navigate these new trends and challenges. Together, communities, politicians, and researchers can create strategies to protect groundwater resources and maintain their sustainability over a long period. Scientists in academia and the larger research community have the opportunity to influence future studies pertaining to groundwater resources by conducting research at the nexus of different sectors and disciplines and by modifying study objectives to address a range of issues (Bhunia et al. 2023). Researchers typically utilize datasets for various purposes beyond convention; these datasets are crucial for tracking developments in the complex and dynamic field of groundwater science. Scientists can assess advancements, track trends, and enhance methods, which fosters an integrative collaborative environment where knowledge exchanged across multiple disciplines and sectors will lead to ongoing improvements in groundwater resource management.

While many in situ remediation technologies are available to remove certain contaminants from groundwater, the emphasis now is on ensuring these procedures align with environmental sustainability objectives. Future thinking relies on applying techniques such as in situ biological remediation, which offers more economical and ecologically friendly options than conventional physical and chemical methods (Syafiuddin et al. 2020). Concerns about groundwater contamination are being thoughtfully addressed by this shift toward more affordable and environmentally friendly solutions while maintaining sustainable environmental standards. As a result, creating a comprehensive framework is essential for evaluating groundwater systems, guaranteeing their stability, and reducing depletion. This strategic approach emphasizes the dedication to a sustainable and peaceful coexistence with our groundwater resources.

9.8.2 Potential Research Areas for Better Understanding and Management

The study of ECs in groundwater research includes several significant fields that complement one another to offer a more complete knowledge and effective control of these contaminants. Creating sophisticated methods for detecting and describing ECs can be one primary area of concentration. Developing robust, cost-effective analytical methods for accurately identifying and measuring pollutants will be essential better to understand these materials' presence in groundwater systems. Moreover, studies examining the frequency, distribution, and temporal patterns of ECs in diverse hydrogeological settings may provide significant insights into the extent of their source origins and occurrence, transport and fate, and impacts on various systems and populations. Thus, understanding the fate and transport mechanisms of ECs in groundwater is an important area of future research. This entails investigating these contaminants' interactions with microbial communities, aquifer characteristics, and soil properties as they move through the subsurface environment. Additionally, predicting the long-term effects of pollutants and developing efficient management strategies depend on understanding their persistence and transformation under various environmental conditions.

Determining the risks of exposure to ECs in groundwater is a major research component. This entails conducting in-depth risk assessments to examine environmental and public health hazards. Exploring the potential combined effects of exposure to different contaminants is necessary to provide a more realistic picture of the overall environmental risk. Research on the movement and fate of pollutants, particularly ECs, aims to shed light on the behavior of these substances and how they persist and migrate through soil and groundwater systems. By concentrating on these research areas, scientists and policymakers can better understand and manage ECs, which will help them create better plans for remediation, prevention, and regulation of groundwater pollution. Innovative technologies are required to clean up groundwater contaminated by recently identified pollutants, and several promising approaches are presently being investigated. Advanced oxidation processes (AOPs) are especially effective at breaking down pollutants, according to Vinayagam et al. (2023). Strong oxidizing agents are used in ozone treatment, UV irradiation, and Fenton's reagent to convert pollutants into less hazardous forms. These methods have successfully been used to break down, transform, and degrade numerous ECs in groundwater and water treatment systems.

Alternatively, resource management strategies can be optimized, contaminant transport can be predicted, and groundwater flow can be simulated with the help of sophisticated modeling techniques and a variety of datasets. Enhancing the skills of professionals and communities involved in groundwater management requires capacity-building and training initiatives. Furthermore, research is being conducted to help determine economic valuation to assess the monetary value of groundwater resources and integrate economic ideas into plans for sustainable management (Qureshi et al. 2012). Moreover, current projects focus on predicting pollutants' fate and transport, providing early alerts for proactive management, and real-time continuous groundwater quality monitoring (Geetha et al. 2023). Developing adaptive management strategies requires a thorough understanding of the effects of land use changes, climate change, urbanization, and agricultural practices on groundwater (Patra et al. 2023). The scientific community and policymakers hope to contribute to a more comprehensive and sustainable groundwater management strategy by focusing on these multifaceted research areas.

9.8.3   Linking to Broader Themes of the Book—the Circular Economy and Lifecycle Management

As a vital component of Earth's hydrological cycle, groundwater is a limited resource that requires a comprehensive and sustainable strategy. The concept of the circular economy strongly emphasizes waste reduction and resource optimization. Managing groundwater entails implementing policies ensuring aquifers are continuously replenished, maintaining good water quality, and using water resources effectively. The tenets of the circular economy promote water recycling and reuse and the incorporation of treated wastewater into groundwater recharge systems (Mannina et al. 2022). This supports closing the loop on water consumption and disposal while improving water systems' general resilience. Furthermore, knowledge of lifecycle management concepts is necessary for understanding the life cycle of water resources, from extraction to consumption to final return to the environment. Groundwater is part of this intricate lifecycle; thus, it is important to consider how to use it sustainably at each stage. To reduce ecological footprints and increase overall system efficiency, lifecycle thinking can be applied to groundwater management by analyzing the environmental effects of extraction, treatment, and disposal. These interrelated themes emphasize the necessity of managing water resources in a coordinated way. Because groundwater is vital to meeting human needs and maintaining ecosystems, it becomes a focal point for implementing circular economy practices and adopting comprehensive lifecycle management strategies. This book seeks to further our understanding of sustainable groundwater resource management—a crucial component of achieving social, economic, and environmental sustainability—by analyzing these themes simultaneously.

Eliminating environmental pollutants is only one aspect of the complex challenge of remediating ECs. This process emphasizes the necessity of considering a substance's whole life cycle, from production to disposal, and is inextricably linked to the larger idea of lifecycle management. Managing ECs in combination with lifecycle management involves a systematic and integrative approach. The first step is to identify the pathways and sources of these contaminants, such as industrial processes, agricultural practices, or other human-related sources. Understanding the full lifecycle makes it possible to create focused remediation plans that address the underlying causes of contamination to stop it from happening again and remove it currently.

Moreover, remedial actions must include a thorough assessment of the environmental fate and transportation of ECs. This means understanding how these pollutants affect ecosystems and pose health risks to humans as they permeate the soil, water, and air. Considering an EC's complete lifecycle, remedial strategies can be modified to address the various forms and transformations throughout the environment. Because the circular economy approach is inextricably linked to lifecycle management, it can significantly impact strategies for EC removal. The circular economy encourages recycling, material reuse, and waste reduction. To reduce the environmental impact of cleanup efforts, this may be considered in the context of remediation by thinking about ways to recover and reuse materials or by-products that have been remediated. Ultimately, successful EC remediation requires an all-encompassing strategy that includes lifecycle management ideas. The full life cycle of these pollutants—from their sources to their possible effects—can be understood to develop remediation strategies that are both efficient and long-lasting, thereby advancing the more general objective of attaining social, economic, and environmental sustainability.

9.9 Conclusion

Emerging contaminants in groundwater are complex, posing significant health and environmental risks, particularly in low-income communities and countries. This complexity requires a comprehensive approach to investigating groundwater contamination and integrating it into broader environmental and public health challenges. The importance of considering both local actions and global consequences, as well as the ever-changing nature of ECs, highlights the need for a comprehensive and collaborative approach to environmental stewardship. A collaborative approach to environmental stewardship is essential as is incorporating 'green' principles and 'circular economics in addressing ECs. This includes developing and using "safe" or "benign" chemicals in manufacturing processes and engaging in more proactive rather than reactive strategies. Such approaches could include the development and use of products and chemicals and pre-market studies, thereby minimizing the introduction of harmful substances into the environment. Sustainable development and environmental management are essential to a sustainable future for all. Rigorous scientific research, innovative policymaking, international cooperation, and community engagement is imperative for the effective management of ECs in groundwater. Thus, there is a need to develop guidelines and practices that protect and manage groundwater sustainably for future generations.

References

Agrawal K, Bhatt A, Chaturvedi V, Verma P (2020) Bioremediation: an effective technology toward a sustainable environment via the remediation of emerging environmental pollutants. In: Emerging technologies in environmental bioremediation. Elsevier, pp 165–196. https://doi.org/10.1016/B978-0-12-819860-5.00007-9

Ashfaq M, Li Y, Rehman MSU et al (2019) Occurrence, spatial variation and risk assessment of pharmaceuticals and personal care products in urban wastewater, canal surface water, and their sediments: a case study of Lahore, Pakistan. Sci Total Environ 688:653–663. https://doi.org/10.1016/j.scitotenv.2019.06.285

Awad YM, Kim S-C, Abd El-Azeem SAM et al (2014) Veterinary antibiotics contamination in water, sediment, and soil near a swine manure composting facility. Environ Earth Sci 71:1433–1440. https://doi.org/10.1007/s12665-013-2548-z

Bao J, Yu W-J, Liu Y et al (2019) Perfluoroalkyl substances in groundwater and home-produced vegetables and eggs around a fluorochemical industrial park in China. Ecotoxicol Environ Saf 171:199–205. https://doi.org/10.1016/j.ecoenv.2018.12.086

Barnes KK, Kolpin DW, Furlong ET et al (2008) A national reconnaissance of pharmaceuticals and other organic wastewater contaminants in the United States — I) groundwater. Sci Total Environ 402:192–200. https://doi.org/10.1016/j.scitotenv.2008.04.028

Bhunia GS, Shit PK, Brahma S (2023) Groundwater conservation and management: recent trends and future prospects. In: case studies in geospatial applications to groundwater resources. Elsevier, pp 371–385

Bouillard JX, R'Mili B, Moranviller D et al (2013) Nanosafety by design: risks from nanocomposite/nanowaste combustion. J Nanoparticle Res 15:1519. https://doi.org/10.1007/s11051-013-1519-3

Brennan NM, Evans AT, Fritz MK et al (2021) Trends in the regulation of Per- and Polyfluoroalkyl substances (PFAS): a scoping review. Int J Environ Res Public Health 18:10900. https://doi.org/10.3390/ijerph182010900

Brion F, De Gussem V, Buchinger S et al (2019) Monitoring estrogenic activities of waste and surface waters using a novel in vivo zebrafish embryonic (EASZY) assay: comparison with in vitro cell-based assays and determination of effect-based trigger values. Environ Int 130:104896. https://doi.org/10.1016/j.envint.2019.06.006

Brusseau ML (2018) Assessing the potential contributions of additional retention processes to PFAS retardation in the subsurface. Sci Total Environ 613–614:176–185. https://doi.org/10.1016/j.scitotenv.2017.09.065

Brusseau ML, Anderson RH, Guo B (2020) PFAS concentrations in soils: background levels versus contaminated sites. Sci Total Environ 740:140017. https://doi.org/10.1016/j.scitotenv.2020.140017

Brusseau ML, Guo B (2022) PFAS concentrations in soil versus soil porewater: Mass distributions and the impact of adsorption at air-water interfaces. Chemosphere 302:134938. https://doi.org/10.1016/j.chemosphere.2022.134938

Bryant JD, Anderson R, Bolyard SC et al (2022) PFAS Experts Symposium 2: key advances in poly- and perfluoroalkyl characterization, fate, and transport. Remediat J 32:19–28. https://doi.org/10.1002/rem.21703

Bunting SY, Lapworth DJ, Crane EJ et al (2021) Emerging organic compounds in European groundwater. Environ Pollut 269:115945. https://doi.org/10.1016/j.envpol.2020.115945

Calafat AM, Kato K, Hubbard K et al (2019) Legacy and alternative per- and polyfluoroalkyl substances in the U.S. general population: paired serum-urine data from the 2013–2014 National Health and Nutrition Examination Survey. Environ Int 131:105048. https://doi.org/10.1016/j.envint.2019.105048

Calenciuc C, Fdez-Sanromán A, Lama G et al (2022) Recent developments in advanced oxidation processes for organics-polluted soil reclamation. Catalysts 12:64. https://doi.org/10.3390/catal12010064

Cao X, Wang C, Lu Y et al (2019) Occurrence, sources and health risk of polyfluoroalkyl substances (PFASs) in soil, water and sediment from a drinking water source area. Ecotoxicol Environ Saf 174:208–217. https://doi.org/10.1016/j.ecoenv.2019.02.058

Chakraborti D, Das B, Murrill MT (2011) Examining India's groundwater quality management. Environ Sci Technol 45:27–33. https://doi.org/10.1021/es101695d

Chakraborti D, Rahman MM, Mukherjee A et al (2015) Groundwater arsenic contamination in Bangladesh—21 years of research. J Trace Elem Med Biol 31:237–248. https://doi.org/10.1016/j.jtemb.2015.01.003

Chakraborty A, Adhikary S, Bhattacharya S et al (2023) Pharmaceuticals and personal care products as emerging environmental contaminants: prevalence, toxicity, and remedial approaches. ACS Chem Health Saf 30:362–388. https://doi.org/10.1021/acs.chas.3c00071

Chen H, Ma LQ, Gao B, Gu C (2013a) Effects of Cu and Ca cations and Fe/Al coating on ciprofloxacin sorption onto sand media. J Hazard Mater 252–253:375–381. https://doi.org/10.1016/j.jhazmat.2013.03.014

Chen W, Xu J, Lu S et al (2013b) Fates and transport of PPCPs in soil receiving reclaimed water irrigation. Chemosphere 93:2621–2630. https://doi.org/10.1016/j.chemosphere.2013.09.088

Conkle JL, Gan J, Anderson MA (2012) Degradation and sorption of commonly detected PPCPs in wetland sediments under aerobic and anaerobic conditions. J Soils Sediments 12:1164–1173. https://doi.org/10.1007/s11368-012-0535-8

Conrad CC, Hilchey KG (2011) A review of citizen science and community-based environmental monitoring: issues and opportunities. Environ Monit Assess 176:273–291. https://doi.org/10.1007/s10661-010-1582-5

Dat ND, Huynh QS, Tran KAT, Nguyen ML (2023) Performance of heterogeneous Fenton catalyst from solid wastes for removal of emerging contaminant in water: a potential approach to circular economy. Results Eng 18:101086. https://doi.org/10.1016/j.rineng.2023.101086

Dewil R, Mantzavinos D, Poulios I, Rodrigo MA (2017) New perspectives for advanced oxidation processes. J Environ Manag 195:93–99. https://doi.org/10.1016/j.jenvman.2017.04.010

Directive (2006) Directive 2006/118/EC of the European Parliament and of the Council of 12 December 2006 on the Protection of Groundwater against Pollution and Deterioration. Off J Eur Union 372

Dolfing J, Van Eekert M, Seech A et al (2007) In Situ Chemical Reduction (ISCR) technologies: significance of low Eh reactions. Soil Sediment Contam Int J 17:63–74. https://doi.org/10.1080/15320380701741438

Dotson A, Westerhoff P, Krasner SW (2009) Nitrogen enriched dissolved organic matter (DOM) isolates and their affinity to form emerging disinfection by-products. Water Sci Technol 60:135–143. https://doi.org/10.2166/wst.2009.333

Du G, Hu J, Huang Z et al (2019) Neonatal and juvenile exposure to perfluorooctanoate (PFOA) and perfluorooctane sulfonate (PFOS): advance puberty onset and kisspeptin system disturbance in female rats. Ecotoxicol Environ Saf 167:412–421. https://doi.org/10.1016/j.ecoenv.2018.10.025

Dulio V, Van Bavel B, Brorström-Lundén E et al (2018) Emerging pollutants in the EU: 10 years of NORMAN in support of environmental policies and regulations. Environ Sci Eur 30:5. https://doi.org/10.1186/s12302-018-0135-3

East A, Anderson RH, Salice CJ (2021) Per- and Polyfluoroalkyl Substances (PFAS) in surface water near US air force bases: prioritizing individual chemicals and mixtures for toxicity testing and risk assessment. Environ Toxicol Chem 40:871–882. https://doi.org/10.1002/etc.4893

Eldawud R, Wagner A, Dong C et al (2018) Carbon nanotubes physicochemical properties influence the overall cellular behavior and fate. NanoImpact 9:72–84. https://doi.org/10.1016/j.impact.2017.10.006

EPA (2017) Technical Fact Sheet—Perfluorooctane Sulfonate (PFOS) and Perfluorooctanoic Acid (PFOA)

Faille J (2010) Vulnérabilité des nappes d'eau souterraine aux pollutions médicamenteuses: aquifer vulnerability to drug pollution (a literature review). Masters Diss Lille Polytech 19

Fick J, Söderström H, Lindberg RH et al (2009) Contamination of surface, ground, and drinking water from pharmaceutical production. Environ Toxicol Chem 28:2522–2527. https://doi.org/10.1897/09-073.1

Francisco LFV, Do Amaral Crispim B, Spósito JCV et al (2019) Metals and emerging contaminants in groundwater and human health risk assessment. Environ Sci Pollut Res 26:24581–24594. https://doi.org/10.1007/s11356-019-05662-5

Freeling F, Alygizakis NA, Von Der Ohe PC et al (2019) Occurrence and potential environmental risk of surfactants and their transformation products discharged by wastewater treatment plants. Sci Total Environ 681:475–487. https://doi.org/10.1016/j.scitotenv.2019.04.445

Gao B (2022) Emerging contaminants in soil and groundwater systems: occurrence, impact, fate and transport. Elsevier, Amsterdam

Gavrilescu M, Demnerová K, Aamand J et al (2015) Emerging pollutants in the environment: present and future challenges in biomonitoring, ecological risks and bioremediation. New Biotechnol 32:147–156. https://doi.org/10.1016/j.nbt.2014.01.001

Gebbink WA, Van Leeuwen SPJ (2020) Environmental contamination and human exposure to PFASs near a fluorochemical production plant: review of historic and current PFOA and GenX contamination in the Netherlands. Environ Int 137:105583. https://doi.org/10.1016/j.envint.2020.105583

Geetha M, Bonthula S, Al-Maadeed S et al (2023) Research trends in smart cost-effective water quality monitoring and modeling: special focus on artificial intelligence. Water 15:3293. https://doi.org/10.3390/w15183293

Gellrich V, Stahl T, Knepper TP (2012) Behavior of perfluorinated compounds in soils during leaching experiments. Chemosphere 87:1052–1056. https://doi.org/10.1016/j.chemosphere.2012.02.011

Giwa A, Yusuf A, Balogun HA et al (2021) Recent advances in advanced oxidation processes for removal of contaminants from water: a comprehensive review. Process Saf Environ Prot 146:220–256. https://doi.org/10.1016/j.psep.2020.08.015

Golet EM, Alder AC, Hartmann A et al (2001) Trace determination of fluoroquinolone antibacterial agents in urban wastewater by solid-phase extraction and liquid chromatography with fluorescence detection. Anal Chem 73:3632–3638. https://doi.org/10.1021/ac0015265

Gu X, Tan Y, Tong F, Gu C (2015) Surface complexation modeling of coadsorption of antibiotic ciprofloxacin and Cu(II) and onto goethite surfaces. Chem Eng J 269:113–120. https://doi.org/10.1016/j.cej.2014.12.114

Gude VG, Maganti A (2021) Desalination of deep groundwater for freshwater supplies. In: Global groundwater. Elsevier, pp 577–583

Guerra F, Attia M, Whitehead D, Alexis F (2018) Nanotechnology for environmental remediation: materials and applications. Molecules 23:1760. https://doi.org/10.3390/molecules23071760

Heffernan AL, Cunningham TK, Drage DS et al (2018) Perfluorinated alkyl acids in the serum and follicular fluid of UK women with and without polycystic ovarian syndrome undergoing fertility treatment and associations with hormonal and metabolic parameters. Int J Hyg Environ Health 221:1068–1075. https://doi.org/10.1016/j.ijheh.2018.07.009

Hepburn E, Madden C, Szabo D et al (2019) Contamination of groundwater with per- and polyfluoroalkyl substances (PFAS) from legacy landfills in an urban re-development precinct. Environ Pollut 248:101–113. https://doi.org/10.1016/j.envpol.2019.02.018

Herrero J, Puigserver D, Nijenhuis I et al (2019) Combined use of ISCR and biostimulation techniques in incomplete processes of reductive dehalogenation of chlorinated solvents. Sci Total Environ 648:819–829. https://doi.org/10.1016/j.scitotenv.2018.08.184

Hewitt SC, Couse JF, Korach KS (2000) Estrogen receptor transcription and transactivation Estrogen receptor knockout mice: what their phenotypes reveal about mechanisms of estrogen action. Breast Cancer Res 2:345. https://doi.org/10.1186/bcr79

Hu XC, Dai M, Sun JM, Sunderland EM (2022) The utility of machine learning models for predicting chemical contaminants in drinking water: promise, challenges, and opportunities. Curr Environ Health Rep 10:45–60. https://doi.org/10.1007/s40572-022-00389-x

Hu XC, Ge B, Ruyle BJ et al (2021) A statistical approach for identifying private wells susceptible to perfluoroalkyl substances (PFAS) contamination. Environ Sci Technol Lett 8:596–602. https://doi.org/10.1021/acs.estlett.1c00264

Huang C, Wu D, Zhang K et al (2022) Perfluorooctanoic acid alters the developmental trajectory of female germ cells and embryos in rodents and its potential mechanism. Ecotoxicol Environ Saf 236:113467. https://doi.org/10.1016/j.ecoenv.2022.113467

Johnson GR, Brusseau ML, Carroll KC et al (2022) Global distributions, source-type dependencies, and concentration ranges of per- and polyfluoroalkyl substances in groundwater. Sci Total Environ 841:156602. https://doi.org/10.1016/j.scitotenv.2022.156602

Karnjanapiboonwong A, Morse AN, Maul JD, Anderson TA (2010) Sorption of estrogens, triclosan, and caffeine in a sandy loam and a silt loam soil. J Soils Sediments 10:1300–1307. https://doi.org/10.1007/s11368-010-0223-5

Khan S, Mu N, Al-Gheethi A, Iqbal J (2021) Engineered nanoparticles for removal of pollutants from wastewater: current status and future prospects of nanotechnology for remediation strategies. J Environ Chem Eng 9:106160. https://doi.org/10.1016/j.jece.2021.106160

Khan S, Mu N, Govarthanan M et al (2022) Emerging contaminants of high concern for the environment: current trends and future research. Environ Res 207:112609. https://doi.org/10.1016/j.envres.2021.112609

Kolpin DW, Furlong ET, Meyer MT et al (2002) Pharmaceuticals, hormones, and other organic wastewater contaminants in U.S. streams, 1999–2000: a national reconnaissance. Environ Sci Technol 36:1202–1211. https://doi.org/10.1021/es011055j

K'oreje KO, Vergeynst L, Ombaka D et al (2016) Occurrence patterns of pharmaceutical residues in wastewater, surface water and groundwater of Nairobi and Kisumu city, Kenya. Chemosphere 149:238–244. https://doi.org/10.1016/j.chemosphere.2016.01.095

Koutros S, Silverman DT, Alavanja MC et al (2016) Occupational exposure to pesticides and bladder cancer risk. Int J Epidemiol 45:792–805. https://doi.org/10.1093/ije/dyv195

Krall AL, Elliott SM, Erickson ML, Adams BA (2018) Detecting sulfamethoxazole and carbamazepine in groundwater: is ELISA a reliable screening tool? Environ Pollut 234:420–428. https://doi.org/10.1016/j.envpol.2017.11.065

Kumah EA, Fopa RD, Harati S et al (2023) Human and environmental impacts of nanoparticles: a scoping review of the current literature. BMC Public Health 23:1059. https://doi.org/10.1186/s12889-023-15958-4

Lapworth DJ, Baran N, Stuart ME, Ward RS (2012) Emerging organic contaminants in groundwater: a review of sources, fate and occurrence. Environ Pollut 163:287–303. https://doi.org/10.1016/j.envpol.2011.12.034

Lapworth DJ, Lopez B, Laabs V et al (2019) Developing a groundwater watch list for substances of emerging concern: a European perspective. Environ Res Lett 14:035004. https://doi.org/10.1088/1748-9326/aaf4d7

Lee BCY, Lim FY, Loh WH et al (2021) Emerging contaminants: an overview of recent trends for their treatment and management using light-driven processes. Water 13:2340. https://doi.org/10.3390/w13172340

Lee SH, Annamalai S, Shin WS (2023) Engineered ball-milled colloidal activated carbon material for advanced oxidation process of ibuprofen: Influencing factors and insights into the mechanism. Environ Pollut 322:121023. https://doi.org/10.1016/j.envpol.2023.121023

Lewis R, Johns L, Meeker J (2015) Serum biomarkers of exposure to perfluoroalkyl substances in relation to serum testosterone and measures of thyroid function among adults and adolescents from NHANES 2011–2012. Int J Environ Res Public Health 12:6098–6114. https://doi.org/10.3390/ijerph120606098

Li D, Zhang L, Zhang Y et al (2019) Maternal exposure to perfluorooctanoic acid (PFOA) causes liver toxicity through PPAR-α pathway and lowered histone acetylation in female offspring mice. Environ Sci Pollut Res 26:18866–18875. https://doi.org/10.1007/s11356-019-05258-z

Li J, He J, Niu Z, Zhang Y (2020) Legacy per- and polyfluoroalkyl substances (PFASs) and alternatives (short-chain analogues, F-53B, GenX and FC-98) in residential soils of China: present implications of replacing legacy PFASs. Environ Int 135:105419. https://doi.org/10.1016/j.envint.2019.105419

Li P, Karunanidhi D, Subramani T, Srinivasamoorthy K (2021) Sources and consequences of groundwater contamination. Arch Environ Contam Toxicol 80:1–10. https://doi.org/10.1007/s00244-020-00805-z

Liu B-W, Wang M-H, Chen T-L et al (2020) Establishment and implementation of green infrastructure practice for healthy watershed management: Challenges and perspectives. Water-Energy Nexus 3:186–197. https://doi.org/10.1016/j.wen.2020.05.003

Liu Z, Lu Y, Song X et al (2019) Multiple crop bioaccumulation and human exposure of perfluo-roalkyl substances around a mega fluorochemical industrial park, China: implication for planting optimization and food safety. Environ Int 127:671–684. https://doi.org/10.1016/j.envint.2019.04.008

Liu Z, Lu Y, Wang T et al (2016) Risk assessment and source identification of perfluoroalkyl acids in surface and ground water: Spatial distribution around a mega-fluorochemical industrial park, China. Environ Int 91:69–77. https://doi.org/10.1016/j.envint.2016.02.020

Lloret L, Hollmann F, Eibes G et al (2012) Immobilisation of laccase on Eupergit supports and its application for the removal of endocrine disrupting chemicals in a packed-bed reactor. Biodegradation 23:373–386. https://doi.org/10.1007/s10532-011-9516-7

Loos R, Locoro G, Comero S et al (2010) Pan-European survey on the occurrence of selected polar organic persistent pollutants in ground water. Water Res 44:4115–4126. https://doi.org/10.1016/j.watres.2010.05.032

Lukač Reberski J, Terzić J, Maurice LD, Lapworth DJ (2022) Emerging organic contaminants in karst groundwater: a global level assessment. J Hydrol 604:127242. https://doi.org/10.1016/j.jhydrol.2021.127242

Luo Y, Guo W, Ngo HH et al (2014) A review on the occurrence of micropollutants in the aquatic environment and their fate and removal during wastewater treatment. Sci Total Environ 473–474:619–641. https://doi.org/10.1016/j.scitotenv.2013.12.065

Manaia CM, Aga DS, Cytryn E et al (2023) The complex interplay between antibiotic resistance and pharmaceutical and personal care products in the environment. Environ Toxicol Chem 5555. https://doi.org/10.1002/etc.5555

Mane MB, Bhandari VM (2022) Developing spherical activated carbons from polymeric resins for removal of contaminants from aqueous and organic streams. Int J Environ Sci Technol 19:10021–10040. https://doi.org/10.1007/s13762-021-03684-6

Manganelli A (2020) Guaraní aquifer system agreement

Mannina G, Gulhan H, Ni B-J (2022) Water reuse from wastewater treatment: the transition towards circular economy in the water sector. Bioresour Technol 363:127951. https://doi.org/10.1016/j.biortech.2022.127951

Masud MAA, Shin WS, Kim DG (2023) Fe-doped kelp biochar-assisted peroxymonosulfate activation for ciprofloxacin degradation: multiple active site-triggered radical and non-radical mechanisms. Chem Eng J 471:144519. https://doi.org/10.1016/j.cej.2023.144519

Miettinen M, Khan SA (2022) Pharmaceutical pollution: a weakly regulated global environmental risk. Rev Eur Comp Int Environ Law 31:75–88. https://doi.org/10.1111/reel.12422

Milinovic J, Lacorte S, Vidal M, Rigol A (2015) Sorption behaviour of perfluoroalkyl substances in soils. Sci Total Environ 511:63–71. https://doi.org/10.1016/j.scitotenv.2014.12.017

Mishra S, Sundaram B (2023) Fate, transport, and toxicity of nanoparticles: an emerging pollutant on biotic factors. Process Saf Environ Prot 174:595–607. https://doi.org/10.1016/j.psep.2023.04.037

Mohamed BA, Hamid H, Montoya-Bautista CV, Li LY (2023) Circular economy in wastewater treatment plants: treatment of contaminants of emerging concerns (CECs) in effluent using sludge-based activated carbon. J Clean Prod 389:136095. https://doi.org/10.1016/j.jclepro.2023.136095

Mompelat S, Le Bot B, Thomas O (2009) Occurrence and fate of pharmaceutical products and by-products, from resource to drinking water. Environ Int 35:803–814. https://doi.org/10.1016/j.envint.2008.10.008

Montagner CC, Vidal C, Acayaba RD (2017) Emerging contaminants in aquatic matrices from Brazil: current scenario and analytical, ecotoxicological and legislational aspects. Quím Nova 40:1094–1110

Moreira FC, Boaventura RAR, Brillas E, Vilar VJP (2017) Electrochemical advanced oxidation processes: a review on their application to synthetic and real wastewaters. Appl Catal B Environ 202:217–261. https://doi.org/10.1016/j.apcatb.2016.08.037

Mukherjee A, Scanlon BR, Aureli A et al (eds) (2021) Global groundwater: source, scarcity, sustainability, security, and solutions. Elsevier, Amsterdam Oxford Cambridge

Mukherjee SG, O'Claonadh N, Casey A, Chambers G (2012) Comparative in vitro cytotoxicity study of silver nanoparticle on two mammalian cell lines. Toxicol in Vitro 26:238–251. https://doi.org/10.1016/j.tiv.2011.12.004

Naidu R, Biswas B (2024) Introduction to inorganic contaminants and radionuclides: global issues and challenges. In: Inorganic contaminants and radionuclides. Elsevier, pp 1–10

Nazifa TH, Kristanti RA, Ike M et al (2020) Occurrence and distribution of estrogenic chemicals in river waters of Malaysia. Toxicol Environ Health Sci 12:65–74. https://doi.org/10.1007/s13530-020-00036-8

Nebot C, Falcon R, Boyd KG, Gibb SW (2015) Introduction of human pharmaceuticals from wastewater treatment plants into the aquatic environment: a rural perspective. Environ Sci Pollut Res 22:10559–10568. https://doi.org/10.1007/s11356-015-4234-z

Nicole W (2013) PFOA and cancer in a highly exposed community: new findings from the c8 science panel. Environ Health Perspect 121:A340. https://doi.org/10.1289/ehp.121-A340

Oh S, Shin WS, Kim HT (2016) Effects of pH, dissolved organic matter, and salinity on ibuprofen sorption on sediment. Environ Sci Pollut Res 23:22882–22889. https://doi.org/10.1007/s11356-016-7503-6

Patra S, Shilky, Kumar A, Saikia P (2023) Impact of land use systems and climate change on water resources: Indian perspectives. In: Rai PK (ed) Advances in water resource planning and sustainability. Springer Nature Singapore, Singapore, pp 97–110

Pepper IL, Brusseau ML, Prevatt FJ, Escobar BA (2021) Incidence of Pfas in soil following long-term application of class B biosolids. Sci Total Environ 793:148449. https://doi.org/10.1016/j.scitotenv.2021.148449

Pierozan P, Jerneren F, Karlsson O (2018) Perfluorooctanoic acid (PFOA) exposure promotes proliferation, migration and invasion potential in human breast epithelial cells. Arch Toxicol 92:1729–1739. https://doi.org/10.1007/s00204-018-2181-4

Pradhan B, Chand S, Chand S et al (2023) Emerging groundwater contaminants: a comprehensive review on their health hazards and remediation technologies. Groundw Sustain Dev 20:100868. https://doi.org/10.1016/j.gsd.2022.100868

Qi Y, Huo S, Hu S et al (2016) Identification, characterization, and human health risk assessment of perfluorinated compounds in groundwater from a suburb of Tianjin, China. Environ Earth Sci 75:432. https://doi.org/10.1007/s12665-016-5415-x

Qureshi ME, Reeson A, Reinelt P et al (2012) Factors determining the economic value of groundwater. Hydrogeol J 20:821–829. https://doi.org/10.1007/s10040-012-0867-x

Raanan R, Harley KG, Balmes JR et al (2015) Early-life exposure to organophosphate pesticides and pediatric respiratory symptoms in the CHAMACOS cohort. Environ Health Perspect 123:179–185. https://doi.org/10.1289/ehp.1408235

Rabølle M, Spliid NH (2000) Sorption and mobility of metronidazole, olaquindox, oxytetracycline and tylosin in soil. Chemosphere 40:715–722. https://doi.org/10.1016/S0045-6535(99)00442-7

Radini S, Marinelli E, Akyol Ç et al (2021) Urban water-energy-food-climate nexus in integrated wastewater and reuse systems: cyber-physical framework and innovations. Appl Energy 298:117268. https://doi.org/10.1016/j.apenergy.2021.117268

Rajput V, Minkina T, Sushkova S et al (2020) ZnO and CuO nanoparticles: a threat to soil organisms, plants, and human health. Environ Geochem Health 42:147–158. https://doi.org/10.1007/s10653-019-00317-3

Ramirez R, Martí V, Darbra RM (2023) Aquatic ecosystem risk assessment generated by accidental silver nanoparticle spills in groundwater. Toxics 11:671. https://doi.org/10.3390/toxics11080671

Re V (2019) Shedding light on the invisible: addressing the potential for groundwater contamination by plastic microfibers. Hydrogeol J 27:2719–2727. https://doi.org/10.1007/s10040-019-01998-x

Reinhart DR, Berge ND, Santra S, Bolyard SC (2010) Emerging contaminants: nanomaterial fate in landfills. Waste Manag 30:2020–2021. https://doi.org/10.1016/j.wasman.2010.08.004

Revelli A, Piane LD, Casano S et al (2009) Follicular fluid content and oocyte quality: from single biochemical markers to metabolomics. Reprod Biol Endocrinol 7:40. https://doi.org/10.1186/1477-7827-7-40

Reyes NJDG, Geronimo FKF, Yano KAV et al (2021) Pharmaceutical and personal care products in different matrices: occurrence, pathways, and treatment processes. Water 13:1159. https://doi.org/10.3390/w13091159

Richardson SD (2010) Environmental mass spectrometry: emerging contaminants and current issues. Anal Chem 82:4742–4774. https://doi.org/10.1021/ac101102d

Roostaei J, Colley S, Mulhern R et al (2021) Predicting the risk of GenX contamination in private well water using a machine-learned Bayesian network model. J Hazard Mater 411:125075. https://doi.org/10.1016/j.jhazmat.2021.125075

Saha S, Narayanan N, Singh N, Gupta S (2022) Occurrence of endocrine disrupting chemicals (EDCs) in river water, ground water and agricultural soils of India. Int J Environ Sci Technol 19:11459–11474. https://doi.org/10.1007/s13762-021-03858-2

Sammut G, Sinagra E, Sapiano M et al (2019) Perfluoroalkyl substances in the Maltese environment – (II) sediments, soils and groundwater. Sci Total Environ 682:180–189. https://doi.org/10.1016/j.scitotenv.2019.04.403

Sarker A, Shin WS, Al Masud MA et al (2023) A critical review of sustainable pesticide remediation in contaminated sites: research challenges and mechanistic insights. Environ Pollut 341:122940. https://doi.org/10.1016/j.envpol.2023.122940

Sassman SA, Lee LS (2005) Sorption of three tetracyclines by several soils: assessing the role of pH and cation exchange. Environ Sci Technol 39:7452–7459. https://doi.org/10.1021/es0480217

Sauvé S, Desrosiers M (2014) A review of what is an emerging contaminant. Chem Cent J 8:15. https://doi.org/10.1186/1752-153X-8-15

Schaefer CE, Choyke S, Ferguson PL et al (2018) Electrochemical transformations of perfluoroalkyl acid (PFAA) precursors and PFAAs in groundwater impacted with aqueous film forming foams. Environ Sci Technol 52:10689–10697. https://doi.org/10.1021/acs.est.8b02726

Schwarzenbach RP, Escher BI, Fenner K et al (2006) The challenge of micropollutants in aquatic systems. Science 313:1072–1077. https://doi.org/10.1126/science.1127291

Sein MM, Zedda M, Tuerk J et al (2008) Oxidation of diclofenac with ozone in aqueous solution. Environ Sci Technol 42:6656–6662. https://doi.org/10.1021/es8008612

Sellaoui L, Gómez-Avilés A, Dhaouadi F et al (2023) Adsorption of emerging pollutants on lignin-based activated carbon: analysis of adsorption mechanism via characterization, kinetics and equilibrium studies. Chem Eng J 452:139399. https://doi.org/10.1016/j.cej.2022.139399

Selwe KP, Thorn JPR, Desrousseaux AOS et al (2022) Emerging contaminant exposure to aquatic systems in the Southern African Development Community. Environ Toxicol Chem 41:382–395. https://doi.org/10.1002/etc.5284

Shahid MK, Kashif A, Fuwad A, Choi Y (2021) Current advances in treatment technologies for removal of emerging contaminants from water – a critical review. Coord Chem Rev 442:213993. https://doi.org/10.1016/j.ccr.2021.213993

Sharma A, Ahmad J, Flora SJS (2018) Application of advanced oxidation processes and toxicity assessment of transformation products. Environ Res 167:223–233. https://doi.org/10.1016/j.envres.2018.07.010

Sharma BM, Bharat GK, Tayal S et al (2016) Perfluoroalkyl substances (PFAS) in river and ground/drinking water of the Ganges River basin: Emissions and implications for human exposure. Environ Pollut 208:704–713. https://doi.org/10.1016/j.envpol.2015.10.050

Shen H, Gao M, Li Q et al (2023) Effect of PFOA exposure on diminished ovarian reserve and its metabolism. Reprod Biol Endocrinol 21:16. https://doi.org/10.1186/s12958-023-01056-y

Shweta N, Samatha S, Keshavkant S (2021) Mechanisms, types, effectors, and methods of bioremediation: the universal solution. In: Microbial ecology of wastewater treatment plants. Elsevier, pp 41–72

Sims JL, Stroski KM, Kim S et al (2022) Global occurrence and probabilistic environmental health hazard assessment of per- and polyfluoroalkyl substances (PFASs) in groundwater and surface waters. Sci Total Environ 816:151535. https://doi.org/10.1016/j.scitotenv.2021.151535

Smallwood AW, DeBord KE, Lowry LK (1992) N, N'Diethyl-m-Toluamide (m-DET): analysis of an insect repellent in human urine and serum by high-performance liquid chromatography. J Anal Toxicol 16:10–13. https://doi.org/10.1093/jat/16.1.10

Sonthithai P, Suriyo T, Thiantanawat A et al (2016) Perfluorinated chemicals, PFOS and PFOA, enhance the estrogenic effects of 17β-estradiol in T47D human breast cancer cells. J Appl Toxicol 36:790–801. https://doi.org/10.1002/jat.3210

Starling AP, Umbach DM, Kamel F et al (2014) Pesticide use and incident diabetes among wives of farmers in the agricultural health study. Occup Environ Med 71:629–635. https://doi.org/10.1136/oemed-2013-101659

Stock F, Reifferscheid G, Brennholt N, Kostianaia E (eds) (2022) The handbook of environmental chemistry, vol 111. Part 1: plastics in the aquatic environment Current status and challanges. Springer, Cham

Sun Z, Zhao M, Chen L et al (2023) Electrokinetic remediation for the removal of heavy metals in soil: limitations, solutions and prospection. Sci Total Environ 903:165970. https://doi.org/10.1016/j.scitotenv.2023.165970

Sutherland DL, Ralph PJ (2019) Microalgal bioremediation of emerging contaminants - opportunities and challenges. Water Res 164:114921. https://doi.org/10.1016/j.watres.2019.114921

Syafiuddin A, Boopathy R, Hadibarata T (2020) Challenges and solutions for sustainable groundwater usage: pollution control and integrated management. Curr Pollut Rep 6:310–327. https://doi.org/10.1007/s40726-020-00167-z

Szczęsna D, Wieczorek K, Jurewicz J (2022) An exposure to endocrine active persistent pollutants and endometriosis — a review of current epidemiological studies. Environ Sci Pollut Res 30:13974–13993. https://doi.org/10.1007/s11356-022-24785-w

Tang ZW, Shahul Hamid F, Yusoff I, Chan V (2023) A review of PFAS research in Asia and occurrence of PFOA and PFOS in groundwater, surface water and coastal water in Asia. Groundw Sustain Dev 22:100947. https://doi.org/10.1016/j.gsd.2023.100947

Taoufik N, Boumya W, Achak M et al (2021) Comparative overview of advanced oxidation processes and biological approaches for the removal pharmaceuticals. J Environ Manag 288:112404. https://doi.org/10.1016/j.jenvman.2021.112404

Tomiyama S, Igarashi T, Tabelin CB et al (2019) Acid mine drainage sources and hydrogeochemistry at the Yatani mine, Yamagata, Japan: a geochemical and isotopic study. J Contam Hydrol 225:103502. https://doi.org/10.1016/j.jconhyd.2019.103502

Torrellas SÁ, García Lovera R, Escalona N et al (2015) Chemical-activated carbons from peach stones for the adsorption of emerging contaminants in aqueous solutions. Chem Eng J 279:788–798. https://doi.org/10.1016/j.cej.2015.05.104

Tratnyek PG, Johnson RL, Lowry GV, Brown RA (2014) IN SITU chemical reduction for source remediation. In: Kueper BH, Stroo HF, Vogel CM, Ward CH (eds) Chlorinated solvent source zone remediation. Springer, New York, pp 307–351

US EPA O (2014) Drinking Water Contaminant Candidate List (CCL) and Regulatory Determination [WWW Document]. https://www.epa.gov/ccl. Accessed 15 Dec 2023

US EPA O (2015) Ground Water Rule [WWW Document]. https://www.epa.gov/dwreginfo/ground-water-rule. Accessed 15 Dec 2023

Umair M, Kanwal T, Loddo V et al (2023) Review on recent advances in the removal of organic drugs by advanced oxidation processes. Catalysts 13:1440. https://doi.org/10.3390/catal13111440

Van Tran T, Bui QTP, Nguyen TD et al (2017) A comparative study on the removal efficiency of metal ions (Cu 2+, Ni 2+, and Pb 2+) using sugarcane bagasse-derived ZnCl 2 -activated carbon by the response surface methodology. Adsorpt Sci Technol 35:72–85. https://doi.org/10.1177/0263617416669152

Vinayagam V, Palani KN, Ganesh S et al (2023) Recent developments on advanced oxidation processes for degradation of pollutants from wastewater with focus on antibiotics and organic dyes. Environ Res 240:117500. https://doi.org/10.1016/j.envres.2023.117500

Vocciante M, Dovì V, Ferro S (2021) Sustainability in electrokinetic remediation processes: a critical analysis. Sustainability 13:770. https://doi.org/10.3390/su13020770

Wagner ED, Hsu K-M, Lagunas A et al (2012) Comparative genotoxicity of nitrosamine drinking water disinfection byproducts in Salmonella and mammalian cells. Mutat Res Toxicol Environ Mutagen 741:109–115. https://doi.org/10.1016/j.mrgentox.2011.11.006

Wang X, Chen M, Gong P, Wang C (2019) Perfluorinated alkyl substances in snow as an atmospheric tracer for tracking the interactions between westerly winds and the Indian Monsoon over western China. Environ Int 124:294–301. https://doi.org/10.1016/j.envint.2018.12.057

Wee SY, Aris AZ (2023) Revisiting the "forever chemicals", PFOA and PFOS exposure in drinking water. NPJ Clean Water 6:57. https://doi.org/10.1038/s41545-023-00274-6

Westerhoff P, Atkinson A, Fortner J et al (2018) Low risk posed by engineered and incidental nanoparticles in drinking water. Nat Nanotechnol 13:661–669. https://doi.org/10.1038/s41565-018-0217-9

Wielsøe M, Kern P, Bonefeld-Jørgensen EC (2017) Serum levels of environmental pollutants is a risk factor for breast cancer in Inuit: a case control study. Environ Health 16:56. https://doi.org/10.1186/s12940-017-0269-6

Wong JKH, Lee KK, Tang KHD, Yap P-S (2020) Microplastics in the freshwater and terrestrial environments: prevalence, fates, impacts and sustainable solutions. Sci Total Environ 719:137512. https://doi.org/10.1016/j.scitotenv.2020.137512

Wong S, Ngadi N, Inuwa IM, Hassan O (2018) Recent advances in applications of activated carbon from biowaste for wastewater treatment: a short review. J Clean Prod 175:361–375. https://doi.org/10.1016/j.jclepro.2017.12.059

Wu S, Li X, Tian Y et al (2021) Excellent photocatalytic degradation of tetracycline over black anatase-TiO2 under visible light. Chem Eng J 406:126747. https://doi.org/10.1016/j.cej.2020.126747

Xiao F, Simcik MF, Halbach TR, Gulliver JS (2015) Perfluorooctane sulfonate (PFOS) and perfluorooctanoate (PFOA) in soils and groundwater of a U.S. metropolitan area: migration and implications for human exposure. Water Res 72:64–74. https://doi.org/10.1016/j.watres.2014.09.052

Xu B, Liu S, Zhou JL et al (2021a) PFAS and their substitutes in groundwater: occurrence, transformation and remediation. J Hazard Mater 412:125159. https://doi.org/10.1016/j.jhazmat.2021.125159

Xu Y, Yu X, Xu B et al (2021b) Sorption of pharmaceuticals and personal care products on soil and soil components: influencing factors and mechanisms. Sci Total Environ 753:141891. https://doi.org/10.1016/j.scitotenv.2020.141891

Xue S, Ke W, Zeng J et al (2023) Pollution prediction for heavy metals in soil-groundwater systems at smelting sites. Chem Eng J 473:145499. https://doi.org/10.1016/j.cej.2023.145499

Yang Y, Ok YS, Kim K-H et al (2017) Occurrences and removal of pharmaceuticals and personal care products (PPCPs) in drinking water and water/sewage treatment plants: a review. Sci Total Environ 596–597:303–320. https://doi.org/10.1016/j.scitotenv.2017.04.102

Yao F, Zhaojun L, Xiaoqing H (2016) Impacts of soil organic matter, iron-aluminium oxides and pH on adsorption-desorption behaviors of oxytetracycline. Res J Biotechnol 11:1

Yin Z, Yu J, Han X et al (2022) A novel phytoremediation technology for polluted cadmium soil: Salix integra treated with spermidine and activated carbon. Chemosphere 306:135582. https://doi.org/10.1016/j.chemosphere.2022.135582

Yu Y, Liu Y, Wu L (2013) Sorption and degradation of pharmaceuticals and personal care products (PPCPs) in soils. Environ Sci Pollut Res 20:4261–4267. https://doi.org/10.1007/s11356-012-1442-7

Zhang T, Lowry GV, Capiro NL et al (2019) *In situ* remediation of subsurface contamination: opportunities and challenges for nanotechnology and advanced materials. Environ Sci Nano 6:1283–1302. https://doi.org/10.1039/C9EN00143C

Zhao Z, Cheng X, Hua X et al (2020) Emerging and legacy per- and polyfluoroalkyl substances in water, sediment, and air of the Bohai Sea and its surrounding rivers. Environ Pollut 263:114391. https://doi.org/10.1016/j.envpol.2020.114391

Zhi D, Yang D, Zheng Y et al (2019) Current progress in the adsorption, transport and biodegradation of antibiotics in soil. J Environ Manag 251:109598. https://doi.org/10.1016/j.jenvman.2019.109598

Zhu X, Jin L, Yang J et al (2017) Perfluoroalkyl acids in the water cycle from a freshwater river basin to coastal waters in eastern China. Chemosphere 168:390–398. https://doi.org/10.1016/j.chemosphere.2016.10.088

Chapter 10
Examining the Potential Spread of Antibiotic Resistance in Groundwater Originating from Artificial Recharge

Brandon Hardiman and Itza Mendoza-Sanchez

Abstract The artificial recharge of groundwater is a topic of growing interest. Artificial recharge utilizes alternative sources of water, including treated wastewater effluent, to augment natural recharge rates and slow water table declines. Artificial recharge can be accomplished through constructed infiltration basins with high infiltration rates to the water table, or injection wells screened at the depth of the aquifer bypassing the vadose zone. One emerging contaminant commonly detected in treated wastewater is antibiotics used to treat bacterial infections in humans and agriculture. Literature has shown that antibiotics can potentially modify the dissemination of antibiotic-resistant bacteria in the environment by augmenting horizontal gene transfer frequencies. Horizontal gene transfer is the exchange of antibiotic resistance, mainly through conjugation, from resistant bacteria to susceptible bacteria. While native subsurface bacteria are often resistant to antibiotics, they pose no direct risk to humans unless pathogens acquire that resistance. The potential modifications to the subsurface microbiome via the artificial recharge of treated wastewater have rarely been explored previously. This chapter aims to give an overview of artificial recharge, wastewater treatment, and bacterial and antibiotic transport in the subsurface influenced by artificial recharge. In addition, an existing artificial recharge site in El Paso, Texas, USA provides a model case for the evaluation of potential dissemination of environmental antibiotic resistance. Significant further research is required to determine the potential risks of spreading environmental antibiotic resistance resulting from the artificial recharge of treated wastewater.

Keywords Treated wastewater · Residual antibiotics · Horizontal gene transfer · El Paso · Texas

B. Hardiman
Water Management and Hydrological Science, Texas A&M University, College Station, TX, USA

I. Mendoza-Sanchez (✉)
Zachary Department of Civil and Environmental Engineering, Texas A&M University, College Station, TX, USA
e-mail: itzamendoza@tamu.edu

© UNESCO 2025
S. Zandaryaa et al. (eds.), *Emerging Pollutants*, Advances in Water Security,
https://doi.org/10.1007/978-3-031-71758-1_10

10.1   Introduction

Environmental antibiotic resistance is a contaminant of growing concern, particularly in environments influenced by treated wastewater effluent containing antibiotic-resistance genes, resistant bacteria and antibiotics (Cacace et al. 2019). Risks resulting from the spread of antibiotic resistant bacteria include the acquisition of resistance to antibiotics by human pathogens via horizontal gene transfer and the exchange of genes conferring resistance to antibiotics which provide the last barrier of defense in the case of infection. In regions with excessive groundwater withdrawals or limited surface water supplies, artificial recharge is one method to increase natural recharge in aquifer systems to help mitigate water table declines, or improve natural groundwater quality (Barnett et al. 2000; Bouwer 2002). Groundwater is a dominant source of drinking water globally and is susceptible to pathogenic contamination from treated wastewater, urban stormwater runoff and agriculture (Li et al. 2015; Regnery et al. 2017b; Sidhu et al. 2020). The purpose of this chapter is to investigate the transport of antibiotics in aquifers subject to artificial recharge and subsequent changes to the subsurface microbiome in developing antibiotic resistance.

Figure 10.1 illustrates a conceptual model of an artificial recharge site using treated wastewater effluent as input into constructed infiltration basins. Wastewater is first treated at a wastewater treatment plant adjacent to the artificial recharge site. Treated wastewater effluent is assumed to be enriched in antibiotic resistance genes, antibiotic resistant bacteria and antibiotic residuals compared to background levels present in the undisturbed groundwater environments as indicated in prior research (Cacace et al. 2019; Andrade et al. 2020). In this conceptual model, we are assuming that the aquifer system is unconfined, and that regional groundwater flow is from left to right. The infiltration basin is constructed with recharge rates exceeding natural levels, with lower hydraulic conductivity and slower regional groundwater flows outside of the basin. In addition to infiltration basins, artificial recharge wells are also commonly used. However, due to costs associated with well maintenance and ease of operation, newer artificial recharge sites typically employ infiltration basins rather than recharge wells where space is not constrained or there is a short distance between the land surface and water table (Pirnie et al. 2011).

Treated wastewater effluent is pumped into an infiltration basin characterized by high hydraulic conductivity to provide rapid vertical recharge into the aquifer. It's assumed that this infiltration basin is kept at a constant level, however, this may not reflect actual recharge conditions which are dependent upon daily fluctuations in wastewater effluent production. In some artificial recharge sites, wastewater effluent is first used for irrigation with any excess then pumped into artificial recharge basins (Pirnie et al. 2011). Important transport processes in the subsurface include the advection, dispersion, sorption and biodegradation of antibiotics in the subsurface, and the attachment and detachment of bacteria.

Once reaching the aquifer, the transport of bacteria is controlled by the direction of regional groundwater flow, with a concentration gradient away from the infiltration basin. However, there are some limitations to the likelihood of bacterial

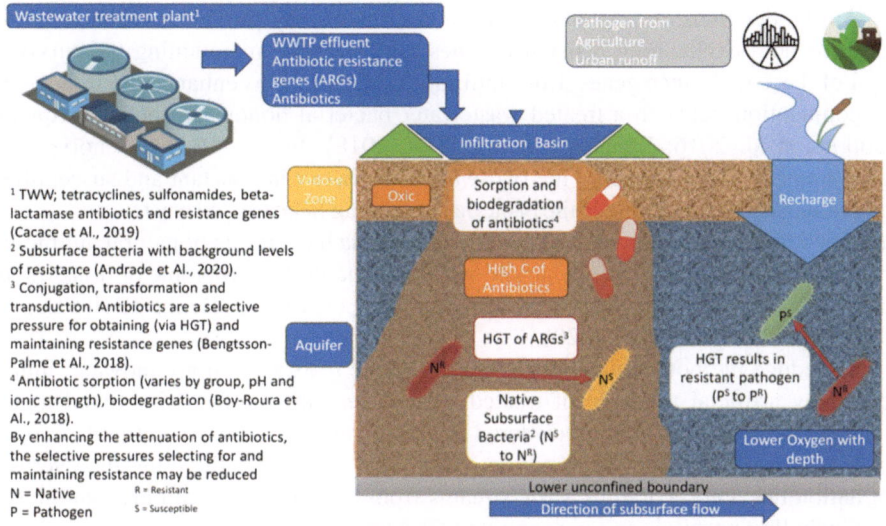

Fig. 10.1 Conceptual model of the artificial recharge of treated wastewater spreading antibiotic resistance in a homogenous, unconfined aquifer. *Not pictured is a recovery well along the flow path from the infiltration basin. *Source* Authors

transport originating from treated wastewater effluent. Wastewater treatment, particularly treatment with UV, has high removal efficiencies for bacteria that significantly reduces the number of viable bacteria able to infiltrate into the subsurface (Lira et al. 2020). Also, bacteria originating from treated wastewater are assumed to have a short lifespan in the subsurface. Infiltration through the vadose zone may further reduce the number of bacteria that can infiltrate into the saturated zone (Hunt and Johnson 2017; Regnery et al. 2017a). Bacteria originating from treated wastewater effluent are not likely to be transported into the subsurface. This chapter will therefore focus on the attenuation of antibiotics in the subsurface and their impact on existing native subsurface bacteria.

Possible sources of antibiotic resistant bacteria in this conceptual model include treated wastewater effluent, natural subsurface bacteria and the exchange of antibiotic-resistance genes between resistant and susceptible bacteria. Of primary concern for the spread of antibiotic resistance in the environment is horizontal gene transfer (HGT), which is the exchange of genetic material through conjugation, transformation, and transduction. Conjugation is the primary method of resistance acquisition by bacteria, where a donor transfers resistance to a recipient unidirectionally via a pilus. Transformation is the direct incorporation of exogenous DNA mediated by proteins. Transduction involves bacteriophages injecting DNA segments into a host to be recombined with chromosomal DNA (Bello-López et al. 2019). We hypothesize that the concentrations of antibiotics present in treated wastewater effluent used in artificial recharge may provide a selective pressure to increase horizontal gene transfer frequencies.

Antibiotics present in treated wastewater effluent may provide a selective pressure for both exchanging resistance genes via HGT and maintaining the survival cost of these resistance genes. Low antibiotic concentrations enhance the frequency of conjugation between a treated wastewater bacterial donor and model recipient (Jutkina et al. 2016; Bengtsson-Palme et al. 2018). In Fig. 10.1, this horizontal gene transfer is illustrated by the two-way arrow between resistant and susceptible native subsurface bacteria. While bacteria originating from treated wastewater are not assumed to be important, native subsurface bacteria have levels of natural antibiotic resistance that may also be exchanged within the natural microbiome or transferred to human pathogens originating from agricultural or urban pollution (Bello-López et al. 2019; Andrade et al. 2020).

To reduce the potential for enhanced antibiotic resistance in groundwater influenced by artificial recharge of treated wastewater, utilization of improved treatment technologies including powdered activated carbon to remove antibiotics from wastewater effluent may be important (Kårelid et al. 2017). Targeting the overuse of antibiotics and separating waste streams from environments with high antibiotic use including hospitals and concentrated feed lots, may be required to prevent high antibiotic concentrations in the wastewater influent (Hocquet et al. 2016). However, further research is needed to determine the risks associated with artificial recharge and the dissemination of antibiotic resistance in the environment.

10.2   The Antibiotic Resistance Problem in Treated Wastewater Effluent

Treated wastewater effluent is a well-established vector of antibiotics in aquatic environments. In addition to antibiotics and trace organic compounds, antibiotic resistance genes and bacteria are frequently detected in treated wastewater effluents (Pallares-Vega et al. 2019; Smyth et al. 2020; Raza et al. 2021). The majority of antibiotics collected in sewage are secreted as unaltered compounds and retain their biological activity after wastewater treatment as conventional treatment processes are not capable of fully removing antibiotics and other pharmaceuticals (Berendonk et al. 2015; Krzeminski et al. 2019; Gomes et al. 2020). Treatment facilities receiving wastewater from hospitals and or animal facilities are widely considered to be at especially high risk for releasing antibiotic compounds and resistance into aquatic environments (Hocquet et al. 2016; Griffin et al. 2019).

The residual level of antibiotics present in wastewater effluent is highly dependent upon the type of wastewater treatment process used. Conventional activated sludge treatment yields the lowest removal efficiency for many antibiotics (Krzeminski et al. 2019). Conventional activated sludge processes remove 80% of fluoroquinolones and tetracyclines prior to discharge. However, macrolides are more persistent and therefore are discharged into aquatic environments (Baquero et al. 2008). In membrane filtration, biodegradation and sorption are the main mechanisms for removal of contaminants, with removal efficacy determined by the hydrophobicity of the chemical. Membrane bioreactors that rely on biofilms can enhance the attenuation of specific contaminants by altering biofilm thickness, diversity and microbial activity. The last steps of wastewater treatment include a disinfection step that employs ozone, UV irradiation or chlorine. Chloride dioxide (ClO_2) is an alternative to chlorine disinfection with greater deactivation strength with lower generation of halogenated disinfection-by-products (Zhong et al. 2019). ClO_2 increased removal of beta-lactam antibiotics, while chlorination reduced the activity of trimethoprim antibiotics. Seasonality may also influence levels of antibiotics in wastewater. In the northern hemisphere, higher prescription rates of antibiotics tends to coincide with higher antibiotics detected in wastewater effluent in the cold fall and winter months. (Caucci et al. 2016; Wang et al. 2021).

Research has established the occurrence and concentration of a variety of antibiotics in conventionally treated wastewater effluent. In an analysis of secondary effluents in four different wastewater treatment plants (WWTPs) in China, fluoroquinolones were detected at the highest concentrations, followed by macrolides, tetracyclines and sulfonamides. The total levels of antibiotics in three of the four treatment facilities had higher antibiotic concentrations in the winter than the spring, further supporting the seasonality of antibiotics observed in wastewater effluent (Wang et al. 2021). Effluents from three conventional WWTPs in New York revealed sulfamethoxazole, trimethoprim, ciprofloxacin, tetracycline and clindamycin ranged in concentrations 0.09–6 ug/L (Batt et al. 2006). Azithromycin and erythromycin were found to be practically unaltered during biological wastewater treatment with low removal efficiencies. In Southern California, sulfamethoxazole was detected in 16 samples of wastewater effluent from four treatment plants (Krzeminski et al. 2019).

Antibiotic resistance genes (ARGs), carried by bacterial communities, encode mechanisms of resistance to various classes of antibiotics and are widely detected in conventionally treated wastewater effluent. While ARGs are often targeted for monitoring antibiotic resistance, the type of bacteria carrying resistance genes are not measured, leading to risk assessment limitations. Culture based methods often yield the best indicators for conducting risk assessment, but they are strongly limited since as little as 1% of total bacterial diversity in the environment can be cultured (Wright 2010). Metagenomic approaches overcome some of these limitations, allowing for evaluation of abundance of resistance genes, mobile genetic elements and human pathogens for risk assessment (Nguyen et al. 2021).

While wastewater treatment is known to reduce the level of pathogenic bacteria in untreated sewage, the overall impact of treatment on non-pathogenic bacterial communities and antibiotic resistance genes is generally less well understood (Lira et al. 2020). In an investigation on the removal of antibiotic resistant bacteria during activated sludge treatment, it was observed that ampicillin and chloramphenicol resistant *E. coli* were reduced by two log units, but the proportion of resistant and susceptible bacteria in the bacterial population remained unchanged after treatment (Krzeminski et al. 2019). In a study of treated wastewater effluent in Dresden, Germany, ARGs measured in wastewater had clear seasonal trends co-occurring with antibiotic prescriptions, with higher levels of ARGs in the Fall and Winter. In treated wastewater samples, eight out of eleven resistance genes were detected at all sampling locations within the plant, conferring resistance to beta lactams, tetracycline, sulfonamides and vancomycin. The genes that encode resistance to sulfonamides (*sul1*, *sul2*) and tetracycline (*tetM*) were most abundant in treated wastewater with 10^7–10^8 gene copies per mL. The reduction in gene copies from inlet to effluent of resistance genes were directly proportional to bacterial removal rates in conventional wastewater treatment plants (Caucci et al. 2016).

Tertiary treatment of wastewater is used in combination with conventional treatment to further reduce pollutants in effluent waters. Tertiary treatment of wastewater is not common, particularly for non-potable or irrigated wastewater reuse in developing countries (Jones et al. 2021). Even in developed countries such as Germany, treatment facilities discharging to surface waters rarely include advanced treatment processes (Karakurt et al. 2019). Beta lactam, macrolide, quinolone, sulfonamide and tetracycline antibiotics have been detected in secondary and tertiary treated wastewater effluents (Nguyen et al. 2021). Powdered activated carbon (PAC) treatment is commonly used for drinking water reuse of treated wastewater, particularly within the context of artificial recharge in developed countries. These systems are often designed to treat wastewater effluent to drinking water standards prior to reuse in artificial recharge (Sheng 2005). PAC tertiary treatment has significant removal efficiencies for culturable antibiotic resistant bacteria present in the effluent (Ravasi et al. 2019). PAC treatment has also shown to be effective in attenuating various antibiotic classes, although pH, charge and solubility of the antibiotic have been identified as important characteristics contributing to point-source antibiotic removal (Kårelid et al. 2017; Phoon et al. 2020; Berges et al. 2021).

Conventional WWTPs, when combined with UV disinfection can increase the removal of antibiotic resistant bacteria but are not likely to significantly reduce the antibiotics present in effluent. Secondary wastewater treatment processes are also effective in reducing the number of antibiotic resistant fecal coliforms in Chicago (Rijal et al. 2009). Carbapenem resistant bacteria and beta lactamase resistant *Escherichia coli* with the potential for mobilization were collected from UV-C treated wastewater effluent from urban wastewater treatment plants in Portugal (Tavares et al. 2020; Araújo et al. 2021). In contrast, metagenomic analysis of UV disinfected WWTP effluent and influent in Portugal revealed that treatment processes effectively removed hosts harboring mobile resistance genes, with high removal efficiencies of *Enterobacteriacae, Firmicutes and Proteobacteria*. Reduction of bacteria from human origin are likely due to enrichment of environmental microorganisms post treatment. (Lira et al. 2020).

10.3 Background Levels of Antibiotic Resistance in Groundwater Bacteria

A literature review of research articles on the occurrence of antibiotic resistance bacteria in groundwater revealed that the majority of bacterial isolates obtained from groundwater were resistant to multiple antibiotics. The occurrence of antibiotic resistance in groundwater was positively correlated with dry periods. Groundwater temperature is another important factor, bacterial proliferation is associated with marginal increases in temperature. Long residence time in groundwater, in combination with exposure to sub-inhibitory concentration of antibiotics may contribute to the detection of resistant bacteria in groundwater environments (Andrade et al. 2020). The exposure to sub-inhibitory levels of antibiotics is considered important for the prevalence of antibiotic resistant bacteria in groundwater. In India, 38% of antibiotics detected in groundwater and surface water were at high enough concentrations to select for antibiotic resistant bacteria (Stanton et al. 2022). Since treated wastewater contains antibiotics at sub-inhibitory concentrations, the artificial recharge of treated wastewater may provide a direct pathway for increasing resistance in native groundwater bacteria.

10.4 Transport of Bacteria from Treated Wastewater Effluent via Managed Aquifer Recharge

While some novel treatment techniques exist that can enhance the removal of organic compounds in managed aquifer recharge sites, these techniques are limited to site-specific approaches (Regnery et al. 2016). The transport of bacteria in the subsurface is determined using advection, dispersion and retention described in colloid filtration theory. Bacteria is known to survive ten to a hundred times longer when attached to particles compared to being suspended in a fluid. In addition, the large size of bacteria compared to other pathogens like viruses means that bacteria are more likely to be removed by filtration shortly after infiltration (Regnery et al. 2017b). The relatively short survival times of bacteria in the subsurface also support these assertions, with the transport of bacteria in the subsurface requiring rapid, preferential flow paths. These assumptions are likely to be true for the majority of groundwater systems with low hydraulic conductivity and relatively slow regional flows outside of an infiltration basin.

Processes affecting microbial attenuation in the subsurface include physical straining and attachment to porous media. Straining is dependent on grain size, bacterial size and shape, saturated water content and clogging, although the presence of macropores can reduce filtering abilities. There are two steps to bacterial adsorption, the first of which is reversable and includes weak interactions between bacteria and porous media through van der Waals forces. The second adsorption step is irreversible adhesion, dependent upon contact time between bacteria and grain media. Bacterial adsorption is dependent upon the presence of organic matter, biofilms, temperature, flow velocity, ionic strength, pH, hydrophobicity, chemotaxis and electrostatic charge (Kristian Stevik et al. 2004). More hydrophobic bacterial strains have been found to increase retention with lower water content in sandy soils (Gargiulo et al. 2008). Similar processes are assumed to occur through the vadose zone with the use of infiltration basins in artificial recharge.

There is evidence for the potential transport of bacteria originating from artificial recharge in relatively rapid flow systems. In a European study, the presence of six antibiotic resistance genes, three bacteria strains known to be human pathogens and fecal indicators were assessed at three different artificial recharge systems. The water recovered from these systems had a wide range of uses from irrigation, street cleaning and consumptive use. Total coliform counts in one specific site (Nardo, Italy) decreased by less than 1 log from input to reclaimed water. The resistance genes *tetO* and *ermB* were detected in all three reclamation sites. The authors conclude that a highly fractured and permeable aquifer only posed partial barriers for bacterial contaminants (Böckelmann et al. 2009). Thus, for artificial recharge systems with rapid, preferential flow pathways, bacterial transport must be accounted for. In an Australian artificial recharge and recovery (ASR) site, recharge with aerobic wastewater effluent resulted in increased bacterial numbers, but reduced microbial diversity in receiving anaerobic groundwater (Ginige et al. 2013). Rather than bacterial transport from the recharged water, the authors hypothesize that redox changes and increasing nitrate, oxygen and organic matter are driving changes in the groundwater microbial community. The importance of bacterial transport in the subsurface is likely to be highly site dependent, in addition to the chemical and microbial composition of the recharge and receiving water.

10.5 Sorption and Biodegradability of Antibiotics in Artificial Recharge

10.5.1 Emerging Contaminants in the Vadose Zone

There are two environmental compartments controlling antibiotic dissemination in the subsurface, namely the vadose zone and the aquifer. In the vadose zone, key transport processes include infiltration, sorption, and biodegradation of antibiotics. Processes affecting antibiotic transport and attenuation in the vadose zone could be key to reducing the risk of contamination in the water table. Many processes are site specific, with transport and removal dependent on the charge and solubility of the antibiotic compound. The soil organic matter, porosity and other properties impact the mobility of antibiotics in the vadose zone (Zhang et al. 2014).

Although infiltration occurring due to irrigation with wastewater have different flow rates compared to infiltration from a recharge basin, investigations on transport of antibiotics during irrigation provide information on the vadose zone compartment. In a paper examining contaminant transport from soil irrigation with treated wastewater in Colorado Springs, soils have been found to accumulate pharmaceuticals in the vadose zone. Acetaminophen, fluoxetine, caffeine, erythromycin and carbamazepine accumulated and persisted in the upper 30 cm of soil at all the irrigation sites. The behavior of other pharmaceutical compounds depended on the organic carbon content of the soil that contributes to retention in the upper soil layers. Erythromycin was detected the most frequently, with the lowest water solubility of all the screened compounds. Water solubility and organic carbon content may affect accumulation of pharmaceuticals in the soil. Carbamazepine and sulfamethoxazole have been found to leach through soil into groundwater when present in irrigated water (Kinney et al. 2006). Most importantly, however, is that sorbed organic chemicals can remain in the solid phase with minimal biodegradation for long periods of time. A study examining agricultural soil and treated wastewater effluent irrigation in China investigated the sorption capacity of four antibiotics in the vadose zone. Column studies suggest that weakly sorbed sulfonamides had a higher recovery rate in the pore water, where trimethoprim was highly sorbed onto agricultural soil and had subsequent lower recovery rates. The high mobility of sulfonamides suggests the possibility for leaching into groundwater that is less likely to occur with trimethoprim. In a review of processes resulting from wastewater reuse in agricultural irrigation on the fate of antibiotics, polarity and ionic state were found to be important to the retention of antibiotic compounds. Polar, ionizable antibiotics tend to remain in the soluble phase. In contrast, non-polar or moderately polar neutral antibiotics are sorbed onto soil particles and desorbed instantaneously until equilibrium with the solute is reached (Christou et al. 2017).

Recharge aquifer treatment techniques use the vadose zone to reduce the concentration of emerging contaminants reaching the aquifer, including antibiotics present in pore water. One study used soil columns to examine the attenuation of emerging contaminants originating from wastewater effluent through the vadose zone. Hydrologic conditions, including repeated drying times to reintroduce oxygen in the vadose zone enhanced antibiotic removal of sulfamethoxazole. Sorption and biodegradation are key processes affecting the attenuation of antibiotics. However, the authors conclude that the low sorption capacity of many emerging contaminants means that biodegradation is the primary process contributing to contaminant attenuation in the vadose zone. A 90% removal rate after 1.3–7.1 days of residence time in soil was observed for sulfamethoxazole. The removal rate of sulfamethoxazole was moderate but positively correlated with greater oxic conditions from longer drying times in the soil column. Sulfamethoxazole removal increased from 56 to 95% with an increase in drying time from 100 to 444 min (Sallwey et al. 2020). These results suggest the importance of the vadose zone for biodegrading emerging contaminants originating from treated wastewater. In addition, the low retention times in the vadose zone in infiltration basins may enhance the transport of antibiotics into groundwater. Enhancing drying times in the vadose zone by alternating between drying periods and infiltration periods in recharge basins may be an important process to enhance the biodegradation of antibiotics.

The contributions of antibiotic concentrations sorbed in soil particles are especially significant as we anticipate that these compounds are retained for long periods of time and may accumulate in the vadose zone with continuous aquifer recharge with treated wastewater. In a study examining tetracycline partitioning between the aqueous and solid phase in agricultural soil, chlortetracycline was found to have the lowest concentration in solution and may continually be released from soil particles. The supply of tetracyclines from the solid to the liquid phase in the vadose zone were limited by desorption rates ranging from 1.26 to 121 $\times$ 10^{-6} (s^{-1}). Soil texture was also important, with high clay and silt soils having the greatest potential for desorption of tetracyclines (Ren et al. 2020). Thus, the prevailing soil texture in an artificial recharge basin may provide conditions for continual release of antibiotic compounds into the pore water.

The presence of preferential flow paths or high hydraulic conductivity in the infiltration basin may reduce the biodegradability of antibiotic compounds. While photodegradation has been identified as an important factor for emerging contaminant removal in constructed wetlands, the high infiltration rates in infiltration basins means that photodegradation prior to infiltration is likely minimal (Krzeminski et al. 2019). In artificial recharge, the combination of low retention times in the vadose zone, and the continual recharge of treated wastewater may further reduce dissolved oxygen in the vadose zone and thus the biodegradability of antibiotic compounds. The polarity and ionic state of antibiotics present in treated wastewater will also affect the attenuation in the vadose zone, with polar and ionizable compounds like sulfonamides having the greatest potential to infiltrate into groundwater. The transport of antibiotics into the water table must also be investigated for the dissemination of antibiotics into groundwater influenced by artificial recharge. While contamination in the vadose zone is likely to be limited to the infiltration basin, infiltration to the water table may have broader implications for regional groundwater supplies down gradient of the recharge basin. The capacity for antibiotics to sorb into sediments will decrease the potential for further infiltration into the water table.

10.5.2 Emerging Contaminants in the Aquifer

The transport of antibiotics is affected by the class of antibiotics, with sulfonamides typically sorbing less and being more mobile than tetracyclines in groundwater (Boy-Roura et al. 2018). Similar to vadose zone processes, the transport of antibiotics is controlled by sorption processes and biodegradation in groundwater. Other important factors include metabolic capacity of the subsurface microbial community, temperature and flow path times between infiltration and reclaimed water (Regnery et al. 2017a). While organic carbon can also provide sorption sites, clay is often the dominant sorbent especially in aquifers with low organic carbon relative to clay content (Regnery et al. 2016). For sulfonamides, coexistence of Fe, Mn and NH4$^+$ ions promote adsorption, while an increase in pH reduces adsorption. Sulfamethoxazole has the greatest potential for mobility into groundwater from the vadose zone, followed by sulfamethazine and sulfamethoxypyridazine due to adsorption capacities (Zuo et al. 2021).

One study investigated the artificial recharge of treated wastewater as a barrier for seawater infiltration in Spain using injection wells. In their study 81 pharmaceuticals were monitored at a well one kilometer downgradient from the injection wells. After one year of injection, a change in redox conditions was observed, in addition to the detection of 11 pharmaceuticals in both treated wastewater and recovered groundwater. The biodegradation of pharmaceuticals was not found to be significant, though dilution of these chemicals was identified as being the primary driver for attenuation. For pharmaceuticals detected in both treated wastewater and recovered water, significant reductions in concentrations were observed. The underlying confined alluvial aquifer is composed of fine sand and gravel with a saturated thickness of 6 m. A volume of 1.6 million cubic meters of treated wastewater was injected to provide a barrier for seawater intrusion. Due to the low clay content, sorption processes were not considered significant to controlling the fate and transport of pharmaceuticals in this region. Treated wastewater was characterized by the presence of oxidized species, supporting the conclusion that treated wastewater effluent is oxic. Changes in aquifer redox conditions from anoxic to oxic were observed after one year of treated wastewater recharge. Thus, we can assume that after the system reaches a steady state, the groundwater geochemistry which is influenced by wastewater effluent from infiltration basins will become primarily oxic (Candela et al. 2016). While the authors identify that sulfamethoxazole is highly resistant to biodegradation in the subsurface, the introduction of oxic conditions through the recharge of treated wastewater means that biodegradation is likely an important factor controlling the fate of these compounds. These results support the assumption that the transport of pharmaceuticals in groundwater can occur if these compounds are not retained in the vadose zone. Desorption of antibiotics in the vadose zone may also release contaminants into the porewater and infiltrate into the water table, with tetracyclines desorbing at high rates in agricultural soils (Ren et al. 2020).

In Colorado, trace organic contaminants were effectively attenuated in a managed aquifer recharge site employing riverbank filtration with subsequent re-aeration, followed by a second artificial recharge and recovery. This sequential artificial recharge process allows for attenuation of a variety of compounds. Sulfamethoxazole was substantially better attenuated under low organic carbon and oxic conditions compared to high organic carbon and anoxic conditions. The charge of the antibiotic is also an important factor, with positively charged trimethoprim being strongly sorbed to negatively charged clay particles. Generally, aquifers with higher clay content increase the overall surface area of particles for antibiotics and other organic contaminants to sorb and be removed from the aqueous phase. At the Colorado artificial recharge site, the carbon rich anoxic subsurface conditions followed by carbon depleted and oxic conditions resulted in higher attenuation of organic contaminants. Redox conditions, combined with subsurface geochemistry, soil microbial composition, temperature, and travel time between the infiltration basin (or injection well) and recovery wells were also identified as potentially significant factors contributing to removal of pharmaceuticals. Sequential redox zonation can increase the attenuation of antibiotics in the saturated zone at artificial recharge and recovery sites (Regnery et al. 2016).

10.6   Horizontal Gene Transfer

One aspect contributing to the problem of antibiotic resistance is horizontal gene transfer. In soils irrigated with treated wastewater, horizontal gene transfer between effluent associated bacteria and soil bacteria was not likely to occur at high frequencies, either due to low transfer frequencies or the environment not supporting the fitness cost of acquired resistance (Marano et al. 2019). The link between horizontal gene transfer frequencies and antibiotic concentrations in groundwater has rarely been described for artificial recharge systems. Conjugation is the main horizontal gene transfer mechanism of bacteria acquiring resistance to antibiotics (Massoudieh et al. 2010). Conjugation requires cell to cell contact to transfer mobilizable plasmids or transposons that contain resistance genes among bacteria (Virolle et al. 2020). Selective pressure of antibiotics present in the environment enhance the maintenance of resistance genes (Zainab et al. 2020; Tang et al. 2022). Due to the necessity of direct contact between cells, natural biofilms in the subsurface may provide a hotspot for horizontal gene transfer due to greater bacterial contact in biofilms required for conjugation compared to the aqueous phase (Abe et al. 2020; Michaelis and Grohmann 2023). Environmental biofilms in the subsurface can form on the surface of soil particles and uptake emerging contaminants present in the aqueous phase (Mendoza-Sanchez and Cunningham 2007). This section will focus on the implications of antibiotic concentrations increasing horizontal gene transfer frequencies in native subsurface bacteria, and the potential for retaining these resistance genes through selective pressures that antibiotics place on environmental bacteria.

Most studies on horizontal gene transfer have been conducted in vitro between two strains of resistant and susceptible bacteria sharing the same genus that are not representative of a bacterial community present in groundwater. One study investigated the conjugative frequency between a complex treated wastewater donor and a model receptor. Antibiotic concentrations many times below the minimum inhibitory concentration were found to promote horizontal gene transfer frequencies (Jutkina et al. 2016). Since native groundwater bacteria have background levels of antibiotic resistance, low concentrations of antibiotics due to the dilution of treated wastewater effluent with groundwater may increase horizontal gene transfer between native groundwater bacteria, furthering the problem of environmental antibiotic resistance.

In contrast, native groundwater bacteria are rarely pathogenic and are unlikely to provide a direct risk to human health even if antibiotic resistance is acquired. The concentrations of antibiotics present in groundwater typically do not pose a significant risk to human health via drinking water from groundwater wells contaminated with antibiotics (Zainab et al. 2020). However, the potential for resistant subsurface bacteria to transfer resistance via horizontal gene transfer to a human pathogen originating from groundwater contamination needs to be investigated.

There is growing evidence that clinical aminoglycoside, vancomycin, beta lacta-mase and quinolone resistance originated from the environmental resistance gene reservoir (Perry and Wright 2013). Horizontal gene transfer from nonpathogenic to pathogenic bacteria has been observed, and is hypothesized as one method through which human pathogens acquire resistance outside of clinical settings (Michaelis and Grohmann 2023). The emergence of multi drug resistant *Aeromonas* strains have been identified in water environments with routes of exposure through inges-tion or bathing with contaminated water, with *Aeromonas* able to receive resistance from phylogenetically distinct bacteria (Bello-López et al. 2019). This highlights the possibility of pathogenic bacteria acquiring resistance from a compatible plasmid genetic element encoding for antibiotic resistance in native bacteria.

Despite the associated survival cost of antibiotic resistance genes in bacteria, once detected in the environment, their presence is often retained for long periods of time at a community level (Andersson and Hughes 2010). High conjugative efficiencies can cause antibiotic resistance genes to persist, despite the lack of selective pres-sure from antibiotics in the environment (Lopatkin et al. 2016). Resistance genes may prolong the survival of bacteria in adverse environmental conditions, especially in biofilms where resistance is maintained through a high frequency of conjuga-tive events (Michaelis and Grohmann 2023). Remediation often focuses on point sources which contribute to environmental antibiotic resistance, including wastew-ater treatment strategies enhancing the attenuation of antibiotics at wastewater treat-ment facilities. Managed aquifer recharge may be key to reduce the risks associated with horizontal gene transfer of antibiotic resistance genes (Regnery et al. 2017b).

10.7 Case Study—El Paso ASR Site

The objective of this case study is to identify a site that has a good description of critical hydrogeological parameters, uses treated wastewater effluent for recharge water and is currently in operation. In a survey of managed aquifer recharge sites globally, 291 artificial recharge sites were identified in the United States as of 2018. This survey includes sites in operation, in the late planning stages prior to initia-tion or closed after some operation period. Out of the 291 sites, 156 used injec-tion wells and 117 used infiltration basins. Among those sites that used reclaimed wastewater as influent for domestic water production, five out of eight used well injection (Stefan and Ansems 2018). While there is a slight majority of injection wells being used for artificial recharge of treated wastewater in the United States, most prior research on emerging contaminant attenuation in ASR settings is in infil-tration basins (Candela et al. 2016; Regnery et al. 2017b). Infiltration basins have numerous benefits over injection wells including lower pumping and maintenance costs. Using the vadose zone may provide additional attenuation of microbes and chemical contaminants, although there may be a greater risk of cross-contamination with pathogens and viruses in an open infiltration basin compared to an injection well (Sheng 2005). Injection wells for aquifer storage and recovery (ASR) are primarily in space limited areas, with further applications in preventing seawater intrusion and improving groundwater quality and storage (Ginige et al. 2013).

10.7.1 Site Description and Hydrological Conditions

The El Paso Water Utility in Texas, USA currently operates both well injection and infiltration basins and is a well-studied example of the use of treated wastewater for aquifer storage and recovery in the United States. As of 2005, 75 million cubic meters of wastewater influent has been recharged into the Hueco Bolson Aquifer since 1985. The purpose of the El Paso ASR site is to preserve falling groundwater levels in the Hueco Bolson aquifer and to improve groundwater quality. Resulting from decades of groundwater overuse in the aquifer, the water table has declined by up to 60 m in some areas. The El Paso AR site is unique in that it is one of the earliest examples of artificial recharge in the United States (Sheng 2005). In the case of the El Paso Water System, infiltration basins with recharge rates of nine feet per day have replaced injection wells that collapsed after 12 years of operation (Pirnie et al. 2011). Ease of maintenance was cited as the reason for the switch from injection wells to infiltration basins, rather than the potential changes in chemical and microbial transport in groundwater.

In the El Paso site, the source water for artificial recharge originates from the Fred Hervey Water Reclamation Plant with a capacity to treat 38,000 cubic meters of wastewater per day from Northeast El Paso. The treatment facility has two processes, with the primary process removing particulates through screening and primary settling in sedimentation tanks. The secondary process treats effluent to drinking water standards using powdered activated carbon treatment, lime treatment, filtration, ozone disinfection and activated carbon filtration to enhance the attenuation of organic contaminants. After secondary treatment of wastewater, the reclaimed water is recharged into the Hueco Bolson aquifer through either injection wells or infiltration basins. The recharged water is recovered in nearby wells and redistributed to the potable water distribution network for municipal and industrial use after two years (Sheng 2005). This potable recovered water is treated to existing drinking water standards and thus, does not quantify the concentrations of pharmaceuticals or antibiotics in the recovered water.

The US Geological Survey has produced reports on the hydrogeologic parameters of the Hueco Bolson aquifer. One study used end member mixing analysis of oxygen and chloride isotopes and tracers to determine the influence of injection on the Hueco Bolson aquifer. End members contributing to groundwater chemistry include injected water, irrigation-affected water, saline groundwater, and freshwater. Injected water was traced as far as 880 m from an injection well. The aquifer consists of alluvial deposits of gravel, sand, clay, and silt ranging in thickness from 30 to 2700 m, with the freshwater zone ranging from 60 to 200 m. The continuous sand and gravel layers are the primary layers of water extraction in the Hueco Bolson aquifer. Hydraulic conductivity ranges from 6.4 to 27 m/day and specific yield ranges from 0.15 to

0.22 (−) respectively. Groundwater velocity in the Hueco Bolson aquifer is typically lower than 0.3 m per day, indicating medium to rapid regional groundwater flow. The highest groundwater velocities, when operated with injection wells, in the ASR site were 0.4 m per day. Recharge sources aside from artificial recharge include natural recharge and irrigation affected water from adjacent corn farms. The transport of trace organic compounds was not considered, however, brominated trihalomethane byproducts of wastewater treatment were not removed during the transport of injected water in the El Paso ASR site (Buszka et al. 1994).

10.7.2 Hypothesis for Contaminant Transport in El Paso ASR Site

Of key importance is the quantification of antibiotics and antibiotic resistant bacteria in wastewater effluent utilized for artificial recharge. While no research has been conducted to determine concentrations of antibiotics in effluent from the Fred Hervey WWTP, wastewater obtained from similar secondary treatment facilities employing activated carbon treatment have detected antibiotics at low concentrations. Despite pharmaceutical removal efficiencies of up to 99.7% at treatment facilities using powdered activated carbon, antibiotics have been found at detectable concentrations, which highlights the necessity for greater monitoring of pharmaceuticals in advanced wastewater treatment facilities (Ravasi et al. 2019; Nguyen et al. 2021). Wastewater treatment to drinking water standards, as is the case in the Fred Hervey WWTP, does not imply effective removal of emerging contaminants. Existing drinking water standards do not address antibiotics and resistance, due to challenges in quantification and risk assessment (Sanganyado and Gwenzi 2019). The National Interim Primary Drinking Water Regulation enforced by the EPA includes regulation on microorganisms, disinfectants, disinfection byproducts, radionuclides, inorganic and organic chemicals but excludes antibiotics and indicators of antibiotic resistance. Europe also lacks regulations on antibiotics and resistance genes in drinking water (Carvalho and Santos 2016). Antibiotics and resistance genes have been detected in treated drinking water, which provides evidence that existing drinking water quality regulations do not adequately address the growing environmental antibiotic resistance problem (Wu et al. 2021).

If the concentrations of antibiotics in wastewater effluents from the Fred Hervey WWTP are found to be high, there is a greater risk of accumulation in the vadose zone and infiltration to the water table and subsequent modifications to the subsurface microbiome. Since antibiotics affect conjugative frequencies even at low concentrations, measurement of low-level antibiotic concentrations in effluent may still contribute to developing antibiotic resistance in the environment. Prior research shows that higher levels of antibiotic concentrations may inhibit conjugation frequencies (Jutkina et al. 2016). However, selective pressures for maintaining resistance genes would increase with higher antibiotic concentrations, particularly if conjugative frequencies are low.

The removal of antibiotic resistant bacteria is also high in treatment facilities that use chlorination, as in the case of the Fred Hervey WWTP. Thus, in the context of the conceptual model, bacterial inputs from wastewater effluent are assumed to be low, which negates the importance of bacterial transport originating from the infiltration basin. In contrast, low removal efficiencies of viable bacteria may introduce the problem of bacterial transport, particularly within the infiltration basin where recharge rates are fast compared to elsewhere in the aquifer. However, this is unlikely to be the case in an advanced WWTP. Recharge using conventional WWTP or untreated wastewater for instance, constitutes a greater risk of bacterial contamination within the infiltration basin.

Aside from wastewater treatment processes contributing to the attenuation of trace organic chemicals, the predominant redox states in the subsurface are known to affect the removal of organic chemicals. In the El Paso ASR wells, aerobic bacteria constitute between 74 and 99% of the microbial population, indicating that the artificial recharge develops oxic aquifer conditions (Buszka et al. 1994). Prior research at other ASR sites also exhibits similar trends from primarily anoxic to primarily oxic groundwater composition after artificial recharge begins. Thus, we can assume that predominant groundwater chemistry comprises of oxic species. Oxic conditions have been shown to increase biodegradability of antibiotics, however, this biodegradability is also dependent upon organic carbon content (Regnery et al. 2016). The highest biodegradability of antibiotics between the infiltration basin and recovery well would likely be from sequential carbon-rich, anoxic groundwater conditions to carbon-depleted and oxic conditions along the flow path. The duration of the flow path from infiltration to recovery of two years is assumed to be sufficient, so the presence of preferential flow paths allowing for rapid transport from infiltration to recovery may reduce antibiotic removal.

Within the context of the El Paso ASR site, effluent is only recharged once irrigation and industrial demands are met. It's assumed that input of wastewater effluent is not constant throughout the year as it is dependent on wastewater production, and irrigation or industrial demands (Pirnie et al. 2011). Enhancing the drying times between recharge and non-recharge periods may further introduce oxic conditions and thus, enhance biodegradation of antibiotic compounds. Biodegradability will be reduced when recharge is constant and drying times are reduced, thereby enhancing the risk of antibiotic infiltration through the vadose zone and into groundwater.

However, of particular concern is whether continual recharge of treated wastewater effluent, even with very small concentrations of antibiotics, can sorb and accumulate in soils. Desorption of antibiotics from the vadose zone may be important to the mobilization of antibiotics into groundwater. Few studies have been performed using wastewater effluent from activated carbon secondary treatment of wastewater on the accumulation of pharmaceuticals in the subsurface. At the El Paso site, the volume of water used for basin infiltration and the continual loading of trace contaminants may further the transport of antibiotics into the vadose zone and thus, the water table. Seasonal trends in antibiotic use may also influence antibiotic concentrations in wastewater effluent, thus there may be greater risk of antibiotic infiltration when effluent antibiotic concentrations are higher in the Winter months (Caucci et al. 2016).

The relatively rapid flow rates in the Hueco Bolson aquifer may enhance bacterial transport in the subsurface, although this is unlikely to originate from the wastewater treatment effluent due to efficient treatment techniques utilized at the Fred Hervey WWTP (Hunt and Johnson 2017). We hypothesize that relatively high regional groundwater flow rates prevalent in the Hueco Bolson aquifer may result in low antibiotic attenuation. The two-year flow path between infiltration basin source and recovery water at El Paso site may not be sufficient alone for the attenuation of pharmaceuticals, as attenuation may also depend on redox zonation and organic carbon content. If regional groundwater flow is slow, this is assumed to increase the likelihood of antibiotic sorption out of the aqueous phase and to the solid phase prior to recovery along the flow path.

10.8 Focus on Monitoring

To better establish the potential risk of antibiotic transport and subsequent changes to the environmental resistance reservoir, future monitoring efforts must focus on sampling of effluent from tertiary WWTPs for antibiotics. Metagenomic analysis of native groundwater and groundwater influenced by artificial recharge would provide additional insights into how the environmental microbiome may be altered by continual recharge of wastewater effluent. Conjugative assays between native groundwater bacteria and a model recipient at environmentally relevant antibiotic concentrations for the ASR site may provide further insights into whether antibiotic resistance is retained or lost in the subsurface. Future research is needed to establish the human risk potential of antibiotic resistance originating from the ASR of treated wastewater effluents.

10.9 Conclusion

This chapter highlights the necessity of investigating artificial recharge as a potential subsurface antibiotic resistance dissemination vector. Of primary concern is the quantification of antibiotic concentrations from advanced wastewater treatment facilities whose effluent is used for artificial recharge. Prior research has reported that antibiotic concentrations and antibiotic resistance genes vary across seasons, and thus robust sampling of effluent across a full year is needed to determine the impact of seasonality on the input of antibiotics. In addition to screening for antibiotics commonly detected in effluents, analysis of last stage antibiotics including carbapenem is necessary, as resistance to these antibiotics pose the greatest risk to human health (Araújo et al. 2021). Attenuation of antibiotics, if detected in advanced WWTP effluent, should focus on late-stage antibiotics where resistance in the environment is not yet widespread.

Aside from wastewater effluents, quantification of antibiotic concentrations in the vadose zone, groundwater and recovered water are needed for existing artificial recharge sites. Robust analysis of redox zonation and organic carbon content throughout the artificial recharge site may be key to enhancing the removal of antibiotic compounds between the source basin and recovery wells. Site specific hydrogeological conditions, including the presence of preferential flow paths that could enhance the risks of pathogenic contamination from sources outside of wastewater effluent must also be quantified. This research is of critical importance particularly within the context of water reuse, antibiotic resistance in the environment, and impacts on human health.

Risk assessment is one of the major challenges associated with antibiotic resistance in the environment. Current indicators of antibiotic resistance include the monitoring of antibiotic resistance genes, but these do not provide any information on the viability of the bacteria or associations with human pathogens. Proposed screening of bacterial groups in the environment includes *E. coli, Klebsiella pneumoniae, Aeromonas, Pseudomonas aeruginosa, Enteroccus faecalis* and *Enterococcus faecium*. Genetic determinants proposed for environmental monitoring include *intI1, sul1, sul2, bla_{CMY-2}, bla_{TEM}, bla_{VIM}, bla_{KPC}, qnrS, aac-(6')-lb-cr, vanA, mecA, ermB, ermF, tetM* and *aph* (Berendonk et al. 2015). Focusing on a consistent set of resistant determinants can enable efficient, high-quality quantification and meaningful comparisons across studies. At a minimum, screening effluents and artificial recharge environments affected by wastewater effluents for these resistance indicators is required.

Horizontal gene transfer is an emerging line of research and future work is required to determine in situ conjugation efficiencies within mixed bacterial communities. Metagenomic analysis of bacteria present in source water, native groundwater within the flow path between source water and recovery wells and the soil may provide insight into the bacterial composition of the recharge water and receiving environment and the potential for mobilization of antibiotic resistance genes. Conjugation assays are needed to determine conjugative frequencies between native subsurface bacteria at antibiotic concentrations determined to be relevant at the recharge site. Additionally, plasmid compatibility between native groundwater bacteria and human pathogens should be characterized to establish the viability of horizontal gene transfer across genera and subsequent human risk assessment.

Acknowledgements This chapter is based upon work supported by the U.S. Geological Survey under Grant No. G21AP10574 through the Texas Water Resources Institute Graduate Student Research program and Faculty Fellows program.

References

Abe K, Nomura N, Suzuki S (2020) Biofilms: hot spots of horizontal gene transfer (HGT) in aquatic environments, with a focus on a new HGT mechanism. FEMS Microbiol Ecol 96:fiaa031. https://doi.org/10.1093/femsec/fiaa031

Andersson DI, Hughes D (2010) Antibiotic resistance and its cost: is it possible to reverse resistance? Nat Rev Microbiol 8:260–271. https://doi.org/10.1038/nrmicro2319

Andrade L, Kelly M, Hynds P et al (2020) Groundwater resources as a global reservoir for antimicrobial-resistant bacteria. Water Res 170:115360. https://doi.org/10.1016/j.watres.2019.115360

Araújo S, Sousa M, Tacão M et al (2021) Carbapenem-resistant bacteria over a wastewater treatment process: Carbapenem-resistant Enterobacteriaceae in untreated wastewater and intrinsically-resistant bacteria in final effluent. Sci Total Environ 782:146892. https://doi.org/10.1016/j.scitotenv.2021.146892

Baquero F, Martínez J-L, Cantón R (2008) Antibiotics and antibiotic resistance in water environments. Curr Opin Biotechnol 19:260–265. https://doi.org/10.1016/j.copbio.2008.05.006

Barnett SR, Howles SR, Martin RR, Gerges NZ (2000) Aquifer storage and recharge: innovation in water resources management. Aust J Earth Sci 47:13–19. https://doi.org/10.1046/j.1440-0952.2000.00760.x

Batt AL, Bruce IB, Aga DS (2006) Evaluating the vulnerability of surface waters to antibiotic contamination from varying wastewater treatment plant discharges. Environ Pollut 142:295–302. https://doi.org/10.1016/j.envpol.2005.10.010

Bello-López JM, Cabrero-Martínez OA, Ibáñez-Cervantes G et al (2019) Horizontal gene transfer and its association with antibiotic resistance in the genus aeromonas spp. Microorganisms 7:363. https://doi.org/10.3390/microorganisms7090363

Bengtsson-Palme J, Kristiansson E, Larsson DGJ (2018) Environmental factors influencing the development and spread of antibiotic resistance. FEMS Microbiol Rev 42:fux053. https://doi.org/10.1093/femsre/fux053

Berendonk TU, Manaia CM, Merlin C et al (2015) Tackling antibiotic resistance: the environmental framework. Nat Rev Microbiol 13:310–317. https://doi.org/10.1038/nrmicro3439

Berges J, Moles S, Ormad MP et al (2021) Antibiotics removal from aquatic environments: adsorption of enrofloxacin, trimethoprim, sulfadiazine, and amoxicillin on vegetal powdered activated carbon. Environ Sci Pollut Res 28:8442–8452. https://doi.org/10.1007/s11356-020-10972-0

Böckelmann U, Dörries H-H, Ayuso-Gabella MN et al (2009) Quantitative PCR monitoring of antibiotic resistance genes and bacterial pathogens in three european artificial groundwater recharge systems. Appl Environ Microbiol 75:154–163. https://doi.org/10.1128/AEM.01649-08

Bouwer H (2002) Artificial recharge of groundwater: hydrogeology and engineering. Hydrogeol J 10:121–142. https://doi.org/10.1007/s10040-001-0182-4

Boy-Roura M, Mas-Pla J, Petrovic M et al (2018) Towards the understanding of antibiotic occurrence and transport in groundwater: findings from the Baix Fluvià alluvial aquifer (NE Catalonia, Spain). Sci Total Environ 612:1387–1406. https://doi.org/10.1016/j.scitotenv.2017.09.012

Buszka PM, Brock RD, Hooper RP (1994) Hydrogeology and selected water-quality aspects of the Hueco Bolson Aquifer at the Hueco Bolson Recharge Project area, El Paso, Texas. USGS

Cacace D, Fatta-Kassinos D, Manaia CM et al (2019) Antibiotic resistance genes in treated wastewater and in the receiving water bodies: a pan-European survey of urban settings. Water Res 162:320–330. https://doi.org/10.1016/j.watres.2019.06.039

Candela L, Tamoh K, Vadillo I, Valdes-Abellan J (2016) Monitoring of selected pharmaceuticals over 3 years in a detrital aquifer during artificial groundwater recharge. Environ Earth Sci 75:244. https://doi.org/10.1007/s12665-015-4956-8

Carvalho IT, Santos L (2016) Antibiotics in the aquatic environments: a review of the European scenario. Environ Int 94:736–757. https://doi.org/10.1016/j.envint.2016.06.025

Caucci S, Karkman A, Cacace D et al (2016) Seasonality of antibiotic prescriptions for outpatients and resistance genes in sewers and wastewater treatment plant outflow. FEMS Microbiol Ecol 92:fiw060. https://doi.org/10.1093/femsec/fiw060

Christou A, Agüera A, Bayona JM et al (2017) The potential implications of reclaimed wastewater reuse for irrigation on the agricultural environment: the knowns and unknowns of the fate of antibiotics and antibiotic resistant bacteria and resistance genes – a review. Water Res 123:448–467. https://doi.org/10.1016/j.watres.2017.07.004

Gargiulo G, Bradford SA, Simunek J et al (2008) Bacteria transport and deposition under unsaturated flow conditions: the role of water content and bacteria surface hydrophobicity. Vadose Zone J 7:406–419. https://doi.org/10.2136/vzj2007.0068

Ginige MP, Kaksonen AH, Morris C et al (2013) Bacterial community and groundwater quality changes in an anaerobic aquifer during groundwater recharge with aerobic recycled water. FEMS Microbiol Ecol 85:553–567. https://doi.org/10.1111/1574-6941.12137

Gomes IB, Maillard J-Y, Simões LC, Simões M (2020) Emerging contaminants affect the micro-biome of water systems—strategies for their mitigation. NPJ Clean Water 3:39. https://doi.org/10.1038/s41545-020-00086-y

Griffin DW, Benzel WM, Fisher SC et al (2019) The presence of antibiotic resistance genes in coastal soil and sediment samples from the eastern seaboard of the USA. Environ Monit Assess 191:300. https://doi.org/10.1007/s10661-019-7426-z

Hocquet D, Muller A, Bertrand X (2016) What happens in hospitals does not stay in hospitals: antibiotic-resistant bacteria in hospital wastewater systems. J Hosp Infect 93:395–402. https://doi.org/10.1016/j.jhin.2016.01.010

Hunt RJ, Johnson WP (2017) Pathogen transport in groundwater systems: contrasts with traditional solute transport. Hydrogeol J 25:921–930. https://doi.org/10.1007/s10040-016-1502-z

Jones ER, Van Vliet MTH, Qadir M, Bierkens MFP (2021) Country-level and gridded estimates of wastewater production, collection, treatment and reuse. Earth Syst Sci Data 13:237–254. https://doi.org/10.5194/essd-13-237-2021

Jutkina J, Rutgersson C, Flach C-F, Joakim Larsson DG (2016) An assay for determining minimal concentrations of antibiotics that drive horizontal transfer of resistance. Sci Total Environ 548–549:131–138. https://doi.org/10.1016/j.scitotenv.2016.01.044

Karakurt S, Schmid L, Hübner U, Drewes JE (2019) Dynamics of wastewater effluent contributions in streams and impacts on drinking water supply via riverbank filtration in Germany—a national reconnaissance. Environ Sci Technol 53:6154–6161. https://doi.org/10.1021/acs.est.8b07216

Kårelid V, Larsson G, Björlenius B (2017) Pilot-scale removal of pharmaceuticals in municipal wastewater: comparison of granular and powdered activated carbon treatment at three wastewater treatment plants. J Environ Manag 193:491–502. https://doi.org/10.1016/j.jenvman.2017.02.042

Kinney CA, Furlong ET, Werner SL, Cahill JD (2006) Presence and distribution of wastewater-derived pharmaceuticals in soil irrigated with reclaimed water. Environ Toxicol Chem 25:317–326. https://doi.org/10.1897/05-187R.1

Kristian Stevik T, Aa K, Ausland G, Fredrik Hanssen J (2004) Retention and removal of pathogenic bacteria in wastewater percolating through porous media: a review. Water Res 38:1355–1367. https://doi.org/10.1016/j.watres.2003.12.024

Krzeminski P, Tomei MC, Karaolia P et al (2019) Performance of secondary wastewater treatment methods for the removal of contaminants of emerging concern implicated in crop uptake and antibiotic resistance spread: a review. Sci Total Environ 648:1052–1081. https://doi.org/10.1016/j.scitotenv.2018.08.130

Li X, Atwill ER, Antaki E et al (2015) Fecal indicator and pathogenic bacteria and their antibiotic resistance in alluvial groundwater of an irrigated agricultural region with dairies. J Environ Qual 44:1435–1447. https://doi.org/10.2134/jeq2015.03.0139

Lira F, Vaz-Moreira I, Tamames J et al (2020) Metagenomic analysis of an urban resistome before and after wastewater treatment. Sci Rep 10:8174. https://doi.org/10.1038/s41598-020-65031-y

Lopatkin AJ, Huang S, Smith RP et al (2016) Antibiotics as a selective driver for conjugation dynamics. Nat Microbiol 1:16044. https://doi.org/10.1038/nmicrobiol.2016.44

Marano RBM, Zolti A, Jurkevitch E, Cytryn E (2019) Antibiotic resistance and class 1 integron gene dynamics along effluent, reclaimed wastewater irrigated soil, crop continua: elucidating potential risks and ecological constraints. Water Res 164:114906. https://doi.org/10.1016/j.wat res.2019.114906

Massoudieh A, Crain C, Lambertini E et al (2010) Kinetics of conjugative gene transfer on surfaces in granular porous media. J Contam Hydrol 112:91–102. https://doi.org/10.1016/j.jconhyd.2009. 10.009

Mendoza-Sanchez I, Cunningham J (2007) Efficient algorithm for modeling transport in porous media with mass exchange between mobile fluid and reactive stationary media. Transp Porous Media 68:285–300. https://doi.org/10.1007/s11242-006-9047-6

Michaelis C, Grohmann E (2023) Horizontal gene transfer of antibiotic resistance genes in biofilms. Antibiotics 12:328. https://doi.org/10.3390/antibiotics12020328

Nguyen AQ, Vu HP, Nguyen LN et al (2021) Monitoring antibiotic resistance genes in wastewater treatment: current strategies and future challenges. Sci Total Environ 783:146964. https://doi. org/10.1016/j.scitotenv.2021.146964

Pallares-Vega R, Blaak H, Van Der Plaats R et al (2019) Determinants of presence and removal of antibiotic resistance genes during WWTP treatment: a cross-sectional study. Water Res 161:319–328. https://doi.org/10.1016/j.watres.2019.05.100

Perry JA, Wright GD (2013) The antibiotic resistance "mobilome": searching for the link between environment and clinic. Front Microbiol 4:138. https://doi.org/10.3389/fmicb.2013.00138

Phoon BL, Ong CC, Mohamed Saheed MS et al (2020) Conventional and emerging technologies for removal of antibiotics from wastewater. J Hazard Mater 400:122961. https://doi.org/10.1016/j. jhazmat.2020.122961

Pirnie ML, ASR Systems L, Jackon, Sjoberg, McCarthy and Wilson L (2011) An assessment of aquifer storage and recovery in Texas

Ravasi D, König R, Principi P et al (2019) Effect of powdered activated carbon as advanced step in wastewater treatments on antibiotic resistant microorganisms. Curr Pharm Biotechnol 20:63–75. https://doi.org/10.2174/1389201020666190207095556

Raza S, Jo H, Kim J et al (2021) Metagenomic exploration of antibiotic resistome in treated wastewater effluents and their receiving water. Sci Total Environ 765:142755. https://doi.org/10.1016/ j.scitotenv.2020.142755

Regnery J, Gerba CP, Dickenson ERV, Drewes JE (2017a) The importance of key attenuation factors for microbial and chemical contaminants during managed aquifer recharge: a review. Crit Rev Environ Sci Technol 47:1409–1452. https://doi.org/10.1080/10643389.2017.1369234

Regnery J, Lee J, Drumheller ZW et al (2017b) Trace organic chemical attenuation during managed aquifer recharge: insights from a variably saturated 2D tank experiment. J Hydrol 548:641–651. https://doi.org/10.1016/j.jhydrol.2017.03.038

Regnery J, Wing AD, Kautz J, Drewes JE (2016) Introducing sequential managed aquifer recharge technology (SMART) – from laboratory to full-scale application. Chemosphere 154:8–16. https://doi.org/10.1016/j.chemosphere.2016.03.097

Ren S, Wang Y, Cui Y et al (2020) Desorption kinetics of tetracyclines in soils assessed by diffusive gradients in thin films. Environ Pollut 256:113394. https://doi.org/10.1016/j.envpol.2019. 113394

Rijal GK, Zmuda JT, Gore R et al (2009) Antibiotic resistant bacteria in wastewater processed by the Metropolitan Water Reclamation District of Greater Chicago system. Water Sci Technol 59:2297–2304. https://doi.org/10.2166/wst.2009.270

Sallwey J, Jurado A, Barquero F, Fahl J (2020) Enhanced removal of contaminants of emerging concern through hydraulic adjustments in soil aquifer treatment. Water 12:2627. https://doi.org/ 10.3390/w12092627

Sanganyado E, Gwenzi W (2019) Antibiotic resistance in drinking water systems: occurrence, removal, and human health risks. Sci Total Environ 669:785–797. https://doi.org/10.1016/j.sci totenv.2019.03.162

Sheng Z (2005) An aquifer storage and recovery system with reclaimed wastewater to preserve native groundwater resources in El Paso, Texas. J Environ Manag 75:367–377. https://doi.org/10.1016/j.jenvman.2004.10.007

Sidhu JPS, Gupta VVSR, Stange C et al (2020) Prevalence of antibiotic resistance and virulence genes in the biofilms from an aquifer recharged with stormwater. Water Res 185:116269. https://doi.org/10.1016/j.watres.2020.116269

Smyth C, O'Flaherty A, Walsh F, Do TT (2020) Antibiotic resistant and extended-spectrum β-lactamase producing faecal coliforms in wastewater treatment plant effluent. Environ Pollut 262:114244. https://doi.org/10.1016/j.envpol.2020.114244

Stanton IC, Tipper HJ, Chau K et al (2022) Does environmental exposure to pharmaceutical and personal care product residues result in the selection of antimicrobial-resistant microorganisms, and is this important in terms of human health outcomes? Environ Toxicol Chem 5498. https://doi.org/10.1002/etc.5498

Stefan C, Ansems N (2018) Web-based global inventory of managed aquifer recharge applications. Sustain Water Resour Manag 4:153–162. https://doi.org/10.1007/s40899-017-0212-6

Tang P-C, Eriksson O, Sjögren J et al (2022) A microfluidic chip for studies of the dynamics of antibiotic resistance selection in bacterial biofilms. Front Cell Infect Microbiol 12:896149. https://doi.org/10.3389/fcimb.2022.896149

Tavares RDS, Tacão M, Figueiredo AS et al (2020) Genotypic and phenotypic traits of blaCTX-M-carrying Escherichia coli strains from an UV-C-treated wastewater effluent. Water Res 184:116079. https://doi.org/10.1016/j.watres.2020.116079

Virolle C, Goldlust K, Djermoun S et al (2020) Plasmid transfer by conjugation in gram-negative bacteria: from the cellular to the community level. Genes 11:1239. https://doi.org/10.3390/gen es11111239

Wang R, Ji M, Zhai H et al (2021) Occurrence of antibiotics and antibiotic resistance genes in WWTP effluent-receiving water bodies and reclaimed wastewater treatment plants. Sci Total Environ 796:148919. https://doi.org/10.1016/j.scitotenv.2021.148919

Wright GD (2010) Antibiotic resistance in the environment: a link to the clinic? Curr Opin Microbiol 13:589–594. https://doi.org/10.1016/j.mib.2010.08.005

Wu J, Cao M, Tong D et al (2021) A critical review of point-of-use drinking water treatment in the United States. NPJ Clean Water 4:40. https://doi.org/10.1038/s41545-021-00128-z

Zainab SM, Junaid M, Xu N, Malik RN (2020) Antibiotics and antibiotic resistant genes (ARGs) in groundwater: A global review on dissemination, sources, interactions, environmental and human health risks. Water Res 187:116455. https://doi.org/10.1016/j.watres.2020.116455

Zhang Y-L, Lin S-S, Dai C-M et al (2014) Sorption–desorption and transport of trimethoprim and sulfonamide antibiotics in agricultural soil: effect of soil type, dissolved organic matter, and pH. Environ Sci Pollut Res 21:5827–5835. https://doi.org/10.1007/s11356-014-2493-8

Zhong Y, Gan W, Du Y et al (2019) Disinfection byproducts and their toxicity in wastewater effluents treated by the mixing oxidant of ClO2/Cl2. Water Res 162:471–481. https://doi.org/10.1016/j.watres.2019.07.012

Zuo R, Liu X, Zhang Q et al (2021) Sulfonamide antibiotics in groundwater and their migration in the vadose zone: a case in a drinking water resource. Ecol Eng 162:106175. https://doi.org/10.1016/j.ecoleng.2021.106175

The opinions expressed in this chapter are those of the author(s) and do not necessarily reflect the views of the UNESCO: United Nations Educational, Scientific and Cultural Organization, its Board of Directors, or the countries they represent.

Part III
Emerging Pollutants and Managing Wastewater and Waste

Chapter 11
Emerging Pollutants—Pitfalls in Their Removal: A Case Study

Hajar Farzaneh, Jayaprakash Saththasivam, and Gordon McKay

Abstract In countries with limited water resources, reusing treated sewage effluent (TSE) is necessary to overcome water scarcity. The quality of the TSE to be reused should be adequate to avoid adverse effects on the environment and its creatures. Pharmaceuticals are among the persistent contaminants to the conventional wastewater treatment process being regularly discharged into sewage effluent from various human activities. In this study, the removal of two widely used pharmaceuticals, ibuprofen (a painkiller) and gemfibrozil (a lipid regulator), from TSE has been investigated using a catalytic ozonation process. Activated carbon was produced from local date pits treated with phosphoric acid at 550 °C with a high surface area of 726 m^2/g to serve as catalysts. The removals were evaluated as single components and in binary systems at different dosages. Very fast and complete removals of both compounds were obtained by the catalytic ozonation process, and the activated carbon reusability showed almost no effect on the removal efficiencies after 10 cycles with almost unchanged phosphate and nitrate concentrations. This work aimed to produce high-quality TSE for possible use in agriculture with high nutrients and low contaminants.

Keywords Emerging pollutants · Wastewater treatment and reuse · Catalytic ozonation · Biochar · Advanced oxidation processes

11.1 Introduction

Global freshwater security is facing intense pressure from the growing population, booming economies, and unprecedented global warming and climate change activities (Haldar et al. 2022). The projection of an additional 20–30% increase in domestic and industrial water demand by 2050 will further intensify the pressure and pose significant challenges in achieving sustainable socioeconomic and human development (Gbandi 2022; Sa'ad et al. 2022). As water resources grow scarcer, the need

H. Farzaneh · J. Saththasivam · G. McKay (✉)
Hamad Bin Khalifa University, Doha, Qatar
e-mail: gmckay@hbku.edu.qa

© UNESCO 2025   253
S. Zandaryaa et al. (eds.), *Emerging Pollutants*, Advances in Water Security,
https://doi.org/10.1007/978-3-031-71758-1_11

for adequately treating and reusing wastewater is gaining traction (Xia et al. 2022). The United Nations (UN), through Sustainable Development Goal 6 on water and sanitation (SDG Target 6.3), has been pushing for a 50% reduction in the amount of untreated wastewater and a significant increase in global recycling and safe reuse (Qadir et al. 2020).

The reuse of wastewater offers feasible solutions in arid and water scarcity regions due to its continuous generation and stable supply. As an alternative water resource, adequately treated wastewater offers many other benefits that include: (i) reducing the freshwater demand for non-potable and industrial applications (Jaramillo and Restrepo 2017), (ii) cheaper alternatives compares to desalination (Jasim et al. 2016), (iii) contains nitrate and phosphate that are beneficial for the growth of plants (Lyu et al. 2016), and (iv) minimize the environmental impacts caused by discharging of wastewater into natural water bodies (Toze 2006). Despite these significant advantages, wastewater needs to be properly treated to remove pathogens, pharmaceuticals and personal care products (PPCPs), and toxic heavy metals that may cause adverse impacts on human health and the environment (Gallen et al. 2018; Bermúdez et al. 2021; Leroy-Freitas et al. 2022; Priya et al. 2022).

PPCPs are widely used all over the world with enormous diversity. More than 3000 pharmaceutical products are used in medicines such as anesthetics, antibiotics, contraceptives, lipid regulators, calmatives, and impotence drugs. These substances are taken by humans during their medical treatment and the excess are discharged into wastewater and aquatic environment either from human bodies after metabolism or by direct disposal in toilets (Ganiyu et al. 2015; Tokumura et al. 2016). PPCPs mainly find their way into aquatic environments through insufficient wastewater treatment before being discharged (Ganiyu et al. 2015; Tokumura et al. 2016). In addition to the municipal wastewater discharge, these PPCPs and endocrine disrupting compounds (EDCs) can enter water systems through inappropriate disposal, landfill leaching, agricultural runoff, drain water, livestock and veterinary wastes. Figure 11.1 illustrates the origins and fate of PPCPs and EDCs in the environment. As a wide range of these chemical compounds are persistent to the conventional sewage treatment process, particularly the micro-organic pollutants, thus using advanced technologies is a viable option. It is suggested that TSE goes through an additional advanced treatment process to minimize the environmental concerns due to the contamination with PPCPs and EDCs.

11.2 Adverse Effects of PPCPs and EDCs

Usually, PPCPs are not biodegradable but they can accumulate in aquatic creatures' bodies which enter the food chain and are then consumed by humans and can adversely affect the environmental species even at concentrations of ng/L and μg/L (Ganiyu et al. 2015). Clofibric acid, diazepam, primidone, gemfibrozil, and carbamazepine are known to be among the most persistent PPCPs (Chen et al. 2006; Hernando et al. 2006; Ebele et al. 2017). Some personal care products are endocrine disrupting compounds (EDCs) (Kasprzyk-Hordern et al. 2009). EDCs are chemicals

Fig. 11.1 Origins and fate of PPCPs and EDCs in the environment. *Source* Authors

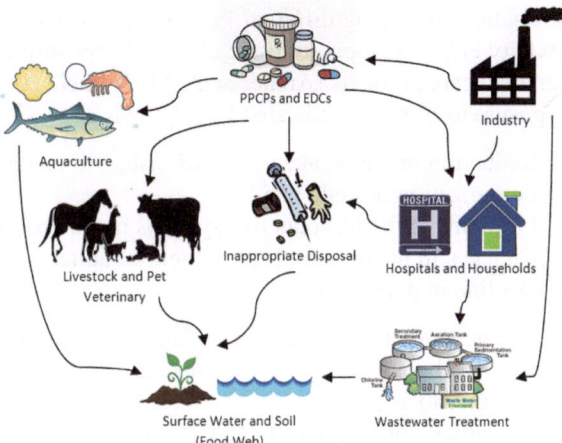

which can alter the hormonal systems of organisms (Esteban et al. 2014), and their disruption endpoints are classified to be estrogenic, androgenic, or thyroidal (Jung et al. 2015). EDCs can cause agonistic effects where they act as a hormone mimic compounds by binding to the hormone receptor sites in a targeted cell and thus activating a response. They can alternatively cause antagonistic effects which can act as hormone blockers and avoid any response by blocking the receptor sites from the natural hormone's interactions (Rahman et al. 2009). The agonistic and antagonistic effects are illustrated in Fig. 11.2. EDCs may reduce fertility, cause feminization and anomalies in the reproductive organs, as well as changes in the sexual behavior of some aquatic creatures such as fish, algae, frogs, etc. (Esteban et al. 2014).

In a recent study by Archer et al. (2017) the effect of exposure to different PPCPs and EDCs was reviewed and briefed as follows. Exposure to bisphenol A, parabens, and UV filters in sunscreens can lead to an increase in the development of breast cancer cells in humans. Concentrations of 100 ng/L of ibuprofen cause decreased egg fertilization in fish and result in decreased fish populations. A similar effect was reported with fish exposure to naproxen at concentrations of 100 ng/L. It is also commonly reported that complex mixtures of different types

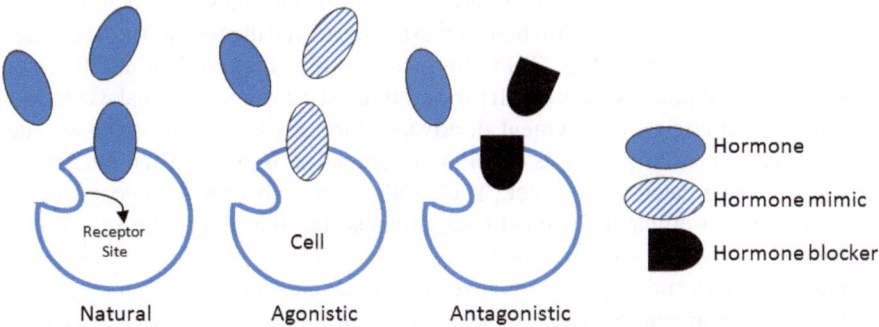

Fig. 11.2 Agonistic and antagonistic responses of EDCs in a cell. *Source* Authors

of organics including different PPCPs and inorganic pollutants may show different toxicity effects compared to the effect of exposure to individual pharmaceutical or personal care products (Archer et al. 2017; Ebele et al. 2017). The general effects of exposure to EDCs as indicated by Esplugas et al. (2007) are:

- Reduction in the breakage of bird, fish, and turtle eggs
- Feminization of male fish
- Problems in the reproductive systems in fish, reptiles, birds, and mammals
- Changes in the immunologic system of aquatic mammals
- Decline in population.

The general effects of EDCs on humans are a reduction of the amount of sperm, an increased possibility of breast, testicular, and prostate cancers, and an incidence of endometriosis (Esplugas et al. 2007; Rahman et al. 2009).

The uptake of a variety of PPCPs by different plants has also been investigated in different studies. Carter et al. (2015) studied the effect of carbamazepine and verapamil on zucchini where they have found that the uptake increases by increasing the pharmaceutical dose. Herklotz et al. (2010) investigated the potential uptake and accumulation of carbamazepine, salbutamol, sulfamethoxazole, and trimethoprim by cabbage. According to their study all these pharmaceuticals were detected in the roots and leaves. These studies show the possibility of PPCPs and EDCs uptake by plants and indicate the importance of crops irrigation using clean water.

Very few PPCPs and EDCs are regulated for a safe drinking water according to the United States EPA to this time. EPA has developed a list of critical EDCs and PPCPs under the final second list of chemicals for tier 1 screening updated June 2014 (USEPA 2014). This list consists of 109 chemicals where 41 are considered as pesticides and the rest are identified under the Safe Drinking Water Act (SDWA) involving the pharmaceuticals erythromycin, lindane, nitroglycerin, and quinoline. This list was developed by taking into account the 85 chemicals considered as drinking water contaminants according to the national primary drinking water regulation and 116 unregulated contaminants listed in the Third Contaminant Candidate List (CCL 3) which may need to be regulated under the SDWA. The CCL 3 includes some pesticides, disinfection byproducts, chemicals used in commerce, waterborne pathogens, pharmaceuticals and biological toxins (USEPA).

Currently, numerous surface waters, ground waters, wastewaters, and drinking waters are contaminated with PPCPs and EDCs worldwide and several attempts have been carried out and currently still being investigated to find the best and most efficient removal method (Ganiyu et al. 2015; Jung et al. 2015; Tokumura et al. 2016). As these harmful compounds are difficult to be eliminated using conventional wastewater treatment technologies, deployment of advanced tertiary treatment processes such as advanced oxidation processes (AOPs) are crucial in improving the quality and standard of the treated wastewater. The AOPs such as ozone/hydrogen peroxide are very fast and efficient for most contaminants, but the cost and production of oxidation by-products are the major issue. For this reason, a combined integrated oxidation and adsorption process is investigated called catalytic ozonation process (COP). This combined method is expected to show high efficiency for the removal of different contaminants and is preferable over other advanced technologies such as membranes. The removal of some PPCPs by COP is briefly explained below.

11.3 Catalytic Ozonation Process

The catalytic ozonation process (COP) is widely studied in the removal of PPCPs from wastewater (Buthiyappan et al. 2016). COP is an advanced oxidation process (AOP) that uses catalysts to promote ozone decomposition, which produces powerful active free radicals needed for the degradation of recalcitrant organic pollutants (Gomes et al. 2017; Mousavi et al. 2017). Catalysts that are commonly used in COP are mainly made of metal oxides (MnO_2, Al_2O_3, TiO_2, ZnO, etc.) (Rekhate and Srivastava 2020; Issaka et al. 2022). A study by Ikhlaq et al. (2015) using γ-Al_2O_3 as a COP catalyst demonstrated notable improvements in the removal of acetic acid and ibuprofen when compared with ozonation alone. Another micropollutant removal study also showed similar efficacy where MnO_2 nanocrystals were successfully used as COP catalysts to remove metoprolol and ibuprofen (He et al. 2021). Porous materials such as activated carbon, carbon nanotubes, and zeolites have also been successfully used as heterogeneous catalysts for COP (Kwong et al. 2008; Restivo et al. 2020; Liu et al. 2021). A study by Gonçalves et al. (2012) showed that the mineralization of sulphamethoxazole, an antibiotic that is difficult to be treated using conventional wastewater treatment technologies, was enhanced using multi-walled carbon nanotubes and activated carbon as catalysts. Li et al. (2020) attributed the efficient removal of ketoprofen using COP catalysts made from peanut shell biochar to the increased active sites and the intense chemical bonds as well as delocalized π electrons.

11.3.1 Case Study: Ibuprofen and Gemfibrozil Removal

In this case study, activated carbon synthesized from waste date pits was assessed as a COP catalyst for the removal of ibuprofen and gemfibrozil from treated sewage effluent (TSE). Waste date pits are available in abundant quantities free of charge in Middle East countries and it is estimated that waste date pits can contribute to the global production of 90,000 tonnes of activated carbon (Farzaneh et al. 2022). Our experimental studies were conducted using activated carbon that was produced from local date pits treated with phosphoric acid at 550 °C. The micropollutant removals were evaluated as single components and in binary systems at different dosages. Most catalytic ozonation processes for ibuprofen removal were studied on pure water using commercial catalysts. For instance, Saeid et al. (2020) used zeolite catalysts and investigated the influence of various support structures and different metals on enhancing the process. Kolosov et al. (2018) investigated the removal of ibuprofen, gemfibrozil, atrazine, and naproxen from synthetic wastewater that mimic the municipal secondary effluent wastewater using commercial catalysts such as zeolite, TiO_2-Al_2O_3, and AlO_2-based catalysts. In all these studies, the catalytic ozonation process was found to be more efficient than ozone alone, proving that catalytic ozonation is a superior process for the removal of pharmaceuticals but the process which can provide catalysts at low cost will be of the greatest advantage to overcome the high cost of ozone. This was the aim of the current work where the catalysts were produced from waste materials (date pits) for the removal of the pharmaceutical compounds at a relatively reduced cost for possible application in agriculture while still providing a high concentration of nutrients in the treated water.

11.3.1.1   Methodology

Chemicals Preparation

Ibuprofen and gemfibrozil purchased from Sigma-Aldrich were used to prepare stock solutions of 40 mg/L each (the maximum concentration is limited by their effective solubility) in Type I ultrapure water with a resistivity of 18.2 MΩ.cm. To increase the dissolution of the pharmaceuticals, the stock solutions were sonicated for one hour at 60 °C. For gemfibrozil, which has lower solubility, 0.8 μL of 1M NaOH was added before sonication. The concentrations of the prepared stock solutions after dissolution were confirmed and validated by HPLC (High Performance Liquid Chromatography) analysis and the calibration lines were produced using ibuprofen and gemfibrozil standards prepared in 100% methanol. The use of methanol in the stock solutions was avoided to eliminate its interference in the oxidation processes.

For the preparation of the dissolved ozone stock solution, a corona discharge ozone generator (BMT 802N, Germany) was used to generate ozone from the oxygen gas. The generated ozone gas was bubbled into ultrapure water in a jacketed column, where the temperature was maintained at 5 °C with water circulation to make a concentrated ozone stock solution of about 40–50 mg/L. The dissolved ozone stock solution was freshly prepared on the date of the experiment and its concentration was accurately measured after one hour using UV-spectrophotometer at 258 nm ($\epsilon_{O_3} = 3150M^{-1}cm^{-1}$) to ensure that the concentration had stabilized before starting the experiments. The concentration of stock ozone solution was measured as follows:

$$C_{ozone} = 16 \times (Abs@258 \text{ nm}) \tag{11.1}$$

where Abs @ 258 nm is the UV light absorbance at the wavelength 258 nm measured using 1 cm quartz cells with a UV–VIS spectrophotometer.

Activated carbon for this work was produced from date pits. Raw date pits were collected from a local company producing seedless date products, then the required quantity was cleaned and dried at 150 °C overnight. The dried date pits were then ground and impregnated with 85% wt. phosphoric acid (BioUltra ≥ 85%, Sigma Aldrich) at a ratio of 2:1 of phosphoric acid to date pits. This ratio was selected based on a literature study showing the high surface area (Jamion et al. 2017). The slurry was mixed at 80 °C for 2 h and then dried overnight at 105 °C. The impregnated date pits were then placed in a muffle furnace (Lindberg Blue M, Thermo Scientific) under a continuous flow of nitrogen gas with the following temperature program:

- Started at room temperature with a ramp at 5 °C/min to 240 °C,
- Followed by a ramp at 2 °C/min to 360 °C,
- Then ramped at 5 °C/min to 550 °C and held for 2 h.

This temperature program was set based on the thermogravimetric analysis (TGA) results from the literature (Danish et al. 2014) where a major weight loss is occurring between the temperatures of 250–350 °C.

The activated carbon was then washed and soaked in water at 80 °C overnight and then rinsed and filtered using filtration paper (Ashless, Whatman) until the pH of the filtrate water was around neutral pH and then dried at 105 °C overnight. The final product was then ground and sieved for a particle size of 75–125 μm using Haver & Boecker standard sieves. The Brunauer–Emmett–Teller (BET) surface area of the produced activated carbon was analyzed using ASAP 2020 Plus, Micromeritics while the CHNS elemental analysis was carried out by Euro EA Elemental Analyzer. The zeta potential analysis was performed using a Zetasizer Nano, Malvern, at different pH values.

Experimental Procedure

The TSE samples were collected from a local wastewater treatment plant and were spiked with ibuprofen and/or gemfibrozil at initial concentrations of 100 μg/L in amber glass bottles. The bottles were sealed with parafilm and were shaken well, then the required dose of activated carbon was added and mixed, and quickly dosed with ozone at specified dosages. The samples were sealed and placed in an incubator shaker (Thermo Scientific, MaxQ 6000) at a controlled temperature. The agitation speed was constant for the entire work in this study at 200 rpm. The samples were kept in the shaker for the desired time and an aliquot from the sample was filtered using a syringe filter at different time intervals for HPLC analysis.

Analytical Procedure

The HPLC analytical method for the analysis of ibuprofen and gemfibrozil and the conditions are similar to the method used in a previous work (Farzaneh et al. 2021). The samples were concentrated by solid-phase extraction (SPE) prior to HPLC analysis. During SPE, the pH of 500 mL samples was adjusted to be between 2 and 2.5 using HCl, then the cartridge (Thermo Scientific SolEx C18) was conditioned using 5 mL methanol and 5 mL ultrapure water at pH 2–2.5. After loading the sample, the cartridge was dried for 15 min. For elution, 10 mL methanol was used, then the collected sample was concentrated by nitrogen gas flow to reduce the volume to 0.5 mL and was then analyzed by HPLC.

Repeatability and Statistical Analysis

To analyze the repeatability of the experiments, all tests were conducted in duplicates and the error bars in all figures express the standard deviation between the runs. To evaluate the differences in the parameters, analysis of variance (ANOVA) was studied using the ANOVA data analysis in Microsoft Excel. The tests, which show higher F-statistic compared to the F critical value, are considered to be statistically different and for P-value less than 0.05, the difference in the experimental data is statistically significant. This statistical analysis was conducted for all the experimental tests in this work to obtain a better conclusion about the different dosages.

For the fitting to the second-order rate of reaction model, the linear regression data analysis was conducted. The correlation coefficient (R^2) was taken into consideration with the F-significance in the ANOVA. The F-significance shows how reliable the model fitting is by giving the number of un-fitted data points to the model. Therefore, the lower the F-significance, the more reliable the model is with less percentage of data that does not fit the model. In addition to this, the 95% confidence intervals were also evaluated for the modeled value of the constant parameter.

11.3.1.2   Characterization of the Produced Date Pits Activated Carbon

The yield of the produced activated carbon from date pits was found to be around 20% of the raw date pits initial weight. The Brunauer–Emmett–Teller (BET) surface area obtained in this study was 726.69 m^2/g, which is comparable to the commercial activated carbon with high pore volume of 0.71 cm^3/g. The use of phosphoric acid as the activation agent helps to break down the fibrous structure of the date pits and creates large pores by the removal of the organic molecules such as lignin and sugar producing large surface area and porous structured activated carbon (Girgis and El-Hendawy 2002; Danish et al. 2014). According to the Carbon, Hydrogen, Nitrogen, and Sulfur (CHNS) elemental analysis, the highest element of the produced activated carbon is carbon at ~60%. High carbon content is typical for activated carbon and shows the number of organic molecules converted to carbon which is the main element that promotes the adsorption process (Maulina and Iriansyah 2018; Zięzio et al. 2020).

The zeta potential analysis of the activated carbon at different pH values showed that the produced activated carbon is negatively charged for the pH range from 3 to 9 and becomes more negative at higher pH values with a slight increase for pH 8 and 9.

11.3.1.3   Evaluation of the Catalytic Ozonation Process

In this section the removal of ibuprofen and gemfibrozil from TSE by catalytic ozonation is investigated, where the produced activated carbon from date pits was used in combination with ozone. Three different dosages of ozone (0.5, 1, and 1.5 mg/L) were combined with different dosages of activated carbon for a contact time of 5 min. The results are shown in Fig. 11.3 for ibuprofen as a single component. Gemfibrozil was completely removed for all the tests and therefore in this section, the results of ibuprofen are only shown. The fast removal of gemfibrozil is due to its fast degradation by ozone (Lee et al. 2013) and might be faster with the catalytic ozonation process. As seen in Fig. 11.3, by increasing both the activated carbon and ozone dosages, the removal efficiency of ibuprofen has increased reaching a maximum removal of 90% using 1.5 mg/L ozone with 7 and 10 mg activated carbon per 100 mL TSE.

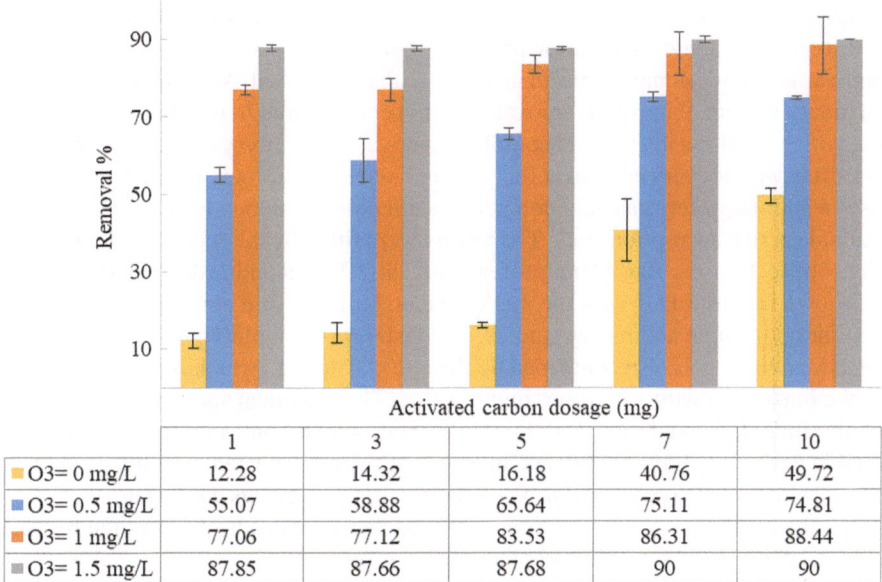

Activated carbon dosage (mg)				
1	3	5	7	10
■ O3= 0 mg/L				
12.28	14.32	16.18	40.76	49.72
■ O3= 0.5 mg/L				
55.07	58.88	65.64	75.11	74.81
■ O3= 1 mg/L				
77.06	77.12	83.53	86.31	88.44
■ O3= 1.5 mg/L				
87.85	87.66	87.68	90	90

Fig. 11.3 Ibuprofen removal as a single component by catalytic ozonation from TSE. Initial ibuprofen conc. = 0.1 mg/L, pH = 8.2, temperature = 20 °C, agitation speed = 200 rpm, contact time = 5 min. *Source* Authors

In this work, the error bars show the standard deviation of duplicate runs for each experiment. As seen in Fig. 11.3, the error bars for the ozone dosages 1 and 1.5 mg/L with the combination of 5, 7, and 10 mg activated carbon are overlapping which shows that the difference between these results is insignificant which was also confirmed by the statistical analysis ANOVA. For 0.5 mg/L ozone dose combined with 7 and 10 mg activated carbon the same values were obtained which shows the result is statistically insignificant. These findings are important to specify the efficiency of the process and more importantly for the selection of the required dosages. When the removal efficiency using a higher dosage is not significant compared to a lower dosage, selecting the lower dosage will be a better choice for cost reduction. According to the results obtained in Fig. 11.3, the optimum combination for cost reduction if the process is to be of only 5 min contact time, is the selection of 1 mg/L ozone and 5 mg/100 mL activated carbon.

To better evaluate the effect of catalytic ozonation on the removal of ibuprofen and gemfibrozil, the removal kinetic was investigated within the first 5 min with 30 s time intervals for ibuprofen and gemfibrozil as single components and in a binary system. For this experiment ozone dosage of 1.5 mg/L was combined with 5 mg activated carbon in 100 mL of spiked TSE. This ozone dose was selected to be compared with ozone alone results obtained in previous work at similar dosage (Farzaneh et al. 2021). Then the activated carbon dose was selected being the optimal dosage for the contact time of 5 min with the selected ozone dosage. As expected, gemfibrozil was completely removed within the first 30 s due to its fast degradation by ozone and might be faster in the catalytic ozonation process. Figure 11.4 shows the result of ibuprofen removal as a single component and in binary with gemfibrozil. As seen in this figure in the binary system, ibuprofen removal was slower compared to the single component test, which confirms the fast and complete removal of gemfibrozil. On the other hand, this effect was not very evident at longer times since gemfibrozil was no longer in the system and ibuprofen was only competing with other compounds in the TSE for the available active adsorption sites. The results were fitted by the second-order kinetics (Fig. 4b) to find the rate of ibuprofen removal. The second-order kinetics results are shown in Table 11.1 with the rate constant, half-life, and correlation coefficient (R^2) for both single and binary studies. The faster removal of ibuprofen in the single component tests was also confirmed by the rate constant, k, which is higher than that for the binary test with ~1.3 times higher value. The statistical analysis by regression for the model showed very low F-significance, which confirms that the model is reliable with close 95% confidence intervals of the rate constants. Comparing these results with the one obtained during ozonation in the previous work (Farzaneh et al. 2021), the catalytic ozonation process showed faster removals of ibuprofen by ~3.6 and ~2.7 times for the single and binary tests, respectively.

Activated Carbon Reusability

The reusability of activated carbon in the catalytic ozonation process was studied by reusing the activated carbon 10 times. After each use, the activated carbon was filtered from the TSE sample, then rinsed with deionized water and dried overnight at 105 °C, and reused in a fresh spiked TSE sample in each cycle. This experiment was carried out for the binary system only with 5 mg of activated carbon combined with 1 mg/L ozone in 100 mL of spiked TSE sample for a contact time of 4 h. In addition to the removal of ibuprofen and gemfibrozil, dissolved organic carbon (DOC) removal was also analyzed for this study. The results as seen in Fig. 11.5, show complete removals of gemfibrozil for all 10 cycles with higher than 95% ibuprofen removal. On the other hand, DOC removal was found to be decreasing by increasing the cycle of activated carbon reuse. This is mainly due to the pore filling of the activated carbon sites by DOC and other contaminants in the TSE which accumulated more and more in the pores in each cycle and the low ozone dosage was not capable of breaking down more organic compounds to free up some sites for adsorption on the surface of activated carbon. This is in addition to the possibility of changes in the activated carbon structure caused by ozone molecules which can adversely affect the adsorption process and the available active sites. As seen in Fig. 11.5, DOC concentration was increased in comparison to the initial DOC concentration in TSE after 10 times reuse of activated carbon. This can be an indication of leaching from organic compounds and/or desorption from the activated carbon particles.

In this work, the main focus for TSE reuse is for irrigation, and therefore the concentrations of the nutrients nitrate and phosphate are important to stay high and unchanged after the treatment. Nitrate and phosphate concentrations were analyzed after each activated carbon reuse cycle and the results are illustrated in Fig. 11.5. As seen in this figure, the removal of nitrate and phosphate was only ~5% by catalytic

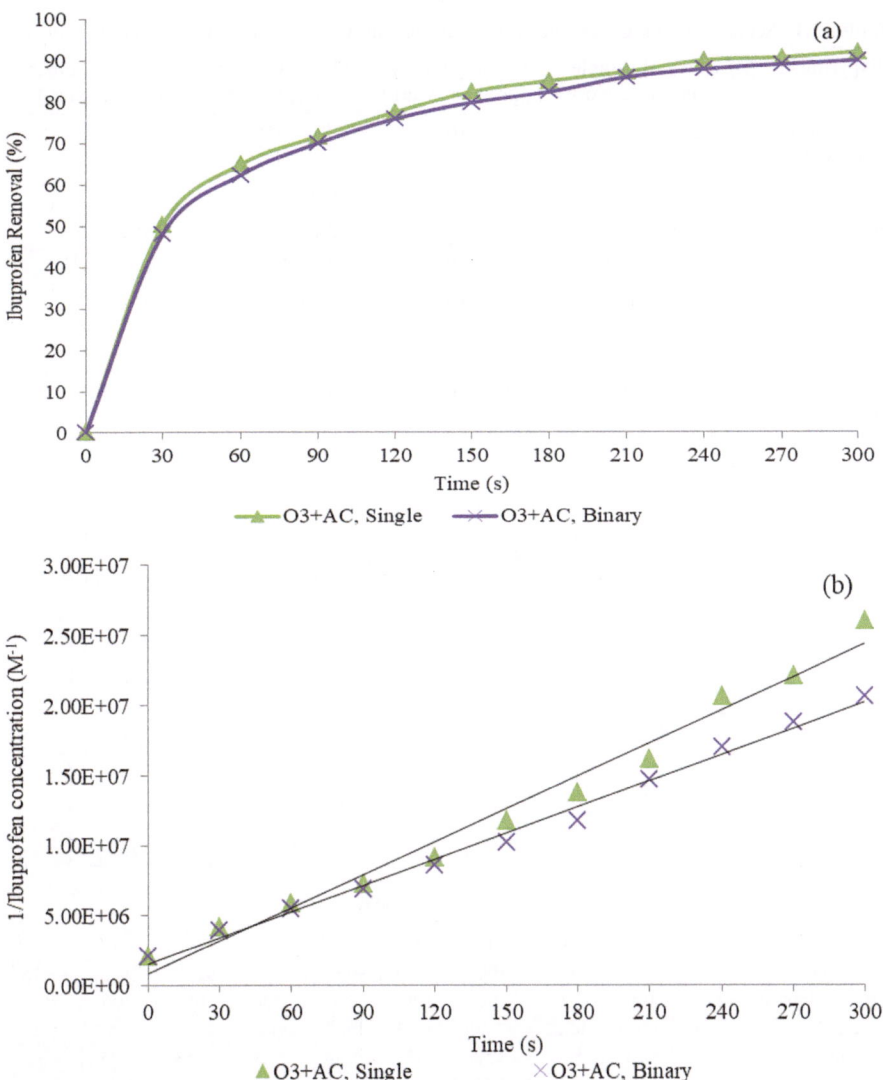

Fig. 11.4 Removal of ibuprofen by catalytic ozonation in single and in a binary system with gemfibrozil. **a** Experimental results and **b** second-order kinetics modelling. Initial ibuprofen conc. = 0.1 mg/L, initial gemfibrozil conc. = 0.1 mg/L, ozone dose = 1.5 mg/L, activated carbon dose = 5 mg/100 mL TSE, pH = 8.2, temperature = 20 °C, agitation speed = 200 rpm. *Source* Authors

ozonation. Ozone does not affect these compounds, but they can be removed through adsorption processes. In this study, the produced activated carbon from date pits is negatively charged, and hence there will be some repulsive effects between these anions and activated carbon particles. This shows that the catalytic ozonation method using the activated carbon produced from date pits is an effective method for the

Table 11.1 Second-order kinetics study for ibuprofen removal by ozonation and catalytic ozonation

Experiment	O_3 only single (previous work)	O_3 only binary (previous work)	O_3 + AC single (current work)	O_3 + AC binary (current work)
Rate constant, k $(M^{-1}s^{-1})$	2.2×10^4	2.3×10^4	7.8×10^4	6.2×10^4
$t_{1/2}$ (s)	93.91	88.79	26.38	33.14
R^2	0.963	0.961	0.982	0.992
F-significance	9.4×10^{-8}	1.2×10^{-7}	3.9×10^{-9}	9.8×10^{-11}
Lower 95% confidence for k $(M^{-1}s^{-1})$	1.9×10^4	1.9×10^4	7×10^4	5.8×10^4
Upper 95% confidence for k $(M^{-1}s^{-1})$	2.5×10^4	2.6×10^4	8.6×10^4	6.6×10^4

Source Authors

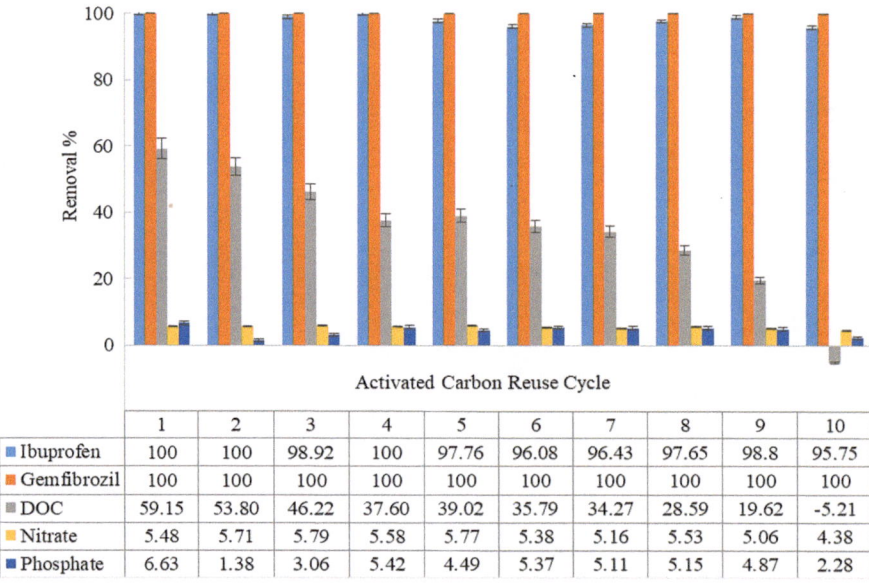

	1	2	3	4	5	6	7	8	9	10
■ Ibuprofen	100	100	98.92	100	97.76	96.08	96.43	97.65	98.8	95.75
■ Gemfibrozil	100	100	100	100	100	100	100	100	100	100
■ DOC	59.15	53.80	46.22	37.60	39.02	35.79	34.27	28.59	19.62	-5.21
■ Nitrate	5.48	5.71	5.79	5.58	5.77	5.38	5.16	5.53	5.06	4.38
■ Phosphate	6.63	1.38	3.06	5.42	4.49	5.37	5.11	5.15	4.87	2.28

Fig. 11.5 Removal percentage of ibuprofen, gemfibrozil, DOC, nitrate, and phosphate by catalytic ozonation at 10 cycles of activated carbon reuse. Initial ibuprofen conc. = 0.1 mg/L, initial gemfibrozil conc. = 0.1 mg/L, initial DOC conc. = 5.474 mg/L, initial nitrate conc. = 19.09 mg/L, initial phosphate conc. = 2.17 mg/L, ozone dose = 1 mg/L, activated carbon dose = 5 mg/100 mL TSE, temperature = 20 °C, initial pH = 8.2, contact time = 4 h, agitation speed = 200 rpm. *Source* Authors

treatment of TSE that can eliminate micropollutants such as pharmaceuticals without affecting the nutrients in TSE, which then can be safely used for irrigation purposes with reduced environmental and health risks.

The production cost of activated carbon was analyzed based on the literature (Ng et al. 2003; Lai and Ngu 2020) considering dosage of 5 mg/100 mL and a plant size for treating 55 MGD of TSE with 10 times reuse of activated carbon. The capital cost including various equipment such as; a hammer mill, chemical tanks, kilns, and a nitrogen generator is estimated to be $2,073,000. Operation and maintenance costs were calculated based on energy, water, phosphoric acid, and transportation costs, resulting in total annual costs of $227,010. The cost of the ozone system and its operation and maintenance are estimated based on a previous work (Farzaneh et al. 2021). By combining the cost of activated carbon production and ozone process, the cost of COP can be estimated. For instance, a combination of 1 mg/L ozone and 5 mg/100 mL activated carbon will give a capital cost of ~$5.2 million and an annual operation and maintenance cost of ~$0.38 million/year. This combination provides a cost-effective solution with high ibuprofen removal efficiency compared to using ozone alone, reducing costs by 19.4%.

11.4 Conclusions

Water scarcity is a serious issue which currently many countries are facing. TSE is known to be a source of water to overcome the water scarcity problem in many countries. Reusing TSE may have adverse effects on the environment and human health in case of inadequate treatment process where some micro organic pollutants can be found in the treated sewage effluents. PPCPs and EDCs are some of these micro organic pollutants found in variety of treated sewage effluents across the world. Among different treatment processes, AOPs have been found to be of the most effective methods for efficient removal of a wide range of PPCPs. To increase the removal efficiency of pharmaceuticals and reduce the number of possible oxidation by-products from wastewater, catalytic ozonation is suggested. In this case study ozone was combined with phosphoric acid-treated activated carbon produced using waste date pits. The produced activated carbon had a high surface area of 726 m^2/g. The combination of the produced activated carbon and ozone in the catalytic ozonation process showed high removals of ibuprofen and gemfibrozil within 5 min contact time. The second-order kinetics study showed ~3.6 times faster removals of ibuprofen compared to the ozone alone. The reusability of activated carbon in the catalytic ozonation process across 10 cycles showed complete removal of gemfibrozil at all cycles whereas ibuprofen removal was decreased by only 5% in the 10th cycle. The removals of phosphate and nitrate throughout the 10 cycles were found to be only ~5%. These findings show that the catalytic ozonation process using the produced activated carbon from date pits is a superior process for the removal of micropollutants from TSE such as pharmaceuticals where its effect on the nutrients phosphate and nitrate is minimal.

Some recommendations for future investigations that were not conducted in the current work are as follows:

- Evaluation of different activation methods for the production of activated carbon for possible further cost reduction such as physical activation using steam.
- Testing the process at scale-up levels to investigate the larger scale treatment process.
- Studying a modified integrated treatment process using a three-phase fluidized bed of ozone, activated carbon, and TSE.
- Analyzing the structure of activated carbon after the COP to see the effect of ozone molecules on the activated carbon particles.

References

Archer E, Petrie B, Kasprzyk-hordern B, Wolfaardt GM (2017) The fate of pharmaceuticals and personal care products (PPCPs), endocrine disrupting contaminants (EDCs), metabolites and illicit drugs in a WWTW and environmental waters. Chemosphere 174:437–446. https://doi.org/10.1016/j.chemosphere.2017.01.101

Bermúdez LA, Pascual JM, Martínez MDMM, Capilla JMP (2021) Effectiveness of advanced oxidation processes in wastewater treatment: state of the art. Water (Switzerland) 13:1–19. https://doi.org/10.3390/w13152094

Buthiyappan A, Abdul Aziz AR, Wan Daud WMA (2016) Recent advances and prospects of catalytic advanced oxidation process in treating textile effluents. Rev Chem Eng 32:1–47. https://doi.org/10.1515/revce-2015-0034

Carter LJ, Williams M, Böttcher C, Kookana RS (2015) Uptake of pharmaceuticals influences plant development and affects nutrient and hormone homeostases. Environ Sci Technol 49:12509–12518. https://doi.org/10.1021/acs.est.5b03468

Chen M, Ohman K, Metcalfe C et al (2006) Pharmaceuticals and endocrine disruptors in wastewater treatment effluents and in the water supply system. Water Qual Res J Canada 41:351–364

Danish M, Hashim R, Ibrahim MNM, Sulaiman O (2014) Optimized preparation for large surface area activated carbon from date (Phoenix dactylifera L.) stone biomass. Biomass Bioenergy 61:167–178. https://doi.org/10.1016/j.biombioe.2013.12.008

Ebele AJ, Abdallah MA, Harrad S (2017) Pharmaceuticals and personal care products (PPCPs) in the freshwater aquatic environment. Emerg Contam 3:1–16. https://doi.org/10.1016/j.emcon.2016.12.004

Esplugas S, Bila DM, Krause LGT, Dezotti M (2007) Ozonation and advanced oxidation technologies to remove endocrine disrupting chemicals (EDCs) and pharmaceuticals and personal care products (PPCPs) in water effluents. J Hazard Mater 149:631–642. https://doi.org/10.1016/j.jhazmat.2007.07.073

Esteban S, Gorga M, Petrovic M et al (2014) Analysis and occurrence of endocrine-disrupting compounds and estrogenic activity in the surface waters of Central Spain. Sci Total Environ 466–467:939–951. https://doi.org/10.1016/j.scitotenv.2013.07.101

Farzaneh H, Loganathan K, Saththasivam J, McKay G (2021) Selectivity and competition in the chemical oxidation processes for a binary pharmaceutical system in treated sewage effluent. Sci Total Environ 765:142704. https://doi.org/10.1016/j.scitotenv.2020.142704

Farzaneh H, Saththasivam J, McKay G, Parthasarathy P (2022) Adsorbent minimization for removal of ibuprofen from water in a two-stage batch process. Processes 10:453. https://doi.org/10.3390/pr10030453

Gallen C, Eaglesham G, Drage D et al (2018) A mass estimate of perfluoroalkyl substance (PFAS) release from Australian wastewater treatment plants. Chemosphere 208:975–983. https://doi.org/10.1016/j.chemosphere.2018.06.024

Ganiyu SO, Van Hullebusch ED, Cretin M et al (2015) Coupling of membrane filtration and advanced oxidation processes for removal of pharmaceutical residues: a critical review. Sep Purif Technol 156:891–914. https://doi.org/10.1016/j.seppur.2015.09.059

Gbandi T (2022) This water is all ours: water demand and civil conflicts. Res Econ 76:120–130. https://doi.org/10.1016/j.rie.2022.06.003

Girgis BS, El-Hendawy ANA (2002) Porosity development in activated carbons obtained from date pits under chemical activation with phosphoric acid. Microporous Mesoporous Mater 52:105–117. https://doi.org/10.1016/S1387-1811(01)00481-4

Gomes J, Costa R, Quinta-Ferreira RM, Martins RC (2017) Application of ozonation for pharmaceuticals and personal care products removal from water. Sci Total Environ 586:265–283. https://doi.org/10.1016/j.scitotenv.2017.01.216

Gonçalves AG, Órfão JJM, Pereira MFR (2012) Catalytic ozonation of sulphamethoxazole in the presence of carbon materials: catalytic performance and reaction pathways. J Hazard Mater 239–240:167–174.https://doi.org/10.1016/J.JHAZMAT.2012.08.057

Haldar K, Kujawa-Roeleveld K, Acharjee TK et al (2022) Urban water as an alternative freshwater resource for matching irrigation demand in the Bengal delta. Sci Total Environ 835:155475. https://doi.org/10.1016/J.SCITOTENV.2022.155475

He Y, Wang L, Chen Z et al (2021) Catalytic ozonation for metoprolol and ibuprofen removal over different MnO2 nanocrystals: efficiency, transformation and mechanism. Sci Total Environ 785:147328. https://doi.org/10.1016/J.SCITOTENV.2021.147328

Herklotz PA, Gurung P, Vanden Heuvel B, Kinney CA (2010) Uptake of human pharmaceuticals by plants grown under hydroponic conditions. Chemosphere 78:1416–1421. https://doi.org/10.1016/j.chemosphere.2009.12.048

Hernando MD, Mezcua M, Fern AR, Barcel D (2006) Environmental risk assessment of pharmaceutical residues in wastewater effluents, surface waters and sediments 69:334–342. https://doi.org/10.1016/j.talanta.2005.09.037

Ikhlaq A, Brown DR, Kasprzyk-Hordern B (2015) Catalytic ozonation for the removal of organic contaminants in water on alumina. Appl Catal B 165:408–418. https://doi.org/10.1016/j.apcatb.2014.10.010

Issaka E, AMU-Darko JNO, Yakubu S et al (2022) Advanced catalytic ozonation for degradation of pharmaceutical pollutants-a review. Chemosphere 289:133208. https://doi.org/10.1016/J.CHEMOSPHERE.2021.133208

Jamion NAB, Hafiff NHBA, Halim NHA et al (2017) Preparation of date seed activation for surfactant recovery. Malays J Anal Sci 21:1045–1053. https://doi.org/10.17576/mjas-2017-2105-06

Jaramillo MF, Restrepo I (2017) Wastewater reuse in agriculture: a review about its limitations and benefits. Sustainability 9:1734. https://doi.org/10.3390/SU9101734

Jasim SY, Saththasivam J, Loganathan K et al (2016) Reuse of treated sewage effluent (TSE) in Qatar. J Water Process Eng 11:174–182. https://doi.org/10.1016/j.jwpe.2016.05.003

Jung C, Son A, Her N et al (2015) Removal of endocrine disrupting compounds, pharmaceuticals, and personal care products in water using carbon nanotubes: a review. J Ind Eng Chem 27:1–11. https://doi.org/10.1016/j.jiec.2014.12.035

Kasprzyk-Hordern B, Dinsdale RM, Guwy AJ (2009) The removal of pharmaceuticals, personal care products, endocrine disruptors and illicit drugs during wastewater treatment and its impact on the quality of receiving waters. Water Res 43:363–380. https://doi.org/10.1016/j.watres.2008.10.047

Kolosov P, Peyot ML, Yargeau V (2018) Novel materials for catalytic ozonation of wastewater for disinfection and removal of micropollutants. Sci Total Environ 644:1207–1218. https://doi.org/10.1016/j.scitotenv.2018.07.022

Kwong CW, Chao CYH, Hui KS, Wan MP (2008) Catalytic ozonation of toluene using zeolite and MCM-41 materials. Environ Sci Technol 42:8504–8509. https://doi.org/10.1021/ES801087F/SUPPL_FILE/ES801087F_SI_001.PDF

Lai JY, Ngu LH (2020) The production cost analysis of oil palm waste activated carbon: a pilot-scale evaluation. Greenh Gases: Sci Technol 10:999–1026. https://doi.org/10.1002/ghg.2020

Lee Y, Gerrity D, Lee M et al (2013) Prediction of micropollutant elimination during ozonation of municipal wastewater effluents: use of kinetic and water specific information. Environ Sci Technol 47:5872–5881. https://doi.org/10.1021/es400781r

Leroy-Freitas D, Machado EC, Torres-Franco AF et al (2022) Exploring the microbiome, antibiotic resistance genes, mobile genetic element, and potential resistant pathogens in municipal wastewater treatment plants in Brazil. Sci Total Environ 842:156773. https://doi.org/10.1016/J.SCITOTENV.2022.156773

Li H, Liu S, Qiu S et al (2020) Catalytic ozonation oxidation of ketoprofen by peanut shell-based biochar: effects of the pyrolysis temperatures 43:848–860. https://doi.org/10.1080/09593330.2020.1807610

Liu ZQ, Huang C, Li JY et al (2021) Activated carbon catalytic ozonation of reverse osmosis concentrate after coagulation pretreatment from coal gasification wastewater reclamation for zero liquid discharge. J Clean Prod 286:124951. https://doi.org/10.1016/J.JCLEPRO.2020.124951

Lyu S, Chen W, Zhang W et al (2016) Wastewater reclamation and reuse in China: opportunities and challenges. J Environ Sci 39:86–96. https://doi.org/10.1016/J.JES.2015.11.012

Maulina S, Iriansyah M (2018) Characteristics of activated carbon resulted from pyrolysis of the oil palm fronds powder. IOP Conf Ser Mater Sci Eng 309:012072. https://doi.org/10.1088/1757-899X/309/1/012072

Mousavi SMS, Dehghanzadeh R, Ebrahimi SM (2017) Comparative analysis of ozonation (O3) and activated carbon catalyzed ozonation (ACCO) for destroying chlorophyll a and reducing dissolved organic carbon from a eutrophic water reservoir. Chem Eng J 314:396–405. https://doi.org/10.1016/j.cej.2016.11.159

Ng C, Marshall W, Rao RM et al (2003) Granular activated carbons from agricultural by-products: process description and estimated cost of production. Bulletin 1–32

Priya AK, Jalil AA, Vadivel S et al (2022) Heavy metal remediation from wastewater using microalgae: recent advances and future trends. Chemosphere 305:135375. https://doi.org/10.1016/J.CHEMOSPHERE.2022.135375

Qadir M, Drechsel P, Jiménez Cisneros B et al (2020) Global and regional potential of wastewater as a water, nutrient and energy source. Nat Resour Forum 44:40–51. https://doi.org/10.1111/1477-8947.12187

Rahman MF, Yanful EK, Jasim SY (2009) Endocrine disrupting compounds (EDCs) and pharmaceuticals and personal care products (PPCPs) in the aquatic environment: implications for the drinking water industry and global environmental health. J Water Health 7:224–243. https://doi.org/10.2166/wh.2009.021

Rekhate CV, Srivastava JK (2020) Recent advances in ozone-based advanced oxidation processes for treatment of wastewater-a review. Chem Eng J Adv 3:100031. https://doi.org/10.1016/j.ceja.2020.100031

Restivo J, Orge CA, Guedes Gorito Dos Santos AS et al (2020) Nanostructured layers of mechanically processed multiwalled carbon nanotubes for catalytic ozonation of organic pollutants. ACS Appl Nano Mater 3:5271–5284. https://doi.org/10.1021/ACSANM.0C00662/SUPPL_FILE/AN0C00662_SI_001.PDF

Sa'ad SF, Wan Alwi SR, Lim JS, Abd Manan Z (2022) The economic study of centralised water reuse exchange system in the industrial park considering wastewater segregation. Comput Chem Eng 164:107863. https://doi.org/10.1016/J.COMPCHEMENG.2022.107863

Saeid S, Kråkström M, Tolvanen P et al (2020) Synthesis and characterization of metal modified catalysts for decomposition of ibuprofen from aqueous solutions. Catalysts 10:1–24. https://doi.org/10.3390/catal10070786

Tokumura M, Sugawara A, Raknuzzaman M et al (2016) Comprehensive study on effects of water matrices on removal of pharmaceuticals by three different kinds of advanced oxidation processes. Chemosphere 159:317–325. https://doi.org/10.1016/j.chemosphere.2016.06.019

Toze S (2006) Reuse of effluent water—benefits and risks. Agric Water Manag 80:147–159. https://doi.org/10.1016/j.agwat.2005.07.010

USEPA (2014) Final second list of chemicals for tier 1 screening, 1–6

USEPA contaminant candidate list 3 - CCL 3. https://www.epa.gov/ccl/contaminant-candidate-list-3-ccl-3

Xia C, Lim X, Yang H et al (2022) Degradation of per- and polyfluoroalkyl substances (PFAS) in wastewater effluents by photocatalysis for water reuse. J Water Process Eng 46:102556. https://doi.org/10.1016/J.JWPE.2021.102556

Zięzio M, Charmas B, Jedynak K et al (2020) Preparation and characterization of activated carbons obtained from the waste materials impregnated with phosphoric acid(V). Appl Nanosci (Switz) 10:4703–4716. https://doi.org/10.1007/s13204-020-01419-6

Chapter 12
Removal of Pharmaceutical Contaminants from Wastewater Using Novel Ceramic Nanomembrane Filters

Edith Nwakaego Chima, Helen M. K. Essandoh, Regina E. Edziyie, Omagbemi Omoloju Yaya, and Nwude O. Micheal

Abstract The inability of the existing conventional wastewater treatment plants to completely eliminate pharmaceutical contaminants from the wastewater is a global environmental issue. The current study is based on the investigation of the laboratory scale ceramic membrane filters (CMFs) made of two Ghanaian clay materials and cassava starch for their performance in improving the pharmaceutical quality of wastewater at different operating parameters. Additionally, the impact of clogging on the developed CMF experiment was considered in this study. The ceramic filters were produced by the paste casting method at varying ratios of soil, starch and grog, mixed with distilled water. The pharmaceutical detection was carried out with a Kromasil C8 column (150 mm × 4.60 mm, 5 μm). The results of the present study showed that the pharmaceutical removal depends on the flow rate, the pH of the feed solution, and the sludge retention age. Codeine, Morphine, paracetamol and tramadol were removed from 76 to 100% at all operating parameter, while ibuprofen and diclofenac were poorly removed at all pH conditions. Their effective removal was achieved at day 21 and 28. Clogging affects the filter performance with no significant change in the flux efficiency. However, cleaning was found to be effective in maintaining filter performance. Considering the results of this study, the CMF can be considered as an efficient and economical treatment method and could be applied in several related industries.

E. N. Chima (✉) · H. M. K. Essandoh
Department of Civil Engineering, College of Engineering, Kwame Nkrumah University of Science and Technology, Kumasi, Ghana
e-mail: edithnwakaego7@gmail.com

R. E. Edziyie
Department of Fisheries and Watershed Management, Kwame Nkrumah University of Science and Technology, Kumasi, Ghana

E. N. Chima · O. O. Yaya · N. O. Micheal
Regional Centre for Integrated River Basin Management, under the auspices UNESCO, Abuja, National Water Resources Institute, Kaduna, Nigeria

© UNESCO 2025   271
S. Zandaryaa et al. (eds.), *Emerging Pollutants*, Advances in Water Security,
https://doi.org/10.1007/978-3-031-71758-1_12

Keywords Ceramic filter · Pharmaceuticals · Removal efficiency · Flow rate ·
Treatment technologies

12.1   Introduction

The risk associated with pharmaceutical pollutants in water has prompted the need
for advanced wastewater treatment technologies in order to enhance their removal
from wastewater and minimize their interaction with the environment (Wang et al.
2018). Residual pharmaceutical compounds have the potential to cause physiological effects on human health and aquatic ecosystems (Souza et al. 2018; Wang et al.
2018). In an aqueous environment, pharmaceuticals occur as a complex pool that
may exerts more toxic effects than single drugs (Patel et al. 2019). Pharmaceuticals
can affect terrestrial animals through the food chain web, while humans can ingest
pharmaceutical residues via the consumption of fish and other types of meat, vegetables, plants, and fruits (Patel et al. 2019). Few research studies have reported their
bioaccumulation and potential toxic effects in both aquatic and terrestrial animals
(Zhou et al. 2009; Liu et al. 2015; Lu et al. 2016; Ko et al. 2018; Tran et al. 2018; Patel
et al. 2019). Nevertheless, previous studies have shown that the present conventional
wastewater treatment plants do not completely eliminate pharmaceutical contaminants from wastewater, as they are not equipped to remove complex compounds in
low concentrations (Couto et al. 2019). This is because majority of human-use pharmaceuticals are excreted as mixtures of the parent compound and metabolites that
are readily cleaved and develop resistance throughout wastewater treatment process,
resulting in their presence in effluents (Mceneff et al. 2014). Additionally, increased
sludge retention time (SRT) which encourages the growth of specialized bacteria are
efficient at accelerating the elimination of pharmaceutical compounds. In most cases,
conventional WWTPs are not designed to operate with an adequate SRT to achieve
this goal (Couto et al. 2019). The presence of pharmaceuticals in the environment is
relevant since some can be activated or transformed chemically by other compounds
and become persistent in the environment (Shraim et al. 2017).

Considering the criticality and dangers associated with these contaminants, great
concerns have been raised regarding their discharge into the environment to protect
human health and the ecosystem (Chonova et al. 2017; Wang et al. 2018), which has
also resulted in the anticipation that in the near future, there will be a proper treatment
method for effective removal of pharmaceuticals and other emerging contaminants
from the wastewater (Khan et al. 2019). Based on the aforementioned, there has
been an accelerated search for alternative methods for proper ways of managing and
treating wastewater (Castillo et al. 2020).

In order to diminish the level of these contaminants in the aqueous environment,
several advanced treatment technologies have been designed and tested for pharmaceutical removal (Chonova et al. 2017; del Álamo et al. 2020; Castillo et al. 2020).
Some of the studied systems include activated carbon filtration and membrane bioreactor (MBR) (Castillo et al. 2020), and ozonation/advanced oxidation processes

(AOPs) (Ajo et al. 2018; Souza et al. 2018). Though high removal efficiencies were achieved in using these systems, there were variations depending on the type of contaminant and treatment methods. According to Casas et al. (2015), MBR has a better removal efficiency for ibuprofen but not for diclofenac and tramadol. A study by Kovalova et al. (2013) found that 86% of pharmaceuticals could be successfully eliminated using activated carbon. Also, Wang et al. (2018) recorded a removal efficiency of 90% on average for different classes of pharmaceuticals. The advanced oxidation process was found to have a removal rate of more than 80% (Souza et al. 2018). Moreover, such treatment methods are sophisticated and expensive to operate (Chonova et al. 2017; Castillo et al. 2020). Despite the high operational and manufacturing costs of the abovementioned treatment technologies, an alternative treatment method with lower costs and that is easy to operate and maintain is needed (Castillo et al. 2020). Therefore, it has become imperative to develop an effective treatment method aimed at removing these contaminants from wastewater with locally available, cost-effective, and easily assessable raw materials.

Clay which forms the basis for ceramic filter production offers the characteristics to be used for this purpose, because it can be easily moulded into any shape that it retains when dry (Ajayi and Lamidi 2015). Naturally occurring clays are widely used for contaminant remediation due to their easy availability, high adsorption capacities, high porosity, high cation exchange capacity, and high specific surface area (Chaari et al. 2020). Various reports have been documented on the use of clay and different burn-out materials for ceramic filter preparation (Hasan et al. 2011; Nair and Mophinkani 2018; Bulta and Micheal 2019; Joseph and Douglas 2021).

Ceramic water filtration is described as a process of making water pass through a permeable ceramic material (Akosile et al. 2020). Ceramic filters have been used over the years as an alternative treatment method for drinking water (Akosile et al. 2020), storm or rain water (Shafiquzzaman et al. 2011), and wastewater (Wei 2015; Tshishonga and Gumbo 2017).

In addition, the ceramic membrane has several advantages, such as steady chemical property, high operating efficiency, high temperature resistance, long lifetime, no chemical additive, high strength, homogenous pore-size, easy installation, and other prominent advantages (Wei 2015). Despite the several advantages of ceramic filters, there is still a limit to their application due to membrane clogging. Thus, pretreatment of influents is required to minimize fouling (Barredo-Damas et al. 2010). Additionally, frequent cleaning is encouraged for its optimal performance.

However, feed characteristics, concentration, volume, and membrane operating parameters such as sludge retention time (SRT), hydraulic retention time (HRT), flow rate, temperature, pH, and pressure can both influence the quantity and quality of the treated water (Joo and Tansel 2015). The specific objectives of the present study are to: (1) evaluate the impact of various operating parameters, including pH, flow rate, and SRT, on the removal performance of the developed ceramic filters; and (2) examine the effect of clogging on the developed ceramic filter as the treatment periods increase.

12.1.1 Overview of Membrane Separation Technology

Membrane technology is an efficient treatment system for the removal of most target compounds, including pharmaceuticals, with removal efficacy a function of polarity, hydrophobicity, chemical nature, molecular weight, and membrane size. The application of membrane technology in wastewater and water treatment is becoming an increasingly attractive solution to water quality challenges (Shon et al. 2013). Membrane filtration is a pressure-driven process in which a membrane acts as a barrier to restrict the passage of pollutants but allows relatively clear water to pass through. Membrane charges and pore size play an important role in membrane function (Abdel-Fatah 2018). Water and wastewater treatment by membrane techniques are technically possible and cost-effective alternative to conventional treatment systems, since they are characterized by high efficacy in removing contaminants which satisfies high environmental requirements. Membranes are thin and porous material layers that eliminate water pollutants by allowing water to pass through at various speeds depending on the size of the membrane pore (Luo et al. 2014).

Reverse osmosis and other membrane filtration techniques, such as nanofiltration, have emerged as potential alternatives for removing micro contaminants. Nanofiltration, which has the advantages of smaller pores (1–10 nm) and low feed water pressure, is essential for the evacuation of emerging pollutants. Micropollutants are typically ineffectively removed during ultrafiltration and microfiltration because the membrane pore diameters are significantly bigger than the molecular sizes of micropollutants. Reverse osmosis and nanofiltration have substantially "tighter" structures than microfiltration and ultrafiltration (Luo et al. 2014). Reverse osmosis, on the other hand, provides almost total evacuation, but the higher energy consumption makes it less desirable (Arman et al. 2021).

Membrane processes, such as pressure-driven filtration technologies like nanofiltration, are regarded as some of the most innovative and very efficient processes. These are regarded as alternate techniques for removing substantial quantities of organic micropollutants, which include pharmaceuticals, personal care products, surfactants, diverse industrial chemicals, etc. (Arman et al. 2021).

In general, size exclusion, membrane adsorption, and charge repulsion are typically used in membrane processes to retain micropollutants. These removal processes are substantially influenced by various variables, including the type of membrane procedure, the characteristics of the membrane, the operating conditions, and the characteristics of the particular micropollutants (Luo et al. 2014). However, the method of membrane filtration has the drawback of fouling, which reduces flux and raises operational expenses (Arman et al. 2021).

12.2   Materials and Methods

12.2.1   Collection of Soil Samples and Preparation

Two soil samples were used in this study: Telekou Bokasso soil (whitish in colour) and Mfensi soil (greenish in colour). Both soil samples and grog were bought from the ceramics department at KNUST, Kumasi, Ghana. The starch was bought from Ayigya Market in Kumasi, in the Ashanti Region of Ghana. The two soil samples were soaked separately in distilled water for 24 h to reduce possible contamination. Thereafter, the supernatant was discarded, and the samples were air dried and ground with a pestle and mortar. The starch and grog were also pulverized. All samples were sieved with a 0.5 mm sieve size to obtain a fine powder with the same particle size. The raw materials utilized in this investigation were chosen for their availability, low cost, and the presence of qualities related to moisture content, plasticity, particle size distribution, swelling ability, fired strength, ion exchange, and surface area features. According to a previous study, these qualities aid in the production of ceramics (Asamoah et al. 2018).

12.2.2   Filter Preparation

The proportion of the two soil samples to burn-out material and grog varies depending on the desired flow rate to be produced (Table 12.1). The dry materials were hand mixed for 5–10 min to achieve an even distribution. Once mixed, about 1000 mL of distilled water was added gradually to the solid mass (Table 12.1). The total mixture was thoroughly stirred to form a smooth, homogeneous mixture. The paste casting method was employed, and the mixture was placed in a round bottom Plaster of Paris mould, and the filters were formed into cylindrical shapes. The filter capacity ranges from 500 to 600 mL of the mixture. Five filters were moulded from each composition, for a total of 20 filters. After 20–30 min, the filters were pressed and air-dried slowly for 10–15 days. However, adequate drying time before firing is required to avoid cracking and deformation (Bulta and Micheal 2019). When ready for firing, different batches of moulded dried filters were sintered for 3–5 h at 900 °C in a temperature-regulated furnace with a 5–10 °C/min heating rate. When the desired temperature has been reached and held for an additional 2 h, the furnace is turned off and allowed to cool overnight. Once cooled, the fired filters were removed and packed until ready for use (Fig. 12.1). Firing is very important in the filter making process as over-firing or under-firing can affect the filtration rate and quality of the filter. During firing, the burn-out/combustible material burns out, leaving small pores, which increases the porosity of the filter. A number of filters were fabricated by variances of burn-out material and clay soil. Table 12.1 shows the mixing ratio of the raw materials used in the production of the ceramic filters.

Table 12.1 Percentage composition of soil materials, grog and starch

Filter code	Number of filters	Flow rate (mL/h)	Telekou Bokasso % by (wt)	Mfensi % by (wt)	Starch % by (wt)	Grog % by (wt)	Firing temperature (°C)
1	5	10	60	30	5	5	900
2	5	50	55	25	15	5	900
3	5	150	50	20	25	5	900
4	5	300	45	15	35	5	900

Source Authors

Plaster of Paris (POP) mould used in forming the filters

Mixing of the raw materials

Pouring of the paste into the POP (a) and formation of the filter (b)

Formed filters ready for pressing

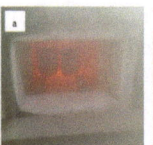
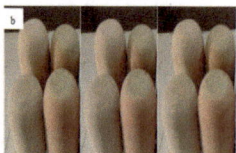

Filters drying at room temperature after pressing (a) and in oven at 105 ± 5 °C (b)

Filters during firing in the furnace (a) and after firing (b)

Fig. 12.1 Ceramic membrane filter (CMF) fabrication processes. *Source* Authors

12.2.3 Ceramic Filtration Experiments

In this present research, a laboratory-scale testing unit was designed for the reduction of diclofenac, ibuprofen, codeine tramadol, paracetamol and morphine in raw wastewater. The ceramic membrane filter was fabricated by the paste casting method using cost-effective and easily available raw materials based on clay soil and cassava starch, and they work by gravity filtration. The unit includes two plastic containers, ceramic filter and a rubber cover (Fig. 12.2). The ceramic filter is cylindrical in shape with one side opened (Fig. 12.3). The open side of the filter was sealed by a

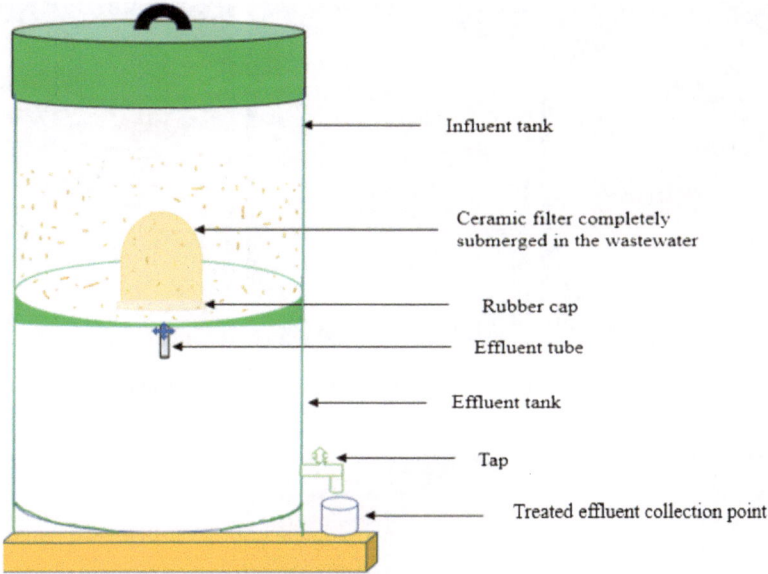

Influent tank

Ceramic filter completely
submerged in the wastewater

Rubber cap

Effluent tube

Effluent tank

Tap

Treated effluent collection point

Fig. 12.2   Diagram of ceramic filtration setup. *Source* Authors

rubber cap with a small opening for the effluent tube (Fig. 12.4). Two plastic buckets housed both the influent and effluent. One of the plastic buckets was manually fed with the raw wastewater that was obtained from the sewer point of KNUST main academic campus, which includes wastewater from lecturer's bungalow, student residence halls, student class rooms, and university laboratories and faculty areas, and the ceramic filter was attached at its bottom, while the second plastic bucket housed the effluent. These wastewater flows by gravity to the university's wastewater treatment facilities. The evaluation of the treatment plant for pharmaceutical removal should be the subject of further study since it was not done at the time this work because the facility was not operating.

To obtain fresh wastewater samples for the analysis, the ceramic filter was submerged vertically on a plastic bucket to ensure an effective use of the entire surface of the ceramic filter. During filtration wastewater passed through the filter by gravitational force and the treated effluent was directed to the effluent tank (a second plastic bucket) through the effluent tube (Fig. 12.3). The ceramic filtration experiment was operated as cross flow membranes and sampling was conducted immediately after treatment. To achieve the objective of this study, the untreated wastewater (1000 mL) obtained from the sewer point of KNUST main academic campus was spiked with 16 ppm of each target compound in addition to its original concentrations. The spiking was important to ensure the presence and high concentrations of all the studied compounds in the wastewater. Prior to spiking, the raw wastewater was filtered using cotton wool. This filtration was necessary to reduce clogging and

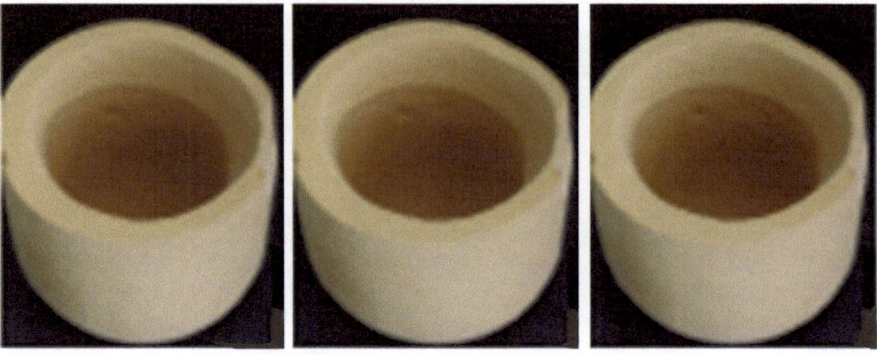

Fig. 12.3 Ceramic filters before covering with the rubber caps. *Source* Authors

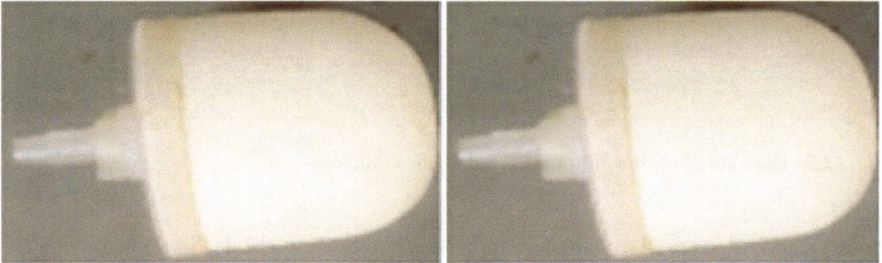

Fig. 12.4 Ceramic filters covered with the rubber caps. *Source* Authors

improve the performance of the filter. The pH of the raw wastewater was found to be 6.5–7.5, while the temperature was between 25.0 and 26.0 °C (Table 12.2).

12.2.4 Effect of Operating Parameters on Pharmaceutical Removal

In this study, the effect of flow rate, pH, and sludge retention time (SRT) on the pharmaceutical removal efficiency was considered. The raw wastewater (1000 mL) was spiked with 16 ppm of the individual pharmaceutical concentrations. The ceramic filtration experiment was conducted, and SPE-HPLC (High Performance Liquid Chromatography) protocols described in Sects. 2.6 and 2.7 below were performed separately for each operating parameter. The percentage removal efficiency (% R) was calculated according to Yadav et al. (2018) as follows:

$$\%\text{Removal} = \frac{C_{\text{inf}} - C_{\text{eff}}}{C_{\text{inf}}} \times 100$$

Table 12.2 Characteristics of the ceramic filter and operating conditions of the ceramic filtration process

Characteristics of the ceramic filter	
Raw materials	Soil materials
Combustible material	Cassava starch
Inner diameter (cm)	4–5
Outer diameter (cm)	7–8
Thickness (cm)	1–2
Shape	Cylindrical
Flow rate (mL/h)	10–300
Operating conditions of the filtration process	
Feed volume (mL)	100–400
Pressure (Kpa)	–
Filtration time (h)	1–10
Configuration	Completely submerged

Source Authors

where: C_{inf} and C_{eff} are the individual contaminant concentrations in both the influent and effluent. 100% removal was assumed when the studied compounds were detected in the influent but not in the effluent, while negative removal efficiency was assumed when the effluent concentration is greater than or equal to that of the influent (Yadav et al. 2018).

12.2.4.1 Effect of Flow Rate

The relationship between the changes in the flow rate of wastewater and removal efficiencies was carried out between four different filters with different flow rates (10, 50, 150, and 300 mL/h), based on recipe combinations ranging from 60:30:5:5, 55:25:15:5, 50:20:25:5, to 45:15:35:5, respectively (Table 12.1). Appendix 3 shows the initial and final concentrations of the individual pharmaceutical in the raw wastewater samples (Fig. 12.5).

12.2.4.2 Effect of pH

In this study, the effect of the solution pH on the removal efficiency was investigated at values of 7.0, 4.5, and 9.0 using a filter with a flow rate of 50 mL/h, since it gave a moderate flow rate. The pH of the solution was adjusted to the desired values using different volumes of 5% (v/v) of NaOH and HPLC-grade acetic acid. The concentrations of the pharmaceutical compounds before and after treatment are described in appendix 2 (Fig. 12.6).

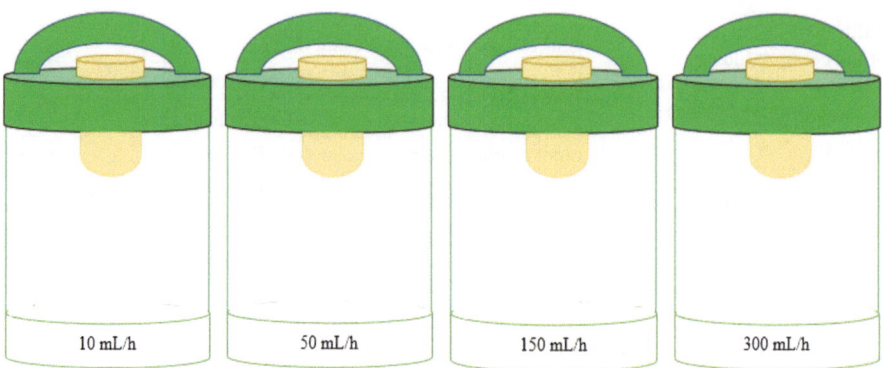

Fig. 12.5 Setup for pharmaceutical removal at different flow rate. *Source* Authors

Fig. 12.6 Setup to determine the effect of sludge retention time using a filter with a 50 mL/h flow rate. Note: The same setup was used to determine different parameters such as pH and sludge retention time. *Source* Authors

12.2.4.3   Effect of Sludge Retention Time

To evaluate the impact of increasing sludge retention time (SRT) on the removal efficiency of the target pharmaceutical contaminants, the spiked wastewater sample was stored for 28 days in the beaker (Appendix 4). About 200–300 mL of the filtered wastewater sample was extracted and analyzed at varying times as follows: 7, 14, 21, and 28 days. This study was carried out with a filter of a flow rate of 50 mL/h under the normal wastewater condition, which falls within the pH range of 6.5–7.5 (Fig. 12.7).

Fig. 12.7 Setup to determine the effect of sludge retention time using a filter with a 50 mL/h flow rate. Note: The same setup was used to determine different parameters such as pH and sludge retention time. *Source* Authors

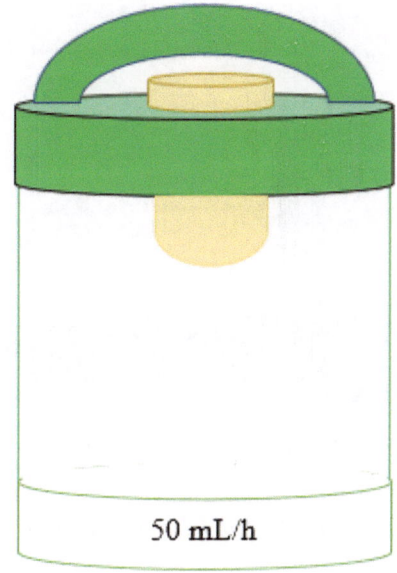

50 mL/h

12.2.5   Materials and Reagents

The following chemicals and reagents were used: Standards of ibuprofen (100.2%), diclofenac (99.1%), paracetamol (100.16%), codeine (99.5%), morphine (99.9%), and tramadol (99.74%) were obtained from Ernest Chemist Ltd., Accra, Ghana. Acetic acid, methanol, acetonitrile, and trifluoroacetic acid were obtained from Fisher Scientific (UK). All reagents were of HPLC-grade. Oasis HLB SPE cartridges (6CC, 500 mg) were obtained from Waters (Milford, MA, USA). Ultrapure water was prepared using the Elga Maxima HPLC ultrapure water system, which was equipped with an ultraviolet radiation. Stock solutions of the target drugs were prepared as follows: approximately 50 mg of the individual reference standard of the target compounds was accurately weighed by an analytical weighing balance. The weighed powder was transferred into six (6) different 50 mL volumetric flasks containing 20 mL of the diluent (methanol and/or water), depending on the solubility of the compounds. The solution was made up to the 50 mL mark on the volumetric flasks with the diluent to make 1000 ppm. The solution was sonicated to ensure complete dissolution and homogeneity. 0.02 mg/mL of the stock solutions were individually pipetted into six (6) separate 10 mL volumetric flasks and were made up to the mark in the volumetric flasks with ultrapure water to produce the working solutions. All the solutions were labelled properly and stored in the refrigerator at 4 °C.

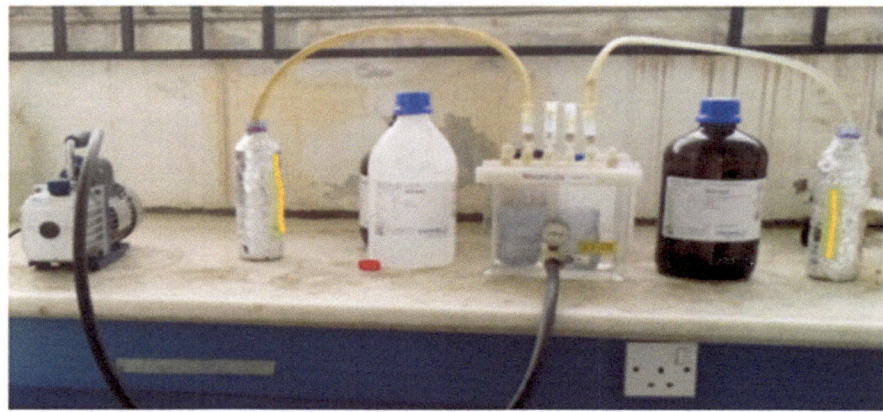

Fig. 12.8 Samples extraction using the solid phase extraction method. *Source* Authors

12.2.6 Solid Phase Extraction

The analytes in the wastewater samples were extracted using Oasis HLB cartridges (6 cc, 500 mg, Waters, Milford, MA, USA). Before extraction, each SPE cartridge was conditioned with 5 mL of methanol followed by 5 mL of distilled water. The prepared samples (about 100–400 mL, depending on the flow rate) were passed through the conditioned cartridges at a rate of 2 mL/min with the aid of an SPE manifold (Supelco Visiprep™, UK) and allowed to dry under vacuum for about 10 min. Each sample was eluted into a centrifuge tube with 3 mL of methanol and water (60:40, v/v). The final extracts were filtered through 0.45 μm polytetrafluoroethylene syringe-less filters (Whatman), evaporated to dryness under a gentle stream of nitrogen and reconstituted to 1 mL with methanol: water (60:40, v/v) in chromatographic vials (Fig. 12.8).

12.2.7 High Performance Liquid Chromatography Analysis

The high-performance liquid chromatography analysis was carried out according to the validated protocol described in our previous study yet to be published. All samples were performed in duplicate (n = 2) and prepared using solid phase extraction protocols. All analyte determinations were carried out by high performance liquid chromatography equipped with an autosampler (Perkin-Elmer Flexar LC Autosampler, N3896, USA). The chromatographic separations were performed on a Kromasil C8 column (150 mm 4.60 mm, 5 μm) at room temperature using a mobile phase containing 0.05% trifluoroacetic acid (pH 2.3) in water and acetonitrile in gradient mode. The mobile phase was delivered at a flow rate of 1 mL/min and a sample

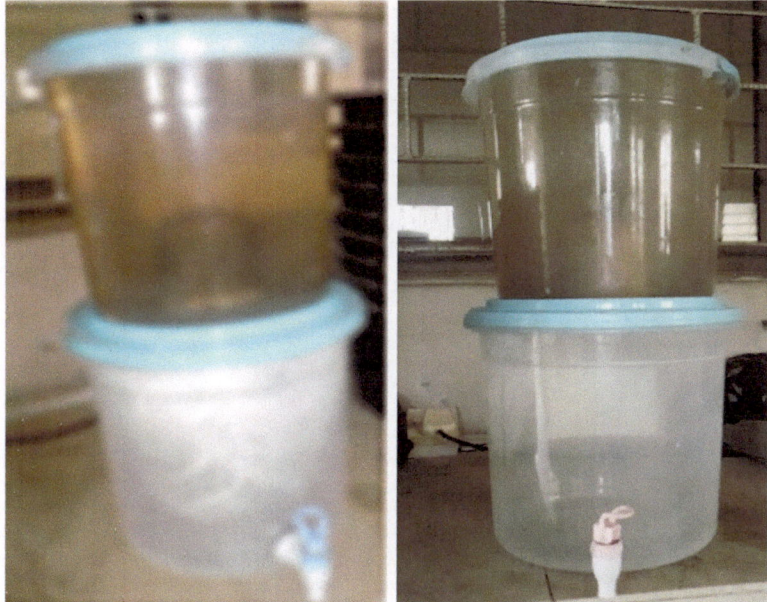

Fig. 12.9 Setup to determine the clogging effect using a filter with a 50 mL/h flow rate. *Source* Authors

injection volume of 20 μL with photo diode array detection at 225 nm. The target compounds were eluted depending on their affinity with the mobile phase.

12.2.8   Effect of Clogging

To determine the membrane clogging effects, the operation periods with the ceramic filtration were extended for four (4) months with non-spiked raw wastewater, and the filter was cleaned after every 8 weeks of operation. The percolated water is measured every week for sixteen weeks (Fig. 12.9).

12.3   Results and Discussion

12.3.1   Sample Analysis

After a successful validation in ultrapure water, a reversed phase stationary phase (150 mm × 4.60 mm, and 5 μm) was considered due to the polar or moderately polar nature of the selected compounds, and also to shorten the time of interaction

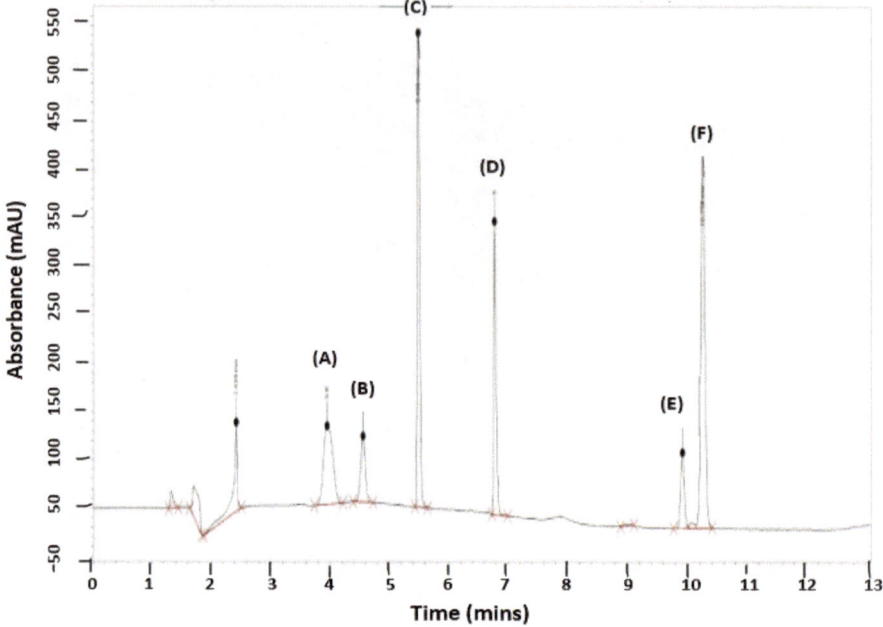

Fig. 12.10 The chromatogram for the HPLC analysis of the six (6) target compounds. *Source* Authors

between the analytes and the stationary phase. This is due to the analytes' increased interaction with the mobile phase and decreased affinity for the stationary phase. The mobile phase was pumped through the column at a flow rate of 1.0 mL/min and an injection volume of 20 μL for every injection. Figure 12.1 shows the corresponding chromatogram and the labels A, B, C, D, E, and F represent morphine (3.953), paracetamol (4.550), codeine (5.488), tramadol (6.769), diclofenac (9.915), and ibuprofen (10.258), while the numbers in the bracket represents their different retention times (Fig. 12.10).

12.3.2 *Effect of Operating Parameter on the Removal Efficiency*

12.3.2.1 Effect of Flow Rate on the Removal Efficiency

From Fig. 12.1, it is indicated that the filters with a lower flow rate treat pharmaceutical contaminants from wastewater samples more efficiently and effectively than filters with a higher flow rate. However, flow rate can be used as an indicator for filter efficacy as water passing at a faster flow rate has less contact time with the filter

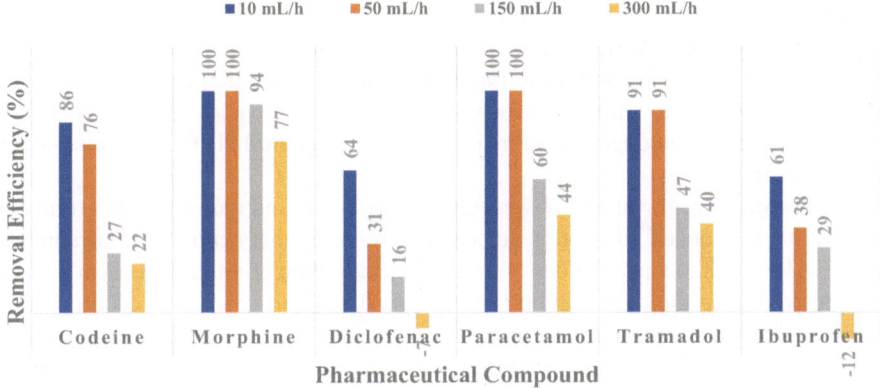

Fig. 12.11 Flow rate and removal efficiency. *Source* Authors

components as compared to when the water is flowing at a slower speed. Nevertheless, increasing the pore size of the ceramic filters may compromise the removal rate during filtration. Of all the filters produced, the results indicated that the filters produced at ratios of 60:30:5:5 and 55:25:15:5 showed better removal efficiency as they could remove between 76 and 100% of most of the pharmaceutical compounds in the wastewater owing to their low flow rate, which resulted from a lower percentage of the constituents of combustible material. Diclofenac and ibuprofen showed a removal percentage of 31–64% with the same filters. There were poor removal rates of all the compounds with filters produced at ratios of 50:20:25:5 and 45:15:35:5, having a flow rate of 150 and 300 mL/h, with negative removal for diclofenac and ibuprofen at a flow rate of 300 mL/h. This actually shows that these filters (flow rates) were not suitable for the treatment of diclofenac and ibuprofen (Fig. 12.11).

12.3.2.2   Effect of pH on the Pharmaceutical Removal Performance

Good removal efficiencies were observed with most of the analytes at the varying pH values as shown in Fig. 12.2. However, diclofenac and ibuprofen showed a significant decrease in removal rates under all conditions, with a negative removal rate for the acidic solution (pH 4.5). It was observed that the most important factor influencing the adsorption process is the pH of the aqueous solutions (Verlicchi et al. 2012; Chaari et al. 2020), because it affects the active sites of the solid surface areas. At basic conditions, as the water passes through the negatively charged ceramic filter, it ionizes the hydroxyl groups on the solid surface area of the ceramic filter, thus giving it a negative charge. This causes the positive ions of the contaminants to be attracted to the negative charges on the surface of the ceramic filter, thus favoring the removal of acidic compounds. Also, the surface areas become positively charged when the pH of the solution is low (acidic condition). In this case, the contaminants' negative ions adhere to the positive charges on the surface of the ceramic filter, facilitating the removal of basic compounds (Luo et al. 2014). However, the increased removal of the basic compounds (morphine, tramadol, codeine, and paracetamol) even for the solution under basic conditions may have resulted from other removal processes rather than adsorption, while their increased removal efficiencies at neutral pH may be due to increased biodegradation by microbes since neutral pH (7.0) is suitable for microbial growth.

The decreased removal rate observed with diclofenac and ibuprofen at neutral pH may be due to reduced adsorption since it was reported that acidic pharmaceuticals like NSAIDs with low pKa values (4.1–4.9) ionize to their conjugate bases at neutral pH, thus reducing their adsorption (Patel et al. 2019). However, low adsorption of ibuprofen has been reported in the previous literature (Couto et al. 2019). At a pH of 9.0, the CMF still produced a lower removal rate for diclofenac and ibuprofen despite the fact that they are acidic compounds, which predicts their increased removal due to increased adsorption between the positive charge of the contaminants and the negative charge on the surface of the ceramic filter. Their poor removal at a pH of 9.0 actually reflects their level of persistency during the CMF treatment processes. Their negative removal from the solution under acidic conditions (pH 4.5) may be due to the strong competition between the hydrogen ion (H^+) of the ceramic filter and positive ions of the contaminants in the solution, resulting in the decreased uptake of the positive ions at the active sites. It could also be due to the deconjugation of glucuronidated and sulfated diclofenac to release active diclofenac and the hydroxyl and carboxyl derivatives of ibuprofen to give active ibuprofen (Patel et al. 2019). Their recalcitrance to CMF treatment processes is of concern since their discharge to the environment has been reported to present some toxic effects (Larsson et al. 2014) (Fig. 12.12).

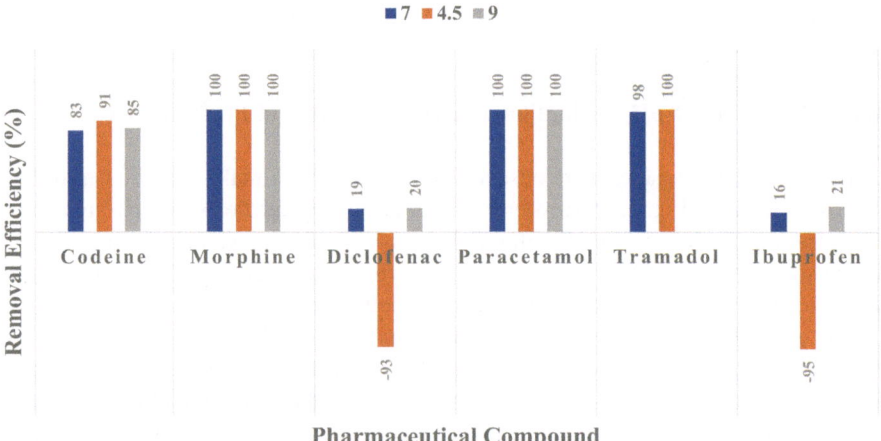

Fig. 12.12 pH and removal efficiency (flow rate: 50 mL/h). *Source* Authors

12.3.2.3   Effect of Sludge Retention Time on the Removal Efficiency

The results obtained from the sludge retention time (SRT) test (Fig. 12.13) showed the removal efficiency of some pharmaceuticals between 81 and 100%. From day 7 to 14, there is a slight improvement in the removal performance of most of the compounds. For example, tramadol was removed from 95 to 98%, while codeine was removed from 81 to 83%. On these days too, morphine and paracetamol were completely removed (100%). The removal rate for ibuprofen and diclofenac at day 7 was low, which implies that the time is not sufficient for the complete degradation of these compounds. It was unlikely that ibuprofen and diclofenac could persist for up to 7 days despite the microbial activities. However, their persistence is of great concern since it may lead to their release into the recipient river and consequently result in exposure to aquatic species, posing a threat to the environment. Nevertheless, a little improvement in the removal rates of diclofenac and ibuprofen was achieved at day 14 (from 68 to 75%). This may be attributed to increasing contact time with the biological organisms causing more biodegradation. From day 21–28, a clear feature common with all the compounds was the increased removal rate, varying from 91 to 100%, suggesting that increasing SRT may be an alternative method to improve the removal efficiency of pharmaceutical contaminants. With a longer SRT, Couto et al. (2019) reported pharmaceutical elimination increased on day 26, although using low SRTs (1.5–5.1 days) had only slight effects on the elimination of several pharmaceutical compounds such as ibuprofen, diclofenac, and others (Luo et al. 2014). A study by Verlicchi et al. (2012) also reported an increased percentage removal of anti-inflammatories and analgesics at increased sludge ages.

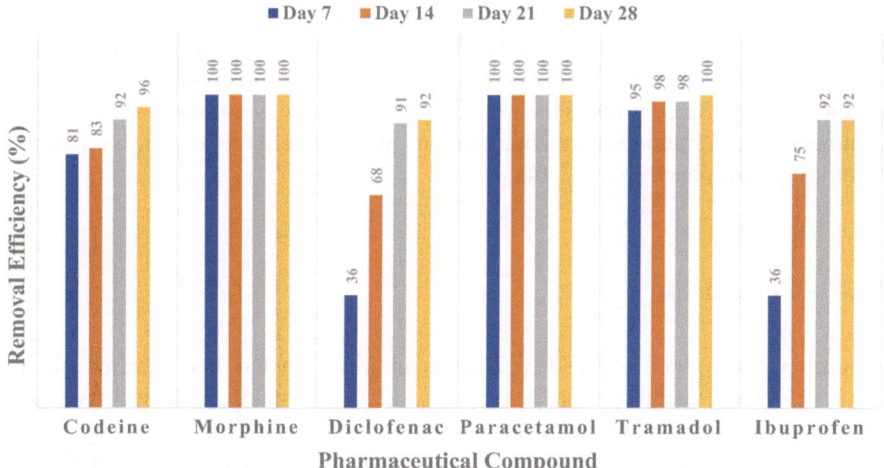

Fig. 12.13 Removal efficiency based on the sludge retention time (flow rate: 50 mL/h). *Source* Authors

12.3.3 Effect of Clogging on Ceramic Membrane Filter Experiment

In order to investigate the long-term clogging behavior of the ceramic membrane filter, the ceramic filtration experiment was operated continuously for four (4) months. The flow rate was measured every week for sixteen (16) weeks. The average flow rate after the first four weeks of operation was about 9.7 L per week, resulting in a flux efficiency of 97–100%. The flow rate was found to reduce considerably as the testing period increased. Increasing the treatment periods from 8 to 12 weeks has led to the drop-in flow rate to an average of 8.2 and flux efficiency of 81–86% (Fig. 12.4). The reduction in the flux of the ceramic filter as the operation period increases can be the result of clogging causing membrane pore blocking. The blocking was the result of the buildup of dirt inside and outside the layers of the ceramic membrane filters. These issues were the main problems associated with the application of the ceramic membrane (Wei 2015).

However, when the filter module was cleaned at week eight (8) using the cleaning procedure of distilled water and a soft brush, at weeks 9–12, the flux returned to 91–93% of the original volume. The result improved after cleaning the filter, which shows that membrane cleaning can restore its standard performance. However, this observation has previously been mentioned by Abdel-Fatah (2018) and Shafiquz-zaman et al. (2011). In Bangladesh, the ceramic filter demonstrated its performance after one year without clogging (Hasan et al. 2011). Although cleaning the filter improves the flow rate, the flow rate never returns to the original values. The flow rate continually diminished even after washing, as can be seen from weeks 13–16 (81–87%) due to subsequent clogging. However, frequent cleaning has been found to be effective in increasing the flow rate, and this can actually be incorporated into the maintenance of ceramic filters. The flow rate also depends on the quality of the influent wastewater. In addition to cleaning the filter modules, to reduce fouling, feed solutions were filtered prior to the treatment using cotton wool, suggesting that pre-treatment of feed solutions could increase flux and maintain the ceramic membrane's optimum performance (Fig. 12.14).

12.4 Conclusion

As earlier highlighted in the text, significant research has been carried out for the development technologies for the removal of pharmaceuticals in wastewater. However, significant gaps still exist, which highlighted the ongoing challenges to ensure the provision of affordable treatment methods. This paper addresses the development of ceramic membrane filters as an alternative treatment method for pharmaceutical removal. It also provides good basis for choosing the best operating parameters for performance evaluation. The study shows that this novel method is applicable

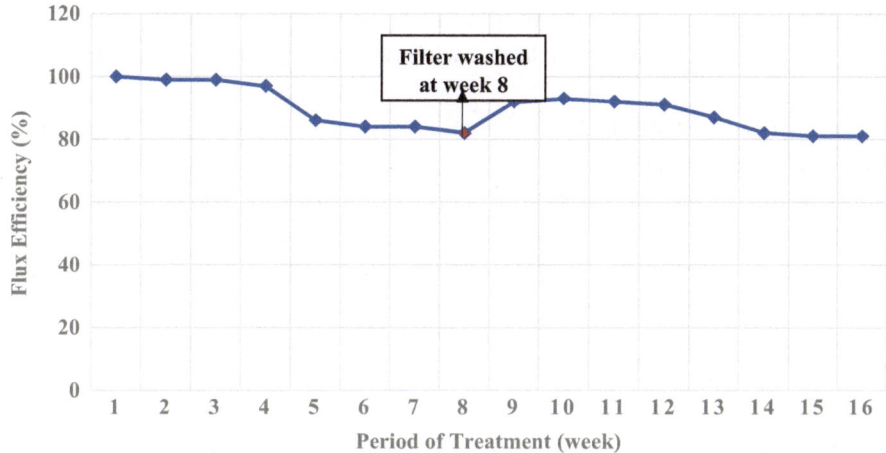

Fig. 12.14 Flux efficiency across four months treatment period (flow rate: 50 mL/h). *Source* Authors

for use for treatment of the selected analgesics. The removal of the selected analgesics from the wastewater is influenced by the operating parameters such as the flow rate of the filter, pH of the feed solution, and sludge retention age. The present study indicated that codeine, tramadol, paracetamol and morphine were removed from 76 to 100% for the different flow rates, pH and sludge retention times. Ibuprofen and diclofenac on the other hand, were resistant to the ceramic filtration process under all the pH conditions and the different flow rates, leading to their incomplete removal. Their effective removal was achieved by employing ceramic membrane filter experiments operated at day 21 and 28, possibly due to increased biodegradation. The persistence of ibuprofen and diclofenac is a major concern, though, as it could result in their release into the environment, endangering the ecosystems. Additionally, further research is recommended on the long-term applications of the developed ceramic filter and its application in the removal of other pharmaceuticals since prediction cannot be made on the rate of removal as pharmaceuticals behave differently during treatment processes.

12.5   Declaration of Interest

The authors declare that they have no known competing financial interests or personal relationships that could have appeared to influence the work reported in this work.

Acknowledgements The authors gratefully acknowledge the support from the National Water Resources Institute, Kaduna, Nigeria. This study was funded by the Regional Water and Environmental Sanitation Centre Kumasi (RWESCK) at the Kwame Nkrumah University of Science and

Table 12.3 Influent and effluent concentrations in the wastewater at different pH

	Codeine	Morphine	Diclofenac	Paracetamol	Tramadol	Ibuprofen
pH 7						
Initial Conc (ppm)	24.37805	16.1041	35.34845	25.2671	65.3652	46.87065
Final Conc (ppm)	4.1498	0	28.6647	0	1.57955	39.474
% removal	83	100	19	100	98	16
pH 4.5						
Initial Conc (ppm)	23.30485	27.9203	22.555	37.2843	28.7853	18.1565
Final Conc (ppm)	2.02155	0	43.5574	0	0	35.464
% removal	91	100	-93	100	100	-95
pH 9						
Initial Conc (ppm)	29.6901	23.65615	21.15475	24.2966	26.80415	17.91575
Final Conc (ppm)	4.3675	0	16.9505	0	bd	14.1174
% removal	85	100	20	100		21

Source Authors

Technology (KNUST), Kumasi, with funding from the Ghana Government through the World Bank under the Africa Centres of Excellence project. We thank you all.

Appendix

See Fig. 12.15 and Tables 12.3, 12.4 and 12.5.

Table 12.4 Table showing the influent and effluent concentrations in the wastewater at different flow rates

	10 mL/h			50 mL/h			150 mL/h			300 mL/h		
	Initial Conc (ppm)	Final Conc (ppm)	% removal	Initial Conc (ppm)	Final Conc (ppm)	% removal	Initial Conc (ppm)	Final Conc (ppm)	% removal	Initial Conc (ppm)	Final Conc (ppm)	% removal
Codeine	24.038	3.403	86	24.038	5.7458	76	24.038	17.632	27	24.038	18.681	22
Morphine	18.747	0.000	100	18.747	0.000	100	18.747	1.123	94	18.747	4.301	77
Diclofenac	28.8	10.507	64	28.836	19.929	31	28.86	24.198	16	28.836	30.867	-7
Paracetamol	30.139	0.000	100	30.139	0.000	100	30.139	11.957	60	30.139	16.791	44
Tramadol	25.035	2.305	91	25.035	2.220	91	25.035	13.298	47	25.035	14.926	40
Ibuprofen	38.965	15.367	61	38.965	24.305	38	38.965	27.497	29	38.965	43.684	-12

Source Authors

Table 12.5 Table showing the pharmaceutical concentrations in the wastewater before and after treatment at different sludge retention times

	Influent Conc (ppm)	Day 7		Day 14		Day 21		Day 28	
		Effluent Conc (ppm)	% removal	Effluent Conc (ppm)	% removal	Effluent Conc (ppm)	% removal	Effluent Conc (ppm)	% removal
Codeine	57.250	10.984	81	9.937	83	4.837	92	2.299	96
Morphine	39.062	0	100	0	100	0	100	0	100
Diclofenac	61.344	39.564	36	19.514	68	5.514	91	4.914	92
Paracetamol	31.838	0	100	0	100	0	100	0	100
Tramadol	48.853	2.510	95	1.208	98	1.098	98	0	100
Ibuprofen	55.926	35.615	36	13.856	75	4.695	92	3.989	92

Source Authors

Fig. 12.15 Color of effluent at different flow rate: 50 (**b**), 150 (**a**) and 300 (**c**) mL/h. *Source* Authors

References

Abdel-Fatah MA (2018) Nanofiltration systems and applications in wastewater treatment: review article. Ain Shams Eng J (Ain Shams University) 9(4):3077–3092. https://doi.org/10.1016/j.asej.2018.08.001

Ajayi BA, Lamidi YD (2015) Formulation of ceramic water filter composition for the treatment of heavy metals and correction of physiochemical parameters in household water. https://doi.org/10.4236/adr.2015.34013

Ajo P, Preis S, Vornamo T, Manttari M, Kallioinen M, Louhi-Kultanen M (2018) Hospital wastewater treatment with pilot-scale pulsed corona discharge for removal of pharmaceutical residues. J Environ Chem Eng (Elsevier) 6(2):1569–1577. https://doi.org/10.1016/j.jece.2018.02.007

Akosile SI, Ajibade FO, Lasisi KH, Ajibade TF, Adewumi JR, Babatola JO, Oguntuase AM (2020) Performance evaluation of locally produced ceramic filters for household water treatment in Nigeria. Sci Afr 7(1–13):e00218. https://doi.org/10.1016/j.sciaf.2019.e00218

Arman NZ, Salmiati S, Aris A, Salim MR, Nazifa TH, Muhamad MS, Marpongahtun M (2021) A review on emerging pollutants in the water environment: existences, health effects and treatment processes. Water (Switzerland) 13(22):1–31. https://doi.org/10.3390/w13223258

Asamoah RB, Nyankson E, Annan E, Agyei-Tuffour B, Efavi JK, Kan-Dapaah K, Apalangya VA, Damoah LNW, Dodoo-Arhin D, Tiburu EK, Kwofie SK, Onwona-Agyeman B, Yaya A (2018) Industrial applications of clay materials from Ghana (a review). Orient J Chem 34(4):1719–1734. https://doi.org/10.13005/ojc/340403

Barredo-Damas S, Alcaina-Miranda MI, Bes-Piá A, Iborra-Clar MI, Iborra-Clar A, Mendoza-Roca JA (2010) Ceramic membrane behavior in textile wastewater ultrafiltration. Desalination 250(2):623–628

Bulta AL, Micheal GAW (2019) Evaluation of the efficiency of ceramic filters for water treatment in Kambata Tabaro zone, southern Ethiopia. Environ Syst Res 8(1):1–15. https://doi.org/10.1186/s40068-018-0129-6

Casas ME, Chhetri RK, Ooi G, Hansen KMS, Litty K, Christensson M, Kragelund C, Andersen HR, Bester K (2015) Biodegradation of pharmaceuticals in hospital wastewater by staged Moving Bed Biofilm Reactors (MBBR). Water Res 83:293–302. https://doi.org/10.1016/j.watres.2015.06.042

Castillo L, Piotrowski P, Farnan J, Tasker TL, Xiong B, Weggler B, Murrell K, Dorman FL, Vanden JP, Burgos WD (2020) Detection and removal of biologically active organic micropollutants from hospital wastewater. Sci Total Environ 700:134469. https://doi.org/10.1016/j.scitotenv

Chaari I, Medhioub M, Jamoussi F, Hamzaoui AH (2020) Acid-treated clay materials (Southwestern Tunisia) for removing sodium leuco-vat dye: characterization, adsorption study and activation mechanism. J Mol Struct 1223(7):128944. https://doi.org/10.1016/j.molstruc.2020.128944

Chonova T, Labanowski J, Bouchez A (2017) Contribution of hospital effluents to the load of micropollutants in WWTP influents. In: The handbook of environmental chemistry (HEC), volume 60, pp 135–152

Couto CF, Lange LC, Amaral MCS (2019) Occurrence, fate and removal of pharmaceutically active compounds (PhACs) in water and wastewater treatment plants—a review. J Water Process Eng 32(8):100927. https://doi.org/10.1016/j.jwpe.2019.100927

del Álamo AC, González CP, Pariente MI, Molina R, Martínez F (2020) Fenton-like catalyst based on a reticulated porous perovskite material: activity and stability for the on-site removal of pharmaceutical micropollutans in a hospital wastewater. Chem Eng J 401(7):126113. https://doi.org/10.1016/j.cej.2020.126113

Hasan MM, Shafiquzzaman MD, Azam MS, Nakajima J (2011) Application of a simple ceramic filter to membrane bioreactor. Desalination 276(1–3):272–277. https://doi.org/10.1016/j.desal.2011.03.062

Joo SH, Tansel B (2015) Novel technologies for reverse osmosis concentrate treatment: a review. J Environ Manag 150:322–335. https://doi.org/10.1016/j.jenvman.2014.10.027

Joseph S, Douglas OE (2021) Filtration rate of ceramic clay filters treated with rice husk. J Sci Technol Educ 9(2):219–224

Khan NA, Ullah S, Ahmed S, Haq I (2019) Recent trends in disposal and treatment technologies of emerging-pollutants-a critical review. Trends Anal Chem 11:115744. https://doi.org/10.1016/j.trac.2019.115744

Ko S, Kotoka F, Kwame F (2018) Assessment and remediation of pollutants in Ghana's Kete-Krachi district hospital effluents using granular and smooth activated carbon. Heliyon 4:e00692. https://doi.org/10.1016/j.heliyon.2018.e00692

Kovalova L, Siegrist H, Von Gunten U, Eugster J, Hagenbuch M, Wittmer A, Moser R, McArdell CS (2013) Elimination of micropollutants during post-treatment of hospital wastewater with powdered activated carbon, ozone, and UV. Environ Sci Technol 47(14):7899–7908. https://doi.org/10.1021/es400708w

Larsson E, Al-Hamimi S, Jönsson JÅ (2014) Behaviour of nonsteroidal anti-inflammatory drugs and eight of their metabolites during wastewater treatment studied by hollow fibre liquid phase microextraction and liquid chromatography mass spectrometry. Sci Total Environ 485–486(1):300–308. https://doi.org/10.1016/j.scitotenv.2014.03.055

Liu J, Lu G, Xie Z, Zhang Z, Li S, Yan Z (2015) Occurrence , bioaccumulation and risk assessment of lipophilic pharmaceutically active compounds in the downstream rivers of sewage treatment plants. Sci Total Environ (Elsevier B.V.) 511:54–62. https://doi.org/10.1016/j.scitotenv.2014. 12.033

Lu MC, Chen YY, Chiou MR, Chen MY, Fan HJ (2016) Occurrence and treatment efficiency of pharmaceuticals in landfill leachate. Waste Manag 55:257–264

Luo Y, Guo W, Ngo HH, Nghiem LD, Hai FI, Zhang J, Liang S, Wang XC (2014) A review on the occurrence of micropollutants in the aquatic environment and their fate and removal during wastewater treatment. Sci Total Environ 473–474:619–641. https://doi.org/10.1016/j.scitotenv. 2013.12.065

Mceneff G, Schmidt W, Quinn B (2014) Pharmaceuticals in the aquatic environment: a short summary of current knowledge and the potential impacts on aquatic biota and humans synthesis report (EPA research report 143) available for download at Prepared for the Environmental Protection Agency. https://www.epa.ie/pubs/reports/research/water/Research142Repo rtFINAL.pdf

Nair CS, Mophinkani K (2018) Effectiveness of locally made ceramic water filters for household water purification. Int J Emerging Technol Adv Eng 75–82

Patel M, Kumar R, Kishor K, Mlsna T, Pittman CU, Mohan D (2019) Pharmaceuticals of emerging concern in aquatic systems: chemistry, occurrence, effects, and removal methods. Chem Rev 119(6):3510–3673. https://doi.org/10.1021/acs.chemrev.8b00299

Shafiquzzaman M, Hasan MM, Nakajima J, Mishima I (2011) Development of a simple, effective ceramic filter for arsenic removal. J Water Environ Technol 9(3):333–347. https://doi.org/10. 2965/jwet.2011.333

Shon HK, Phuntsho S, Chaudhary DS, Vigneswaran S, Cho J (2013) Nanofiltration for water and wastewater treatment - a mini review. Drinking Water Eng Sci 6(1):47–53. https://doi.org/10. 5194/dwes-6-47-2013

Shraim A, Diab A, Alsuhaimi A, Niazy E, Metwally M, Amad M, Sioud S, Dawoud A (2017) Analysis of some pharmaceuticals in municipal wastewater of Almadinah Almunawarah. Arab J Chem King Saud Univ 10:S719–S729. https://doi.org/10.1016/j.arabjc.2012.11.014

Souza FS, Da Silva VV, Rosin CK, Hainzenreder L, Arenzon A, Pizzolato T, Jank L, Feris LA (2018) Determination of pharmaceutical compounds in hospital wastewater and their elimination by advanced oxidation processes. J Environ Sci Health - Part A Toxic/hazard Subst Environ Eng 53(3):213–221. https://doi.org/10.1080/10934529.2017.1387013

Tran NH, Reinhard M, Gin KYH (2018) Occurrence and fate of emerging contaminants in municipal wastewater treatment plants from different geographical regions-a review. Water Res (Elsevier Ltd.) 133:182–207. https://doi.org/10.1016/j.watres.2017.12.029

Tshishonga M, Gumbo J (2017) The use of ceramic water filters in improving the microbial quality of drinking water. In: 9th international conference on advances in science, engineering, technology and waste management (ASETWM), 3–6 December 2017.https://doi.org/10.17758/eares.eap 1117029

Verlicchi P, Al Aukidy M, Zambello E (2012) Occurrence of pharmaceutical compounds in urban wastewater: Removal, mass load and environmental risk after a secondary treatment-a review. Sci Total Environ (Elsevier B.V.) 429:123–155. https://doi.org/10.1016/j.scitotenv.2012.04.028

Wang Y, Wang X, Li M, Dong J, Sun C, Chen G (2018) Removal of pharmaceutical and personal care products (PPCPs) from municipal wastewater with integrated membrane systems, MBR-RO/ NF. Int J Environ Res Public Health 15(2):1–12. https://doi.org/10.3390/ijerph15020269

Wei Y (2015) Application of inorganic ceramic membrane in wastewater treatment: membrane fouling control and cleaning procedure (Ism3e):289–291. https://doi.org/10.2991/ism3e-15.201 5.72

Yadav MK, Short MD, Gerber C, Van Den AB, Aryal R (2018) Occurrence, removal and environmental risk of markers of five drugs of abuse in urban wastewater systems in South Australia. Environ Sci Pollut Res 33(26):33816–33826

Zhou JL, Zhang ZL, Banks E, Grover D, Jiang JQ (2009) Pharmaceutical residues in wastewater treatment works effluents and their impact on receiving river water. J Hazard Mater 166:655–661. https://doi.org/10.1016/j.jhazmat.2008.11.070

Chapter 13
Emerging Pollutants in Wastewater: A Challenge for Water Reuse

Sabrine Hattab, Chayma Alaya, and Mohamed Banni

Abstract Although water represents two-third of our planet, we are facing a serious water shortage. The water crisis is the most pervasive, serious and invisible dimension of the earth's ecological devastation. Wastewater reuse has been considered as a promising solution to cope with the problem of water deficiency around the word. Both at the domestic and industrial levels, recycling and reuse of wastewater could have several sustainable benefits. The recycled water can be reused in the original process, reused in another process, or used in another sector or application. The implementation of wastewater reuse should take into account all the various types of affected stakeholders, accounting in addition for the external costs and benefits derived from the reuse decision. The objective of this chapter is to analyze the potential of wastewater reuse mainly for agricultural purposes stressing the toxicological aspects due to the potential presence of contaminants in these waters.

Keywords Wastewater · Climate change · Water scarcity · Toxic compounds

13.1 Introduction

Water is a precious resource for all forms of life and one of the most important raw materials exploited by man for his subsistence. Most of water needs are met by rainwater, which is accumulated in the surface or stored as groundwater. These water resources are becoming increasingly inaccessible (Goel 2006; Kummu et al. 2016). Some have attributed this water shortage to climate change, thus the greenhouse

S. Hattab · C. Alaya · M. Banni (✉)
Laboratory of Agrobiodiversity and Ecotoxicology, Higher Institute of Agronomy, University of Sousse, Sousse, Tunisia
e-mail: m_banni@yahoo.fr

S. Hattab
Regional Research Centre in Horticulture and Organic Agriculture, Chott-Mariem, Sousse, Tunisia

M. Banni
Higher Institute of Biotechnology, University of Monastir, Monastir, Tunisia

© UNESCO 2025
S. Zandaryaa et al. (eds.), *Emerging Pollutants*, Advances in Water Security,
https://doi.org/10.1007/978-3-031-71758-1_13

effect, rising temperatures, drought and lack of rainfall are major problems facing planet Earth today. This scarcity is also due to the growing demand for fresh water for agricultural, municipal, industrial and agricultural uses (Ganjegunte et al. 2017). Agriculture is one of the sectors that are highly dependent on water; about 70% of the fresh water withdrawn in the world is used for agricultural irrigation. The effects would be much greater in regions where water is scarce, such as semi-arid and arid regions (Nechifor and Winning 2017).

Under these conditions, it is therefore imperative to look for other alternative water sources, such as treated wastewater and desalinated seawater, to supplement conventional freshwater sources. Reusing treated wastewater could be an alternative option and a great way to alleviate water scarcity. Thus, many countries have decided to turn wastewater into irrigation resources to meet urban demand and alleviate water shortages, especially in areas where water is scarce (Yi et al. 2011). However, due to the nature of this water, its use for irrigation can pose problems. Health risks, increased salinity and toxicity risks are among the main concerns. Thus long-term irrigation with treated wastewater can lead to changes in soil characteristics, the accumulation of toxic substances and pollutants (resistant pathogens, antibiotics, metallic trace elements, organic compounds and microplastics, etc.) and several changes in microbial properties of irrigated soils as well as altered biological soil fertility (Klay et al. 2010; Hidri et al. 2014; Bedbabis et al. 2014; Mkhinini et al. 2018b; Lüneberg et al. 2018; Boughattas et al. 2021). These hazardous substances can be accumulated by soil fauna, often by ingestion (Hirano and Tamae 2011; Mkhinini et al. 2018a; Boughattas et al. 2021). As a consequence, they can provoke crucial alterations and disrupt the functional roles of soil organisms and their related metabolic activities (Pelosi et al. 2014; Boughattas et al. 2016).

13.2 Conventional Waters

13.2.1 State of Water Distribution in the World

The total amount of water on earth is estimated at over 332 million cubic miles (Gleick 1993). Of this total volume of water, only 8.3 million cubic miles are fresh water. Nearly 70% of this freshwater is trapped in glaciers and ice caps, the remaining 2.5 million cubic kilometers of water is 96% groundwater, and 0.032% of the total water occurs in the form of fresh surface water (lakes and rivers), ice and snow or other minor sources. Lakes and rivers account for less than 21% of surface fresh water. The total freshwater resources of the planet are estimated at around 43,750 cubic miles per year. However, humans use about 54% of this accessible runoff (FAO 2003; Gerbens-Leenes et al. 2008). In addition to its physical and quantitative dimensions, the perception of water resources also implies qualitative extent. Water is divided into two types of resources: non-renewable and renewable water resources. Surface water and groundwater, such as average river flow on an annual basis, are considered renewable water resources, while deep aquifers, which do not have a significant rate of renewal at scale Human time, are considered non-renewable water resources (FAO 2003).

13.2.2 Water and Climate Change

The scientific evidence for the global warming issue is now unequivocal, and scientists agree on the role of human activities in this warming. Anthropogenic emissions of greenhouse gases (GHGs), such as carbon dioxide, methane and nitrous oxide, have increased dramatically since pre-industrial times (Field and Barros 2014; OMM 2019). Climate change thus manifests itself through the frequency and magnitude of extreme weather events such as heat waves. Indeed, the average temperature on the surface of the globe has increased by almost 0.9 °C since the nineteenth century (Cheng et al. 2019). In fact, temperature change has a direct impact on the Earth's water budget (Field and Barros 2014; Bates et al. 2008; Schewe et al. 2014). Evaporation from the land surface has increased due to the global trend of increasing air temperatures in all regions, resulting in decreased water availability in different seasons (GIEC 2018), as well as the degradation of the physical, chemical and biological properties of freshwater lakes and rivers. Indeed, this change has a negative impact on many freshwater species and on the composition of communities and habitats (Field and Barros 2014). In addition, the rainfall deficit caused by climate change weakens soil moisture, river flow and groundwater recharge (Changnon 1987). Other risks are added, such as pollution. In fact, water pollutant concentrations and pathogen contamination are higher during droughts (UNESCO 2020).

13.3 Treated Wastewater

13.3.1 Definition and Field of Use

Wastewater is liquids and solids suspended in water that are discharged into sewers and represent the waste of community life (Sonune and Ghate 2004). They usually contain a high load of oxygen-demanding wastes, pathogens, organic matter, plant nutrients, chemicals inorganic, minerals and sediments. They may also contain toxic compounds (Tchobanoglous et al. 1991). Wastewater can be (Sonune and Ghate 2004):

- Domestic: wastewater discharged from homes, commercial establishments and similar facilities.
- Industrial: wastewater in which industrial waste predominates. In recent years, wastewater treatment has been seen as a real alternative to help reduce the water deficit. Treated wastewater (EUTs) are defined as "water that has received at

least secondary treatment and disinfection base and reused after discharge from a sewage treatment plant servant" (Hashem and Qi 2021). EUTs are generally used as a resource to meet urban demand and address water shortages (Ofori et al. 2021).

The known sources of Contaminants of Emerging Concern (CECs) include runoff and leakage from landfills, agriculture, hospitals, industrial and domestic wastewater (UNEP/MAP and Plan Bleu 2020). CECs span natural and artificial chemical substances and their by-products, antibiotic resistant genes (ARGs) and, more recently, the SARS-CoV-2 virus (Pastorino and Ginebreda 2021). Conventional wastewater treatment plants (WWTPs) are not designed to remove these contaminants; hence the removal is limited even when a tertiary treatment is in place. Treated wastewater reuse in agriculture, among the highest water consuming human activities, is one of the most debated solutions to tackle water scarcity. However, this approach is also considered as a source of contamination mainly with CECs (e.g, microplastics, pharmaceuticals) as well as pathogens. Nowadays, Europe's as well as Euro-Mediterranean long-term policy increasingly adopts a more systemic perspective that links the environment, human health and well-being and encourages solutions, simultaneously providing social and economic benefits and building resilience. The EU Mission to restore our ocean, seas and waters by 2030, the EU Biodiversity Strategy, the EU Action Plan Towards Zero Pollution for Air, Water and Soil, the Convention for the protection of the Mediterranean Sea against pollution, the proposed European Climate Law, and the UN Decade of Ocean Science for Sustainable Development all aim at protecting and restore marine and freshwater ecosystems and biodiversity. Actually, we are in need of innovative approaches with main target to provide response to prevent the spread of emerging pollutants and pathogens from treated wastewaters, and improve conditions for increasing water reuse for irrigation purposes. Innovative solutions are in need to mobilize both scientists and local communities for a sustainable and healthy environment. Urgent innovative solutions are in need to promote local actions and social innovations in co-creation with societal stakeholders, communities, scholars and citizens, to raise awareness of water quality issues and combat microplastic and emerging pollution at source directly, positively affecting society, environment, human health and economy.

13.3.2 Composition of Wastewater

13.3.2.1 Organic Compounds

Xenobiotic organic compounds are part of the composition of wastewater. Actually, trace organic compounds that cause new concerns include a wide range of chemicals, encompassing pharmaceuticals products and their metabolites, endocrine disruptors, personal care products, etc. (Fatta-Kassinos et al. 2011). Recent investigations reported the presence of these chemicals in wastewater effluents. Disposal of unwanted drugs and human excretion are some of the pathways by which these compounds enter the wastewater stream (Tong et al. 2011). In addition, analyzes by gas chromatography and mass spectrometry samples of filtered runoff water from agricultural fields irrigated by treated wastewater (EUTs) detected the presence of many non-target compounds presenting possible toxicological importance, including pesticides' metabolites, chemical pesticide adjuvants, plasticizers, flame retardants, ingredients for personal care chemicals and pharmaceuticals. Although the toxicity of most of these chemicals is poorly evaluated, some of them may have slight but long-term toxicological effects. (Pedersen et al. 2003).

13.3.2.2 Metallic Trace Elements

Contamination of water by metallic trace elements (MTEs) is a form of water pollution and becomes a severe hazard to all living beings. (Tchounwou et al. 2012). Miscellaneous industries such as textiles, leather, mining, fertilizers, pesticides, metallurgy, paper mills and surface treatment generate and discharge into the environment wastewater containing a variety of types of ETMs (Jaishankar et al. 2014). ETMs mainly present in wastewater are Pb, Cd, Cr, Zn, Cu, Fe, Ni, and the Mg. These MTEs are therefore not biodegradable in nature and when consumed, they easily accumulate in biota (Huang et al. 2014; Mahmud et al. 2016; Mkhinini et al. 2018a). This waste water containing ETMs is directly or indirectly discharged into lakes, streams, rivers or oceans, notably in developing countries (Kumar et al. 2017; Owlad et al. 2009).

13.3.2.3 Microplastics

Microplastics (MPs) are plastics less than 5 mm in size, produced deliberately, i.e. primary, (Browne et al. 2011) or secondary MPs, resulting weathering of bigger plastics (Andrady 2011), which can have harmful effects on biota and ecosystems (Van Cauwenberghe et al. 2015). Microplastics are present in wastewater (Browne et al. 2011) and can become more harmful by adsorbing toxic agents, such as pathogenic organisms (McCormick et al. 2014) and pharmaceuticals (Ziajahromi et al. 2017). Despite the importance of wastewater as a source of microplastics, a gap of knowledge still exists in the scientific literature world regarding their potential impact if reused for agriculture purposes. Microplastics present in wastewater come mainly from domestic discharges (clothing, cosmetics and cleaning products...) (Carney Almroth et al. 2018). In sewage treatment plants, the most of the microplastics are eliminated, retained in the solid fraction which can contaminate terrestrial ecosystems when applied as fertilizer, and some remain in the final effluent that contaminates aquatic ecosystems (Prata 2018). Non-domestic sources of microplastics in wastewater can comprise: pre-production pellets lost during transport or manufacturing (Sheavly and Register 2007), plastic particles used in airbrushing to clean paint and motors (Gregory 1996), waste polystyrene used for filling or shipping, ...

13.3.2.4   Pathogens

Wastewater flows contain different types of pathogens, including bacteria, viruses, protozoa and helminths (Cai and Zhang 2013; Chahal et al. 2016), which can penetrate from many sources: human and animal fecal waste or water contaminated with feces from other domestic uses (Hill 2003; Gerardi and Zimmerman 2004).Although most of these pathogens are eliminated by the various types of wastewater treatment, the huge volumes of water rejected into the environment transport a considerable load of micro-organisms that vary according to various factors such as geographical area, season and type of treatment process (Kitajima et al. 2014; Newton and McClary 2019). Bacteria are considered as the most diverse group of human pathogens in wastewater with enteric pathogenic bacteria representing the majority of these germs, including Escherichia spp, Salmonella spp, Shigella spp, Vibrio cholerae, Klebsiella spp, Yersinia spp, Leptospira spp, Legionella pneumophila, Aeromonas hydrophila, Pseudomonas and Mycobacterium spp (Stevik et al. 2004; Maynard et al. 2005; Cai and Zhang 2013; Varela and Manaia 2013). The rise in the use of antibiotics in several countries has led to the increased presence of multi-resistant bacteria in wastewater (Bitton 2005; Bouki et al. 2013). Viruses are another important group of water-borne human pathogens, with the main germs present in wastewater consisting of enteric viruses such as rotavirus, adenovirus, hepatitis A, astrovirus, norovirus and various enteroviruses (Ashbolt 2004; Cai and Zhang 2013). Industrial waste from slaughterhouses can also contribute zoonotic viruses to wastewater, such as hepatitis E, sopaviruses and animal adenoviruses (Wyn-Jones et al. 2011). These viruses can be inadequately eliminated by the secondary treatment processes used for bacteria (Ottoson et al. 2006). Also, parasites are one of the most frequently identified pathogens in wastewater (Hatam-Nahavandi et al. 2015). The use of partially or untreated wastewater can lead to the transmission of these organisms. In fact, the infectious stages of parasites—eggs, oocytes and cysts—are resistant stages due to the presence of solid external layers that protect them from various physical and chemical destructions and can make them resistant to different types of treatment processes. Pathogenic parasites can therefore persist and survive in wastewater for a longer period than bacteria and viruses (Tomass and Kidane 2012; Hogan et al. 2013). The presence of parasites in wastewater has been demonstrated in several African cities. Studies have shown that wastewater in Africa contains a variety of pathogenic parasites such as protozoa and helminths. Although wastewater treatment systems eliminate most helminths, they still cannot eliminate protozoa (Zacharia et al. 2018).

13.4   Effects of Treated Wastewater

13.4.1   Impacts on Human Health

Unlike drinking water intended for domestic use, treated wastewater can contain organic pollutants, toxic substances and even pathogens as well as microplastics (Yi et al. 2011). These Hazardous and pathogens may provoke health risk to consumers if not adequately managed (Khalid et al. 2018). The ingestion of contaminated food and inhalation are the main ways by which pathogens and pollutants can enter the human body (Yiet al. 2011). Wastewater is a main source of human exposure to pathogenic microorganisms such as bacteria, viruses and protozoa. Non-treated wastewater contains different types of bio-contaminants such as Salmonella spp, *Escherichiacoli*. These pathogens can persist in large amounts even in treated wastewater, particularly if advanced filtration treatment such that the membranes or disinfection are not included in the treatment protocol. A high occurrence of enteric viruses has been spotted in wastewater effluent waste that has undergone secondary treatment, with adenovirus being the most common (Schlindwein et al. 2010). Besides, treated wastewater may contain persistent substances such as MTEs and non-biodegradable contaminants, which can bio-concentrate in the environment matrixes and enter the food chain (Cherfi et al. 2015). Although the concentrations of MTEs in the wastewaters used for irrigation are generally low, long-term irrigation practices may pose an environmental risk and as a consequence a human health risk (Kim et al. 2015). The effects of MTEs on human health are well documented; some can cause cancer, while others are harmful to the nervous, circulatory, immune, endocrine or, as well as for vital organs (Cherfi et al. 2015). There may also be a risk of a synergistic effect with other pollutants, such as residual pharmaceuticals and antibiotics (Becerra-Castro et al. 2015).

13.4.2   Impact of Treated Wastewater on the Terrestrial Ecosystem

Several studies have revealed the impact of wastewater on soil, plants/crops and water. Studies on soils revealed that the use of treated wastewater (TWW) for irrigation can affect soil parameters and cause a decrease in pH, an increase in salinity, levels of potassium (K), phosphorus (P), manganese (Mn) and iron (Fe), an accumulation of nutrients and MTEs, a disturbance of electrical conductivity, as well as changes in the structure and texture of the soil, the rate of adsorption of sodium and changes in the soil microbial community (Mohammad and Mazahreh 2003; Klay et al. 2010; Bedbabis et al. 2014; Ibekwe et al. 2018). Additionally, other studies on plants have shown that the use of treated wastewater has resulted in the accumulation and translocation of ETM in fruits or edible parts of plants (Kalavrouziotis et al. 2012). MTEs can affect photosynthesis and in general plant metabolism, and opening of stomata, which are directly linked to plant growth (Parveen et al. 2015) and can also modify microbial communities or cause phytotoxicity (Becerra-Castro et al. 2015). In addition to metallic trace elements, microplastics present in wastewater can have negative effects on terrestrial ecosystems. To our knowledge, direct studies on particles from EUT irrigation are very limited. Nevertheless, the irrigation of plants with TWW may not present any direct risk for plants, because direct uptake of particles by plants is unlikely (Hurley and Nizzetto 2018). However, TWW irrigation can promote the bio-concentration of particles in agricultural soils and have a negative impact on soil integrity and wildlife. The plastic particles can also promote the presence of other contaminants (persistent organic pollutants (POPs), metals, etc. in the soil due to their hydrophobic surface which allows them to adsorb other contaminants (Blair et al. 2017; Horton et al. 2017; He et al. 2018). This may depreciate the soil quality and induce toxicity in soil biota and ultimately plants uptake of contaminants.

13.5   Diagnosis of Terrestrial and Aquatic Ecosystems: Biomonitoring Approach

Environmental and biological monitoring is becoming a crucial issue to evaluate the extent of environmental contamination in the current state of the quality of the ecosystem (Sariana et al. 2020). Thus, biomonitoring is a novel approach for assessing the exposure of the environment and humans to natural and anthropogenic chemicals. This technique takes advantage of the fact that untaken chemicals may provoke changes at cellular and tissue levels reflecting this exposure. The marker can be either the chemical substance itself, or the product of its degradation, or a biological change in organisms. The results of these measurements offer insights on the quantities of chemicals to which organisms were exposed, as well as on the resultant effects induced. Because of the reliability between the selected organisms and the corresponding hazardous, the biomonitoring can provide robust data on potential effects, reflecting the corresponding degree of harmfulness in the environment. (Zhang et al. 2004).

13.5.1   Bioindicator Approach

Bioindicators are defined as organisms (or communities of organisms) that may reflect the quality of the environment (or part of the environment) (Markert et al. 2003) In the context of environmental monitoring assessments. Environmentalists have established a comprehensive set of criteria for bioindicators are considered valid biological indicators (Zaghloul et al. 2020) Thus, a perfect bioindicator should have the following characteristics: it can accumulate high amounts of hazardous without dying, it lives in a sedentary style, reflecting thus clearly spotted pollution, it is sufficiently abundant and widely distributed toallow repeated sampling and comparison, its service life is sufficiently long to allow comparison between different ages, it is easy to sample and to maintain in the laboratory, it and a good dose–effect relationship can be observed in it (Zhou et al. 2008).

The use of biomarkers in the study of the effects of environmental contaminants on bioindicators such as earthworms for soils our mussels for marine ecosystems has been shown to be very revealing of the organism/population dynamic response to stress (Jänsch et al. 2005; Maity et al. 2008; Hattab et al. 2015; Boughattas et al. 2016) Thus, using a set of cytotoxic, genotoxic, biochemical and transcriptomic parameters, earthworms could contribute to the ecological assessment of any soil additive or pollutant to maintain soil fertility and ecosystem functionality (Boughattas et al. 2022).

13.5.1.1 Biomarkers of Cytotoxicity

Lysosomal membrane stability

The lysosomal system has been revealed as a specific target for the toxic effects of pollutants at the subcellular level, (Moore 1990), and the pathological modifications of lysosomes have been particularly crucial in revealing negative environmental effects on numerous organisms (Moore 1980; Giamberini and Pihan 1997), when exposed to pollutants, one of the characteristic pathological alterations is a decrease in the integrity of the lysosomal membrane (Moore 1988). The lysosomal membrane stability test is considered as cytotoxicity test and a widely used biomarker in bioindicators such as mussels, worms, fish after exposure to contaminants (Svendsen et al. 2004; Boughattas et al. 2022).

Biomarkers of oxidative stress

Biomarkers of oxidative stress can be divided into two groups: molecules that are affected by interactions with reactive oxygen species (ROS) in the environment, such as DNA, lipids (including phospholipids), proteins and carbohydrates, and the second category consists on molecules of the antioxidant system that react in response to increased redox stress. Among these changes, some are known to have direct effects on the function of the molecule (e.g, inhibition of enzyme function), others can only reflect the extent of oxidative stress in the environment (Ho et al. 2013).

Enzymatic biomarkers

The cell can apply a defense mechanism against oxidative stress either by transforming ROS in less toxic compounds or by scavenging them. The antioxidant defense system consists of some enzymes, some proteins and some low-weight molecules (Nandi et al. 2019).

Glutathione-S-transferase (GST) represents a family of multifunctional enzymes involved in numerous intracellular transport and biosynthetic pathways (George and Buchanan 1990). It is a phase II enzyme known to be directly implicated in the detoxification of various organic xenobiotics (Booth et al. 2000) as well as in the metabolism of lipophilic organic toxins, but it also plays a pivot role in the cellular protection against oxidative stress (Hattab et al. 2015; Maity et al. 2018; Mkhinini et al. 2018a).

Catalase (CAT) is a well-known antioxidant enzyme, its activity generally increased in organisms exposed to oxidative stress (Boughattas et al. 2016). This enzyme is in charge of the neutralization of hydrogen peroxide, thus maintaining an optimal level of molecule in the cell, which is also essential for signaling processes cellular (Nandi et al. 2019). Thus, catalase breaks down two molecules of hydrogen peroxide (the disproportionation of H2O2 usually produced by SOD) into one molecule of oxygen and two molecules of water (Deisseroth and Dounce 1970).

Biomarkers of lipid peroxidation

Lipids are sensitive targets of oxidation due to their molecular structure abundant in reactive double bonds. The most studied markers of the peroxidation of lipids are isoprostanes (IsoP), Thiobarbituric acid (T Bars) and Malonedialdehyde (MDA) (Porter et al. 1995).

Free radicals initiate the process of lipid peroxidation in an organism. Malonedialdehyde (MDA) is one of the end products of polyunsaturated fatty acid peroxidation in the cells. An increase in free radicals leads to overproduction of MDA. The level of MDA is generally known as a marker of oxidative stress as well as to assess the state of lipid peroxidation membranes (Gaweł et al. 2004).

13.5.1.2 Biomarkers of Neurotoxicity

Acetylcholinesterase (AChE) is an enzyme crucial for the correct transmission of nervous impulses. Its inhibition is directly associated with the mechanisms of toxic action of some pesticides and metals (Booth et al. 2000; Boughattas et al. 2016). ACHE is known as the main cholinesterase in worms, insects as well as in mussels and fish (Rault et al. 2007), where it is the key enzyme in nervous system that ends nerve impulses by catalyzing the hydrolysis of the neurotransmitter; acetylcholine into choline and acetate (Caselli et al. 2006).

13.5.1.3 Genotoxicity Biomarkers

The micronucleus test (MNi) is a very useful parameter for biomonitoring exercises in the field. A better understanding of the mechanisms of MNi production allows realistic conclusions to be drawn about their biological significance. It has been demonstrated that a disruption of the mitotic apparatus as well as impaired function topoisomerase may be involved in the formation of micronuclei. Furthermore, the role of DNA conformational changes induced by state alterations cytosine methylation and cellular DNA repair capacity has been proved. The fate of cells containing micronuclei is not well known: cells can theoretically be cytostatic and the formation of micronuclei may therefore be a means for the body to withdraw genetic damages (Stopper and Müller 1997). The MNi frequency test has been shown to be a robust and useful biomarker for assessing DNA alteration in a set of bioindicators such as earthworms, mussels, rats after exposure to genotoxic contaminants (Boughattas et al. 2016; Khalil 2016; Mkhinini et al. 2018a).

13.6 Conclusion

Treated wastewaters reuse in agriculture may constitute an interesting alternative to conventional waters in a climate change context. This option can become a real solution under certain extreme condition such as arid and semi-arid regions. However, it is mandatory to improve the treatment processes aiming to minimize potential contamination loads. Emerging contaminants may pose a critical problem in such unconventional waters and pollutants mitigation strategies should be taken into consideration in wastewater reuse.

References

Andrady AL (2011) Microplastics in the marine environment. Mar Pollut Bull 62:1596–1605. https://doi.org/10.1016/j.marpolbul.2011.05.030

Ashbolt NJ (2004) Microbial contamination of drinking water and disease outcomes in developing regions. Toxicology 198:229–238. https://doi.org/10.1016/j.tox.2004.01.030

Bates B, Kundzewicz Z, Wu S (2008) Climate change and water. Intergovernmental Panel on Climate Change Secretariat

Becerra-Castro C, Lopes AR, Vaz-Moreira I, Silva EF, Manaia CM, Nunes OC (2015) Wastewater reuse in irrigation: a microbiological perspective on implications in soil fertility and human and environmental health. Environ Int 75:117–135. https://doi.org/10.1016/j.envint.2014.11.001

Bedbabis S, Ben Rouina B, Boukhris M, Ferrara G (2014) Effect of irrigation with treated wastewater on soil chemical properties and infiltration rate. J Environ Manag 133:45–50. https://doi.org/10.1016/j.jenvman.2013.11.007

Bitton G (2005) Wastewater microbiology. Wiley, New York

Blair RM, Waldron S, Phoenix V, Gauchotte-Lindsay C (2017) Micro- and nanoplastic pollution of freshwater and wastewater treatment systems. Springer Sci Rev 5:19–30. https://doi.org/10.1007/s40362-017-0044-7

Booth L, Heppelthwaite V, Mcglinchy A (2000) The effect of environmental parameters on growth, cholinesterase activity and glutathione S-transferase activity in the earthworm (Apporectodea caliginosa). Biomarkers 5:46–55. https://doi.org/10.1080/135475000230532

Boughattas I, Hattab S, Boussetta H, Sappin-Didier V, Viarengo A, Banni M, Sforzini S (2016) Biomarker responses of Eisenia andrei to a polymetallic gradient near a lead mining site in North Tunisia. Environ Pollut 218:530–541. https://doi.org/10.1016/j.envpol.2016.07.033

Boughattas I, Hattab S, Zitouni N, Mkhinini M, Missawi O, Bousserrhine N, Banni M (2021) Assessing the presence of microplastic particles in Tunisian agriculture soils and their potential toxicity effects using Eisenia andrei as bioindicator. Sci Total Environ 796:148959. https://doi.org/10.1016/j.scitotenv.2021.148959

Boughattas I, Zitouni N, Hattab S, Mkhinini M, Missawi O, Helaoui S, Mokni M, Bousserrhine N, Banni M (2022) Interactive effects of environmental microplastics and 2,4-dichlorophenoxyacetic acid (2,4-D) on the earthworm Eisenia andrei. J Hazard Mater 424:127578. https://doi.org/10.1016/j.jhazmat.2021.127578

Bouki C, Venieri D, Diamadopoulos E (2013) Detection and fate of antibiotic resistant bacteria in wastewater treatment plants: a review. Ecotoxicol Environ Saf 91:1–9. https://doi.org/10.1016/j.ecoenv.2013.01.016

Browne MA, Crump P, Niven SJ, Teuten E, Tonkin A, Galloway T, Thompson R (2011) Accumulation of microplastic on shorelines woldwide: sources and sinks. Environ Sci Technol 45:9175–9179. https://doi.org/10.1021/es201811s

Cai L, Zhang T (2013) Detecting human bacterial pathogens in wastewater treatment plants by a high-throughput shotgun sequencing technique. Environ Sci Technol 47:5433–5441. https://doi.org/10.1021/es400275r

Carney Almroth BM, Åström L, Roslund S, Petersson H, Johansson M, Persson N-K (2018) Quantifying shedding of synthetic fibers from textiles; a source of microplastics released into the environment. Environ Sci Pollut Res 25:1191–1199. https://doi.org/10.1007/s11356-017-0528-7

Caselli F, Gastaldi L, Gambi N, Fabbri E (2006) In vitro characterization of cholinesterases in the earthworm Eisenia andrei. Comput Biochem Physiol Part C Toxicol Pharmacol 143:16–421. https://doi.org/10.1016/j.cbpc.2006.04.003

Chahal C, van den Akker B, Young F, Franco C, Blackbeard J, Monis P (2016) Pathogen and particle associations in wastewater. Adv Appl Microbiol 97:63–119. https://doi.org/10.1016/bs.aambs.2016.08.001

Changnon SA (1987) Detecting drought conditions in Illinois. Illinois State Water Survey, Champaign (2204 Griffith Dr, Champaign 61820)

Cheng L, Abraham J, Hausfather Z, Trenberth KE (2019) How fast are the oceans warming? Science 363:128–129. https://doi.org/10.1126/science.aav7619

Cherfi A, Achour M, Cherfi M, Otmani S, Morsli A (2015) Health risk assessment of heavy metals through consumption of vegetables irrigated with reclaimed urban wastewater in Algeria. Process Saf Environ Prot 98:245–252. https://doi.org/10.1016/j.psep.2015.08.004

Deisseroth A, Dounce AL (1970) Catalase: physical and chemical properties, mechanism of catalysis, and physiological role. Physiol Rev 50:319–375. https://doi.org/10.1152/physrev.1970.50.3.319

FAO (Food and Agriculture Organization of the United Nations) (2003) Review of world water resources by country

Fatta-Kassinos D, Kalavrouziotis IK, Koukoulakis PH, Vasquez MI (2011) The risks associated with wastewater reuse and xenobiotics in the agroecological environment. Sci Total Environ 409:3555–3563. https://doi.org/10.1016/j.scitotenv.2010.03.036

Field CB, Barros VR (2014) Climate change 2014 – impacts, adaptation and vulnerability: global and sectoral aspects. Cambridge University Press, Cambridge

Ganjegunte G, Ulery A, Niu G, Wu Y (2017) Effects of treated municipal wastewater irrigation on soil properties, switchgrass biomass production and quality under arid climate. Ind Crops Prod 99:60–69. https://doi.org/10.1016/j.indcrop.2017.01.038

Gaweł S, Wardas M, Niedworok E, Wardas P (2004) Malondialdehyde (MDA) as a lipid peroxidation marker. Wiadomosci Lek Wars Pol 1960 57:453–455

George SG, Buchanan G (1990) Isolation, properties and induction of plaice liver cytosolic glutathione-S-transferases. Fish Physiol Biochem 8:437–449. https://doi.org/10.1007/BF00003400

Gerardi MH, Zimmerman MC (2004) Wastewater pathogens. Wiley, New York

Gerbens-Leenes W, Hoekstra AY, van der Meer TH (2008) The water footprint of bioenergy and other primary energy carriers

Giamberini L, Pihan J (1997) Lysosomal changes in the hemocytes of the freshwater mussel Dreissena polymorpha experimentally exposed to lead and zinc. Dis Aquat Organ 28:221–227. https://doi.org/10.3354/dao028221

GIEC (2018) Summary for policymakers of IPCC special report on global warming of 1.5°C approved by governments

Gleick PH (1993) Water in crisis: a guide to the world's fresh water resources. Oxford University Press, Oxford

Goel PK (2006) Water pollution: causes, effects and control. New Age International

Gregory MR (1996) Plastic 'scrubbers' in hand cleansers: a further (and minor) source for marine pollution identified. Mar Pollut Bull 32:867–871. https://doi.org/10.1016/S0025-326X(96)000 47-1

Hashem MS, Qi X (2021) Treated wastewater irrigation—a review. Water 13:1527. https://doi.org/10.3390/w13111527

Hatam-Nahavandi K, Mahvi AH, Mohebali M, Keshavarz H, Mobedi I, Rezaeian M (2015) Detection of parasitic particles in domestic and urban wastewaters and assessment of removal efficiency of treatment plants in Tehran, Iran. J Environ Health Sci Eng 13:4. https://doi.org/10.1186/s40201-015-0155-5

Hattab S, Boughattas I, Boussetta H, Viarengo A, Banni M, Sforzini S (2015) Transcriptional expression levels and biochemical markers of oxidative stress in the earthworm Eisenia andrei after exposure to 2,4-dichlorophenoxyacetic acid (2,4-D). Ecotoxicol Environ Saf 122:76–82. https://doi.org/10.1016/j.ecoenv.2015.07.014

He D, Luo Y, Lu S, Liu M, Song Y, Lei L (2018) Microplastics in soils: analytical methods, pollution characteristics and ecological risks. TrAC Trends Anal Chem 109:163–172. https://doi.org/10.1016/j.trac.2018.10.006

Hidri Y, Fourti O, Eturki S, Jedidi N, Charef A, Hassen A (2014) Effects of 15-year application of municipal wastewater on microbial biomass, fecal pollution indicators, and heavy metals in a Tunisian calcareous soil. J Soils Sediments 14:155–163. https://doi.org/10.1007/s11368-013-0801-4

Hill VR (2003) Prospects for pathogen reductions in livestock wastewaters: a review. Crit Rev Environ Sci Technol 33:187–235. https://doi.org/10.1080/10643380390814532

Hirano T, Tamae K (2011) Earthworms and soil pollutants. Sensors 11:11157–11167. https://doi.org/10.3390/s111211157

Ho E, Karimi Galougahi K, Liu C-C, Bhindi R, Figtree GA (2013) Biological markers of oxidative stress: applications to cardiovascular research and practice. Redox Biol 1:483–491. https://doi.org/10.1016/j.redox.2013.07.006

Hogan JN, Daniels ME, Watson FG, Oates SC, Miller MA, Conrad PA, Shapiro K, Hardin D, Dominik C, Melli A, Jessup DA, Miller WA (2013) Hydrologic and vegetative removal of Cryptosporidium parvum, Giardia lamblia, and Toxoplasma gondii Surrogate microspheres in coastal wetlands. Appl Environ Microbiol 79:1859–1865. https://doi.org/10.1128/aem.03251-12

Horton AA, Walton A, Spurgeon DJ, Lahive E, Svendsen C (2017) Microplastics in freshwater and terrestrial environments: evaluating the current understanding to identify the knowledge gaps and future research priorities. Sci Total Environ 586:127–141. https://doi.org/10.1016/j.scitot env.2017.01.190

Huang Y, Li J, Chen X, Wang X (2014) Applications of conjugated polymer based composites in wastewater purification. RSC Adv 4:62160–62178. https://doi.org/10.1039/C4RA11496E

Hurley RR, Nizzetto L (2018) Fate and occurrence of micro(nano)plastics in soils: knowledge gaps and possible risks. Curr Opin Environ Sci Health 1:6–11. https://doi.org/10.1016/j.cocsh.2017.10.006. Micro and nanoplastics (TAP Rocha-Santos (eds))

Ibekwe AM, Gonzalez-Rubio A, Suarez DL (2018) Impact of treated wastewater for irrigation on soil microbial communities. Sci Total Environ 622–623:1603–1610. https://doi.org/10.1016/j.scitotenv.2017.10.039

Jaishankar M, Tseten T, Anbalagan N, Mathew BB, Beeregowda KN (2014) Toxicity, mechanism and health effects of some heavy metals. Interdiscip Toxicol 7:60–72. https://doi.org/10.2478/intox-2014-0009

Jänsch S, Amorim MJ, Römbke J (2005) Identification of the ecological requirements of important terrestrial ecotoxicological test species. Environ Rev 13:51–83

Kalavrouziotis IK, Koukoulakis P, Kostakioti E (2012) Assessment of metal transfer factor under irrigation with treated municipal wastewater. Agric Water Manag 103:114–119. https://doi.org/10.1016/j.agwat.2011.11.002

Khalid S, Shahid M, Natasha, Bibi I, Sarwar T, Shah AH, Niazi NK (2018) A review of environmental contamination and health risk assessment of wastewater use for crop irrigation with a focus on low and high-income countries. Int J Environ Res Public Health 15:895. https://doi.org/10.3390/ijerph15050895

Khalil AM (2016) Physiological and genotoxic responses of the earthworm Aporrectodea caliginosa exposed to sublethal concentrations of AgNPs. J Basic Appl Zool Invertebr Parasitol 74:8–15. https://doi.org/10.1016/j.jobaz.2015.12.004

Kim HK, Jang TI, Kim SM, Park SW (2015) Impact of domestic wastewater irrigation on heavy metal contamination in soil and vegetables. Environ Earth Sci 73:2377–2383. https://doi.org/10.1007/s12665-014-3581-2

Kitajima M, Iker BC, Pepper IL, Gerba CP (2014) Relative abundance and treatment reduction of viruses during wastewater treatment processes–identification of potential viral indicators. Sci Total Environ 488–489:290–296. https://doi.org/10.1016/j.scitotenv.2014.04.087

Klay S, Charef A, Ayed L, Houman B, Rezgui F (2010) Effect of irrigation with treated wastewater on geochemical properties (saltiness, C, N and heavy metals) of isohumic soils (Zaouit Sousse perimeter, Oriental Tunisia). Desalination 253:180–187. https://doi.org/10.1016/j.desal.2009.10.019

Kumar M, Gogoi A, Kumari D, Borah R, Das P, Mazumder P, Tyagi VK (2017) Review of perspective, problems, challenges, and future scenario of metal contamination in the urban environment. J Hazard Toxic Radioact Waste 21:04017007. https://doi.org/10.1061/(ASCE)HZ.2153-5515.0000351

Kummu M, Guillaume JHA, de Moel H, Eisner S, Flörke M, Porkka M, Siebert S, Veldkamp TIE, Ward PJ (2016) The world's road to water scarcity: shortage and stress in the 20th century and pathways towards sustainability. Sci Rep 6:38495. https://doi.org/10.1038/srep38495

Lüneberg K, Schneider D, Siebe C, Daniel R (2018) Drylands soil bacterial community is affected by land use change and different irrigation practices in the Mezquital Valley, Mexico. Sci Rep 8:1413. https://doi.org/10.1038/s41598-018-19743-x

Mahmud HNME, Obidul Huq AK, Binti Yahya R (2016) The removal of heavy metal ions from wastewater/aqueous solution using polypyrrole-based adsorbents: a review. RSC Adv 6:14778–14791. https://doi.org/10.1039/C5RA24358K

Maity S, Banerjee R, Goswami P, Chakrabarti M, Mukherjee A (2018) Oxidative stress responses of two different ecophysiological species of earthworms (Eutyphoeus waltoni and Eisenia fetida) exposed to Cd-contaminated soil. Chemosphere 203:307–317. https://doi.org/10.1016/j.chemosphere.2018.03.189

Maity S, Roy S, Chaudhury S, Bhattacharya S (2008) Antioxidant responses of the earthworm Lampito mauritii exposed to Pb and Zn contaminated soil. Environ Pollut Barking Essex 1987 151:1–7. https://doi.org/10.1016/j.envpol.2007.03.005

Markert BA, Breure AM, Zechmeister HG (2003) Definitions, strategies and principles for bioindication/biomonitoring of the environment (Chap 1). In: Markert BA, Breure AM, Zechmeister HG (eds) Trace metals and other contaminants in the environment, bioindicators & biomonitors, pp 3–39. Elsevier. https://doi.org/10.1016/S0927-5215(03)80131-5

Maynard C, Berthiaume F, Lemarchand K, Harel J, Payment P, Bayardelle P, Masson L, Brousseau R (2005) Waterborne pathogen detection by use of oligonucleotide-based microarrays. Appl Environ. Microbiol 71:8548–8557. https://doi.org/10.1128/AEM.71.12.8548-8557.2005

McCormick A, Hoellein TJ, Mason SA, Schluep J, Kelly JJ (2014) Microplastic is an abundant and distinct microbial habitat in an urban river. Environ Sci Technol 48:11863–11871. https://doi.org/10.1021/es503610r

Mkhinini M, Boughattas I, Alphonse V, Livet A, Bousserrhine N, Banni M (2018a) Effect of treated wastewater irrigation in East Central region of Tunisia (Monastir governorate) on the biochemical and transcriptomic response of earthworms Eisenia andrei. Sci Total Environ 647:1245–1255. https://doi.org/10.1016/j.scitotenv.2018.07.449

Mkhinini M, Boughattas I, Boussetta H, Alphonse V, Livet A, Giusti-Miller S, Banni M, Bousserrhine N (2018b) Short term treated wastewater reuse impact on soil microbial biomass, bacterial functional diversity and enzymatic activities in the presence of earthworms Eisenia Andrei, pp 301–303. https://doi.org/10.1007/978-3-319-70548-4_96

Mohammad MJ, Mazahreh N (2003) Changes in soil fertility parameters in response to irrigation of forage crops with secondary treated wastewater. Commun Soil Sci Plant Anal 34:1281–1294. https://doi.org/10.1081/CSS-120020444

Moore M (1980) Cytochemical determination of cellular responses to environmental stressors in marine organisms

Moore MN (1988) Cytochemical responses of the lysosomal system and NADPH-ferrihemoprotein reductase in molluscan digestive cells to environmental and experimental exposure to xenobiotics. Mar Ecol Prog Ser 46:81–89

Moore MN (1990) Lysosomal cytochemistry in marine environmental monitoring. Histochem J 22:187–191. https://doi.org/10.1007/BF02386003

Nandi A, Yan L-J, Jana CK, Das N (2019) Role of catalase in oxidative stress- and age-associated degenerative diseases. Oxid Med Cell Longev 2019:9613090. https://doi.org/10.1155/2019/9613090

Nechifor V, Winning M (2017) Projecting irrigation water requirements across multiple socio-economic development futures – a global CGE assessment. Water Resour Econ 20:16–30. https://doi.org/10.1016/j.wre.2017.09.003

Newton RJ, McClary JS (2019) The flux and impact of wastewater infrastructure microorganisms on human and ecosystem health. Curr Opin Biotechnol Energy Biotechnol Environ Biotechnol 57:145–150. https://doi.org/10.1016/j.copbio.2019.03.015

Ofori S, Puškáčová A, Růžičková I, Wanner J (2021) Treated wastewater reuse for irrigation: pros and cons. Sci Total Environ 760:144026. https://doi.org/10.1016/j.scitotenv.2020.144026

OMM (Organisation météorologique mondiale) (2019) Déclaration de l'OMM sur l'état du climat mondial en 2018

Ottoson J, Hansen A, Björlenius B, Norder H, Stenström TA (2006) Removal of viruses, parasitic protozoa and microbial indicators in conventional and membrane processes in a wastewater pilot plant. Water Res 40:1449–1457. https://doi.org/10.1016/j.watres.2006.01.039

Owlad M, Aroua MK, Daud WAW, Baroutian S (2009) Removal of hexavalent chromium-contaminated water and wastewater: a review. Water Air Soil Pollut 200:59–77. https://doi.org/10.1007/s11270-008-9893-7

Parveen T, Hussain A, Someshwar Rao M (2015) Growth and accumulation of heavy metals in turnip (Brassica rapa) irrigated with different concentrations of treated municipal wastewater. Hydrol Res 46:60. https://doi.org/10.2166/nh.2014.140

Pastorino P, Ginebreda A (2021) Contaminants of Emerging Concern (CECs): occurrence and fate in aquatic ecosystems. Int J Environ Res Public Health 18:13401. https://doi.org/10.3390/ijerph182413401

Pedersen JA, Yeager MA, Suffet IH (Mel) (2003) Xenobiotic organic compounds in runoff from fields irrigated with treated wastewater. J Agric Food Chem 51:1360–1372. https://doi.org/10.1021/jf025953q

Pelosi C, Barot S, Capowiez Y, Hedde M, Vandenbulcke F (2014) Pesticides and earthworms. A review. Agron Sustain Dev 34:199–228. https://doi.org/10.1007/s13593-013-0151-z

Porter NA, Caldwell SE, Mills KA (1995) Mechanisms of free radical oxidation of unsaturated lipids. Lipids 30:277–290. https://doi.org/10.1007/BF02536034

Prata JC (2018) Microplastics in wastewater: state of the knowledge on sources, fate and solutions. Mar Pollut Bull 129:262–265. https://doi.org/10.1016/j.marpolbul.2018.02.046

Rault M, Mazzia C, Capowiez Y (2007) Tissue distribution and characterization of cholinesterase activity in six earthworm species. Comp Biochem Physiol B Biochem Mol Biol 147:340–346. https://doi.org/10.1016/j.cbpb.2007.01.022

Sariana LG, Berame JS, Mariano M, Lascano JP, Macasinag ML, Alam Z (2020) Environmental biomonitoring of terrestrial ecosystems in the philippines: a critical assessment and evaluation. IAMURE Int J Ecol Conserv 32:1–1

Schewe J, Heinke J, Gerten D, Haddeland I, Arnell NW, Clark DB, Dankers R, Eisner S, Fekete BM, Colón-González FJ, Gosling SN, Kim H, Liu X, Masaki Y, Portmann FT, Satoh Y, Stacke T, Tang Q, Wada Y, Wisser D, Albrecht T, Frieler K, Piontek F, Warszawski L, Kabat P (2014) Multimodel assessment of water scarcity under climate change. Proc Natl Acad Sci 111:3245–3250. https://doi.org/10.1073/pnas.1222460110

Schlindwein AD, Rigotto C, Simões CMO, Barardi CRM (2010) Detection of enteric viruses in sewage sludge and treated wastewater effluent. Water Sci Technol J Int Assoc Water Pollut Res 61:537–544. https://doi.org/10.2166/wst.2010.845

Sheavly SB, Register KM (2007) Marine debris & plastics: environmental concerns, sources, impacts and solutions. J Polym Environ 15:301–305. https://doi.org/10.1007/s10924-007-0074-3

Sonune A, Ghate R (2004) Developments in wastewater treatment methods. Desalination 167:55–63. https://doi.org/10.1016/j.desal.2004.06.113

Stevik TK, Aa K, Ausland G, Hanssen JF (2004) Retention and removal of pathogenic bacteria in wastewater percolating through porous media: a review. Water Res 38:1355–1367. https://doi.org/10.1016/j.watres.2003.12.024

Stopper H, Müller SO (1997) Micronuclei as a biological endpoint for genotoxicity: a mini review. Toxicol in Vitro 11:661–667. https://doi.org/10.1016/S0887-2333(97)00084-2. International workshop on in vitro toxicology

Svendsen C, Spurgeon DJ, Hankard PK, Weeks JM (2004) A review of lysosomal membrane stability measured by neutral red retention: is it a workable earthworm biomarker? Ecotoxicol Environ Saf 57:20–29. https://doi.org/10.1016/j.ecoenv.2003.08.009. Ecological, physiological, and physiochemical factors in earthworm ecotoxicology. The third international workshop on earthworm ecotoxicology

Tchobanoglous G, Burton FL, Eddy M (1991) Wastewater engineering: treatment, disposal, and reuse. McGraw-Hill

Tchounwou PB, Yedjou CG, Patlolla AK, Sutton DJ (2012) Heavy metal toxicity and the environment. In: Luch A (ed) Molecular, clinical and environmental toxicology, vol 3. Environmental toxicology, experientia supplementum, pp 133–164. Springer, Basel. https://doi.org/10.1007/978-3-7643-8340-4_6

Tomass Z, Kidane D (2012) Parasitological contamination of wastewater irrigated and raw manure fertilized vegetables in Mekelle City and Its Suburb, Tigray, Ethiopia. Momona Ethiop J Sci 4:77–89. https://doi.org/10.4314/mejs.v4i1.74058

Tong AYC, Peake BM, Braund R (2011) Disposal practices for unused medications around the world. Environ Int 37:292–298. https://doi.org/10.1016/j.envint.2010.10.002

UNESCO (2020) The United Nations world water development report 2020: water and climate change. United Nations

United Nations Environment Programme/Mediterranean Action Plan and Plan Bleu (2020) State of the Environment and Development in the Mediterranean

Van Cauwenberghe L, Devriese L, Galgani F, Robbens J, Janssen CR (2015) Microplastics in sediments: a review of techniques, occurrence and effects. Mar Environ Res Part Oceans: Implic Safe Mar Environ 111:5–17. https://doi.org/10.1016/j.marenvres.2015.06.007

Varela AR, Manaia CM (2013) Human health implications of clinically relevant bacteria in wastewater habitats. Environ Sci Pollut Res Int 20:3550–3569. https://doi.org/10.1007/s11356-013-1594-0

Wyn-Jones AP, Carducci A, Cook N, D'Agostino M, Divizia M, Fleischer J, Gantzer C, Gawler A, Girones R, Höller C, de Roda Husman AM, Kay D, Kozyra I, López-Pila J, Muscillo M, Nascimento MSJ, Papageorgiou G, Rutjes S, Sellwood J, Szewzyk R, Wyer M (2011) Surveillance of adenoviruses and noroviruses in European recreational waters. Water Res 45:1025–1038. https://doi.org/10.1016/j.watres.2010.10.015

Yi L, Jiao W, Chen X, Chen W (2011) An overview of reclaimed water reuse in China. J Environ Sci China 23:1585–1593. https://doi.org/10.1016/s1001-0742(10)60627-4

Zacharia A, Outwater AH, Ngasala B, Deun RV (2018) Pathogenic parasites in raw and treated wastewater in Africa: a review. Resour Environ 8:232–240

Zaghloul A, Saber M, Gadow S, Awad F (2020) Biological indicators for pollution detection in terrestrial and aquatic ecosystems. Bull Natl Res Cent 44:127. https://doi.org/10.1186/s42269-020-00385-x

Zhang Z, Pang K, Cotton P, Wong C, Ranjan S (2004) Rapid simulation of correlated defaults and the valuation of basket default swaps. In: Probability, finance and insurance. World Scientific, pp 150–163. https://doi.org/10.1142/9789812702715_0009

Zhou Q, Zhang J, Fu J, Shi J, Jiang G (2008) Biomonitoring: an appealing tool for assessment of metal pollution in the aquatic ecosystem. Anal Chim Acta 606:135–150. https://doi.org/10.1016/j.aca.2007.11.018

Ziajahromi S, Neale PA, Rintoul L, Leusch FDL (2017) Wastewater treatment plants as a pathway for microplastics: development of a new approach to sample wastewater-based microplastics. Water Res 112:93–99. https://doi.org/10.1016/j.watres.2017.01.042

Part IV
A Circular Economy Approach: Lifecycle Management of Emerging Pollutants

Part IV

Chapter 14
Management of Emerging Pollutants with a Circular Economy Approach: Lessons from Developed Countries and a Case Study in Northern Cyprus

Farhad Bolouri, Hüseyin Gökçekuş, and Vahid Nourani

Abstract Emerging pollutants (EPs) represent contemporary challenges arising from the advancement of medicines, industries, and related factors, posing threats to both human and environmental well-being. This research project employs extensive library sources, including books, academic papers, and authoritative reports, to assess the impact and effectiveness of policies implemented in developing countries for managing EPs through a circular economy framework. Subsequently, it endeavors to tailor practical solutions suitable for the specific circumstances in Northern Cyprus, drawing insights from the experiences of the countries examined during the literature review. The importance of identifying and comprehensively documenting EPs cannot be overstated, as this information is pivotal for informed decision-making regarding their management. Therefore, it is imperative for Northern Cyprus to address this issue systematically, emphasizing the careful collection of data on EPs across the entire region. Collaborative data exchange with Southern Cyprus is a viable proposition to gather comprehensive data for the entirety of Cyprus, enabling the implementation of sound management decisions. Embracing a circular economy perspective is of paramount significance in the decision-making process for managing EPs. This entails exploring opportunities to utilize EPs in the production of related products, aiming to minimize environmental harm—an approach already evident in practices such as recycling car batteries and certain computer components.

Keywords Wastewater management · Circular economy · Emerging pollutants · Hazardous materials · Northern Cyprus

F. Bolouri (✉) · H. Gökçekuş · V. Nourani
Civil Engineering Department, Energy, Environment, and Water Research Center, Faculty of Civil and Environmental Engineering, Near East University, Nicosia, Türkiye
e-mail: farhad.bolouri@neu.edu.tr

V. Nourani
Center of Excellence in Hydroinformatics, Faculty of Civil Engineering, University of Tabriz, Tabriz, Iran

© UNESCO 2025   317
S. Zandaryaa et al. (eds.), *Emerging Pollutants*, Advances in Water Security,
https://doi.org/10.1007/978-3-031-71758-1_14

14.1 Introduction

The concept of emerging pollutants (EPs) is in accordance with the definition provided by the NORMAN Network, an international organization dedicated to facilitating the sharing of information regarding newly identified environmental substances. NORMAN Network also promotes the validation and standardization of measurement techniques and monitoring instruments in this field (Bunke et al. 2019). As of the present moment, the NORMAN Network group has identified approximately 970 EPs over the past decade (NORMAN 2016). EPs can infiltrate the environment through a range of sources, which encompass anthropogenic actions like water treatment, fumigation, and agriculture. Additionally, heavy metals can originate from diverse sources, including vehicle emissions, the chemical industry, coal combustion, municipal waste, pharmaceutical contamination, and dust deposition. Particulate matter and nitrogen oxides can enter the environment via ambient air, and, organic micropollutants are found in water sources, which are often linked to biomass combustion, pharmaceuticals, organic compounds, and pesticides (Patil et al. 2019; Bauherr et al. 2020; Gondhalekar et al. 2021; O'Flynn et al. 2021).

Exposure to EPs presents several potential health risks to humans, including: (i) Metabolic Syndromes: Unbalanced diets, compounded by unhealthy lifestyles and environmental contaminant exposure, can contribute significantly to the prevalence of metabolic syndromes such as diabetes, heart disease, and obesity (Huang and Fang 2020); (ii) Developmental Issues: Children, in their crucial growth and development phase with immature organ and system development, constitute a vulnerable population when it comes to environmental pollution. Exposure to environmental hazards during childhood can serve as a critical risk factor for adult-onset diseases (Xiaoming 2021); (iii) Endocrine Disruption: Endocrine-disrupting chemicals (EDCs) have the potential to disrupt the body's endocrine system, leading to adverse effects on development, reproduction, neurology, and immunity (Klánová and Barouki 2019); (iv) Dyslipidemia: Emerging chemicals and complex chemical mixtures have the capacity to induce dyslipidemia, an abnormal lipid profile (e.g., cholesterol and fat) in the bloodstream (Klánová and Barouki 2019); (v) Heavy Metal Toxicity: Exposure to heavy metals carries the potential for chronic and harmful health impacts on local inhabitants, encompassing developmental issues, neurological damage, and cancer (Yousaf et al. 2016)."

A circular economy is an approach designed to encourage sustainable resource management and minimize waste (Ala'a et al. 2021; Sharma et al. 2021; Richter 2022). Transitioning to a circular economy has some benefits like economic benefits with implementing circular economy principles can create new business opportunities, foster job creation, and enhance competitiveness in the global market. Also, social benefits because circular economy principles contribute to social equity by addressing issues such as equitable resource distribution and effective waste management (Gutberlet and Carenzo 2020; Lazaridou et al. 2021; Purchase et al. 2021; Ohiomah and Sukdeo 2022; Niwalkar et al. 2023). The circular economy encompasses diverse practices, such as designing products for durability and repairability

(Bigerna et al. 2021), recycling and upcycling textiles (Manickam and Duraisamy 2019), converting waste to energy through biomass (Hoang et al. 2022), implementing closed-loop systems for plastic recycling (Sherwood 2020), obtaining Cradle-to-Cradle Certification for sustainable production (Kopnina 2018), promoting car-sharing and ride-sharing initiatives (Pouri and Hilty 2020), utilizing renewable energy systems (Olabi 2019), managing e-waste through recycling programs (Murthy and Ramakrishna 2022), exploring eco-friendly packaging alternatives (Meherishi et al. 2019), and implementing water recycling and reuse systems (Giakoumis et al. 2020).

The underlying hypothesis of this research is that the policies employed in advanced nations can be adapted and tailored to suit the localized and practical requirements of developing nations like Northern Cyprus. Northern Cyprus is part of the island of Cyprus. This area has various environmental problems such as air pollution and of course water scarcity. Like other regions of the world, it is affected by climate change, which makes the mentioned problems more acute. In the meantime, since its economy depends on things like education and universities and of course tourism, it can be hoped that with solutions like the circular economy, practical solutions can be taken to reduce the mentioned environmental problems (İşçioğlu 2013; Shirkhani et al. 2021).

Considering the harm of EPs to human health and the environment, as well as the lack of sufficient attention to this matter even in developed countries, and considering the positive approach of the circular economy in solving such environmental problems; In this research, the situation of developed countries in the face of EPs is investigated and then an attempt is made to provide suggestions for Northern Cyprus.

14.2 Methodology

In this research, using library sources (books, papers, and authoritative reports), the policies implemented in developed countries for the management of EPs with a circular economy approach have been reviewed. Then, according to the conditions of Northern Cyprus and using the experiences gained from the countries investigated in the section of library studies, practical solutions for Northern Cyprus have also been presented. To do this, the "circular economy approach", "circular economy principles" and "circular economy principles applied to wastewater" have been discussed in more detail in this section. Then in the results and discussion section, also the topic in question in four subsections "emerging pollutants in developed countries", "current regulations in developed countries regarding emerging pollutants", "challenges in regulating emerging pollutants in developed countries", "potential solutions for regulating "emerging pollutants" have been examined and finally, in the fifth sub-section under the title "circular economy and potential solutions for emerging pollutants in Northern Cyprus", practical suggestions have been given for Northern Cyprus according to the previous four sub-sections.

14.2.1 Circular Economy Approach

The concept of a circular economy centers on the deceleration, closure, and narrowing of resource cycles through the advocacy of sustainable production, consumption, and waste management techniques. It has garnered considerable attention across diverse sectors, encompassing business, construction, water and wastewater management, as well as agriculture (Ala'a et al. 2021; Sharma et al. 2021; Richter 2022). The circular economy stands apart from a conventional economic model by prioritizing sustainable resource management and waste reduction. In the traditional linear economy, resources are extracted, utilized, and discarded as waste. In contrast, the circular economy strives to extend the lifespan of resources, extract their maximum value, and promote the recovery and regeneration of products and materials when they reach the end of their life cycle. This approach can result in decreased input expenses, savings in waste management costs, and enhanced product quality for consumers, improved waste collection systems, and a reduction in both waste landfills and greenhouse gas emissions (Soliwoda et al. 2020; Fidanchevski et al. 2022; Gorokhova et al. 2023; Nandi et al. 2023).

14.2.2 Circular Economy Principles

Some commonly embraced principles within the circular economy framework encompass: (i) Minimizing waste and pollution through design: This principle centers on the elimination of waste and pollution through the creation of products and processes that minimize waste and emissions; (ii) Extending the life of products and materials: This principle strives to prolong the use of products and materials by advocating for practices like reuse, repair, and recycling; (iii) Promoting natural system rejuvenation: This principle encourages the restoration of ecosystems and the utilization of renewable resources to regenerate natural systems; (iv) Closing material loops: This principle focuses on closing material loops by emphasizing the recovery and reuse of materials and the reduction in the consumption of virgin resources; (v) Encouraging collaboration and innovation: This principle aims to stimulate collaboration and innovation by endorsing the sharing of knowledge, resources, and expertise among stakeholders (Domenech and Borrion 2022; Huang et al. 2022; Rodríguez-Espíndola et al. 2022; Vural Gursel et al. 2022).

14.2.3 Circular Economy Principles Applied to Wastewater

Circular economy principles find application in various aspects of wastewater management. Below are instances from the search results: (i) Microalgae-Based Wastewater Treatment: Circular economy principles can be incorporated into wastewater treatment through microalgae-based methods. Microalgae exhibit the ability to sequester carbon dioxide while recovering nutrients from sewage. Achieving this objective necessitates the development of advanced microalgae-based systems with enhanced efficiency and process control (González-Camejo et al. 2021); (ii) Waste Material for Nanoparticle Synthesis: Waste materials, such as Olive Mill Wastewater (OMWW), can serve as reducing agents in nanoparticle synthesis, aligning with circular economy principles. This approach reduces reliance on toxic substances and encourages the utilization of waste materials (De Matteis et al. 2023); (iii) Livestock Production: Circular economy principles can be applied to enhance livestock production in Sub-Saharan Africa. This involves improving livestock feeding practices, including dual-purpose animal breeding, and utilizing crop residues as livestock feed (Duncan et al. 2023); (iv) Urban Housing: In urban housing, circular economy principles can be implemented by designing more sustainable and resource-efficient homes. This approach addresses concerns related to housing demand, affordability, and waste generation (Marchesi et al. 2020); (v) Agriculture: Circular economy principles are relevant to agriculture through the promotion of waste material reuse and recycling, waste reduction in production, and designs that facilitate disassembly. Examples include using biosolids as a nutrient source and crop residues as livestock feed (Aggarwal and Mahajan 2021); (vi) Valorization of Agro-Industrial Biowaste: Agro-industrial biowaste can be transformed into green nanomaterials for wastewater treatment, aligning with green chemistry and circular economy principles. This approach reduces waste and supports material recovery and reuse (Omran and Baek 2022). Circular economy principles in wastewater management involve viewing wastewater as a valuable resource. Essential elements, such as nutrients, are extracted for agricultural purposes, and advanced technologies enhance treatment processes for efficiency and resource recovery. The treated water meets quality standards for safe reuse in irrigation or industrial applications, reducing reliance on external sources. Sustainable energy practices are promoted through processes like biogas production via anaerobic digestion. Flexible decentralized treatment systems and public engagement initiatives raise awareness about water conservation. Green infrastructure, including constructed wetlands, aids natural filtration. Regulatory support provides incentives for industries to adopt water-efficient technologies. Implementing these principles not only reduces environmental impact but also promotes sustainable water use, energy efficiency, and resource conservation. Collaboration among governments, industries, and communities is essential for successful implementation.

14.3   Results and Discussion

In this part, the items "Emerging pollutants in developed countries", "Current regulations in developed countries regarding emerging pollutants", "Challenges in regulating emerging pollutants in developed countries" and "Potential solutions for regulating emerging pollutants" are reviewed and then "Circular economy and potential solutions for emerging pollutants in Northern Cyprus" is presented.

14.3.1   Emerging Pollutants in Developed Countries

While the search results are not exhaustive in listing all EPs in Europe, they cite various instances of substances found in the environment, such as: (i) Heavy metals in urban soils, originating from sources like vehicle emissions, the chemical industry, coal combustion, municipal solid waste, drug contamination, and dust sedimentation (Oladipupo 2017); (ii) Particulate matter and nitrogen oxides in ambient air, recognized by the WHO and the European Environmental Agency as significant pollutants impacting human health (Dalecká and Bartošková 2023); (iii) Organochlorine pollutants (OCs) detected in fish, with potential effects on parasite infection modulation and consumer health risks (Monnolo et al. 2023); (iv) Nanoparticle heavy metal pollutants in water, classified as EPs and subject to environmental monitoring and surveillance concerns (Faye et al. 2012).

Some instances of EPs observed in the United States are as; (i) Airborne Contaminants: The continuous expansion in the array and complexity of pollutants present in the ambient air, facilitated by real-time and automated analyses, has led to the identification of emerging airborne pollutants (Kim and Brown 2011); (ii) Hazardous Chemicals: Evolving understanding of how pollutants interact within the environment, coupled with advances in detection technologies capable of identifying minute quantities of toxic chemicals, necessitates adjustments to existing US environmental regulations (Binder, 1984); (iii) Particulate Matter: Fine particulate matter, denoted as PM2.5, has been definitively linked to a range of adverse health effects in the human population, spanning from aggravated allergies and chronic illness development to premature mortality (Zhang et al. 2022); (iv) Blood Transfusion Proteins: Investigations in proteomics have unveiled that red blood cells contain approximately 3000 gene products, some of which may be categorized as EPs (DeSimone and Vinchi 2021).

EPs in Canada can be mentioned: (i) Porphyrins and Persistent Organic Pollutants: A research investigation carried out on surf scoters in British Columbia detected exposure to both legacy and emerging persistent organic pollutants, referring to toxic substances that endure in the environment and accumulate within the food chain (Wilson et al. 2010); (ii) Acid Rain: Canada has identified acid rain as an emerging pollutant of concern. Surveillance of acid rain control initiatives has unveiled instances of ongoing lake acidification, posing potential adverse impacts on aquatic ecosystems (Jeffries et al. 2003).

Some instances of EPs observed in Australia are as: (i) Microplastics: These tiny plastic particles are prevalent in the environment, including water bodies and soil, and are recognized as EPs in Australia. They have the potential to adversely impact aquatic ecosystems and human health (Jain and Dhabas 2020); (ii) Pharmaceuticals: Pharmaceuticals are designated as EPs in Australia due to their presence in the environment, primarily through wastewater, with detections in surface water and groundwater (Wong 2012); (iii) Flame Retardants: Chemicals added to products to decrease flammability, flame retardants are considered EPs in Australia and can pose risks to human health and the environment (Reichwaldt and Ghadouani 2016); (iv) Per- and Polyfluoroalkyl Substances (PFAS): PFAS, a group of chemicals used in various industrial and consumer goods, are recognized as EPs in Australia. They have the potential to harm both human health and the environment (Vasseur 2010); (v) Pesticides: EPs in Australia, pesticides enter the environment via runoff and have been identified in surface water and groundwater (Grosshandlera et al. 2017).

14.3.2 Current Regulations in Developed Countries regarding Emerging Pollutants

The current regulations concerning EPs in Europe are characterized by a lack of clear definition. The strategy employed by the Water Framework Directive (WFD) to combat chemical pollution combines measures related to pollution sources with environmental quality standards, using the latter as a safeguard for regulating emissions, discharges, and losses. Substances of concern are managed using a two-tiered approach, addressing them at both the European level (priority substances) and the river basin level (river basin-specific pollutants). These levels are interconnected through data sources and procedural foundations, with the assumption that priority substances represent a subset of river basin-specific pollutants that exceed local risk thresholds and necessitate Europe-wide measures. Environmental quality standards serve a triple purpose in water policy: they establish risk thresholds, guide management strategies for contaminated areas, and indicate a favorable chemical status if concentrations of all priority substances adhere to the corresponding environmental quality standards. The prioritization of chemicals under the Water Framework Directive relies on evidence related to the inherent hazards of the substances, particularly their aquatic ecotoxicity and human toxicity through aquatic exposure routes. This evidence is gathered through monitoring widespread environmental contamination and other factors that suggest the potential for such contamination, including production volume, usage patterns, and usage volume of the substances (Heiss and Küster 2015). Nevertheless, there remains a need for governments to compile a list of localized priority pollutants to support the overarching enhancement of water ecological environmental quality (Fytianos et al. 2021).

In the United States, the Environmental Protection Agency (EPA) has introduced a suite of initiatives and programs aimed at addressing emerging pollutants (EPs), which include the Emerging Contaminants Program and the Endocrine Disruptor Screening Program. These initiatives share a common objective of identifying and regulating EPs that have the potential to endanger both human health and the environment. Furthermore, comprehensive research efforts have been undertaken to assess the presence of EPs in the nation's drinking water, offering suggested thresholds for these emerging contaminants. These research outcomes underscore the imperative need for measures aimed at curbing chemical contamination in drinking water. Interventions encompass expanding sewage treatment facilities, upgrading tertiary treatment processes, and implementing more stringent controls to reduce pesticide applications (Zini and Gutterres 2021; Kang et al. 2022; Castro and Obusan 2023).

Research endeavors have been conducted to evaluate the presence of EPs within the Australian environment and propose recommended limits for these emerging contaminants. These studies underscore the urgency of implementing measures to mitigate chemical contamination within the environment. Viable interventions may encompass expanding and enhancing sewage treatment systems, alongside the regulation and reduction of pesticide applications (Brack 2019; Serafimov et al. 2021; Benedetti et al. 2022).

14.3.3 Challenges in Regulating Emerging Pollutants in Developed Countries

The regulation of EPs in developed countries comes with several challenges, including: (i) Inadequate legislation: Existing laws related to substance supply and utilization are insufficient and primarily designed for regulating individual or groups of compounds (Schärer 2009); (ii) Absence of explicit workplace regulations: There's a lack of explicit regulations regarding chemical exposure in occupational settings (Lovas et al. 2020); (iii) Lack of a cohesive and strategic policy framework: Many regions worldwide, including advanced environmental policy areas like the European Union (EU), still lack a cohesive and strategic policy framework (Panagiotakis and Dermatas 2022); (iv) High costs associated with addressing problems: The risks and challenges posed by contaminated sites, along with the high costs involved in mitigating these issues, underscore the need for a comprehensive approach centered around integrated pollution prevention and control (Weber et al. 2008); (v) Challenges in adopting more efficient and cost-effective methods for screening and prioritizing chemicals, as well as addressing increasingly complex issues like life-stage susceptibility and genetic variations in the human population (Kramer et al. 2009).

In fact, in the end, it can be assessed that due to the lack of sufficient regulations regarding EPs, and on the one hand, serious challenges in the path of these regulations, and on the other hand, the focus of countries on registering primary regulations to control them, it can be said that the principles of the circle economy have not been considered as it should be in the preparation of relevant regulations.

14.3.4   Potential Solutions for Regulating Emerging Pollutants

In Europe, various potential strategies have been proposed to regulate emerging pollutants (EPs). These strategies encompass: (i) Developing a unified and strategic policy framework: The absence of a unified and strategic policy framework is noticeable not only in Europe but also in many regions worldwide, including advanced environmental authorities like the European Union (EU). The establishment of such a framework could effectively address the regulatory challenges associated with EPs (Panagiotakis and Dermatas 2022); (ii) Embracing the 'polluter pays principle': The 'polluter pays principle' holds that parties responsible for pollution should be held accountable for the environmental harm they cause and bear the associated costs. A comprehensive application of this principle could facilitate the remediation of contaminated areas and the prevention of future contamination incidents (Weber et al. 2008); (iii) Implementing an integrated approach to pollution prevention and control: Embracing a comprehensive strategy rooted in integrated pollution prevention and control measures could help address the risks and challenges posed by contaminated sites, as well as the substantial costs associated with their remediation and prevention (Weber et al. 2008).

In the United States, a federal law governing the quality of public drinking water; Safe Drinking Water Act (SDWA) obliges the EPA to establish standards for contaminants in drinking water and mandates that public water systems monitor and treat their water to meet those standards (Weinmeyer et al. 2017); (iv) Collaboration: Addressing EPs in the United States requires collaboration among government agencies, industry stakeholders, and the public. This entails the sharing of information and resources, as well as joint efforts to develop and implement effective solutions (Richardson 2003).

In Australia, potential strategies for addressing EPs include: (i) National Pollutant Inventory (NPI): The NPI serves as a comprehensive national repository of emissions data encompassing various pollutants, including emerging ones. This openly accessible database offers insights into the types and quantities of pollutants discharged by diverse industries and facilities (Weng et al. 2012; Tang and Mudd 2015); (ii) Environmental Protection Act (EP Act): The EP Act constitutes federal legislation responsible for regulating the environmental impact of activities. Under this law, companies are obligated to secure environmental approvals before initiating specific activities, with provisions for penalties in cases of non-compliance (Cole 2008); (iii) the Australian and New Zealand Guidelines for Fresh and Marine Water Quality play a crucial role by offering guidance regarding permissible pollutant levels in water bodies. These guidelines serve as reference points for regulators in establishing water quality standards and assist industries in managing their environmental footprint (Warne et al. 2014).

14.3.5 Circular Economy and Potential Solutions for Emerging Pollutants in Northern Cyprus

According to the review of the management status of developed countries regarding EPs, it can be concluded that due to the status of identifying these pollutants and the importance of the available data on these items to make decisions about how to manage them. It is essential to address this issue in Northern Cyprus as well and to collect EPs data in the entire region carefully for consecutive years even data exchange with Southern Cyprus can be proposed to collect data for Cyprus so that the correct management decisions are applied. Considering that there are various methods to encourage businesses and industries to embrace circular economy principles for the reduction of EPs, some suggestions can also be made for Northern Cyprus. One approach involves providing financial incentives and tax breaks to facilitate a smoother transition towards circular practices (Cohen et al. 2022). The support of public authorities through policy incentives and instruments can play a significant role in bolstering the adoption of circular approaches within companies (Demko-Rihter et al. 2023). Another effective strategy is the implementation of educational programs aimed at raising awareness about the advantages of circular practices (Ncube et al. 2023). Simultaneously, regulatory frameworks can be developed to facilitate the shift towards sustainable production and consumption patterns (Ncube et al. 2023). Furthermore, forward-thinking industries stand to benefit from innovative circular value creation (De Aguiar Hugo et al. 2023). Embracing circular economy principles not only allows businesses to curtail their environmental impact but also enhances the efficient utilization of resources, potentially leading to cost reductions and increased profitability (De Aguiar Hugo et al. 2023).

According to the circular economy approach, it is very important that during the decision-making process for the management of EPs, the nature of the circular economy is also considered and, for example, EPs are used in the manufacture of related products in order to cause less damage to the environment; exactly what is done in the case of car batteries, some computer parts, etc. Implementing circular economy principles offers a pathway to mitigate the impact of EPs through the promotion of resource efficiency and waste reduction. One effective strategy involves integrating sustainable practices during the initial stages of new product development (Jugend et al. 2020). Furthermore, businesses and industries can be motivated to embrace circular economy principles for the purpose of reducing EPs. This can be achieved by providing financial incentives and tax benefits, launching educational initiatives to highlight the advantages of circular practices, and establishing regulatory frameworks to facilitate the shift towards sustainable production and consumption patterns (Salvioni and Almici 2020; Ghosh et al. 2022). Innovative industries can also capitalize on the potential for novel circular value creation (Jugend et al. 2020). Moreover, the strategic use of digital technologies in conjunction with circular economy practices can enhance supply chain resilience and sustainability, serving as dynamic capabilities (Cherrafi et al. 2022). Finally, adopting a circular economy approach in the management of biosolids presents opportunities to enhance safe nutrient recycling and diminish the impact of emerging organic pollutants (De Amorim Júnior et al. 2022).

14.4 Conclusion

This study aims to assess the implementation of circular economy principles in managing the life cycle of EPs in advanced nations, including the United States and European countries. Subsequently, it seeks to explore the adaptability of the policies and approaches from these developed countries to the specific context of Northern Cyprus. While the issue of EPs in European surface waters is on the rise, measures are being employed to proactively identify and mitigate new risks, with a focus on water prioritization for control and remediation efforts. There are also suggestions for the establishment of a European-wide data infrastructure to facilitate the efficient collection and exchange of extensive screening data, enhancing cost-effective assessment. In the United States, a significant influx of EPs into sewage systems poses threats to the well-being of citizens. American scientists emphasize the importance of acquiring dependable data through a variety of tests and further investigations. In this context, considering Northern Cyprus as a developing nation, it is advisable to prioritize the collection of precise and scientifically rigorous data as a foundation for subsequent screening and assessment efforts. Since the new concept of EPs and the laws and regulations in this regard have not yet reached the stage of evolution and have not been implemented in developed countries, it was not possible to present statistics and analyze these cases in this research. Considering that these laws are evolving and being implemented in developed countries, it is suggested that more attention be paid to statistics and figures in future research, and the analysis of this part is also done and modeled for the future.

References

Aggarwal V, Mahajan R (2021) Applying circular economy principles to agriculture: selected case studies from the Indian context. In: Challenges and Opportunities of Circular Economy in Agri-Food Sector: Rethinking Waste, pp 227–243

Ala'a H, Osman AI, Kumar PSM, Jamil F, Al-Haj L, Al Nabhani A, Rooney DW (2021) Circular economy approach of enhanced bifunctional catalytic system of CaO/CeO2 for biodiesel production from waste loquat seed oil with life cycle assessment study. Energy Convers Manag 236:114040

Bauherr S, Larsberg F, Petrich A, Sperber HS, Klose-Grzelka V, Luckner M, Schwarzer R (2020) Macropinocytosis and clathrin-dependent endocytosis play pivotal roles for the infectious entry of Puumala virus. J Virol 94(14):10–1128

Benedetti M, Giuliani ME, Mezzelani M, Nardi A, Pittura L, Gorbi S, Regoli F (2022) Emerging environmental stressors and oxidative pathways in marine organisms: current knowledge on regulation mechanisms and functional effects. Biocell 46(1):37

Bigerna S, Micheli S, Polinori P (2021) New generation acceptability towards durability and repairability of products: circular economy in the era of the fourth industrial revolution. Technol Forecast Soc Chang 165:120558

Binder G (1984) New responses to emerging environmental problems in USA. Environ Conserv 11(4):367–367

Brack W (2019) Solutions for present and future emerging pollutants in land and water resources management. Policy briefs summarizing scientific project results for decision makers. Environ Sci Europe 31(1):1–3

Bunke D, Moritz S, Brack W, Herráez DL, Posthuma L, Nuss M (2019) Developments in society and implications for emerging pollutants in the aquatic environment. Environ Sci Eur 31(1):1–17

Castro AE, Obusan MCM (2023) Microbial quality and emerging pollutants in freshwater systems of Mega Manila, Philippines: a scoping review. Urban Water J 1–13

Cherrafi A, Chiarini A, Belhadi A, El Baz J, Benabdellah AC (2022) Digital technologies and circular economy practices: vital enablers to support sustainable and resilient supply chain management in the post-COVID-19 era. TQM J 34(7):179–202

Cohen J, Rosado L, Gil J (2022) How is the construction sector addressing the Circular Economy? Lessons from current practices and perceptions in Argentina. IOP Conf Ser: Earth Environ Sci 1078(1):012008. IOP Publishing

Cole D (2008) Creative sentencing: using the sentencing provisions of the South Australian Environment Protection Act to greater community benefit. Environ Plan Law J 25(1):13–21

Dalecká A, Bartošková A (2023) Invited commentary on "Interactions between long-term ambient particle exposures and lifestyle on the prevalence of hypertension and diabetes: insight from a large community-based survey." J Epidemiol Commun Health 77(7):419–420

De Aguiar Hugo A, de Nadae J, da Silva Lima R (2023) Consumer perceptions and actions related to circular fashion items: perspectives of young Brazilians on circular economy. Waste Manag Res 41(2):350–367

De Amorim Júnior SS, de Souza Pereira MA, Morishigue M, da Costa RB, de Oliveira Guilherme D, Magalhães Filho FJC (2022) Circular economy in the biosolids management by nexus approach: a view to enhancing safe nutrient recycling—pathogens, metals, and emerging organic pollutants concern. Sustainability 14(22):14693

De Matteis V, Griego A, Scarpa E, Cascione M, Singh J, Rizzello L (2023) Size effect of silver nanoparticles derived from olive mill wastewater in THP-1 cell lines. Appl Sci 13(10):6033

Demko-Rihter J, Sassanelli C, Pantelic M, Anisic Z (2023) A framework to assess manufacturers' circular economy readiness level in developing countries: an application case in a Serbian Packaging Company. Sustainability 15(8):6982

DeSimone RA, Vinchi F (2021) Screening out the exposome to improve transfusion quality. Hema-Sphere 5(7)

Domenech T, Borrion A (2022) Embedding circular economy principles into urban regeneration and waste management: framework and metrics. Sustainability 14(3):1293

Duncan AJ, Ayantunde A, Blummel M, Amole T, Padmakumar V, Moran D (2023) Applying circular economy principles to intensification of livestock production in Sub-Saharan Africa. Outlook Agric 00307270231199116

Faye CB, Dutouquet C, Amodeo T, Frejafon E, Delalain P, Aguerre-Chariol O, Delépine-Gilon N (2012) Detection of nanoparticle heavy metal Pollutants in water by laser-induced break-down spectroscopy (LIBS). In: Three international conference on safe production and use of nanomaterials

Fidanchevski E, Šter K, Mrak M, Kljajević L, Žibret G, Teran K, Merta I (2022) The valorization of selected quarry and mine waste for sustainable cement production within the concept of circular economy. Sustainability 14(11):6833

Fytianos G, Ioannidou E, Thysiadou A, Mitropoulos AC, Kyzas GZ (2021) Microplastics in Mediterranean coastal countries: a recent overview. J Mar Sci Eng 9(1):98

Giakoumis T, Vaghela C, Voulvoulis N (2020) The role of water reuse in the circular economy. In: Advances in chemical pollution, environmental management and protection, vol 5, pp 227–252. Elsevier

Ghosh A, Bhola P, Sivarajah U (2022) Emerging associates of the circular economy: analyzing interactions and trends by a mixed methods systematic review. Sustainability 14(16):9998

Gondhalekar D, Hu HY, Chen Z, Tayal S, Bekchanov M, Sauer J, Drewes JE (2021) The emerging environmental economic implications of the urban Water–Energy–Food (WEF) Nexus: water reclamation with resource recovery in China, India, and Europe. In: Oxford Research Encyclopedia of Environmental Science

González-Camejo J, Ferrer J, Seco A, Barat R (2021) Outdoor microalgae-based urban wastewater treatment: recent advances, applications, and future perspectives. Wiley Interdiscip Rev Water 8(3):e1518

Gorokhova T, Shpatakova O, Toponar O, Zolotarova O, Pavliuk S (2023) Circular economy as an alternative to the traditional linear economy: case study of the EU. Revista De Gestão Social e Ambiental 17(5):e03385–e03385

Grosshandlera W, Averilla J, Clearya T, Weinertb D (2017) Recent developments in building fire detection technologies

Gutberlet J, Carenzo S (2020) Waste pickers at the heart of the circular economy: a perspective of inclusive recycling from the Global South

Heiss C, Küster A (2015) In Response: a regulatory perspective on prioritization of emerging pollutants in the context of the Water Framework Directive. Environ Toxicol Chem 34(10):2181–2183

Hoang AT, Varbanov PS, Nižetić S, Sirohi R, Pandey A, Luque R, Ng KH (2022) Perspective review on municipal solid waste-to-energy route: characteristics, management strategy, and role in circular economy. J Clean Prod 359:131897

Huang Y, Fang M (2020) Nutritional and environmental contaminant exposure: a tale of two co-existing factors for disease risks. Environ Sci Technol 54(23):14793–14796

Huang Y, Shafiee M, Charnley F, Encinas-Oropesa A (2022) Designing a framework for materials flow by integrating circular economy principles with end-of-life management strategies. Sustainability 14(7):4244

İşçioğlu D (2013) ICT and E-government applications in Northern Cyprus. Kıbrıs Araştırmaları Dergisi 17(41):13–21

Jain SK, Dhabas PC (2020) Renewable energy sources. Microgrids. U.S. Energy Information Administration (EIA)

Jeffries DS, Brydges TG, Dillon PJ, Keller W (2003) Monitoring the results of Canada/USA acid rain control programs: some lake responses. Environ Monit Assess 88:3–19

Jugend D, Camargo Fiorini PD, Pinheiro MAP, da Silva HMR, Pais Seles BMR (2020) Building circular products in an emerging economy: an initial exploration regarding practices, drivers and barriers: case studies of new product development from medium and large Brazilian companies. Johnson Matthey Technol Rev 64(1):59–68

Kang Y, Chen Y, Zhou M, Xu Y, Feng H, Chen R (2022) Identification of regional industrial priority pollutants in surface water: a field study in Taihu Lake Basin. Front Environ Sci 10:2365

Kim KH, Brown RJ (2011) Emerging measurement techniques for airborne pollutants. Sci World J 11:2599–2601

Klánová J, Barouki R (2019) Towards a more reliable assessment of health risks associated with the population exposure to bisphenols. J Epidemiol Commun Health 73(11):988–989

Kopnina H (2018) Circular economy and cradle to cradle in educational practice. J Integr Environ Sci 15(1):119–134

Kramer MG, Firestone M, Kavlock R, Zenick H (2009) The future of toxicity testing for environmental contaminants. Environ Health Perspect 117(7):A283–A283

Lazaridou DC, Michailidis A, Trigkas M (2021) Exploring environmental and economic costs and benefits of a forest-based circular economy: a literature review. Forests 12(4):436

Lovas S, Varga O, Ádám B (2020) Chemical pollutants in closed spaces of transportation and storage of non-dangerous goods: a combined qualitative approach

Manickam P, Duraisamy G (2019) 3Rs and circular economy. In: Circular economy in textiles and apparel, pp 77–93. Woodhead Publishing

Marchesi M, Tweed C, Gerber D (2020) Applying circular economy principles to urban housing. IOP Conf Ser: Earth Environ Sci 588(5):052065. IOP Publishing

Meherishi L, Narayana SA, Ranjani KS (2019) Sustainable packaging for supply chain management in the circular economy: a review. J Clean Prod 237:117582

Monnolo A, Clausi MT, Del Piano F, Santoro M, Fiorentino ML, Barca L, Ferrante MC (2023) Do organochlorine contaminants modulate the parasitic infection degree in Mediterranean Trout (Salmo trutta)? Animals 13(18):2961

Murthy V, Ramakrishna S (2022) A review on global E-waste management: urban mining towards a sustainable future and circular economy. Sustainability 14(2):647

Nandi S, Hervani AA, Helms MM, Sarkis J (2023) Conceptualizing Circular economy performance with non-traditional valuation methods: lessons for a post-pandemic recovery. Int J Log Res Appl 26(6):662–682

Ncube A, Mtetwa S, Bukhari M, Fiorentino G, Passaro R (2023) Circular economy and green chemistry: the need for radical innovative approaches in the design for new products. Energies 16(4):1752

Niwalkar A, Indorkar T, Gupta A, Anshul A, Bherwani H, Biniwale R, Kumar R (2023) Circular economy-based approach for green energy transitions and climate change benefits. Proc Indian Natl Sci Acad 89(1):37–50

NORMAN (2016) NORMAN network group. http://www.norman-network.net/?q=node

O'Flynn D, Lawler J, Yusuf A, Parle-McDermott A, Harold D, Mc Cloughlin T, White B (2021) A review of pharmaceutical occurrence and pathways in the aquatic environment in the context of a changing climate and the COVID-19 pandemic. Anal Methods 13(5):575–594

Ohiomah I, Sukdeo N (2022) Challenges of the South African economy to transition to a circular economy: a case of remanufacturing. J Remanufacturing 12(2):213–225

Olabi AG (2019) Circular economy and renewable energy. Energy 181:450–454

Oladipupo AJ (2017) Soil metal distribution under different land uses of emerging mega cities in Southwest Nigeria and the associated ecological risk. Jordan J Earth Environ Sci 8(1)

Omran BA, Baek KH (2022) Valorization of agro-industrial biowaste to green nanomaterials for wastewater treatment: approaching green chemistry and circular economy principles. J Environ Manag 311:114806

Panagiotakis I, Dermatas D (2022) New European Union soil strategy: a potential worldwide tool for sustainable waste management and circular economy. Waste Manag Res 40(3):245–247

Patil SS, Bhagwat RV, Kumar V, Durugkar T (2019) Megaplastics to nanoplastics: emerging environmental pollutants and their environmental impacts. In: Environmental contaminants: ecological implications and management, pp 205–235

Pouri MJ, Hilty LM (2020) Digitally enabled sharing and the circular economy: towards a framework for sustainability assessment. In: Advances and new trends in environmental informatics: ICT for sustainable solutions, pp 105–116. Springer International Publishing

Purchase CK, Al Zulayq DM, O'Brien BT, Kowalewski MJ, Berenjian A, Tarighaleslami AH, Seifan M (2021) Circular economy of construction and demolition waste: a literature review on lessons, challenges, and benefits. Materials 15(1):76

Reichwaldt ES, Ghadouani A (2016) Can mussels be used as sentinel organisms for characterization of pollution in urban water systems? Hydrol Earth Syst Sci 20(7):2679–2689

Richardson SD (2003) Disinfection by-products and other emerging contaminants in drinking water. TrAC Trends Anal Chem 22(10):666–684

Richter JL (2022) A circular economy approach is needed for electric vehicles. Nat Electron 5(1):5–7

Rodríguez-Espíndola O, Cuevas-Romo A, Chowdhury S, Díaz-Acevedo N, Albores P, Despoudi S, Dey P (2022) The role of circular economy principles and sustainable-oriented innovation to enhance social, economic and environmental performance: evidence from Mexican SMEs. Int J Prod Econ 248:108495

Salvioni DM, Almici A (2020) Transitioning toward a circular economy: the impact of stakeholder engagement on sustainability culture. Sustainability 12(20):8641

Schärer M (2009) Evaluation of surface waters in Switzerland and Europe. Environ Sci

Serafimov V, Stets O, Shkolyk A (2021) Seaports in the BRI: challenges, solutions and emerging regulations. Lex Portus 7:14

Sharma HB, Vanapalli KR, Samal B, Cheela VS, Dubey BK, Bhattacharya J (2021) Circular economy approach in solid waste management system to achieve UN-SDGs: solutions for post-COVID recovery. Sci Total Environ 800:149605

Sherwood J (2020) Closed-loop recycling of polymers using solvents: Remaking plastics for a circular economy. Johnson Matthey Technol Rev 64(1):4–15

Shirkhani S, Fethi S, Alola AA (2021) Tourism-related loans as a driver of a small island economy: a case of Northern Cyprus. Sustainability 13(17):9508

Soliwoda M, Wieliczko B, Kulawik J (2020) Circular economy vs. sustainability of agribusiness. Zagadnienia Ekonomiki Rolnej/Probl Agric Econ 1

Tang M, Mudd GM (2015) The pollution intensity of Australian power stations: a case study of the value of the National Pollutant Inventory (NPI). Environ Sci Pollut Res 22:18410–18424

Vasseur P (2010) Foreword ISTA14 special issue. Environ Toxicol 5(25):429–430

Vural Gursel I, Elbersen B, Meesters KP, van Leeuwen M (2022) Defining circular economy principles for biobased products. Sustainability 14(19):12780

Warne MSJ, Batley GE, Braga O, Chapman JC, Fox DR, Hickey CW, Van Dam R (2014) Revisions to the derivation of the Australian and New Zealand guidelines for toxicants in fresh and marine waters. Environ Sci Pollut Res 21:51–60

Weber R, Gaus C, Tysklind M, Johnston P, Forter M, Hollert H, Zennegg M (2008) Dioxin- and POP-contaminated sites—contemporary and future relevance and challenges: overview on background, aims and scope of the series. Environ Sci Pollut Res 15:363–393

Weinmeyer R, Norling A, Kawarski M, Higgins E (2017) The Safe Drinking Water Act of 1974 and its role in providing access to safe drinking water in the United States. AMA J Ethics 19(10):1018–1026

Weng Z, Mudd GM, Martin T, Boyle CA (2012) Pollutant loads from coal mining in Australia: discerning trends from the National Pollutant Inventory (NPI). Environ Sci Policy 19:78–89

Wilson LK, Harris ML, Trudeau S, Ikonomou MG, Elliott JE (2010) Properties of blood, porphyrins, and exposure to legacy and emerging persistent organic pollutants in surf scoters (Melanitta perspicillata) overwintering on the south coast of British Columbia, Canada. Arch Environ Contam Toxicol 59:322–333

Wong MH (ed) (2012) Environmental contamination: health risks and ecological restoration. CRC Press

Xiaoming SHI (2021) Actively promoting research on health effects of children's multi-source environmental exposure. 中国学校卫生 42(9):1285–1287

Yousaf B, Liu G, Wang R, Imtiaz M, Zia-ur-Rehman M, Munir MAM, Niu Z (2016) Bioavailability evaluation, uptake of heavy metals and potential health risks via dietary exposure in urban-industrial areas. Environ Sci Pollut Res 23:22443–22453

Zhang R, Johnson NM, Li Y (2022) Establishing the exposure–outcome relation between airborne particulate matter and children's health. Thorax 77(4):322–323

Zini LB, Gutterres M (2021) Chemical contaminants in Brazilian drinking water: a systematic review. J Water Health 19(3):351–369

The opinions expressed in this chapter are those of the author(s) and do not necessarily reflect the views of the UNESCO: United Nations Educational, Scientific and Cultural Organization, its Board of Directors, or the countries they represent.

Chapter 15
Reverse Logistics in the Disposal of Empty Pesticide Packaging—A Case Study in the Municipality of Ibirama, Santa Catarina, Brazil

Maria Pilar Serbent, Camila Schwarz Pauli, Anderson Fozina Krüger, and Willian Jucelio Goetten

Abstract Brazil is an important player in the food production sector. Among other environmental problems associated with the development of this activity, conventional agriculture generates a type of waste that should be highlighted due to its pollution potential and/or its hygienic aspect: Pesticide packaging. These materials must be returned to the place where the products were purchased, which is called reverse logistics and is a tool of the National Solid Waste Law. This research shows the results of the third stage of the university extension project entitled Agriculture and Socio-Environmental Responsibility in the Alto Vale de Itajaí, Santa Catarina, in which, firstly, the health aspects of rural properties in Ibirama were identified, including general knowledge about the risks associated with pesticides and the destination of empty containers, then environmental education activities were carried out with young people and adolescents in the region and, finally, a situational diagnosis of the destination of empty pesticide containers was carried out. Thus, the objective of this study was to evaluate the impact of the campaign for the collection of empty pesticide containers organized by the Municipality of Ibirama (27°03′25″ south 49°31′04″ west), in collaboration with the Agricultural Association of the Itajaí River Basin, for the years 2019, 2020, 2021 and 2022. The municipality has around 550 registered farmers. In the collection organized by the municipality of Ibirama

M. P. Serbent (✉) · W. J. Goetten
Santa Catarina State University, Ibirama, Brazil
e-mail: mariapilar.serbent@udesc.br

M. P. Serbent
Chemical Research and Technology Center UTN-CONICET, Córdoba, Argentina

C. Schwarz Pauli
Department of Environment, Ibirama City Hall, Brazil

A. Fozina Krüger
Economic Development and Environment, Ibirama City Hall, Brazil

W. J. Goetten
Intermunicipal Sanitation Regulatory Agency (ARIS), Florianópolis, Brazil

© UNESCO 2025
S. Zandaryaa et al. (eds.), *Emerging Pollutants*, Advances in Water Security,
https://doi.org/10.1007/978-3-031-71758-1_15

333

for 2019, 52 farmers participated and 1,056 plastic containers (855 washable and 201 non-washable), seventy-seven cardboard containers, 854 lids and eleven glass bottles were collected. In 2020, 77 farmers participated in the itinerant collection and 1,674 plastic containers, 15 kg of flexible packaging, 60 kg of cardboard and 1,674 lids were collected. In 2021, 58 farmers participated in itinerant acceptance and 1311 washed plastic packaging, 6.73 kg flexible packaging, 34 kg cardboard and 1386 caps were collected. In 2022, 72 farmers participated in itinerant acceptance and 1649 washed plastic packaging, 26.77 kg flexible packaging, 82 kg cardboard and 1596 caps were collected. Although the number of farmers did not increase in the period covered (2019–2022), the number of farmers who took part in the packaging collection campaigns has increased over the years. The disposal of empty packaging must be conducted correctly and in accordance with environmental legislation to avoid pollution of the environment, reduce health risks for the population and ensure the quality of life for generations. Therefore, monitoring the process of reverse logistics, as carried out in this work, is essential.

Keywords National solid waste policy · Reverse logistics · Final disposal of the empty pesticide containers

15.1 Introduction

In 2020, Brazilian agribusiness fed eight hundred million people (Contini and Aragão 2021), making the country one of the largest producers in the world and therefore, making the country massively use pesticides for its production (Pignati et al. 2017). Terrestrial and aquatic environmental compartments may be the fate of products used in agriculture, considering that only 5% or less of the applied amount has been shown to reach target organisms, resulting in soil and water contamination (Javaid et al. 2016).

In addition to the issue of environmental quality, humans can also be affected by the adverse effects of pesticides, either through the sale of pesticides in agricultural establishments, or through occupational exposure to which thousands of agricultural workers are exposed, or indirectly through the ingestion of contaminated food or exposure in a contaminated environment (Friedrich et al. 2022; Marques and Silva 2021; Pereira et al. 2019) since the effects may include poisoning by inhalation (inhalation), skin contact (dermal), or ingestion, with various short- and long-term effects.

In areas of notable agricultural development, a social commitment to the environment is essential, with clear support and participation from governments, educational institutions, and non-governmental organizations working together on the proper management of solid waste. At the national level, the National Solid Waste Policy (PNRS), established by Federal Law No. 12,305 in 2010, emphasizes the importance of environmental education to promote, among other things, the prevention, reduction, reuse, and recycling of solid waste (Brazil 2010). Due to Brazil's importance in food production, the prevailing agricultural production methods have led to an increased generation of a type of waste that should be highlighted for its pollution potential and/or hygienic aspect: pesticide packaging. These materials must be returned to the place where the products were purchased, which is called reverse logistics and is a tool of the PNRS.

Pesticide packaging is classified as Class I—Hazardous Solid Waste—by the Brazilian Association of Technical Standards (ABNT) through the NBR 10.004/2004, and therefore must be sent to an environmentally sound final destination, since the chemical substances contained in pesticides have the capacity to alter the environment by affecting the natural chain, contaminating the soil, water, and air, in addition to harming public health (Brazil 2004).

Brazilian legislation, through Federal Law No. 9.974, has introduced the concept of shared responsibility among the actors of the agricultural chain in the process of acceptance and final disposal of empty pesticide packaging, establishing the specific roles of each (Brazil 2000). Authorities at the local, state, and federal levels monitor the fulfillment of the legal obligations of each link in the chain and share with the distribution channels and the processing industry the responsibility for educating and regularly raising awareness among farmers of the importance of participating in take-back logistics. In this link of the chain, the farmer is responsible for returning all empty pesticide product packaging. The return can be made at the distributor's designated packaging acceptance or collection points, on the invoice, or at the farm where the pesticide was purchased.

The publication of Law No. 14.026/2020, known as the New Sanitary Framework, brought changes to the PNRS. Among other things, it dealt with the deadline for the promotion of environmentally sound solid waste disposal throughout the country. Another recent regulatory aspect in Brazil was the publication of the first Reference Standard for Basic Sanitation, issued by the National Water and Sanitation Agency (ANA) and approved by Resolution No. 79 of June 14, 2021, which establishes the regulation, structure, and parameters for charging for the provision of public waste management services throughout the country (Abar 2022).

In some regions there are also ambulatory collections organized by resellers. The itinerant reception is a practice instituted by the Sistema Campo Limpo. The Sistema Campo Limpo is the name of the Brazilian reverse logistics program for empty pesticide packaging, in which the National Institute for Empty Packaging Processing (INPEV) acts as an intelligence hub. It covers all regions of the country and is based on the concept of shared responsibility enacted by Federal Law No. 9.974. To classify and facilitate reverse logistics of pesticide packaging, it is divided into two broad groups: washable and non-washable packaging. Washable packaging is rigid (plastic and metal) and is used for packaging liquid formulations that are diluted in water. Non-washable packaging is used for packaging products that are not sprayed with water, as well as all flexible packaging and secondary packaging. The regulation imposed by Federal Law No. 9.974 and its regulations extends to all pesticide use in the country, including use on small farms where subsistence agriculture is practiced. In the state of Santa Catarina, family farms predominate, accounting for 78% of

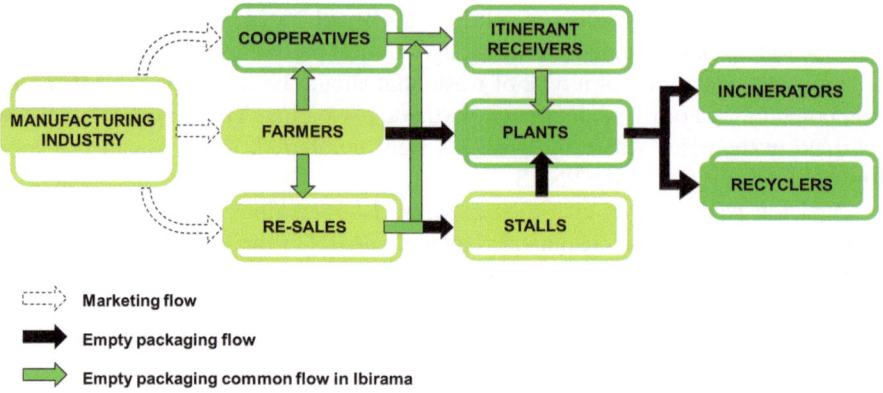

Fig. 15.1 Process of taking back pesticide packaging after use in the municipality of Ibirama, Santa Catarina. *Source* Authors

production (Santa Catarina 2019a). This reality is also reflected in the Alto Vale do Itajaí region, where the municipality of Ibirama is located.

According to the report of the Brazilian Forestry Agency, by July 31, 2019, 1195 rural plots were registered in Ibirama (Brazil 2019), and by February 4, 2023, 1078 plots were registered in the system of unification of the municipalities of the Alto Vale do Itajaí (AMAVI), by the City Hall of Ibirama (Amavi 2023). In this municipality, small producers can return their empty pesticide packaging to the collection points that are set up during the annual campaigns for this purpose (Figure 15.1). The campaigns consist of setting up temporary structures in places far from the collection points to facilitate the return of the material by the farmers. It should be noted that the Federal Law No. 12.305, in its article 33, provides for the mandatory implementation of the take-back system for the distributors and traders of pesticides for their packaging, the companies that sell these products in the municipality of Ibirama say that they do not implement the take-back system because there is no suitable place for the deposit.

Triple washing consists of cleaning pesticide packaging after use on land. All packaging that does not go through this procedure that does not have holes in the bottom and where the lids are not separated from the packaging, as well as packaging of group B, except for cardboard, is destined for incineration.

According to data from the Information System for Monitoring Water Quality for Human Consumption (SISAGUA) from 2014 to 2017, Santa Catarina is among the three states with the highest pesticide contamination in water supplies (Brazil 2018). This shows the importance of knowledge about the correct application of pesticides in the field and the correct implementation of triple washing, which contributes to efficient backward logistics of these residues without environmental damage to water resources, as well as the promotion of sustainable alternatives that avoid the use and/ or dependence on these products.

Although it is not the responsibility of the municipality to manage the waste of backward logistics, Art. 41—Item XIV of the Municipal Master Plan (Ibirama 2008) establishes a program to raise awareness about the use of pesticides and the disposal of packaging. In addition, item 6.1 and 11.1 of the Municipal Sanitation Plan (Ibirama 2016) mentions pesticide packaging as part of the reverse logistics and hazardous waste management program. This initiative is promoted by the Municipality of Ibirama in collaboration with the Association of Agricultural Companies of the Itajaí River Basin (AABRI). This municipal action is crucial because in 2018, the municipality of Ibirama ranked third among the municipalities with the most confirmed cases of pesticide poisoning (11) in the state of Santa Catarina (Santa Catarina 2019b).

The objective of this study was to describe and evaluate the impact of the campaign organized by the municipality of Ibirama together with the Association of Farmers of the Itajaí River Basin for the collection of empty pesticide containers in 2019, 2020, 2021 and 2022.

15.2   Materials and Methods

The first phase of this study was conducted in a rural community in the municipality of Ibirama, located in Alto Vale do Itajaí, in the central region of the Santa Catarina state (Figure 15.2). The municipality has an area of 246.7 km^2 with an estimated population of 19,096 inhabitants (Brazil 2020). The main economic activities in the municipality include the textile sector, local trade, and tourism, with a focus on agriculture, where tobacco and corn are the main crops. It is estimated that 235,184 liters of pesticides (16l/ha) are consumed in Ibirama per year (Pignati et al. 2017).

The project collaborated with the community of Ibirama and focused on the disposal of pesticide packaging, as it is a type of waste with pollution potential and health significance. To this end, the annual campaign, which took place in August 2019, August 2020, August 2021, and December 2022 and included six collection points in different neighborhoods, was accompanied by municipal officials: Centro, Dalbérgia, Rafael Baixo, Rafael Alto, Rio Sellin and Ribeirão das Pedras. The campaigns are publicized on the website of the City Hall of Ibirama, on social media and in places where pesticides are sold and collected. This campaign evaluated compliance with INPEV recommendations and standards for reception, control (separation into washable and non-washable packaging, rigid packaging—for seed processing—and secondary packaging—collection boxes—and flexible packaging), processing (in reception points) and storage. Table 15.1 shows the results of the campaign in the period 2019–2022.

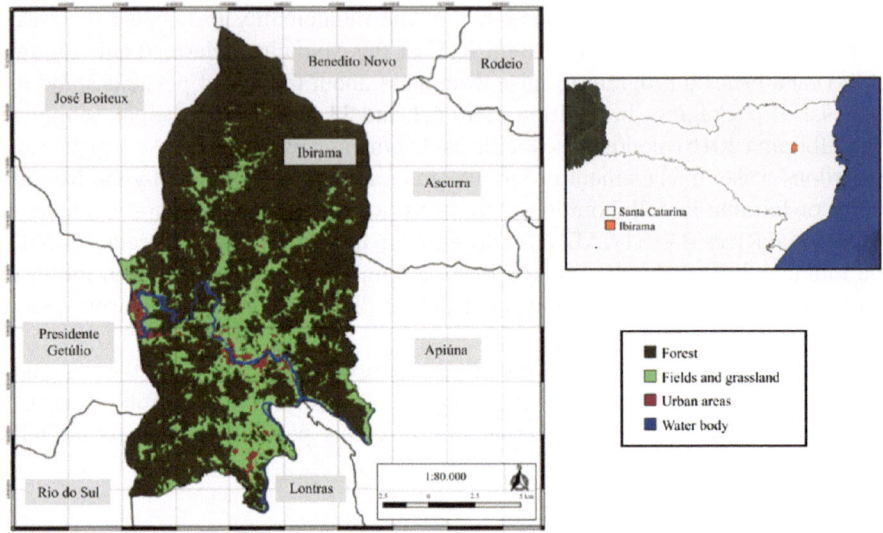

Fig. 15.2 Location of Ibirama, Santa Catarina, with details on land use and occupation. *Source* Ibirama City Hall, Brazil, and Authors

Table 15.1 Situation diagnosis of the fate of empty pesticide packaging in Ibirama

Year	2019	2020	2021	2022
Registered farmers in Ibirama	627	585	612	531
Farmers participating in the campaign	52 (8.29%)*	77 (13.16%)*	58 (9.48%)*	72 (13.56%)*
Washable packaging (units)	855	1674	1311	1649
Non-washable packaging	201 (units)	15 kg	N/A	N/A
Cardboard	77 (units)	60 kg	34 kg	82 kg
Tampas (units)	854	1674	1386	1596
Glass containers (units)	11	N/A	N/A	N/A
Flexible packaging (kg)	N/A	N/A	6.73	26.77

N/A: unaccounted
*percentage of farmers
Source Authors

15.3 Results and Discussion

According to the list of establishments that distribute agrochemical products and providers of crop protection services registered at Santa Catarina, there are few places in the Itajaí region registered to sell agrochemicals for Ibirama (4), Apiúna (1), Ascurra (4), Benedito Novo (2), Lontras (2), Presidente Getúlio (5), Rio do Sul

(4), Rodeio (1), José Boiteux (2), Dona Emma (2), Vitor Meireles (1) and Witmarsum (1) (Santa Catarina 2019c).

However, farmers indicated that there is an illegal market for the sale of pesticides, without the technical support of authorized professionals to purchase these products. The farm workers also expressed that they use the wastewater produced by the triple washing to prepare syrup for the crops.

The empty packaging was transported to a collection point connected to INPEV (Fig. 15.3), maintained by AABRI and located in the municipality of Aurora, Santa Catarina, 31 km from Ibirama (as the crow flies). The reverse logistics system for empty pesticide packaging in this reception point is a pioneer in ambulatory reception of pesticide packaging nationwide, since it represents about 20% of the total pesticide packaging collected in the state of Santa Catarina (SC) and has a positive return from rural producers (França et al. 2018).

Comparing the last collections in the municipality of Ibirama, it can be noted that in 2021 there was a decrease in pesticide packaging collected, due to the covid-19 pandemic, while in 2022 the collection increased again, showing the importance of the correct disposal of packaging. As a positive consequence of the work conducted by AABRI over the last decade, a 28% decrease in the amount collected for incineration was recorded in 2017 (França et al. 2018). At the national level, the Campo-Limpo system processed 44,261 tons of empty pesticide packaging, representing 94% of the total number of products of this type sold in the country (Abrelpe 2020).

Although the number of farmers did not increase during the period in question (2019–2022), the number of farmers who took part in the packaging collection campaigns increased over the years.

Fig. 15.3 Reception center connected to the INPEV in the neighboring municipality of Aurora, Santa Catarina. *Source* Authors

Fig. 15.4 Acceptance of empty packaging in the municipality of Ibirama, Santa Catarina. *Source* Authors

Glass packaging is not currently produced by the processing industry, so the presence of this type of packaging is due to old rural estates that have kept this packaging for a long time and are returned by farmers due to itinerant income. Glass packaging, washable packaging that has a certain amount of residue, and non-washable packaging are sent directly to incinerators by the collection points, since recycling is not possible. Although incineration reduces the volume and hazardousness of the waste, it releases toxic pollutants into the atmosphere through the degradation of the liquid or solid product residues by physical–chemical reactions. Thus, it would be necessary to propose ways to treat the wastewater generated after washing the packages, since they are not always used in the syrups used in the cultures.

15.4   Conclusion

The present work summarizes the results of an experience related to the socio-environmental problem in the municipality of ibirama of the Alto Vale de Itajaí region, characterized by agricultural activity within family farms. This chapter brings together the results of an experiment in the Alto Vale de Itajaí, Santa Catarina, Brazil, a region characterized by agricultural activity. The campaigns aimed to encourage the return of empty pesticide containers after use as a way of mitigating the socio-environmental problems arising from conventional agriculture. However, water quality and soil analysis were not part of the scope of the work presented.

Even if it is not the responsibility of the municipality, this institution has an interest in promoting awareness about the use of pesticides and the disposal of packaging. Although the number of farmers did not increase in the period covered (2019–2022), the number of farmers who took part in the packaging collection campaigns has increased over the years. The activities developed in recent years need to be improved to reach a wider audience through environmental education actions linked to different institutions.

Acknowledgements To Rosdalva Schroder from the Health Surveillance of the Secretariat of Regional Development in Ibirama. To Jean and Natália, former research fellows of UDESC.

References

Abar (2022) Associação Brasileira de Agências Reguladoras. O papel das agências reguladoras na implementação do novo marco legal de saneamento básico em relação aos serviços de limpeza urbana e manejo de resíduos sólidos (Hoos CG et al (eds)). https://www.cnmp.mp.br/portal/images/CMA/residuos/53_3.Opapeldasagenciasregulad orasnaimplementacaodonovomarco-ABAR.pdf. Accessed 08 March 2024

Abrelpe (2020) Associação Brasileira de Empresas de Limpeza Pública e Resíduos Especiais. Panorama dos resíduos sólidos no Brasil. https://abrelpe.org.br/panorama-2020/. Accessed 28 June 2021

Amavi (2023) Associação dos Municípios do Alto Vale do Itajaí. Propriedades: CAR. https://www.amavi.org.br/sistemas/car/cons_propriedade. Accessed 04 Feb 2023

BRAZIL (2000) Subchefia para Assuntos Jurídicos. Lei N°. 9.974, de 06 de junho de 2000. http://www.planalto.gov.br/ccivil_03/LEIS/L9974.htm. Accessed 06 Sept 2019

BRAZIL (2004) ABNT, Associação Brasileira de Normas Técnicas. NBR 10004. Resíduos sólidos: classificação

BRAZIL (2010) Lei N°. 12.305, de 2 de agosto de 2010. http://www.planalto.gov.br/ccivil_03/_ato 2007-2010/2010/lei/l12305.htm. Accessed 08 Sept 2019

BRAZIL (2018) Ministério da Saúde. SISAGUA, Sistema de Informação de Vigilância da Qualidade da Água para Consumo Humano. https://app.rios.org.br/index.php/s/ljppVjrP37ak8HE. Accessed 09 Sept 2019

BRAZIL (2019) Serviço Florestal Brasileiro. CAR, CADASTRO AMBIENTAL RURAL. Módulo de Relatórios. http://www.florestal.gov.br/modulo-de-relatorios. Accessed 06 Sept 2019

BRAZIL (2020) Estimativas da população residente no Brasil e Unidades da Federação. https://ftp. ibge.gov.br/Estimativas_de_Populacao/Estimativas_2020/POP2020_20210331.pdf. Accessed 28 June 2021

Contini E, Aragão A (2021) O Agro Brasileiro alimenta 800 milhões de pessoas. Embrapa, Brasilia. https://www.embrapa.br/documents/10180/26187851/Popula%C3%A7%C3%A3o+ alimentada+pelo+Brasil/5bf465fc-ebb5-7ea2-970d-f53930b0ec25?version=1.0&download= true. Accessed 02 Nov 2023

França I, Campos de Sá L, Dalpian P (2018) Logística reversa de embalagens vazias de agrotóxicos: o caso de sucesso da central de recebimento de embalagens vazias de Aurora/SC. In: VI Simpósio da Ciência do Agronegócio. https://www.ufrgs.br/cienagro/wp-content/uploads/ 2018/10/Log%C3%ADstica-reversa-de-embalagens-vazias-de-agrot%C3%B3xicos-o-caso-de-sucesso-da-central-de-receb.pdf. Accessed 09 Sept 2019

Friedrich K, do Monte Gurgel A, Sarpa M, Bedor CNG, de Siqueira MT, Gurgel IGD, da Silva Augusto LG (2022) Toxicologia crítica aplicada aos agrotóxicos – perspectivas em defesa da vida. Saúde Debate 46:293–315. https://doi.org/10.1590/0103-11042022e220

Ibirama (2008) Lei Complementar N°. 73, de 22 de dezembro de 2008. https://www.ibirama.sc.gov. br/download.php?id=15. Accessed 06 Sept 2019

Ibirama (2016) PLANO MUNICIPAL DE SANEAMENTO BÁSICO. Plano Municipal de Saneamento Básico Consolidado. http://ibirama.sc.gov.br/download.php?id=1343. Accessed 09 Sept 2019

Javaid MK, Ashiq M, Tahir M (2016) Potential of biological agents in decontamination of agricultural soil. Scientifica 2016:1598325. https://doi.org/10.1155/2016/1598325

Marques JMG, Silva MV (2021) Estimativa de ingestão crônica de resíduos de agrotóxicos por meio da dieta. Rev Saúde Pública 55:36. https://doi.org/10.11606/s1518-8787.2021055002197

Pereira RA, Maciel Lima Costa C, Maciel Lima E (2019) O impacto dos agrotóxicos sobre a saúde humana e o meio ambiente. Rev Extensão 3:29–37

Pignati W, Lima FANS, Lara SS, Corrêa MLM, Barbosa JR, Leão LHC, Pignatti MG (2017) Distribuição espacial do uso de agrotóxicos no Brasil: uma ferramenta para a vigilância em saúde. Cien Saude Colet 22:3281–3293. https://doi.org/10.1590/1413-812320172210.17742017

Santa Catarina (2019a) Epagri, Empresa De Pesquisa Agropecuária E Extensão Rural De Santa Catarina. Agricultura familiar responde por metade do faturamento da agropecuária catarinense. https://www.epagri.sc.gov.br/index.php/2019/11/01/agricultura-familiar-responde-por-metade-do-faturamento-da-agropecuaria-catarinense/. Accessed 04 Sept 2023

Santa Catarina (2019b) VSPEA, Vigilância em Saúde de Populações Expostas a Agrotóxicos. Informativo VSPEA 01/2019. http://www.vigilanciasanitaria.sc.gov.br/phocadownload/Noticias/2019/aBRIL/informativo%20vspea%2001.pdf. Accessed 08 Sept 2019

Santa Catarina (2019c) Cidasc, Companhia Integrada De Desenvolvimento Agrícola De Santa Catarina. Secretaria De Estado Da Agricultura E Da Pesca. Relação dos Estabelecimentos que comercializam produtos agrotóxicos e prestadores de serviços fitossanitários registrados em Santa Catarina. https://sigen.cidasc.sc.gov.br//ConsultaEstabelecimentoAx/ConsultaEmpresa. Accessed 09 Sept 2019

Chapter 16
Microalgae as Bio-based Circular Solutions for Harmful Algal Bloom (HAB) in Lake Tegel, Berlin, Germany

Kei Namba and Armin Dolatimehr

Abstract The proliferation of Harmful Algal Blooms (HABs) in freshwater systems has emerged as a pressing environmental concern in recent years. Lake Tegel in Berlin, Germany, serves as a notable case study. This research elucidates the multifaceted drivers behind HABs, including the exacerbating roles of anthropogenic climate change, wastewater intrusion, and eutrophication. Of particular concern are potent toxins like anatoxin-a, which have been linked to neurotoxicity and faunal mortalities. This study proposes adopting a circular economy paradigm to tackle Harmful Algal Blooms (HABs) challenges. By leveraging biotechnological advancements, the research suggests detoxifying affected aquatic systems and converting these toxins into commercially viable byproducts, such as biofertilizers. This investigation underscores the imperative of a holistic understanding of HAB etiology, the potential efficacy of microalgal bioremediation, and the broader socioeconomic implications of integrating circular economy principles into environmental remediation strategies.

Keywords Harmful Algal Bloom (HAB) · Lake Tegel · Climate change · Circular economy · Microalgae · Bioremediation · Biofertilizer · Entrepreneurship

K. Namba (✉)
Technische Universität Berlin, Berlin, Germany
e-mail: k.namba@tu-berlin.de

A. Dolatimehr
Independent Researcher, Berlin, Germany

© UNESCO 2025

343

S. Zandaryaa et al. (eds.), *Emerging Pollutants*, Advances in Water Security,
https://doi.org/10.1007/978-3-031-71758-1_16

16.1 Background

16.1.1 Problem Statement

The research aims two-fold: First, to investigate the impacts of Harmful Algal Bloom (HAB) on freshwater bodies, as their full impacts are currently unknown. Second, to contribute to providing a solution to the problem using a circular economy approach: we produce resources out of HAB as byproducts because of toxin removal and pollutants treatment.

Although this study focuses on the HAB problem of Lake Tegel in Berlin, Germany, marine and freshwater HABs are a growing human and environmental risk on a global scale. The 2019 Intergovernmental Panel for Climate Change assessment report concluded that 'the occurrence of HABs, their toxicity, and risk on natural and human systems are projected to continue to increase with warming and rising CO_2 in the 21st century (Hallegraeff et al. 2021). At the global level, GlobalHAB, "Global Harmful Algal Blooms," was established as a new scientific programme on HABs cosponsored by the Intergovernmental Oceanographic Commission (IOC) of UNESCO and the Scientific Committee on Oceanic Research (SCOR) that will operate for ten years from 2016 to 2025 (Intergovernmental Panel on Climate Change (IPCC) 2019). Understanding the biodiversity and biogeography of harmful organisms and their toxins is fundamental to identifying trends in HAB occurrence, enabling their prediction and management, and, when possible, mitigating their impacts. Furthermore, correct taxonomic identification of harmful species is crucial to achieving this objective (Berdalet et al. 2017).

In Lake Tegel, Tychonema sp. was present in significant concentrations and was the primary cause of anatoxin-A. The neurotoxicity of anatoxin-a caused the death of several dogs exposed to this water body (Fastner et al. 2018). Identification of high anatoxin-a concentration in the dead animal bodies also revealed the necessity of further investigation around anatoxin-a and its metabolites (Fastner et al. 2018). Tychonema is a toxic filamentous bacteria identified as the leading producer of anatoxin-a and homoanatoxin-a in freshwater (Salmaso et al. 2016; Shams et al. 2015). Due to its mixotrophic lifestyle, this cyanobacterium has been able to grow successfully and rapidly in aquatic environments (Evseev et al. 2023). The Anatoxin group comprises anatoxin-a, homoanatoxin-a, and dihydroanatoxin-a. These toxins attach to acetylcholine receptors in the synapses between neurons and muscle tissue to imitate acetylcholine (Christensen and Khan 2020). Consequently, muscles get overstimulated, disrupting the respiratory system and leading to death (Colas et al. 2021) (Fig. 16.1).

HABs, also known as cyanobacteria in freshwater due to rising temperature caused by anthropogenic factors, climate change, intrusion of wastewater or agricultural runoff into freshwater, and eutrophication, produce harmful toxins and negatively affect people's health and the environment. The occurrence of HAB can be anthropogenic and considered a disaster. In the case of the mass fish die-off at the Oder

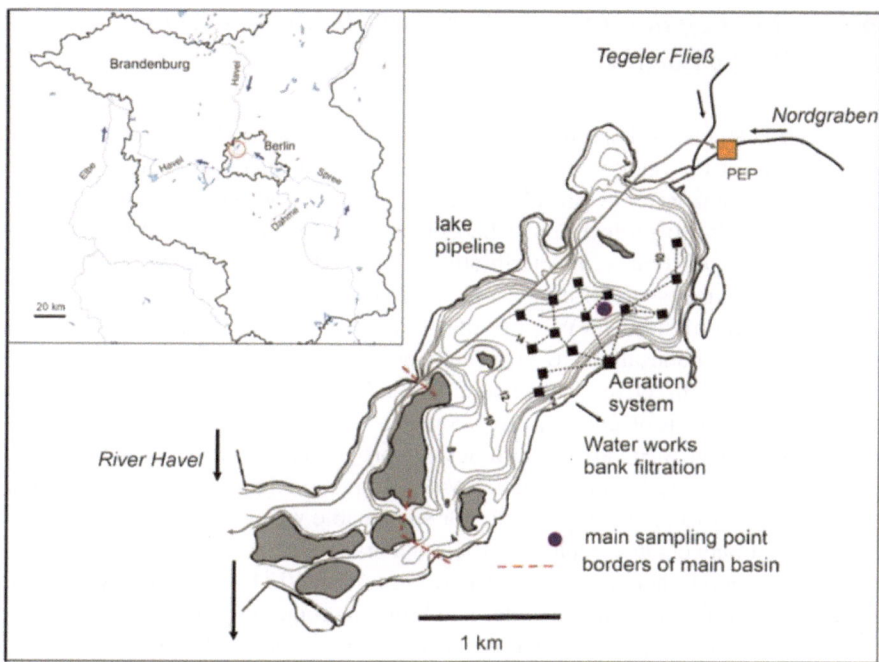

Fig. 16.1 Lake Tegel, its characteristics, tributaries, and position of the phosphorus elimination plant (PEP) (from the Berlin Waterbody Atlas). *Source* Chorus et al. 2020

River between Germany and Poland in 2022, the accident's root cause was a rapid rise in salinity.

However, the key finding of the investigations by experts was that human activities caused it (Umweltbundesamt 2022). The impacts of human activities, including both climatic and non-climatic factors, on HABs are well documented (Gobler 2020). International Commission for the Protection of the Odra against Pollution (ICPO) (Die Internationale Kommission zum Schutz der Oder gegen Verunreinigungen (IKSO) was founded by the Czech Republic, Poland, and Germany based on the Convention on the Protection of the Oder. The Convention entered into force following its ratification on April 26, 1999. In response to the fish dying off in the Oder River in August 2022, the Commission has tightened its International Warning and Alarm Plan for the Oder (IWAPO) (Mkoo n.d.). The counties bordering the river, the city of Frankfurt (Oder), and the Brandenburg Ministry of Agriculture, Environment, and Climate Protection (MLUK) also published recommendations and ordinances to mitigate the consequences of the disaster (Ministerium für Landwirtschaft, Umwelt und Klimaschutz Brandenburg n.d.).Tackling the problem of HAB directly contributes to meeting the goals of the EU directives on water quality.

16.1.2 EU Regulations and Policy and Implementation at the Local Level

The Water Framework Directive (Wasserrahmenrichtlinie 2000/60/EG), set out in 2000, is central to water management in EU member states. It requires member States to use their River Basin Management Plans (RBMPs) and Programmes of Measures (PoMs) to protect and, where necessary, restore water bodies to reach good status [OBJ] [OBJ] to prevent deterioration. Good status means both good chemical and good ecological status (European Commission 2023).

In the context of the EU's Water Framework Directive, the German Environment Agency (Umweltbudesamt, UBA) formulates recommendations as guidelines for implementing EU directives. The Federal Water Act (Wasserhaushaltsgesetz, WHG) is the core of water protection law and connects water use approval to preserve the water balance. The WHG also determines that programs of measures and river basin management plans, according to the EU Water Framework Directive and Flood Directive regulated by the general principle for sustainable water resource management, to prevent climate change impacts (Article 6, Para, 1 No.5 WHG).

In Berlin, good chemical status, according to the WFD, is not achieved due to exceedances of, for example, mercury and bromine-containing organic compounds. Low runoff due to drought, for example, during heavy rains, overflows from the sewer system into surface waters are also a significant problem. In addition, many barriers in the form of weirs and barrages make it impossible for fish and other aquatic life to pass. Many watercourses need more diversity due to built-up banks and beds (at the bottom), straightening, and a need for riparian woodland. In the past, this led to a severe impoverishment of species (BUND Berlin n.d.-a).

Due to the interdependent nature of water, regional and inter-state collaboration for emerging pollutants is necessary. In 2022, Berlin and Brandenburg agreed on a joint strategy to improve the protection of water bodies and to deal with anthropogenic trace substances from wastewater treatment plants, which is an essential component of the Masterplan Water Berlin as described in the Position paper on the strategy for dealing with anthropogenic trace substances from wastewater treatment plants (Senatsverwaltung für Mobilität, Umwelt, Verkehr und Klimaschutz Berlin (SenMVKU) 2022).

In the context of the WFD, the state of Brandenburg has created the Water Development Concept (Gewässerentwicklungskonzept, GEK) and the Nutrient Reduction Concept (Nährstoffreduzierungskonzept, NRK). The NRK envisages equipping all large-scale wastewater treatment plants in the Berlin-Brandenburg Spree-Havel region with an advanced treatment stage for phosphorus elimination by 2027. This is not the place to discuss the plan for Berlin to upgrade its wastewater treatment plant at Lake Tegel (Landesamt für Umwelt Brandenburg, LfU n.d.).

There are other directives related to the water quality. The Bathing Water Directive (76/160/EWG) poses the requirements governing the quality of water bodies used for bathing. In Germany, the Act on the Prevention and Control of Infectious Diseases in Humans (Protection Against Infection Act -Infektionsschutzgesetz IfSG) regulates the quality of swimming and bathing pool water used for swimming and bathing pools (Bundesgesundheitsministerium n.d.).

To implement the goals of the WFD in Berlin and protect water bodies and biodiversity, a CSO network called "Wassernetz Berlin" was launched in 2023 to promote citizens' active participation (BUND Berlin n.d.-b). In Berlin, there is a civil society-led project, "Flussbad Berlin," with the aim to transform the Spree Canal in the Mitte district of Berlin into a clean and swimmable body of water. They have created a platform involving government, science, and art experts through their activities (Flussbad Berlin n.d.).

Another EU directive related to water pollution is the Nitrate Directive (91/676/EWG). The Nitrates Directive mandates that farmers observe good agricultural practices and prepare action programs to reduce nitrate runoff. Authorities must monitor the effectiveness of these measures. In Germany, lawmakers transposed the Directive into national law through the Fertilisers Ordinance (Düngeverordnung), which regulates the use of fertilizers and the storage of sludge, among other things (UNECE 2021).

The Directive obliges Germany to prevent exceeding the quality standard of fifty milligrams of nitrate per liter. Since 2008, monitoring points in Germany have increasingly exceeded the 50 mg/l nitrate threshold each year. On June 21, 2018, the European Court of Justice found Germany guilty of violating the EU Nitrates Directive. Agriculture is the most critical source of high nitrate concentrations in groundwater (Umweltbundesamt (UBA) n.d.).

The revised Urban Wastewater Directive (91/271/EWG) in 2024 is one of the key deliverables under the EU's zero-pollution action plan. It requires the collection of wastewater from households and small plants and the reduction of the organic load. Moreover, removing at least 75% of phosphorus and nitrogen requires wastewater treatment in urban wastewater (Council of the European Union 2024).

Although the authorities have warned the public, they have not yet taken sufficient action. Scientists are currently investigating the causes and effects of the problem (Tagesspiegel 2017) (BZ Berlin 2023). Müggel have had these HAB problems since 2017. Lake Tegel is the second largest water body in Berlin. In 2021, the Berlin State Office for Health and Social Affairs (Berliner Landesamt für Gesundheit und Soziales, LAGeSo) imposed a bathing ban at Lake Tegel due to contamination, causing a threat to public health (Berliner Zeitung 2022).

16.1.3 Methodology

We have reviewed scientific articles and policy documents related to toxin removal, water chemistry, circular economy, and microalgae cultivation. As a first step, we investigate the root causes of the HAB problem, followed by ways to remove the toxins as we clean up the lake and treat them. As a second step, we will investigate the potential application of microalgae cultivation for toxin removal and policy environment for broader applications (i.e., bio fertilizer).

16.2 Potential Causes

16.2.1 Algal Bloom-Producing Microorganisms

HABs have an impact on freshwater bodies in Canada, and Lake Erie is a prime example (Engstrom et al. 2022; Hushchyna et al. 2023). This lake, one of the Great Lakes in North America, has been repeatedly affected by these blooms (Ho and Michalak 2015). The leading cause of these blooms is the runoff of nutrients, especially phosphorus (Watson et al. 2016). Apart from disrupting the lake's ecosystem, these blooms also pose health risks and economic challenges to the communities that depend on its waters. In California, USA, algal blooms have significantly impacted the Clear Lake water body, leading to fish deaths and illnesses in mammals (Jang and Otim 2023). The issue has escalated, with these blooms now affecting private drinking water intakes in the Clear Lake area, raising concerns about water safety and public health (Solomon et al. 2022). To effectively manage blooms in various environments, it is crucial to understand the factors that influence them. These factors encompass the strategies algae adopt to maximize growth, acquire and utilize resources efficiently, and adapt to changes in availability. Algae also employ dynamics and reproductive mechanisms for propagation and dispersal. Additionally, they have evolved adaptations to minimize losses caused by damage, pH extremes, temperature variations, salinity changes, desiccation risks, sinking tendencies, advective losses, and threats from predators or pathogens. By understanding these mechanisms, we can gain insights into the intricate nature of HABs in different freshwater ecosystems. This knowledge enables us to develop practical management approaches for each ecosystem (Watson et al. 2015).

16.2.2 Different HAB-Producing Genera

The presence of Anatoxin-a in lakes and its association with specific algal genera such as Anabaena (Wiltsie et al. 2018), Cuspidothrix (Kim et al. 2022; Napiórkowska-Krzebietke et al. 2023), Dolichospermum (Kramer et al. 2022), Microcoleus (Stielow and Ballantine 2003), Oscilatoria (Zhang et al. 2020), Phormidium (Minasyan et al. 2018), Planktothrix (Derot et al. 2020), Raphidiopsis (Burford et al. 2022), Sphaerospermopsis (Budzyńska et al. 2019) is a critical in understanding HABs in freshwater. Anatoxin-a, a potent neurotoxin, is produced by various cyanobacteria and poses significant risks to aquatic ecosystems and public health. Several algal genera, such as Anabaena, Aphanizomenon, Dolichospermum, Raphidiopsis, and Sphaerospermopsis, which are commonly found in lake environments, primarily produce cylindrospermopsin, a potent and increasingly observed cyanobacterial toxin. These cyanobacteria are known for their adaptability to various environmental conditions, often flourishing in nutrient-rich waters that lead to eutrophication. The presence of Cylindrospermopsin in lakes is particularly concerning due to its cytotoxic and hepatotoxic effects, posing significant risks to aquatic ecosystems and human health. Anabaena and Aphanizomenon, for instance, are notorious for forming extensive blooms that can result in high concentrations of this toxin Anabaena (Dreher et al. 2022), Aphanizomenon (de Figueiredo et al. 2022). Numerous studies have identified Dolichospermum and Raphidiopsis as critical contributors to the cylindrospermopsin load in freshwater systems. Dolichospermum (de Figueiredo et al. 2022) and Raphidiopsis (Stefanova et al. 2020) both play significant roles. Sphaerospermopsis, although less commonly reported, also plays a role in producing this toxin Sphaerospermopsis (Kim et al. 2020). The proliferation of these cyanobacteria and the subsequent release of Cylindrospermopsin in lakes necessitate vigilant monitoring and effective management strategies to mitigate potential health risks and preserve water quality. Another famous cyanotoxin in water bodies is Microcystin, with persistent cyanobacteria blooms produced by the cyanobacterial genus Microcystis (Jacquemin et al. 2023), which is a widespread and critical concern in lake ecosystems globally. Microcystis thrives in nutrient-rich, stagnant waters, often leading to extensive algal blooms that can dominate the phytoplankton community in lakes. These blooms are visually unappealing and pose significant ecological and health risks due to microcystin production.

Saxitoxin, responsible for paralytic shellfish poisoning, originates from several types of algae, including the marine species complex Alexandrium tamarense and several cyanobacteria found in freshwater and brackish environments like Cuspidothrix, Dolichospermum, Microseira, and Raphidiopsis. Alexandrium tamarense is well-known in the marine ecosystem for its significant bloom events, often termed 'red tides.' These events can lead to saxitoxin accumulating in shellfish, which poses a considerable threat to both sea life and humans who consume these shellfish (Lyu et al. 2019). In freshwater and brackish water systems, genera like Cuspidothrix and Dolichospermum can produce saxitoxin under certain environmental conditions, contributing to the complexity and danger of HABs (González-Madina et al. 2019). These cyanobacteria can thrive in various aquatic environments, and their ability to produce saxitoxin is a significant concern for water quality and public health, particularly in areas where water bodies are used for recreation or as drinking water sources.

Brevetoxin, a neurotoxic compound primarily associated with HABs in marine environments, is produced by algal genera such as Karenia and Chloromorum. Karenia, particularly Karenia, is well-known for its blooms in coastal waters, often referred to as "red tides," which can lead to widespread fish kills, marine mammal and bird mortalities, and respiratory issues in humans (Turley et al. 2022). These blooms, characterized by their reddish discoloration of seawater, release brevetoxins into the environment, causing significant ecological and health impacts. Chloromorum, although less commonly reported than Karenia, also contributes to brevetoxin production in marine ecosystems (Anderson et al. 2021). Lastly, Domoic acid, another neurotoxin, is produced by the diatom genus Pseudo-nitzschia in marine environments. Blooms of Pseudo-nitzschia, often linked to nutrient-rich waters, pose significant threats to marine ecosystems, public health, and fisheries, highlighting the need for vigilant monitoring and management (Sandoval-Belmar et al. 2023).

16.2.3 Microorganisms for Resource Recovery

HABs present a unique opportunity to create valuable products, such as biofuels, functional foods, pharmaceuticals, and UV-absorbing compounds. Efficient harvesting of these blooms can aid in extracting excess nutrients from the environment, turning a potential ecological problem into a resource. Algae, with their robust bioresource potential, are increasingly recognized for their role in environmental remediation and as a renewable energy source. Their rapid growth and high carbohydrate and lipid production make them particularly valuable (Kim et al. 2015). Furthermore, algae's secondary metabolites are gaining importance in the food, pharmaceutical, and cosmetic industries (Vu et al. 2020). Developing large-scale cultivation systems with optimized conditions is crucial to harnessing these benefits effectively. Research explores approaches such as composting and mechanization to manage and repurpose algae, particularly in regions like the Great Lakes, to tackle secondary environmental pollution (Kim et al. 2015).

In laboratory simulations of HABs, Chlamydomonas reinhardtii, and Chlorella vulgaris have been identified as effective strains for phosphate removal and recovery in high-phosphate environments like lakes or ponds, transforming phosphate into a valuable resource (Pandhal et al. 2017). Additionally, the biomass accumulations of Heterosigma akashiwo during HABs significantly deplete nitrogen oxides and phosphates, highlighting the diverse roles of different algal species in nutrient balances within aquatic ecosystems (Lemley et al. 2021).

Exploring methods to enhance phosphorus uptake, researchers have found that wall-breaking agents like hydrogen peroxide and sodium hydroxide effectively release phosphorus from microalgae. This phosphorus is then absorbed by magnesium-enriched hydrochar, with a remarkable release efficiency of up to 90.5% under optimal conditions (Deng et al. 2019).

Further studies on nutrient recovery and removal from synthetic wastewater using algae have shown that Chlorella vulgaris ACUF_809 can achieve a phosphorus removal efficiency of 32%. In comparison, nitrate removal efficiency in all reactors exceeded 93%. The coupled cultivation of C. pyrenoidosa and C. vulgaris under mixotrophic conditions with ammonium acetate supplementation is a promising solution for simultaneous nitrate and phosphate removal from phosphorus-rich wastewaters (Moreno Osorio et al. 2020).

Lastly, the significant biomass and survival duration of Microcystis in sediment layers may explain the prolonged recovery of lakes affected by these blooms, even after the reduction of external nutrient inputs (Brunberg and Boström 1992). This comprehensive understanding of algae's role in resource recovery highlights their potential to address environmental challenges and create sustainable solutions.

16.2.4 Occurrence of Genus Tychonema Producing Anatoxin-A

Over the past few decades, the HAB has significantly evolved globally (Anderson 2009). The proliferation of autotrophic and heterotrophic microorganisms is the leading cause of this phenomenon. These blooms result from natural processes within the water bodies or anthropogenic activities (Boyd 2020; Sellner et al. 2003). Among more than five thousand microorganism species in water bodies, cyanobacteria may result in visible scum and affect thermal stratification, which can lead to significant economic losses for aquaculture, fisheries, and tourist as well as adverse effects on the environment and human health (Boyd 2020; Hallegraeff 1993). Therefore, regular monitoring and detection of these microorganisms can prevent human and natural disasters.

16.2.5 Pharmaceutical Micropollutants

Due to demographic change and the increasing use of medicines, pharmaceutical micropollutants from Berlin's wastewater are increasingly concerned about water quality. Pharmaceuticals in the aquatic environment pose a significant risk of harming ecosystems and drinking water abstraction. This situation is evident in shallow dimictic Lake Tegel, where drinking Water is produced through bank filtration and artificial groundwater recharge. Simultaneously, a considerable anthropogenic micro-pollutant load originates from a municipal wastewater treatment plant (Schimmelpfennig et al. 2016). Pharmaceutical contamination in the lakes in Berlin is not a new phenomenon. Massmann et al. (2008) studied the behavior of residues of phenazone-type pharmaceuticals during bank filtration at a field site in Berlin. The concentrations of the pharmaceutical residues in the shallow, young bank filtrate (travel times < one month) was correlated to the prevailing hydrochemical conditions at the field site.

Although Berlin's water utility (Berliner Wasserbetriebe, BWB) is upgrading all six sewage treatment plants primarily to eliminate phosphorus and nitrogen from the wastewater, the utility is now constructing a new ozone plant at the Schönerlinde wastewater treatment plant northeast of the capital. Scheduled for operation in 2024, the new plant aims to remove trace substances from the Water.

BWB is aware that kidney damage caused by medications has already been noticed in fish in Lake Tegel. Since the purified wastewater from the Schönerlinde sewage treatment plant flows into Lake Tegel, which in turn feeds the second-largest waterworks of Berliner Wasserbetriebe, construction of the capital's first ozone plant will begin there (ND Aktuell 2021).

16.2.6   The Impacts of Climate Change

Negative consequences of climate change, such as rising temperature, and stormwater runoff into water freshwater bodies, are considered to increase the risks of HAB s (Gobler 2020; Glibert 2020; Neerugatti et al. 2022).

Ladwig et al. (2018) studied Lake Tegel in Berlin as an example of adaptive urban lake management in response to climate change. Their study suggests that Lake Tegel's annual stratification patterns will change due to climate change. The winter stratification will decrease, whereas the summer stratification will intensify. Additional nutrient-free discharges by an active elimination plant can mitigate an increase in the stability of the summer stratification period. Lake Tegel could shift from a dimictic seasonal mixing type to a monomictic one. These physical changes will affect the lake's water quality. An increased summer stratification period combined with a higher buoyancy frequency will function as a favorable habitat for forming cyanobacteria blooms. Nonetheless, they argue that the elimination plant acts as a necessary "life-support system" for Lake Tegel because the discharges from the plant function as a buffer against nutrient-rich waters from the River Havel.

A study by the Leibniz Institute of Freshwater Ecology and Inland Fisheries (IGB) has found out that climate change impacts HAB issues at Lake Müggel in Berlin: summer water temperatures have risen by about two degrees Celsius over the past 40 years. At the same time, the periods during which the lake was frozen shortened (Berliner Zeitung 2016; IGB 2021).

Another possible cause could be stormwater runoff into freshwater bodies. Nutrient-driven HAB is common in anthropogenically altered water bodies. Stormwater runoff is one of the largest sources of nutrients for freshwater bodies. Stormwater retention ponds in urban and suburban areas are likely environments for harmful cyanobacteria blooms and were thus targeted for an in-depth investigation assessing taxonomic composition, bloom morphological composition, toxicity, and impact of nutrients and other environmental drivers (Grogan et al. 2023).

Heavy rainfall could increase the risk of HAB. Furthermore, the sealed surface on the streets in Berlin absorbs rainwater. As heavy rainfall is expected to occur more frequently due to the changing climate, investing in heavy rain management by transitioning into blue-green infrastructure (e.g., Berlin's sponge city principle) could also reduce the risks of HAB.

The relationship between heavy rainfall and Harmful Algal Blooms (HABs) is complex and multifaceted, as evidenced by numerous studies across different regions. In Florida, the impact of rainfall from hurricanes and El Niño events leads to higher nutrient loads in water bodies like the Indian River Lagoon and St. Lucie Estuary, fueling HABs. This is particularly evident in Lake Okeechobee, where significant HABs create an indirect risk for the St. Lucie Estuary through nutrient-rich discharges during high rainfall periods, and the absence of HABs in the lake can result in lower bloom intensities in the estuary due to rapid water turnover from discharges (Philips et al. 2020). A study on Chaohu Lake, China, using a remote sensing-based approach for monitoring HABs, shows that these blooms typically occur from May to November, peaking in the summer months, and are more frequent in the western region. The study identifies that higher temperatures and light rain are conducive to HAB growth, with wind being a primary factor in promoting their expansion. This demonstrates that environmental and meteorological factors like temperature, rainfall, and wind significantly influence HAB dynamics (Ma et al. 2021). The impact of changing rainfall patterns on toxic cyanobacterial blooms was discussed in another study, which notes that increased frequency and intensity of rainfall events, interspersed with more extended drought periods, can exacerbate these blooms. Heavy rainfall contributes to more significant nutrient input and eutrophication in waterbodies, fostering conditions favorable for cyanobacterial growth. Conversely, significant storm events may disrupt blooms temporarily, underlining the complexity of these dynamics and the need for further research to understand the relationship between rainfall patterns and HABs (Reichwaldt and Ghadouani 2012). Researchers have developed innovative methods to enhance the accuracy of predicting the likelihood of HABs. These models utilize advanced data-driven techniques, including neural network trimming, as demonstrated in a case study conducted in Western Xiamen Bay, China. Conventional models often face challenges due to spatial heterogeneity and the unique behaviors of distinct species, which can undermine their predictive accuracy.

Conventional models frequently encounter challenges due to spatial heterogeneity and the unique behaviors of distinct species, which can compromise their predictive accuracy. Integrated modeling approaches, including applications in various aquatic environments, underscore the importance of combining different paradigms and techniques for effective HAB prediction (Chen et al. 2006). The study on the dynamics of cyanobacteria blooms in urban lakes, such as Lake Pampulha, reveals more complex relationships, particularly in urban areas where stormwater runoff plays a significant role. Efforts to reduce total phosphorous (TP) levels, even by as much as half, fall short of achieving the water quality goal of maintaining a maximum concentration of 60 μg chla L-1. This situation highlights the inherent challenges in mitigating cyanobacteria biomass solely through TP reduction. Furthermore, the study sheds light on the consequences of increased imperviousness in urban catchment areas, leading to an escalation in runoff volume, Total Suspended Solids (TSS), TP, and nitrate (NO3-) loads that flow into the lake. This increase, particularly notable at the onset of the wet season, contributes to a significant growth in cyanobacteria biomass due to the additional influx of nutrients from urban runoff (Silva et al. 2019).

In conclusion, stormwater runoff significantly impacts freshwater ecosystems, particularly influencing cyanobacteria such as Tychonema. When stormwater flows into freshwater bodies, it often carries a variety of pollutants, including nutrients like nitrogen and phosphorus. These excessive nutrients can facilitate the eutrophication process and blooms. Moreover, stormwater can also introduce sediment and other pollutants that alter the Water's physical and chemical properties, further impacting Tychonema's growth and distribution. The effect of stormwater on Tychonema and similar organisms is a growing concern, highlighting the need for effective stormwater management practices to protect freshwater ecosystems. While the impact of stormwater on freshwater ecosystems is a well-recognized environmental issue, there needs to be more specific research focusing on how stormwater affects Tychonema, a genus of cyanobacteria, in these environments. This gap highlights the need for more targeted studies to understand the unique interactions between stormwater pollutants and Tychonema, which could be crucial for managing and mitigating the effects of urban runoff on freshwater biodiversity and water quality.

16.3 Potentials for a Solution—Applications of Microalgae for Many Purposes

Microalgae is not just toxic and pollutant but can be used for many good purposes. Microalgae are photosynthetic organisms that produce valuable metabolites under different conditions, such as extreme temperatures, high salinity, osmotic pressure, and ultraviolet radiation. It reduces carbon dioxide, phosphorus, and nitrogen in freshwater ecosystems and promotes biodiversity. It is also a sustainable source of nutrients, protein, animal feed, and food additives, or it can be used as biofertilizer and soil stabilizer, bioplastics, biofuel, and hydrogen.

A circular economy approach maximizes resource efficiency, while decarbonization is one of many microalgae cultivation benefits. Despite many benefits and potential applications of macro and microalgae (e.g., nutrients, raw materials, carbon capture, biofuels), there are many technical, regulatory, and socio-economic challenges for applications. Other challenges could be consumers' acceptance of food applications, safety, and contamination issues.

Nature-based solutions (NbS) is an approach to managing climate change impacts in urban areas by using organisms, soils and sediments, and landscape features to reduce climate change hazards; they hold promise as being more flexible, multifunctional, and adaptable to an uncertain and non-stationary climate future than traditional approaches (Hobbie and Grimm 2020). Microalgae cultivation can not only be used to produce high-value products for human consumption; it can mitigate wastewater treatment plants' carbon footprint as a tertiary treatment, thus serving as a nature-based solution (de Lima Barizão et al. 2023). Below, describe the potentials of microalgae that provide multiple solutions to problems beyond Water.

16.3.1 Microalgae's Bioremediation Potential

Bioremediation is a method that harnesses the power of living organisms to eliminate or neutralize pollutants from areas. Microalgae have shown potential for bioremediation due to their ability to absorb and store different toxins from the environment. Their fast growth rate and ability to address contaminants, such as metals and organic pollutants make them a promising option for environmental restoration initiatives.

16.3.2 Types of Toxins Microalgae Can Remediate

Microalgae, as primary producers, utilize photosynthesis to convert energy into chemical forms (Masojídek et al. 2013). Their cellular structures, including cell walls and vacuoles, exhibit an affinity for various environmental toxins, enabling them to capture and concentrate these substances from their surroundings. Metals are particularly concerning among these environmental toxins. Microalgae employ biosorption mechanisms, where specific functional groups like carboxyl, hydroxyl, and amino groups on their cell wall polysaccharides and proteins bind to these metals, regulating the biosorption process (Vijayaraghavan and Yun 2008). In addition to extracellular adsorption, intracellular uptake mechanisms facilitate the transport and storage of pollutants within cellular structures such as vacuoles. This unique ability allows microalgae to reduce metal concentrations in aquatic systems (Renu and Singh 2017). Beyond metals, organic pollutants also pose significant challenges to water systems. Microalgae have a role in breaking down and eliminating organic pollutants from water systems. They possess metabolic pathways that enable them to absorb and transform contaminants, including pharmaceuticals, pesticides, and hydrocarbons (Xiong et al. 2018). Microalgae can degrade or convert pollutants in the presence of bacterial consortia (Muñoz and Guieysse 2006).

Another significant environmental concern is nutrient pollution in aquatic habitats. Microalgae can mitigate excess nitrogen and phosphorus compounds in water bodies. Microalgae are adept at absorbing these nutrients rapidly during their growth and metabolic activities, thereby reducing the risks associated with eutrophication (Wang et al. 2010).

16.3.3 Biomass Application

Microalgae reduce nutrient levels in the water and convert these nutrients into valuable biomass that can be harvested and used for various purposes, such as producing biofuels or animal feed (Chen et al. 2010). The efficiency of removal by microalgae can vary depending on the characteristics of distinct species, the availability of light, and the prevailing physical and chemical conditions in their environment (Cho et al. 2015).

16.3.4 Mechanisms of Toxin Removal

Adsorption is a process where harmful substances gather on the surface of cells without entering the cell membrane. This process depends on the chemical interactions between toxin molecules and specific functional groups on the cell walls, such as carboxyl, hydroxyl, and amino groups (Vijayaraghavan and Yun 2008). These interactions can take forms such as van der Waals forces, hydrogen bonding, and ionic interactions. These interactions hinge on the toxin's chemical structure and its surrounding environment's pH level. As we move from external to internal interactions, absorption provides another mode of toxin interaction. Unlike absorption, absorption entails microalgae actively taking in and integrating toxins into their structure. Unlike absorption, absorption entails taking in and integrating toxins into the structure of microalgae. This internalization happens through transport mechanisms across the cell membrane, facilitated by protein carriers or occurring through passive diffusion. Compartmentalized organelles such as vacuoles separate them from the cell's core metabolic activities (De Schamphelaere and Janssen 2004). Beyond mere internalization, microalgae also can transform these substances. Metabolic transformation refers to the process by which microalgae modify toxins biochemically, making them less harmful or completely harmless. This occurs through the machinery in microalgal cells, which can convert toxins into different compounds through oxidation, reduction, hydrolysis, or conjugation. These biotransformations play a role in detoxifying pathways that allow microalgae to thrive in environments with elevated levels of potentially harmful substances. Merging the mechanisms of adsorption and absorption, we encounter the concept of biosorption.

Biosorption involves the ability of microalgae to both adsorb and absorb toxins. This process is especially effective for metals as the metal ions interact with biological molecules in microalgal cells like polysaccharides, proteins, and lipids. The effectiveness of biosorption depends on factors such as the type of algae, concentration of toxins, pH levels, and temperature. It can successfully remove various environmental pollutants (Volesky 2001).

16.3.5   Adsorption and Absorption Mechanisms

Adsorption is a process where harmful substances gather on the surface of cells without entering the cell membrane. This process depends on the chemical interactions between toxin molecules and specific functional groups on the cell walls, like carboxyl, hydroxyl, and amino groups (Vijayaraghavan and Yun 2008). These interactions can take forms such as van der Waals forces, hydrogen bonding, and ionic interactions. These interactions depend on both the toxin's chemical structure and its surrounding environment's pH level. As we transition from external to internal interactions, absorption emerges as another significant mode of toxin interaction. Unlike absorption, absorption involves microalgae taking in and incorporating toxins into their structure. This internalization occurs through transport mechanisms operating across the cell membrane, either facilitated by protein carriers or passive diffusion. Once inside the cell, toxins can reside in the cytoplasm or become compartmentalized within organelles such as vacuoles, sequestering them from the cell's core metabolic activities (De Schamphelaere and Janssen 2004).

Understanding adsorption and absorption processes further underscores the importance of environmental factors that influence these mechanisms. As discussed in a recent study, adsorption depends significantly on the physicochemical environment, including pH, temperature, and redox reactions (Bhatt et al. 2022). The adsorption phenomenon relies on the surrounding physicochemical environment like pH, temperature, and redox reactions, as discussed in a study available (Bhatt et al. 2022). This contextual dependence is crucial in determining how different substances interact with microalgae in diverse environmental conditions.

In line with these concepts, research has shown that altering the physicochemical properties of microalgae can enhance their ability to purify water from contaminants. Alkaline modification was applied to Scenedesmus obliquus biomass, enhancing its capacity to absorb tramadol and other pharmaceutical compounds. Experimental findings highlighted that this modified algal biomass effectively captured tramadol, reaching a remarkable 91% elimination rate in a short span of 45 min. The underlying mechanism of this adsorption involves the interaction of hydrophilic elements. Specifically, the amino and carbonyl structures in the pharmaceuticals interact with the hydroxyl and carbonyl functional groups on the modified biomass's surface. Ali et al. (2018) gave more details about the study.

Further illustrating the versatility of microalgae in environmental remediation, another species, Microcystis aeruginosa, has shown promise in antibiotic remediation. This microalgae promoted the hydrolysis of tetracycline under conditions of increased pH and inhibited abiotic photolytic reactions by the shading effect to the water column compared with control experiments. This research is described by Pan et al. 2021. Expanding beyond microalgae, the integration of advanced materials in environmental cleanup efforts is also gaining traction. We designed a hybrid magnetic-ZnO-based photocatalyst platform for efficient sunlight-driven removal of cyanobacteria and mineralization of cyanotoxins (specifically anatoxin-A). The photocatalyst's effectiveness in eliminating two types of microalgae with different morphology and toxicity—Spirulina platensis Paracas and Anabaena flos-aquae—was evaluated. The hybrid photocatalyst Ni@ZnO@ZnS-Spirulina was synthesized through a multistep process using Spirulina as a biotemplate. Furthermore, these photocatalysts demonstrated effective degradation of anatoxin-A under artificial sunlight conditions. More details on this innovation can be found in (Serrà et al. 2020).

16.3.6   Metabolic Transformation and Biosorption

Biosorption involves the ability of microalgae to both adsorb and absorb toxins, a particularly effective process for metal ions. These metal ions interact with biological molecules in microalgal cells, such as polysaccharides, proteins, and lipids. The effectiveness of biosorption varies depending on factors like the type of algae, the concentration of toxins, pH levels, and temperature, and it has proven to be a successful method for removing a range of pollutants from the environment (Volesky 2001). This concept of biosorption, which combines the mechanisms of adsorption and absorption, is further complemented by the microalgae's capacity for metabolic transformation. In this process, microalgae biochemically modify toxins to become less harmful or completely harmless. This transformation occurs through various processes within the microalgal cells, including oxidation, reduction, hydrolysis, or conjugation, thereby playing a vital role in the detoxification pathways that enable microalgae to survive in environments with elevated levels of potentially harmful substances.

Supporting this concept, specific microalgal species have demonstrated remarkable abilities in removing contaminants. For instance, Nostoc ellipsosporum excels in eliminating heavy metals, achieving reduction percentages of over 95% for all tested metal ions. Similarly, Anabaena variabilis has shown significant efficiency in reducing levels of various heavy metals like chromium, lead, iron, copper, and molybdenum. Chlorella vulgaris is another species that exhibits strong heavy metal removal capabilities. These findings, detailed in a study at (Ghazal et al. 2018), highlight the potential of microalgae in environmental remediation. In another example, Chlorella sp. Cha-01, Chlamydomonas sp. Tai-03, and Mychonastes sp. YL-02 were observed to effectively remove cephalosporin antibiotics through a combination of hydrolysis, photolysis, and biosorption, as described in research available at (Guo et al. 2016).

Further demonstrating the versatility of microalgae in biotransformation, studies involving Scenedesmus obliquus and Chlorella pyrenoidosa have shown significant success. These freshwater microalgae have effectively transformed over 95% of progesterone within five days. Additionally, Scenedesmus obliquus achieved near-complete transformation for norgestrel, whereas Chlorella pyrenoidosa was less effective, leaving about 40% untransformed. This research, detailed at (Peng et al. 2014), illustrates the diverse capabilities of microalgae in processing and neutralizing different substances, marking them as valuable tools in environmental management and pollution control.

16.3.7 Factors Affecting Toxin Removal by Microalgae

The effectiveness of removing toxins using microalgae depends significantly on the strain that is used. Different microalgae strains have metabolic pathways, cell wall structures, and uptake mechanisms, which can significantly impact their ability to bind and transform toxins. Therefore, it is crucial to understand each strain's attributes to optimize the toxin removal process. Comparative studies have shown that different strains of microalgae vary in their capacity to absorb toxins emphasizing the importance of selecting the strain based on the type of contaminants present (Wang and Chen 2009). Beyond microalgae strain, environmental factors play a significant role in toxin removal. Factors like pH, temperature, and light intensity influence the effectiveness of microalgae in eliminating toxins. For example, the pH level can affect how toxins are ionized and the charge distribution on the surface of microalgae cells, which in turn impacts how toxins are absorbed.

Meanwhile, temperature affects metabolic rates and cellular uptake mechanisms. Similarly, light intensity indirectly influences energy processes involved in toxin removal through its effect on activity. Considering these interactions, assessing conditions that maximize toxin removal efficiency becomes crucial. Biotic factors also add another layer of complexity to the equation. The ability of microalgae to remove toxins can be influenced by the presence of organisms in the system. For example, bacteria can form relationships with microalgae, resulting in a combination called algal bacterial consortia. This partnership can enhance the removal process by utilizing metabolic activities, where bacteria break down organic toxins to make them easier for microalgae to absorb.

On the other hand, when competing with types of algae, nutrient availability may be affected, potentially impacting the efficiency of toxin removal. The interactions between organisms in cultures have diverse effects on toxin removal outcomes (Muñoz and Guieysse 2006). As we consider the adaptability and potential of microalgae, they have a vast array of applications. Microalgae have become increasingly promising in various real-world applications, from producing biofuels to treating wastewater. Their fast growth rates and ability to capture carbon dioxide make them solutions for carbon capture and bioenergy generation. Moreover, their biomass is rich in lipids and has been used to create biodiesel, biogas, and bioethanol. We have witnessed the shift from small-scale laboratory studies to applications using large-scale photobioreactors and open pond systems (Chisti 2007).

The use of microalgae for bioremediation has become increasingly important in dealing with areas contaminated with heavy metals, organic pollutants, and excessive nutrients. Their natural ability to adsorb, absorb, and transform contaminants has been harnessed to restore environments ranging from mining waste to runoff. Field experiments have proven the effectiveness of microalgae in reducing the impacts of eutrophication in freshwater systems and eliminating substances from wastewater. These specific applications highlight the potential of microalgae as an alternative to traditional remediation methods, emphasizing their crucial role in ecological restoration (Bilanovic et al. 1988).

One of the crucial factors for microalgae is pH value; microalgae cultivation leads to an increase in pH value, primarily due to CO_2 assimilation during photosynthesis and nitrogen absorption, where nitrate ions are reduced to ammonia, producing OH − ions. This pH elevation significantly influences microalgal bioprocesses, particularly in pathogen elimination, as discussed in a study (Chai et al. 2021). Additionally, research with Desmodesmus pleiomorphus has demonstrated enhanced cadmium removal at a lower pH, specifically at pH 4.0, indicating pH's critical role in microalgae's heavy metal removal capabilities. This finding is elaborated in the research available (Monteiro et al. 2010). Therefore, pH manipulation in microalgal systems emerges as a potential strategy for improving bioremediation efficiency.

16.3.8 Microalgae-Bacterial Interactions

The symbiotic interaction between microalgae and bacteria in cultivation systems, mainly when grown in domestic wastewater, is pivotal for sustainable environmental management. Studies, such as one detailed by (Dao et al. 2018), demonstrate how bacteria facilitate microalgal growth by producing indole acetic acid (IAA), with microalgae potentially secreting substances to stimulate bacterial activity further. This mutualistic exchange, extending to the provision of essential nutrients and supporting biofuel production, is explored in another study (Yao et al. 2019). Moreover, as highlighted in a review by (Nagarajan et al. 2022), microalgae and bacteria interactions enhance nutrient recycling in wastewater treatment, offering a greener alternative to traditional methods. This synergy between microalgae and bacteria holds significant promise for advancing circular economy and sustainability in wastewater treatment and bioenergy production.

16.4 Challenges and Limitations

Scaling up microalgae cultivation from small-scale experiments to operational levels presents significant challenges. As systems grow, ensuring optimal growth conditions, such as uniform light distribution, becomes increasingly complex. Closed photobioreactors encounter challenges regulating temperature, efficiently transferring CO_2, and preventing biofouling. In contrast, open pond systems must contend with issues such as contamination and water loss through evaporation. Furthermore, the economics of scaling up are only sometimes straightforward, as larger systems bring increased complexity and potential costs. Striking a balance between productivity and operational costs on a large scale remains challenging (Chisti 2007). Building on the topic of scalability, the post-cultivation processes also present their own sets of challenges.

Extracting microalgae from cultivation systems is a step that demands a significant amount of energy, often comprising around 30% of the overall production cost. Traditional methods like centrifugation or flocculation for harvesting can be both energy-intensive and expensive. Therefore, it is vital to develop cost-effective techniques, such as bioflocculation or the utilization of gravity settlers to ensure the feasibility of the microalgae industry (Uduman et al. 2010). Following the extraction phase, subsequent challenges revolve around the properties of the extracted biomass. After harvesting, challenges arise when it comes to managing the biomass. The high-water content in the biomass requires methods to remove the Water before further processing. However, traditional drying techniques can harm components within the biomass. Additionally, if not appropriately managed, storing biomass can result in degradation and a loss of product quality. It is crucial to develop optimized approaches for drying, preserving, and storing biomass to fully utilize its derived products (Rodolfi et al. 2009).

While technical challenges exist throughout the cultivation and processing stages, ensuring the purity of the microalgae culture is another vital consideration. The growth of specific types of microalgae can be negatively affected when unwanted species invade open pond systems. These intruder species have the potential to outperform the desired microalgae, affecting productivity and potentially introducing harmful substances. It is challenging to develop strains with growth advantages or optimize cultivation conditions to support the desired microalgae. Additionally, selecting high-yielding strains for desired products remains a bottleneck in the industry.

16.4.1 Microalgae Production

There are three ways in which microalgae can be produced: in open systems, bioreactors, photobioreactors, or fermenters. Microalgae is unicellular and comp rises 30–70% protein, 10–30% carbohydrates, and 10–50% lipids. In open system production, there are ponds, lakes, and oceans.

In the next section, regulatory and socio-economic issues and challenges related to algae production are elaborated in the context of the EU and Germany.

The European Union (EU) hosts the EU4Algae Platform, a platform for collaboration among European algae stakeholders in the context of the EU Algae Initiative and the EU blue bioeconomy report, with an overview of the latest micro and macro algae cultivation systems developments and how seaweed can transform regional economies (Berliner Zeitung, European Commission 2016, 2023).

The Commission adopted the Algae Initiative to unlock the potential of algae in the EU. The Initiative proposes twenty-three actions to create opportunities for the industry, enabling it to grow into a robust, sustainable, and regenerative sector capable of meeting the growing EU demand.

However, despite algae being an untapped resource that can produce food, feed, pharmaceuticals, bioplastics, fertilizers, and biofuels with a limited carbon and environmental footprint, microalgae products remain niche in Europe and require further development.

EU's Algae initiative sets out how the EU can increase the sustainable production, safe consumption, and innovative use of algae and algae-based products. The Initiative will help in achieving the objectives of the European Green Deal, the transition to a green, circular, and carbon–neutral EU, and a post-Covid recovery (European Commission 2021a, b).

However, in addition to technical challenges regarding microalgae cultivation, there are regulatory and socio-economic barriers, including high production costs, low-scale production, limited knowledge of the markets, consumers' needs, environmental impacts of algae cultivation, and fragmented industrial and research networks. To overcome those hindrances, market support, the entire business environment, social acceptance, and closing knowledge, research, and innovation gaps are required.

From the regulatory point of view, the EU's Novel Food Regulation No 258/98 is a significant barrier for the microalgae industry for food applications (European Union 1997). The new regulation from 2018 includes the requirement that Member States should no longer conduct the assessment. Instead, the European Food Safety Authority (EFSA) should conduct it.

Lähteenmäki-Uutela et al. (2021) argue that the novel food regulation applies to macroalgae foods previously used as food, and organic macroalgae are a specific regulatory category. The maximum levels of heavy metals may be a barrier for macroalgae foods, feeds, and fertilizers.

The EU project IntegraSea (2019–2021) studied the potential for off-shore seaweeds to suppress toxic microalgae by nutrient takeout and thus protect shellfish farms, using high-value native seaweeds and potentially producing seaweed extracts for 'natural' products from the harvests (EUMOFA 2022).

The abilities of macroalgal and microalgal systems to remove undesirable chemical and biological contaminants should continue to be validated by establishing pilot or demonstration plants at scale-up levels—microalgae as adsorbents and metabolizers of pollutants in the input streams and seaweeds as suppressors, by nutrient competition, of toxic microalgae in marine eutrophication and for other aquaculture activities.

The aims of circularity and carbon reduction are being tackled on land by linkage of nutrient outputs from other production, processing, and wastes-management industries to microalgal and macroalgal biomass production, which is feasible in the current state of technological advancement, given efficient management of design, including modular processing units, linkage of sources of inputs to algal production units, including effective co-location and permissive or at least flexible regulations and permit systems (EABA 2021).

The major challenges include:

- to understand how to process potentially harmful components out, such as heavy metals and arsenic, spoilage organisms, microalgal toxins, ang excessive iodine, and make the safety of the end-products inherent and implicit;
- to re-define wastes so they can be used as inputs;
- to ensure that Life Cycle Assessments (LCAs) and other econometric and functional analyses are detailed enough and contain accurate data to be trustworthy. The national and EU-wide definitions of wastes limit the use of algae in the Circular Economy as nutrient and CO_2 absorbers.

The absolute change would be a recognition that it does not matter what the origin is of the algal biomass or the inputs into the process provided that the process has been validated as producing something safe or the outputs/end-products have monitorable criteria for acceptance, which means they are independent of the sources of inputs.

16.4.2 Circular Economy Principles for Algae Cultivation

In contrast to linear economic models based on a 'take-make-consume-throw away' pattern, the circular economy is about extending the lifecycle of products and minimizing waste through reusing, repairing, refurbishing, and recycling existing materials and products. It implies reducing waste to a minimum. When a product reaches the end of its life, its materials are kept within the economy wherever possible, thanks to recycling. These can be productively used repeatedly, thereby creating further value.

Circular economy approaches towards water include purifying wastewater while recovering energy at sewage treatment plants (e.g., biogas) and using nutrients recovered from wastewater for agriculture (e.g., phosphorus fertilizer).

Linear- and fossil-dependent economic activities, urbanization, and population growth have increased the discharge of nutrients (nitrogen and phosphorus) to water bodies and carbon dioxide (CO_2) to the atmosphere. Considering the cost of conventional nitrogen production processes, declining global organic phosphorus reserves, and the burden of climate change on economies, leaking nutrients and CO_2 are both environmental and economic challenges. When integrated, the nutrient removal, CO_2 sequestration, and biofuel production processes could address the linear economy's challenges around waste management and fossil-based resource dependence (Calicioglu and Demirer 2022).

Examples of implementing circular economy principles for micro-algae cultivation include algae-based bioplastics using materials such as polyhydroxyalkanoates (PHAs), polylactic acid (PLA), alginate, carrageenan, and ulvan (Bin et al. 2024) and bio-based aliphatic (degradable) polyesters (Rosenboom et al. 2022).

Creating more efficient and sustainable products from the start would help to reduce energy and resource consumption, as it is estimated that more than 80% of a product's environmental impact is determined during the design phase. Toxin removal of HAB in freshwater could provide an alternative solution to improve water quality and produce resources as byproducts in the process.

16.4.3 Reduction of Raw Material Dependence

Some EU countries are dependent on other countries for their raw materials. Recycling raw materials mitigates the risks associated with supply, such as price volatility, availability, and import dependency (European Parliament 2023).

Phosphorus insecurity is an emerging global issue (Walsh et al. 2023), and the current phosphorate life cycle is linear (Barquet et al. 2020). The EU is entirely dependent on imported phosphate rock (Eurostat n.d.). In light of this, the European Commission has added phosphorus (European Commission n.d.-c). Moreover, phosphate rock is on the list of 20 Critical Raw Materials for which supply security is at risk and economic importance is high. Phosphate rock is identified as non-substitutable and of high economic importance. European policies to address raw materials' criticality and dependency include improving the efficiency of materials use and recycling, waste policy, and international cooperation to address supply security (Phosphorus Platform n.d.).

The EU's Commission proposal for a Critical Raw Materials Act (CRM Act) is a comprehensive response to critical raw materials supply disruption risks and the structural vulnerabilities of EU critical raw materials supply chains. It is expected to ensure EU access to a secure and sustainable supply of critical raw materials, enabling Europe to meet its climate and digital objectives, keeping EU industrial competitiveness, and ensuring the well-functioning of the single market (European Commission n.d.). From securing critical raw materials locally and achieving a circular economy, removing HAB toxin could solve several problems.

The Federal Cabinet adopted the National Bioeconomy Strategy in Germany on January 15, 2020. The report also recognized the multi-functionality of microalgae as sustainable sources of food, medicine, biofuels, and fertilizer (Bioökonomie.de n.d.). Germany, Spain, and Italy are the top three countries for European microalgae production. Spirulina producers are in France, Italy, Germany, and Spain.

However, as discussed in the previous section, algae production in Europe remains scattered and limited by technological, regulatory, and market-related barriers. However, the European algae sector has considerable potential for sustainable development as long as the acknowledged economic, social, and environmental challenges are addressed (European Commission 2021a, b).

Another area for improvement could be the ownership of water and land resources. As lakes in Berlin are publicly owned, obtaining permission from authorities to take samples from the lake is required. Nevertheless, there is an example of a Finnish algae producer where access to water and land has been less complicated. Finland's coastal and inland water areas have traditionally been under private ownership in conjunction with possession of land. Individual owners manage most of these water areas jointly (Salmi and Varjopuro 2001).

16.4.4 Administrative Barriers

Administrative issues are certainly the most challenging factors to experiment with to remove toxins from Lake Tegel and cultivate algae. Although we approached Berlin's water utility (BWB) and the Berlin Senat, as well as Berlin's rainwater agency, for permission to experiment on toxin removal, it was not successful, and we could not get any positive response from the authorities. The unfamiliarity of authorities to work with innovators and access to water resources make it hard to realize the project. Berlin has become a hub for startups over the past 20 years. However, Berlin has room for improvement in turning startups into fast-growing companies. Potential measures include providing fast-track administrative procedures in all business-related matters, improving recruitment processes for highly qualified employees, and further expanding the knowledge transfer between research institutes and fast-growing companies (Kritikos 2016). Berlin's administration has been putting much effort into developing a startup ecosystem in the city and created a Startup Agenda 2022–2026, bringing together actors from government, industry, and communities (Berlin Partner 2022). Nevertheless, gaining support from local authorities to pursue innovation projects would be critical to proceed with the concept (Berlin's Rainwater Agency n.d.). Therefore, fostering collaboration between government and innovators, transferring knowledge from science to business, and eliminating bureaucracy in administration would be essential and urgent.

16.5 Discussion

The first part of this Chapter deals with several root causes of the HAB problem at Lake Tegel. Although there are several root causes, including stormwater runoff, rising temperature due to climate change, pharmaceutical micropollutants, and high salinity, it is clear that human activities such as demographic changes, increasing use of medicines, and wastewater discharge are key driving factors for the occurrence of toxic HABs. There is yet to be an effective measure to tackle this problem, even though civil society organizations are promoting awareness of the problems.

The second part of the Chapter proposed microalgae cultivation as a potential solution to the problem, highlighting numerous benefits through the circular economy approach. The capability of macroalgal and microalgal systems to remove undesirable chemical and biological contaminants from freshwater resources should garner more attention. Biofertilizer, particularly as a byproduct of toxin removal, holds significant appeal. Given the global phosphorus shortage at the EU level, local fertilizer production is urgently needed to establish a self-reliant system. There is a growing demand for transitioning to locally produced fertilizer.

In this research, we also identified regulatory and socio-economic challenges. While treating toxic algae in freshwater bodies and cultivating microalgae are technically feasible, several challenges persist, including the need for greater social acceptance and awareness and high production costs. Therefore, this Chapter aims to address research and innovation gaps in microalgae cultivation to combat harmful algal blooms in freshwater bodies and generate valuable resources such as phosphorus as a byproduct.

References

Ali MEM et al (2018) Removal of pharmaceutical pollutants from synthetic wastewater using chemically modified biomass of green alga Scenedesmus obliquus. Ecotoxicol Environ Saf 151:144–152. https://doi.org/10.1016/j.ecoenv.2018.01.012

Anderson DM (2009) Approaches to monitoring, control, and management of harmful algal blooms (HABs). Ocean and Coast Manag 52(7):342–347. https://doi.org/10.1016/j.ocecoaman.2009.04.006

Anderson DM et al (2021) Marine harmful algal blooms (HABs) in the United States: history, current status and future trends. Harmful Algae 102:101975. https://doi.org/10.1016/j.hal.2021.101975

Barquet K, Järnberg L, Rosemarin A, Macura B (2020) Identifying barriers and opportunities for a circular phosphorus economy in the Baltic Sea region. Water Res 171:115433

Berdalet E, Montresor M, Reguera B, Roy S, Yamazaki H, Cembella A, Raine R (2017) Harmful algal blooms in fjords, coastal embayments, and stratified systems: recent progress and future research. Oceanography 30(1):46–57

Berlin Partner (2022) Startup agenda 2022–2026. https://www.berlin-partner.de/en/news/detail/startup-agenda-2022-2026

Berliner Zeitung (2016) Studie: Klimawandel hat Auswirkungen auf den Berliner Müggelsee. https://www.berliner-zeitung.de/zukunft-technologie/studie-klimawandel-hat-auswirkungen-auf-den-berliner-mueggelsee.11307

Berliner Zeitung (2022) Bisherige Art der Blaualgen-Bekämpfung kann Seen giftiger machen. https://www.berliner-zeitung.de/gesundheit-oekologie/bisherige-art-der-blaualgen-bekaempfung-kann-seen-giftiger-machen-li.229855

Berlin's Rainwater Agency (n.d.) Although innovation and decentralized rainwater management are encouraged through Berlin's rainwater agency, staff shortage within the rainwater agency was the main reason for their inability to gain support. https://regenwasseragentur.berlin

Bhatt P et al (2022) Microalgae-based removal of pollutants from wastewaters: occurrence, toxicity and circular economy. Chemosphere 306:135576. https://doi.org/10.1016/j.chemosphere.2022.135576

Bilanovic D, Shelef G, Sukenik A (1988) Flocculation of microalgae with cationic polymers—effects of medium salinity. Biomass 17(1):65–76

Bin Abu Sofian ADA, Lim HR, Manickam S, Ang WL, Show PL (2024) Towards a sustainable circular economy: algae-based bioplastics and the role of internet-of-things and machine learning. ChemBioEng Rev 11(1):39–59

Bioökonomie.de (n.d.) Politikstrategie Deutschland. https://biooekonomie.de/themen/politikstrategie-deutschland

Boyd CE (2020) Microorganisms and water quality. In: Boyd CE (ed) Water quality: an introduction, pp 233–267. Springer International Publishing. https://doi.org/10.1007/978-3-030-23335-8_12

Brunberg A-K, Boström B (1992) Coupling between benthic biomass of Microcystis and phosphorus release from the sediments of a highly eutrophic lake. In: Hart BT, Sly PG (eds) Sediment/water interactions. Springer, Netherlands, pp 375–385

Budzyńska A, Rosińska J, Pełechata A, Toporowska M, Napiórkowska-Krzebietke A, Kozak A, ... Pawlik-Skowrońska B (2019) Environmental factors driving the occurrence of the invasive cyanobacterium Sphaerospermopsis aphanizomenoides (Nostocales) in temperate lakes. Sci Total Environ 650:1338–1347

BUND Berlin (n.d.-a) EU-Wasserrahmenrichtlinie. https://www.bund-berlin.de/themen/stadtnatur/stadtwasser/eu-wasserrahmenrichtlinie/

BUND Berlin (n.d.-b) Wassernetz Berlin. https://www.bund-berlin.de/stadtnatur/stadtwasser/wassernetz-berlin/

Bundesgesundheitsministerium (n.d.) Schwimm- und Badebeckenwasser. https://www.bundesgesundheitsministerium.de/service/begriffe-von-a-z/s/schwimm-und-badebeckenwasser.html

BUND (2024) EU-Wasserrahmenrichtlinie. https://www.bund-berlin.de/themen/stadtnatur/stadtw asser/eu-wasserrahmenrichtlinie/

Burford MA et al (2022) Effects of terrestrial dissolved organic matter on a bloom of the toxic cyanobacteria, Raphidiopsis Raciborskii. Harmful Algae 117:102269. https://doi.org/10.1016/ j.hal.2022.102269

BZ Berlin (2023) Blaualgen-Alarm am Tegeler See. https://www.bz-berlin.de/berlin/reinickendorf/ blaualgen-alarm-am-tegeler-see

Calicioglu O, Demirer GN (2022). Role of microalgae in circular economy. In: Integrated wastewater management and valorization using algal cultures, pp 1–12. Elsevier

Chai WS et al (2021) Multifaceted roles of microalgae in the application of wastewater biotreatment: a review. Environ Pollut 269:116236. https://doi.org/10.1016/j.envpol.2020.116236

Chen P, Min M, Chen Y, Wang L, Li Y, Chen Q, Wang C, Wan Y, Wang X, Cheng Y (2010) Review of biological and engineering aspects of algae to fuels approach. Int J Agric Biol Eng 2(4):1–30

Chen Q et al (2006) Hydroinformatics techniques in eco-environmental modelling and management. J Hydroinf 8(4):297–316. https://doi.org/10.2166/hydro.2006.011

Chisti Y (2007) Biodiesel from microalgae. Biotechnol Adv 25(3):294–306

Cho D-H, Ramanan R, Heo J, Lee J, Kim B-H, Oh H-M, Kim H-S (2015) Enhancing microalgal biomass productivity by engineering a microalgal–bacterial community. Biores Technol 175:578–585

Christensen VG, Khan E (2020) Freshwater neurotoxins and concerns for human, animal, and ecosystem health: a review of anatoxin-a and saxitoxin. Sci Total Environ 736:139515. https:// doi.org/10.1016/j.scitotenv.2020.139515

Chorus I, Köhler A, Beulker C, Fastner J, van De Weyer K, Hegewald T, Hupfer M (2020) Decades needed for ecosystem components to respond to a sharp and drastic phosphorus load reduction. Hydrobiologia 847(21):4621–4651

Colas S, Marie B, Lance E, Quiblier C, Tricoire-Leignel H, Mattei C (2021) Anatoxin-a: overview of a harmful cyanobacterial neurotoxin from the environmental scale to the molecular target. Environ Res 193:110590. https://doi.org/10.1016/j.envres.2020.110590

Council of the European Union (2024) Urban wastewater: Council and Parliament reach a deal on new rules for more efficient treatment and monitoring. https://www.consilium.europa.eu/en/ press/press-releases/2024/01/29/urban-wastewater-council-and-parliament-reach-a-deal-on-new-rules-for-more-efficient-treatment-and-monitoring/

Dao G-H et al (2018) Enhanced microalgae growth through stimulated secretion of indole acetic acid by symbiotic bacteria. Algal Res 33:345–351. https://doi.org/10.1016/j.algal.2018.06.006

De Schamphelaere KAC, Janssen CR (2004) Effects of dissolved organic carbon concentration and source, pH, and water hardness on chronic toxicity of copper to Daphnia magna. Environ Toxicol Chem: Int J 23(5):1115–1122

de Figueiredo DR et al (2022) Bacterioplankton community shifts during a spring bloom of Apha-nizomenon gracile and Sphaerospermopsis aphanizomenoides at a temperate shallow lake. Hydrobiology 1(4):499–517

de Lima Barizão AC, de Oliveira Gomes LE, Brandão LL, Sampaio IC, de Moura IV, Gonçalves RF, de Oliveira JP, Cassini ST (2023) Microalgae as tertiary wastewater treatment: energy production, carbon neutrality, and high-value products. Algal Res 72:103113

Deng Y et al (2019) Optimization and mechanism studies on cell disruption and phosphorus recovery from microalgae with magnesium modified hydrochar in assisted hydrothermal system. Sci Total Environ 646:1140–1154. https://doi.org/10.1016/j.scitotenv.2018.07.369

Derot J, Yajima H, Jacquet S (2020) Advances in forecasting harmful algal blooms using machine learning models: a case study with Planktothrix rubescens in Lake Geneva. Harmful Algae 99:101906. https://doi.org/10.1016/j.hal.2020.101906

Dreher TW et al (2022) 7-epi-cylindrospermopsin and microcystin producers among diverse Anabaena/Dolichospermum/Aphanizomenon CyanoHABs in Oregon, USA. Harmful Algae 116:102241. https://doi.org/10.1016/j.hal.2022.102241

EABA (2021) Algae as novel food in Europe. https://naff.eaba-association.org/files/1678271864_ 72766c9383de1a456235aa09c5492aac_algae-as-novel-food-in-europe.pdf

Engstrom CB et al (2022) Seasonal development and radiative forcing of red snow algal blooms on two glaciers in British Columbia, Canada, summer 2020. Remote Sens Environ 280:113164. https://doi.org/10.1016/j.rse.2022.113164

European Commission (2021) Algae production industry in Europe. https://knowledge4policy.ec. europa.eu/bioeconomy/algae-production-industry-europe_en

European Commission (2021) Blue bioeconomy: towards a strong and sustainable EU algae sector. https://ec.europa.eu/info/law/better-regulation/have-your-say/initiatives/12780-Blue-bioeconomy-towards-a-strong-and-sustainable-EU-algae-sector_en

European Commission (2023) EU blue bioeconomy report out. https://oceans-and-fisheries.ec.eur opa.eu/news/eu-blue-bioeconomy-report-out-2023-01-13_en

European Commission (n.d.) Critical raw materials. In: Single market economy. https://single-mar ket-economy.ec.europa.eu/sectors/raw-materials/areas-specific-interest/critical-raw-materials_ en

European Commission (n.d.) Water framework directive. https://environment.ec.europa.eu/topics/ water/water-framework-directive_en

European Parliament (2023) Circular economy: definition, importance, and benefits. https://www. europarl.europa.eu/topics/en/article/20151201STO05603/circular-economy-definition-import ance-and-benefits

European Union (1997) Regulation (EC) No 258/97 of the European Parliament and of the Council of 27 January 1997 concerning novel foods and novel food ingredients. https://eur-lex.europa. eu/legal-content/en/ALL/?uri=CELEX%3A31997R0258

Eurostat (2012) Agri-environmental indicator: Risk of pollution by phosphorus. https://ec.eur opa.eu/eurostat/statistics-explained/index.php?title=Archive:Agri-environmental_indicator_-_ risk_of_pollution_by_phosphorus&oldid=105042#:~:text=The%20EU%20is%20almost%20e ntirely,ends%20up%20in%20animal%20manure

Evseev P, Tikhonova I, Krasnopeev A, Sorokovikova E, Gladkikh A, Timoshkin O, Miroshnikov K, Belykh O (2023) Tychonema sp. BBK16 characterisation: lifestyle, phylogeny and related phages. Viruses 15(2):442

EUMOFA (2022) Blue bioeconomy report 2022. https://eumofa.eu/documents/20178/84590/blue+ bioeconomy+report+2022+final.pdf/eb889d94-74a6-2c15-e136-4d2204118c6a?t=167344185 5108

Fastner J, Beulker C, Geiser B, Hoffmann A, Kröger R, Teske K, Hoppe J, Mundhenk L, Neurath H, Sagebiel D (2018) Fatal neurotoxicosis in dogs associated with tychoplanktic, anatoxin-a producing Tychonema sp. in mesotrophic lake Tegel, Berlin. Toxins 10(2):60

Flussbad Berlin (n.d.). https://www.flussbad-berlin.de

Ghazal FM et al (2018) The use of microalgae in bioremediation of the textile wastewater effluent. Nat Sci 16:98–104

Glibert PM (2020) Harmful algae at the complex nexus of eutrophication and climate change. Harmful Algae 91:101583

Gobler CJ (2020) Climate change and harmful algal blooms: insights and perspective. Harmful Algae 91:101731

González-Madina L et al (2019) Drivers of cyanobacteria dominance, composition and nitrogen fixing behavior in a shallow lake with alternative regimes in time and space, Laguna del Sauce (Maldonado, Uruguay). Hydrobiologia 829(1):61–76. https://doi.org/10.1007/s10750-018-3628-6

Grogan AE, Alves-de-Souza C, Cahoon LB, Mallin MA (2023) Harmful algal blooms: a prolific issue in urban stormwater ponds. Water 15(13):2436

Guo W-Q et al (2016) Removal of cephalosporin antibiotics 7-ACA from wastewater during the cultivation of lipid-accumulating microalgae. Biores Technol 221:284–290. https://doi.org/10. 1016/j.biortech.2016.09.036

Hallegraeff GM (1993) A review of harmful algal blooms and their apparent global increase. Phycologia 32(2):79–99

Hallegraeff G, Enevoldsen H, Zingone A (2021) Global harmful algal bloom status reporting. Harmful Algae 102:101992

Hobbie SE, Grimm NB (2020) Nature-based approaches to managing climate change impacts in cities. Philos Trans R Soc B 375(1794):20190124

Ho JC, Michalak AM (2015) Challenges in tracking harmful algal blooms: a synthesis of evidence from Lake Erie. J Great Lakes Res 41(2):317–325. https://doi.org/10.1016/j.jglr.2015.01.001

Hushchyna K et al (2023) Multicollinearity and multi-regression analysis for main drivers of cyanobacterial harmful algal bloom (CHAB) in the Lake Torment, Nova Scotia, Canada. Environ Model & Assess 28(6):1011–1022. https://doi.org/10.1007/s10666-023-09907-z

IGB (2021) Leibnitz-Institut für Gewässerökologie und Binnenfischerei. Wie der Klimawandel unsere Gewässer verändert 3 Fragen an 3 Forschende, NA

Intergovernmental Panel on Climate Change (IPCC) (2019) Special Report on the Ocean and Cryosphere in a Changing Climate (SROCC). https://www.ipcc.ch/srocc/

Jacquemin SJ et al (2023) Exploring long-term trends in microcystin toxin values associated with persistent harmful algal blooms in Grand Lake St Marys. Harmful Algae 122:102374. https://doi.org/10.1016/j.hal.2023.102374

Jang K, Otim O (2023) Harmful algae blooms: an analysis of recent spatiotemporal trends on California's inland waterbodies. In: Environmental science: processes & impacts [Preprint]

Kim JK et al (2015) Potential applications of nuisance microalgae blooms. J Appl Phycol 27(3):1223–1234. https://doi.org/10.1007/s10811-014-0410-7

Kim Y-J, Park H-K, Kim I-S (2020) Invasion and toxin production by exotic nostocalean cyanobacteria (Cuspidothrix, Cylindrospermopsis, and Sphaerospermopsis) in the Nakdong River, Korea. Harmful Algae 100:101954. https://doi.org/10.1016/j.hal.2020.101954

Kim Y-J, Park H-K, Kim I-S (2022) Assessment of the appearance and toxin production potential of invasive nostocalean cyanobacteria using quantitative gene analysis in Nakdong River, Korea. Toxins 14(5):294

Kramer BJ, Hem R, Gobler CJ (2022) Elevated CO2 significantly increases N2 fixation, growth rates, and alters microcystin, anatoxin, and saxitoxin cell quotas in strains of the bloom-forming cyanobacteria, Dolichospermum. Harmful Algae 120:102354. https://doi.org/10.1016/j.hal.2022.102354

Kritikos AS (2016) Berlin: a hub for startups but not (yet) for fast-growing companies. DIW Econ Bull 6(29/30):339–345

Kumar M, Seth K, Choudhary S, Kumawat G, Nigam S, Joshi G, … Harish (2023) Toxicity evaluation of iron oxide nanoparticles to freshwater cyanobacteria Nostoc ellipsosporum. Environ Sci Pollut Res Int 30(19):55742–55755

Landesamt für Umwelt Brandenburg (n.d.) Nährstoffreduzierungskonzept. https://lfu.brandenburg. de/lfu/de/aufgaben/wasser/fliessgewaesser-und-seen/gewaesserbelastungen/naehrstoffreduzier ungskonzept/#

Lähteenmäki-Uutela A, Rahikainen M, Camarena-Gómez MT, Piiparinen J, Spilling K, Yang B (2021) European Union legislation on macroalgae products. Aquacult Int 29:487–509

Lemley DA, Adams JB, Largier JL (2021) Harmful algal blooms as a sink for inorganic nutrients in a eutrophic estuary. Mar Ecol Prog Ser 663:63–76

Lyu Y et al (2019) Optimized culturing conditions for an algicidal bacterium Pseudoalteromonas sp. SP 48 on harmful algal blooms caused by Alexandrium tamarense. MicrobiologyOpen 8(8):e00803

Ma J et al (2021) Spatio-temporal variations and driving forces of harmful algal blooms in Chaohu Lake: a multi-source remote sensing approach. Remote Sens 13(3):427

Maritime Forum (n.d.) EU4Algae. https://maritime-forum.ec.europa.eu/theme/blue-economy-and-fisheries/blue-economy/eu4algae_en

Masojídek J, Torzillo G, Koblížek M (2013). Photosynthesis in microalgae. In: Handbook of microalgal culture: applied phycology and biotechnology, pp 21–36

Massmann G, Dünnbier U, Heberer T, Taute T (2008) Behaviour and redox sensitivity of pharmaceutical residues during bank filtration–Investigation of residues of phenazonetype analgesics. Chemosphere 71(8):1476–1485

Minasyan A et al (2018) Diversity of cyanobacteria and the presence of cyanotoxins in the epilimnion of Lake Yerevan (Armenia). Toxicon 150:28–38. https://doi.org/10.1016/j.toxicon.2018.04.021

Ministerium für Landwirtschaft, Umwelt und Klimaschutz Brandenburg (n.d.). Situation an der Oder. https://mluk.brandenburg.de/mluk/de/umwelt/wasser/situation-an-der-oder/

Mkoo (n.d.). https://www.mkoo.pl/index.php?mid=23&lang=EN

Monteiro CM, Castro PML, Malcata FX (2010) Cadmium removal by two strains of desmodesmus pleiomorphus cells. Water Air Soil Pollut 208(1):17–27. https://doi.org/10.1007/s11270-009-0146-1

Moreno Osorio JH et al (2020) Nutrient removal efficiency of green algal strains at high phosphate concentrations. Water Sci Technol 80(10):1832–1843. https://doi.org/10.2166/wst.2019.431

Muñoz R, Guieysse B (2006) Algal–bacterial processes for the treatment of hazardous contaminants: a review. Water Res 40(15):2799–2815

Nagarajan D et al (2022) Microalgae-based wastewater treatment – microalgae-bacteria consortia, multi-omics approaches and algal stress response. Sci Total Environ 845:157110. https://doi.org/10.1016/j.scitotenv.2022.157110

Napiórkowska-Krzebietke A, Dunalska JA, Bogacka-Kapusta E (2023) Ecological Implications in a Human-Impacted Lake—a case study of cyanobacterial blooms in a recreationally used water body. Int J Environ Res Public Health 20(6):5063

ND Aktuell (2021) Wasserschutz: Lebenselixier in Gefahr. https://www.nd-aktuell.de/artikel/115 9469.wasserschutz-lebenselixier-in-gefahr.html

Neerugatti KRE, Veldurthi NK, Heo, J (2022). Emerging pollutants in water bodies: a cause -and -effect analysis. In: Nano-enabled technologies for water remediation, pp 23–38. Elsevier

Pan M et al (2021) Mitigating antibiotic pollution using cyanobacteria: removal efficiency, pathways and metabolism. Water Res 190:116735. https://doi.org/10.1016/j.watres.2020.116735

Pandhal J et al (2017) Harvesting environmental microalgal blooms for remediation and resource recovery: a laboratory scale investigation with economic and microbial community impact assessment. Biology 7(1):4

Peng F-Q et al (2014) Biotransformation of progesterone and norgestrel by two freshwater microalgae (Scenedesmus obliquus and Chlorella pyrenoidosa): transformation kinetics and products identification. Chemosphere 95:581–588. https://doi.org/10.1016/j.chemosphere.2013.10.013

Phlips EJ et al (2020) Hurricanes, El Niño and harmful algal blooms in two sub-tropical Florida estuaries: direct and indirect impacts. Sci Rep 10(1):1910. https://doi.org/10.1038/s41598-020-58771-4

Phosphorus Platform (n.d.) Phosphate rock in EU critical raw materials list. https://www.phosphorusplatform.eu/scope-in-print/news/359-phosphate-rock-in-eu-critical-raw-materials-list

Reichwaldt ES, Ghadouani A (2012) Effects of rainfall patterns on toxic cyanobacterial blooms in a changing climate: between simplistic scenarios and complex dynamics. Water Res 46(5):1372–1393. https://doi.org/10.1016/j.watres.2011.11.052

Renu AM, Singh K (2017) Heavy metal removal from wastewater using various adsorbents: a review. J Water Reuse Desal 7(4):387–419

Rodolfi L, Chini Zittelli G, Bassi N, Padovani G, Biondi N, Bonini G, Tredici MR (2009) Microalgae for oil: Strain selection, induction of lipid synthesis and outdoor mass cultivation in a low-cost photobioreactor. Biotechnol Bioeng 102(1):100–112

Rosenboom JG, Langer R, Traverso G (2022) Bioplastics for a circular economy. Nat Rev Mater 7(2):117–137

Salmaso N, Cerasino L, Boscaini A, Capelli C (2016). Planktic Tychonema (Cyanobacteria) in the large lakes south of the Alps: phylogenetic assessment and toxigenic potential. FEMS Microbiol Ecol 92(10):fiw155. https://doi.org/10.1093/femsec/fiw155

Salmi P, Varjopuro R (2001) Private water ownership and fisheries governance in Finland

Sandoval-Belmar M et al (2023) A cross-regional examination of patterns and environmental drivers of Pseudo-nitzschia harmful algal blooms along the California coast. Harmful Algae 126:102435. https://doi.org/10.1016/j.hal.2023.102435

Schimmelpfennig S, Kirillin G, Engelhardt C, Dünnbier U, Nützmann G (2016) Fate of pharmaceutical micro-pollutants in Lake Tegel (Berlin, Germany)—from entry to sediment. Environ Sci Pollut Res 23(4):3520–3531

Sellner KG, Doucette GJ, Kirkpatrick GJ (2003) Harmful algal blooms: causes, impacts and detection. J Ind Microbiol Biotechnol 30(7):383–406. https://doi.org/10.1007/s10295-003-0074-9

Senatsverwaltung für Umwelt, Verkehr und Klimaschutz Berlin (n.d.) Positionspapier Spurenstoffe. https://www.berlin.de/sen/uvk/_assets/umwelt/wasser-und-geologie/masterplan-wasser/positionspapier-spurenstoffe.pdf

Serrà A et al (2020) Efficient magnetic hybrid ZnO-based photocatalysts for visible-light-driven removal of toxic cyanobacteria blooms and cyanotoxins. Appl Catal B 268:118745. https://doi.org/10.1016/j.apcatb.2020.118745

Shams S, Capelli C, Cerasino L, Ballot A, Dietrich DR, Sivonen K, Salmaso N (2015) Anatoxin-a producing Tychonema (Cyanobacteria) in European waterbodies. Water Res 69:68–79

Sharma KK, Schuhmann H, Schenk PM, Madl T (2013) An operating model for optimizing microalgae growth in large-scale open ponds. Biores Technol 128:444–451

Silva TFG et al (2019) Impact of urban stormwater runoff on cyanobacteria dynamics in a tropical urban lake. Water 11(5):946

Solomon GM et al (2022) Notes from the field: harmful algal bloom affecting private drinking water intakes—clear lake, California, June–November 2021. Morb Mortal Wkly Rep 71(41):1306

Stefanova K et al (2020). Pilot search for cylindrospermopsin-producers in nine shallow Bulgarian waterbodies reveals nontoxic strains of Raphidiopsis raciborskii, R. mediterranea and Chrysosporum bergii. Biotechnol & Biotechnol Equip 34(1):384–394

Stielow S, Ballantine DL (2003) Benthic cyanobacterial, Microcoleus lyngbyaceus, blooms in shallow, inshore Puerto Rican Seagrass habitats, Caribbean Sea. Harmful Algae 2(2):127–133. https://doi.org/10.1016/S1568-9883(03)00007-6

Suresh Kumar K, Dahms HU, Won EJ, Lee JS, Shin KH (2015 Mar) Microalgae—A promising tool for heavy metal remediation. Ecotoxicol Environ Saf 113:329–352. https://doi.org/10.1016/j.ecoenv.2014.12.019. Epub 2014 Dec 19. PMID: 25528489

Tagesspiegel (2017) So kann man sich vor Blaualgen schützen. https://www.tagesspiegel.de/berlin/so-kann-man-sich-vor-blaualgen-schutzen-5498260.html

Turley BD et al (2022) Relationships between blooms of Karenia brevis and hypoxia across the West Florida Shelf. Harmful Algae 114:102223. https://doi.org/10.1016/j.hal.2022.102223

Uduman N, Qi Y, Danquah MK, Forde GM, Hoadley A (2010) Dewatering of microalgal cultures: a major bottleneck to algae-based fuels. J Renew Sustain Energy 2(1)

Umweltbundesamt (n.d.) Indicator: nitrate in groundwater. https://www.umweltbundesamt.de/en/data/environmental-indicators/indicator-nitrate-in-groundwater#assessing-the-development

Umweltbundesamt (2022) Fish die-off: salt discharges caused mass. https://www.umweltbundesamt.de/en/press/pressinformation/fish-die-off-salt-discharges-caused-mass

UNECE (2021) Germany: revision of targets, protocol on water and health. https://unece.org/sites/default/files/2021-07/Germany_Revision%20of%20targets_Protocol%20on%20Water%20and%20Health_15%20July%202021.pdf

Volesky B (2001) Detoxification of metal-bearing effluents: biosorption for the next century. Hydrometallurgy 59(2–3):203–216

Vu HP, Nguyen LN, Zdarta J, Nga TT, Nghiem LD (2020) Blue-green algae in surface water: problems and opportunities. Curr Pollut Rep 6:105–122

Walsh M, Schenk G, Schmidt S (2023) Realising the circular phosphorus economy delivers for sustainable development goals. npj Sustain Agric 1(1):2

Wang L, Min M, Li Y, Chen P, Chen Y, Liu Y, Wang Y, Ruan R (2010) Cultivation of green algae chlorella sp. in different wastewaters from municipal wastewater treatment plant. Appl Biochem Biotechnol 162(4):1174–1186. https://doi.org/10.1007/s12010-009-8866-7

Watson SB et al (2015) Harmful algal blooms (Chap. 20). In Wehr JD, Sheath RG, Kociolek JP (eds) Freshwater algae of North America, Second edn, pp 873–920. Academic. https://doi.org/10.1016/B978-0-12-385876-4.00020-7

Watson SB et al (2016) The re-eutrophication of Lake Erie: harmful algal blooms and hypoxia. Harmful Algae 56:44–66. https://doi.org/10.1016/j.hal.2016.04.010

Wiltsie D et al (2018) Algal blooms and cyanotoxins in Jordan lake, North Carolina. Toxins 10(2):92

Xiong JQ, Kurade MB, Jeon BH (2018) Can microalgae remove pharmaceutical contaminants from water? Trends Biotechnol 36(1):30–44

Yao S et al (2019) Microalgae–bacteria symbiosis in microalgal growth and biofuel production: a review. J Appl Microbiol 126(2):359–368

Zhang W et al (2020) Physiological differences between free-floating and periphytic filamentous algae, and specific submerged macrophytes induce proliferation of filamentous algae: a novel implication for lake restoration. Chemosphere 239:124702. https://doi.org/10.1016/j.chemosphere.2019.124702

Chapter 17
Spatio-Temporal Assessment of Chlorine Residuals in the Water Distribution System of Dhaka City

Md. Mezanur Rahaman, Atikur Rahman, and Tanvir Ahmed

Abstract Most Dhaka City dwellers in Bangladesh depend on potable water supply from Production Tube wells (PTWs) under the Dhaka Water Supply and Sewerage Authority (DWASA). Water is a means of transmitting disease and causes illness by being contaminated with pathogenic microorganisms and dissolved pollutants. Chlorination is usually done in large networks to ensure acceptable water quality. Although a chlorination system is adopted in a particular District Metered Area (DMA) of Dhaka city, it is controlled by a limited number of potential chlorine injection points, typically the PTWs of such DMAs. The present study assesses the effectiveness of the chlorination system adopted in a particular DMA of Dhaka city. For that, residual chlorine was measured and predicted at selected locations within the network over space and time under various operating conditions using EPANET. The residual chlorine model was calibrated and validated using time-varying residual chlorine concentration data collected at several control points within the DMA. The measured bulk residual chlorine decay rate (k_b) was -0.2975 h^{-1}. The calibrated wall residual chlorine decay rate constant (k_w) was -0.0065 ft/s. Model simulation of the baseline scenario indicated that more than 50% of pipes do not get any chlorine under the present chlorination scheme. The additional six injection points increased the chlorine coverage over 86% of the pipes. EPANET is a tool that may be used to control the concentration of residual chlorine and assess the effectiveness of the existing chlorination system in different DMAs under DWASA. Chlorine by-products are categorized as emerging pollutants, with their presence in water resources linked to endocrine disruption and other adverse biological effects.

Md. M. Rahaman (✉)
TA Consultant, Asian Development Bank, Bangladesh Residence Mission, Dhaka, Bangladesh
e-mail: mezuiipf@gmail.com

A. Rahman
Cooperative Agricultural Research Center, College of Agriculture, Food, and Natural Resources, Prairie View A&M University, Prairie View, TX, USA

T. Ahmed
Department of Civil Engineering, Bangladesh University of Engineering and Technology, Dhaka, Bangladesh

© UNESCO 2025
S. Zandaryaa et al. (eds.), *Emerging Pollutants*, Advances in Water Security,
https://doi.org/10.1007/978-3-031-71758-1_17

Keywords Water chlorination · Dhaka WASA · Bulk chlorine decay · Wall chlorine decay · EPANET

17.1 Introduction

Proper understanding, characterization, and prediction of water quality behavior in drinking water distribution systems are critical to meeting regulatory requirements and customer-oriented expectations. Drinking water is a vehicle for disease transmission, and the risk of getting a waterborne disease increases with the level of contamination by pathogenic microorganisms. The provision of an adequate supply of safe water is one of the eight components of primary health care identified by the International Conference on Primary Health Care in Alma-Ata in 1978. The main objective of any water distribution system is to make water available to the consumer at proper quality and pressure, with acceptable quality in terms of flavor, odor, appearance, and sanitary security. However, preserving the water quality throughout the distribution system is one of the most challenging technological issues for suppliers (Bello et al. 2019). It is not an easily attainable objective because of the distribution system's complexity and dynamism concerning building materials, element diversity, and other properties variability.

Drinking water has been disinfected since the beginning of the twentieth century when it was discovered that microbiological contamination risks by water-borne diseases, namely cholera or typhoid fever, decreased when disinfectants were used (Minear and Amy 1996. Deterioration of water quality in distribution networks dramatically impacts human health and public acceptance of tap water (Mostafa et al. 2013). UV radiation, Ozone, and chlorine can be applied for disinfection. The first two agents do not generate significant Disinfection By-Products (DBPs) (Chowdhury et al. 2009) nor interact significantly with the water they are supposed to disinfect. On the other hand, chlorine disinfection presents the advantages of efficiency and durability (Castro et al. 2010). Chlorine is the most widely used disinfectant in drinking water systems worldwide due to its germicidal potency, economy, and efficiency (Mostafa et al. 2013). Chlorine is relatively easy to handle, the capital costs of installation are low, and it is cost-effective and simple to dose, measure, and control (Freese and Nozaic 2004; Sadiq and Rodriguez 2004; Warton et al. 2006a, b).

Residual chlorine should be maintained through network pipes to prevent contamination and microbial re-growth. As chlorine travels through the pipes in the distribution system, it can react with a variety of materials both within the bulk water and from the pipe wall (Rossman et al. 1994). A reaction between the bulk of the water and the pipe wall means that the residual concentration of chlorine decreases as water travels through the distribution network. In the bulk of the water, free chlorine is mainly consumed by reactions with natural organic matter (NOM) and other reactive substances. In contrast, the consumption of residual chlorine at the pipe wall is primarily associated with various reactions at the attached bio-film and the corrosion surface of the pipe wall (Kima et al. 2014). The hydraulic conditions are also a critical factor governing the decay of residual chlorine within transmission and distribution pipes (Stoianov and Aisopou 2014). The satisfactory maintenance of residual chlorine concentration at a customer's tap is one of the most important criteria for ensuring good water quality in distribution systems (Kima et al. 2014; Dongare et al. 2023).

Dhaka Water Supply and Sewerage Authority (DWASA) has divided Dhaka City into approximately 100 District Metered Areas (DMAs) to improve the water supply network of Dhaka City. Although there is a protocol for chlorination at the pump stations and a schedule for monitoring residual chlorine, the adoption of these protocols has been found to be absent in most DMAs. Chlorine is often administered arbitrarily, and its effectiveness in the water distribution system remains largely unknown. Presence of free chlorine indicates that most dangerous organisms in the water have been removed, and it is safe to drink. Even the concentration of residual chlorine within a water distribution network may vary depending on the location and time. In this case, a water quality (residual chlorine) model may be used as a necessary tool to predict the spatial and temporal variation of residual chlorine concentration throughout the specified water distribution network.

Although chlorine is a well-known and widely used chemical in various industries and water treatment processes, over-application can be a significant environmental and human health risk. Chlorine by-products are categorized as emerging pollutants, with their presence in water resources linked to endocrine disruption and other adverse biological effects (Levi 2009). The contamination of water by chlorine and its by-products is a growing global concern, particularly in urbanized and industrialized areas where chlorine is extensively used for disinfection (Morin-Crini et al. 2022). The recent increase in chlorine levels in water bodies underscores the need for updated water treatment measures to address this emerging pollutant and protect water quality and ecosystem health (Neerugatti et al. 2022).

This Chapter aims to evaluate the effectiveness of the existing chlorinating system introduced in a designated DMA for Dhaka city which makes chlorine measurements at targeted sites within the network. The results of a simulation and the various calibration parameters are set out, in order to enhance existing practices for residual chlorine within the distribution system over time and space.

17.2   Chemistry of Chlorine Equilibrium

The disinfectant capabilities of chlorine depend on its chemical form in water, which in turn depends on pH, temperature, organic content of water, and other water quality factors. In water treatment applications, chlorine is typically applied as compressed gas under pressure (i.e., dissolved in water at the point of application) or as solutions of either sodium hypochlorite or solid calcium hypochlorite. Essentially, the three forms are chemically equivalent because of the rapid equilibrium between dissolved molecular gas and the dissociation products of hypochlorite compounds (Haas 1999).

The following Eqs. (17.1–17.4) demonstrate the reactions of the three compounds in water:

$$H_2 + Cl_2 = HOCl + H^+ + Cl^- \tag{17.1}$$

$$NaOCl = Na^+ + OCl^- \tag{17.2}$$

$$Ca(OCl)_2 = Ca^{2+} + 2OCl^- \tag{17.3}$$

$$HOCl = H^+ + OCl^- \tag{17.4}$$

HOCl and OCl are oxidants and effective germicides, particularly against bacteria and viruses, with some effectiveness against protozoa and endospores. *HOCl* is the stronger and more effective of the two species (Brown et al. 2011a, b). Chlorine combines with various reducing agents and organic compounds, thus increasing the chlorine demands that must be satisfied before chlorine is available for disinfection.

17.2.1 Chlorine Demand

Chlorine demand is the amount of chlorine in the solution that is used up or inactivated after a period of time, which is not available for use as a germicide. When chlorine is added to water, not all of it is available to act against contaminants. Some are deactivated by sunlight. Some are consumed by reactions with other chemicals in the water or by out-gassing as Cl_2. More commonly, it is used up directly by disinfection of the pathogens already present in the water or by combining with ammonia (NH_3) and ammonium (NH_4^+) (by-products of living bacteria) to form various chloramines (Deborde and Gunten 2008).

17.2.2 Total Chlorine

When water's chlorine demand is accounted for, the remaining chlorine concentration is called total chlorine. Total chlorine comprised (a) combined chlorine and (b) free available chlorine.

17.2.2.1 Combined Chlorine

The term combined chlorine usually refers to residual chlorine that has combined with NH_3 or NH_4^+ to form mono-chloramine (NH_2Cl), di-chloramine $(NHCl_2)$ or tri-chloramine (NCl_3). The presence and concentration of these combined forms depend on a number of factors including the ratio of chlorine to ammonia–nitrogen, chlorine dose, temperature, pH and alkalinity. The Eqs. (17.5–17.7) demonstrate the reactions of the three compounds in water which contain chlorine:

$$NH_3 + Cl_2 = NH_2Cl + HCl \qquad (17.5)$$

$$NH_2Cl + Cl_2 = NHCl_2 + HCl \qquad (17.6)$$

$$NHCl_2 + Cl_2 = NCl_3 + HCl \qquad (17.7)$$

In addition to chlorinating ammonia, chlorine also reacts to oxidize ammonia to chlorine-free products (e.g., Nitrogen gas and Nitrate), as shown by the Eqs. (17.8–17.9).

$$3Cl_2 + 2NH_3 = N_2 + 6H^+ + 6Cl^- \qquad (17.8)$$

$$4Cl_2 + NH_3 = NO_3^- + 9H^+ + 8Cl^- \qquad (17.9)$$

Combined chlorine is noteworthy here because chloramines are oxidizers used as germicides, though they have reduced potentiality; therefore, disinfecting power is lower than other chlorine species such as $HOCl$, OCl^- or ClO_2.

17.2.2.2 Free Available Chlorine

Free Available Chlorine (FAC) is any residual chlorine available after the chlorine demand is met to react with new sources of bacteria or other contaminants. The relative proportion of these "free chlorine" forms is pH- and temperature-dependent. Some species of FAC that might be present are molecular chlorine (Cl_2), hypochlorous acid (HOCl), hypochlorite (OCl^-), and trichloride (Cl_3^-). A complex formed by molecular chlorine and the chloride ions (Cl^-) (Deborde and Gunten 2008). The whole of the chlorine addition process can be represented in Fig. 17.1.

Pure water will have no combined chlorine or chlorine demand because it has no pollutants. Thus, the free chlorine concentration will be equal to the concentration of chlorine initially added. In natural waters, especially surface water supplies such as rivers, organic material will demand chlorine, and nitrates will form combined chlorine. Thus, the free chlorine concentration will be less than the chlorine concentration initially added.

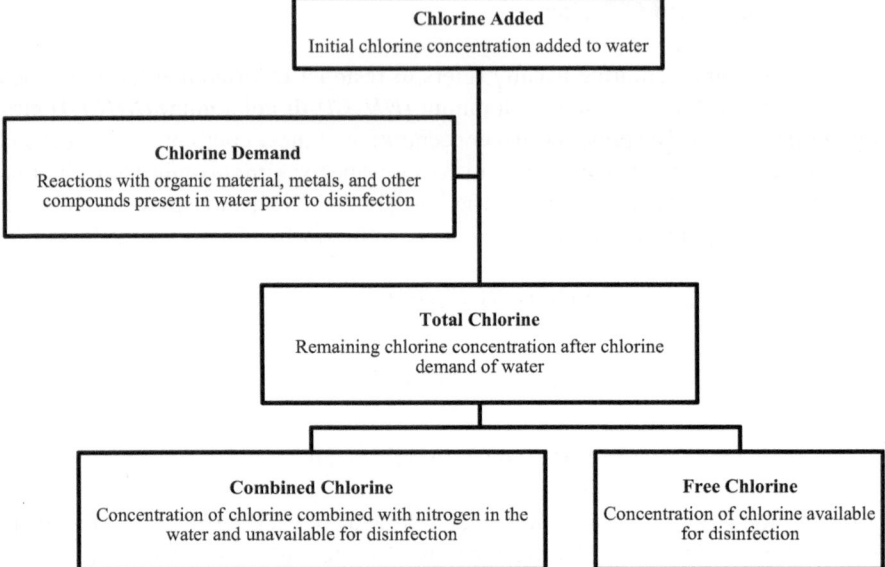

Fig. 17.1* Chlorine addition flow chart. *Source* Taleb et al. 2020

17.2.3 The Importance of pH

pH significantly changes the relative effectiveness of chlorine as a disinfectant. Different species of chlorine ions are more prevalent at different pH levels. Under typical water treatment conditions of pH 6–9, *HOCl* and *OCl*- are the main chlorine species. Figure 17.2 shows that chlorine hydrolysis into *HOCl* is almost complete at pH ≤ 4. Dissociation of *HOCl* into *OCl⁻* begins around pH 5.5 and increases drastically.

This is important because *HOCl* and *OCl⁻* do not have the same effectivity as disinfectants. *HOCl* can be 80–100% more effective disinfectant than *OCl⁻*.

Fig. 17.2* Distribution of free chlorine species in aqueous solutions. *Source* Deborde et al. 2008

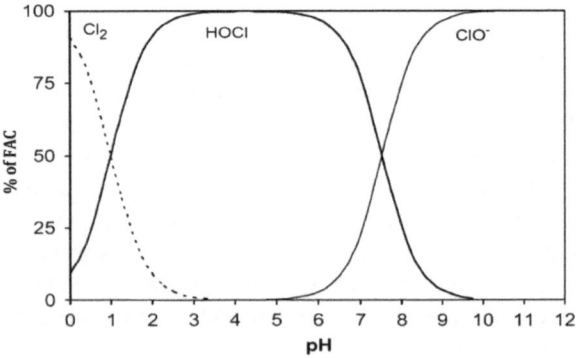

Optimum disinfection occurs at pH 5–6.5, where *HOCl* is the prevailing species of free chlorine present. As pH rises above that level, the ratio shifts to primarily *OCl⁻*. At pH 7.5, the ratio is about even. When the pH value increases to 8 or higher, *OCl⁻* is the dominant species. Therefore, assuming the concentration of Cl_2 species is constant, the higher the pH of the solution rises above 5.5, the lower the FAC's oxidation capability and disinfecting power (Deborde and Gunten 2008).

17.2.4 Mechanisms of Chlorine Decay

As chlorine is a potent oxidizing agent, it may also react with the material of pipe walls in distribution systems and attached biofilm and miscellaneous accumulated sediment. The consumption of chlorine within the bulk aqueous phase is often referred to as the bulk decay of chlorine, while that due to biofilms and at the distribution pipe wall is known as the wall decay. This is illustrated in Fig. 17.3 (Rossman 2000). In the figure, free chlorine *(HOCl)* reacts with natural organic matter *(NOM)* in the bulk phase. It is also transported through a boundary layer at the pipe wall to oxidize iron *(Fe)* released from pipe wall corrosion. Bulk fluid reactions can also occur within tanks (Vasconcelos et al. 1997). The impact of flow velocity on the loss of chlorine residual in unlined metallic pipes could be higher than in old unlined metal pipes; the loss of chlorine residual increases with velocity (Stoianova and Aisopou 2014).

Some chlorine is also lost through natural evaporation. If, ideally, the chlorinated water is pure and the material of the pipes is inert, the only mechanism leading to the decay would be that of natural evaporation, especially in particular areas of the distribution system, namely reservoirs and other free surface flow (Castro et al. 2010).

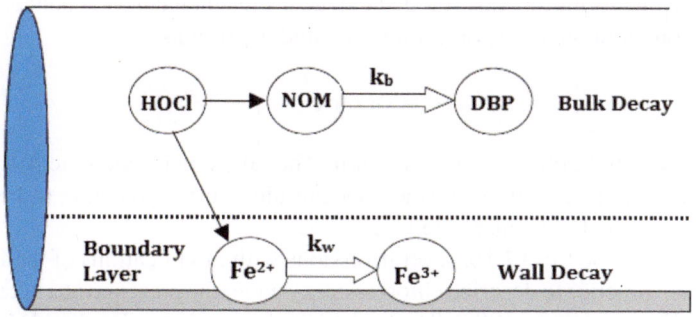

Fig. 17.3 Reaction zones within a distribution pipe. *Source* Rossman 2000

17.2.5 Mathematical Modeling of Chlorine Decay

Mathematical modeling of chlorine decay along the water supply system is a problem whose solution is not yet absolutely mastered. The data should be associated with the hydraulics, such as flow rate, flow pattern, networking, contour, pipe sizing, retention time, system age, and distances. The reaction pathways of chlorine are still relatively unknown due to the site-specific and heterogeneous nature of the natural organic matter *(NOM)*. The chlorine decay rate is typically rapid immediately after the dose and relatively slower after some time.

Several assumptions are considered for developing a mathematical model.

(i)   The disappearance of chlorine flowing through a pipe is governed by first-order kinetics.

(ii) This disappearance is due to reactions within the bulk flow and at sites along the pipe wall (or close to the wall).

(iii) These rates of reactions can be different, with the overall rate of the wall reaction also being affected by the rate at which chlorine can be transported from the bulk flow to the pipe wall.

Decay simulation conducted within EPANET considers the phenomena of chlorine reaction with chemical species in bulk fluid and with pipe walls. The contribution of bulk fluid is introduced into the software using a first-order kinetics constant k_b. The contribution of wall reactions is introduced into the software through another constant k_w, the meaning of which is more complex. Usually, calibrated values of k_b and k_w are used in the model to match the operating conditions.

17.2.5.1 Bulk Chlorine Decay

This kinetic law forms an equation that calculates residual chlorine concentration in water c throughout the transportation time t and is given as:

$$c = c_0.exp(-k_b t) \tag{17.10}$$

where, c_0 = initial chlorine concentration. The adjustment coefficient, k_b, can be determined by adjusting the chlorine concentration curve over time to Eq. (17.10) by recurring to the least square method.

In addition to the Eq. 17.10, several versions with modifications of the first-order models are in practice to describe chlorine decay under the presence of a heterogenous reductant (Haas and Karra 1984). The limited first-order decay (Eq. 17.11) kinetic law assumes that a fraction of the initial chlorine concentration c_0 remains unchanged, and only the remainder decays exponentially according to a first-order law. The parallel first-order decay (Eq. 17.12) assumes two components to the reaction, each decaying according to a first-order law but with different decay rate constants.

$$c = c^* + (c_o - c^*).exp(-k_b t) \tag{17.11}$$

$$c = c_o.z.exp(-k_{bfast}.t) + c_o(1 - z).exp(-k_{bslow}.t) \qquad (17.12)$$

It was found that the parallel first-order decay model yielded the most accurate results due to the presence of slow and fast-reacting components in water. However, the applicability of the results to potable water is questionable, and the initial chlorine dose was up to 10 mg/L. Furthermore, the approach appears to be inconsistent with references to free and combined chlorine.

There has been a study of the chlorine decay modeling in sand-filtered water (prior to contact tank chlorination but after pre-chlorination) at the Macao water treatment plant in South China. It has been found that chlorine decay can be split into two phases: an initial consumption during the first hour (representing the time spent in the contact tank and the remainder of the WTW) and long-term chlorine consumption after 1-h in the distribution system (Zhang et al. 1992; Brown et al. 2011a, b).

Decaying of a substance while moving through a pipe can be generally described as an nth power function of concentration as shown in Eq. (17.13):

$$r = k_b c^n \qquad (17.13)$$

where r is the rate of reaction (mass/volume/time) (Nagwan et al. 2013), k_b is the reaction constant (concentration raised to the power of $[1 - n]$ divided by time), c is the reactant concentration (mass/volume), and n is the reaction order. Chlorine decay is adequately modeled by a simple first-order reaction ($n = 1$, $k_b < 0$) and hence the equation becomes:

$$r = k_b c \qquad (17.14)$$

The USEPA Water Treatment Plant model described chlorine decay by dividing the time into three stages: an initial rapid reaction period ($t < 5$ min), a second-order reaction (5 min $< t < 5$ h), and a third period defined by a first-order reaction ($t = 5$ h). Later, bulk decay was represented through a two-stage approach at the WTW and a standard linear model in distribution systems (Clark 1998).

17.2.5.2 Wall Chlorine Decay

Aside from bulk decay, chlorine also decays due to interactions with pipe and tank walls and fittings. This 'wall decay', which includes corrosive reactions with the wall material itself, with adhering biofilms and with accumulated sediments, is primarily a function of the mass transfer of chlorine from the bulk water to the pipe walls, pipe material, diameter, age, inner coating and the presence of attached biofilms (Vieira et al. 2004; Warton et al. 2006a, b; Al-Jasser 2007).

The value of the wall reaction rate constant, k_w (mass/area/time), is influenced by the factor of mass transfer between the bulk of flow and the wall interface and the amount of wall area available for reaction. The factor is represented by a mass transfer coefficient (k_f) (length/time) which depends on the molecular diffusivity of the traced substance. Chlorine diffusivity is equal to 1.44×10^{-5} cm^2/s in water at 25 °C (AWWARF 1996). For a particular chemical species, k_f is a function of pipe diameter, flow velocity, and temperature (Edwards et al. 1979).

The EPANET is automatically adjusted to account for mass transfer between the bulk flow and the wall based on the molecular diffusivity of the reactant under study and the Reynolds number of the flow. In the case of zero-order kinetics, which is recommended by the EPANET program manuals, the wall reaction rate cannot be greater than the mass transfer rate, resulting in Eq. 17.15:

$$r = Min\,(k_w \times k_f \times c) \times (2/R) \qquad (17.15)$$

where r is the rate of reaction (mass/volume/time) and R is the pipe radius (length) (Nagwan et al. 2013).

The value of k_f can be determined using the Eqs. 17.16–17.20.

$$k_f = Sh\frac{D}{d} \qquad (17.16)$$

$$Sh = 0.023R^{0.83}Sc^{0.33} \text{ for } R > 2300 \qquad (17.17)$$

$$Sh = 3.65 + \frac{0.0668(\frac{d}{L})(R*Sc)}{1 + 0.04 * [(\frac{d}{L})(R*Sc)]^{\frac{2}{3}}} \qquad (17.18)$$

$$R = \frac{ud}{v} \qquad (17.19)$$

$$Sc = \frac{v}{d} \qquad (17.20)$$

where Sh = Sherwood number, R = Reynolds number, Sc = Schmidt Number, D = molecular diffusivity of chlorine in water, u = flow velocity in the pipe, v = kinematic viscosity of water, d = pipe diameter, and L = pipe length (Edwards et al. 1979).

17.2.5.3 Mass-Conservation Equation for Chlorine Decay

The one-dimensional conservation-of-mass equation for a dilute concentration of total free chlorine in water flowing through a section of a pipe is

$$\frac{\partial c}{\partial t} = u\frac{\partial c}{\partial x} - k_b \times c - \frac{k_f}{r_h}(c - c_w) \qquad (17.21)$$

where C = chlorine concentration in the bulk flow, t = time, x = distance along the pipe, r_h= hydraulic radius of pipe (one-half the pipe radius), and c_w= chlorine concentration at the pipe wall (Rossman et al. 1993).

The term on the left side of (17.21) represents the rate of change of chlorine concentration at different pipe sections. The first term on the equation's right-side accounts for chlorine's advective flux through the section. The second term represents chlorine decay within the bulk flow; the third term accounts for the transport of chlorine from the bulk flow to the pipe wall and the subsequent reaction. The inverse of the hydraulic radius represents the specific surface area available for reaction (i.e., the pipe-wall area per unit of pipe volume) (Rossman et al. 1993).

Assuming that the reaction of chlorine at the pipe wall is first-order with respect to the wall concentration c_w and that it proceeds at the same rate as chlorine is transported to the wall results in the following mass balance for chlorine at the wall:

$$k_f (c - c_w) = k_w \times c_w \tag{17.22}$$

Solving Eq. (17.22) for c_w and substituting it into (17.21) gives the following equation, which describes the time variation of chlorine along a single pipe.

$$\frac{\partial c}{\partial t} = u \frac{\partial c}{\partial x} - k_b \times c - \frac{k_w \times k_f \times c}{r_h(k_w + k_f)} \tag{17.23}$$

For a drinking water distribution system, the mass-conservation equation for the ith pipe can be expressed as

$$\frac{\partial c_i}{\partial t} = u \frac{\partial c}{\partial x} - k_i \times c_i \tag{17.24}$$

The subscript 'i' indicates the ith pipe in the network, and $= k_i$ an overall decay constant containing the bulk decay constant, the hydraulic radius, and the mass-transfer coefficient.

The overall decay constant is as follows:

$$k_i = k_b + \frac{k_w \times k_f}{r_h(k_w + k_f)} \tag{17.25}$$

A known hydraulics system (which may change over time) can be solved with a known initial condition for chlorine throughout the network at time 0 and a boundary condition at the head end junction of each pipe i where $x_i = 0$. Assuming that complete and instantaneous mixing occurs at pipe junctions, this boundary condition can be expressed with the following conservation-of-mass equation:

$$C_{i|x=0} = \frac{\sum q_p \times c_{p|x=L} + M}{\sum q_p + S_i} \tag{17.26}$$

The summation is made over all the pipes p that have flow q_p into the head junction of pipe 'i'; $M =$ any external mass flow of chlorine introduced at the head of the pipe 'i' and $S_i =$ any external flow of water introduced at the head of the pipe 'i'. To solve for the chlorine concentration within pipe 'i', one must know the concentrations in all pipes flowing into pipe 'i'.

Storage tanks can be modeled as thoroughly mixed, variable-volume reactors. Equations (17.23) and (17.24) represent a coupled set of differential/algebraic equations over all pipes in the network. Under a set of known time-varying hydraulic conditions, they can be solved using an explicit discretization technique called the Discrete Volume Element Method (DVEM) (Rossman et al. 1993).

Within each time period, when hydraulic conditions are constant, DVEM divides each pipe into a number of segments. After reactions are completed for all pipes, the resulting mixture concentration at each junction node is computed and then released into the head end segments of pipes with flow leaving the node. This sequence of steps is repeated until the time when a new hydraulic condition occurs. This method has been incorporated into a general-purpose distribution-system simulation computer code called EPANET (Rossman 2000, USPEA).

17.3 Case Study: Assessing the Effectiveness of a Chlorination System in DWASA

17.3.1 Selection and Location of the Study Area

The DMA 505 water distribution network was selected to assess the effectiveness of the present residual chlorine system in that network. It is one of the 10 DMAs under Zone-5 of DWASA. The DMA area is characterized by medium population density. The water is supplied from six deep production tube wells (PTWs). The network is located in the north part of the Gulshan model town of Dhaka City Corporation and in between the geographic coordinates of 23°47′41.29″N and 90°24′51.34″E and 23°46′49.21″N and 90°25′0.44″E. It serves Gulshan 1 and 2 and parts of Baridhara of Dhaka city, over 1.47 square kilometers. Figure 17.4 shows the water distribution network's layout with PTWs of DMA 505. The DMA network comprises 348 junctions, 419 pipes, six pumps, six valves, and 12 reservoirs.

17.3.2 Water Quality Parameter of Production Tube Wells

The water samples from six PTWs were collected, and quality parameters were measured. The tested parameters were pH, alkalinity, hardness, color, ammonia, total coliform, fecal coliform, organic constituents, carbon dioxide, iron, arsenic, chloride ions, and manganese.

Layout of Existing Water Distribution Network of DMA 505

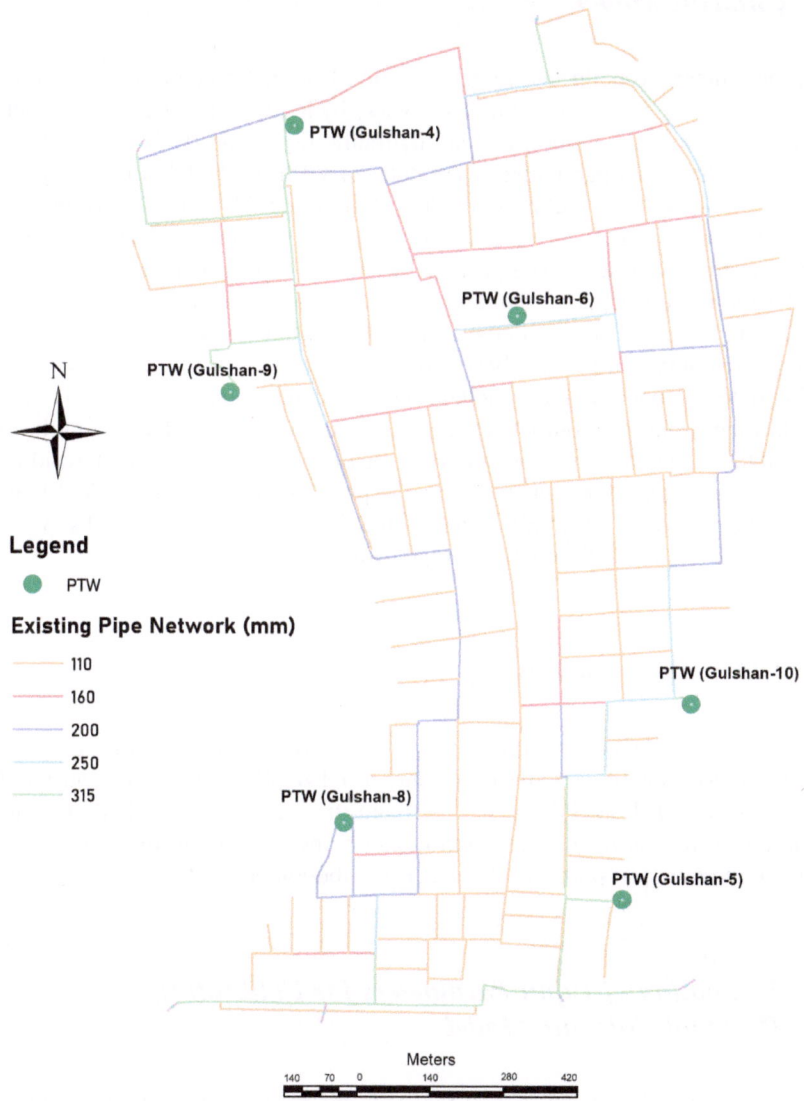

Fig. 17.4 Layout of water distribution network with 6 PTWs. *Source* Drawn by the authors based on the shapefile of DMA 505 Water Distribution Network

17.3.3 Input Parameters for Hydraulic Model and Residual Chlorine Model

The input parameters to build the model collected from the commissioning report were a layout of the water distribution network, hydraulic base demand of each node with respect to time pattern, and six hydraulic flow-head (Q-H) data for six pump curves. Other input parameters were determined from the laboratory experiment and field observations, including model calibration. EPANET assumes that all pipes are full at all times. Flow direction is from the end of the higher hydraulic head to that at the lower head. The principal hydraulic input parameters for nodes include elevation, base demand, demand categories, and demand pattern, as well as for pipes: start and end nodes, diameter, length, roughness coefficient, and status (open, closed, or contains a check valve) were also collected from the commissioning report. Computed outputs for nodes include actual demand, total head, pressure, and residual chlorine. Computed outputs for pipes include flow rate, velocity, unit head loss, friction factor, average reaction rate (over the pipe length), and average residual chlorine as a water quality parameter (over the pipe length). The Hazen-Williams formula has been used in EPANET to determine the hydraulic head loss by water flowing in a pipe due to friction with the pipe walls.

17.3.4 Chlorination Scheme of DMA 505

The chlorination experiment was conducted in July 2017. At the time of the experiment, chlorination was being carried out only at PTW 505 Gulshan-10 and PTW 505 Gulshan-6 out of the 6 PTWs. It was found that the higher residual chlorine concentrations were near the source of chlorination, medium at intermediate points, and absent at the remotest points of the water distribution network.

17.3.5 Estimation of Input Parameters for Calibrating Residual Chlorine Model

There are mainly two input parameters for calibrating the residual chlorine model in a water distribution network using EPANET 2.0, the bulk water decay constant, k_b and the wall decay constant, k_w. From laboratory tests on bottles, the bulk decay coefficient kb was developed. EPANET 2.0 simulations were carried out to calibrate k_w with the field observed data.

17.3.5.1   Residual Chlorine Bulk Decay Rate Constant (k_b)

The residual chlorine decay rate constant of bulk flow (k_b) was determined by analyzing the collected samples from six PTWs. The samples were analyzed in the laboratory by placing a sample of chlorinated water in a series of non-reacting glass bottles and analyzing the contents of each bottle at different points in time. Standard method 4500—Cl G (APHA 2005) was used to characterize samples' free residual chlorine concentration. After analyzing the residual chlorine concentration, the bulk decay rate constant (k_b) was determined using Eq. 17.10.

17.3.5.2   Chlorine Wall Decay Constant (k_w)

The wall decay rate constant, k_w, was derived from calibrating the residual chlorine model data obtained from the EPANET 2.0 and observed in the area of DMA 505. Two intermediate points INP2/J505-161 and INP3/J505-182, positioned between the PTW 505 Gulshan-6 and the PTW 505 Gulshan-10, were selected to monitor change in residual chlorine and record the observed data. Then, a trial and error method was applied to calibrate the observed data using EPANET 2.0. After successful calibration, the wall reaction rate constant, k_w (mass/area/time), was obtained. The mass transfer between the bulk of flow and the wall interface, as well as the amount of wall space available for reaction, has an influence on it. The factor is represented by a mass transfer coefficient (k_f) (length/time) which depends on the molecular diffusivity of the traced substance. Chlorine diffusivity is equal to $0.144 m^2/s$ in water at 25 °C (Awwarf 1996). For a particular chemical species, k_f is a function of pipe diameter, flow velocity, and temperature (Edwards et al. 1979).

17.3.5.3   Field Observation of Residual Chlorine

Water samples from six PTWs and 37 distribution network locations of DMA 5 were collected to analyze the residual chlorine concentration. Residual chlorine levels were measured using the HACH total chlorine test kit (Harp 2002) at the sampling site.

17.4   Results

17.4.1   Determination of Residual Chlorine Bulk Decay Rate Constant (k_b)

Figure 17.5a–d show the residual chlorine decay with respect to time for the four selected PTWs. The results of four representative PTWs are reported. The values of k_b are $-0.350\,h^{-1}$, $-0.450\,h^{-1}$, $-0.180\,h^{-1}$, and $-0.210\,h^{-1}$ for PTW 505 Gulshan-8, Gulshan-5, Gulshan-9, and Gulshan-6, respectively, using the chlorine decay curves as shown in figures. The average observed value of k_b is $-0.2975\,h^{-1}$ of this study area.

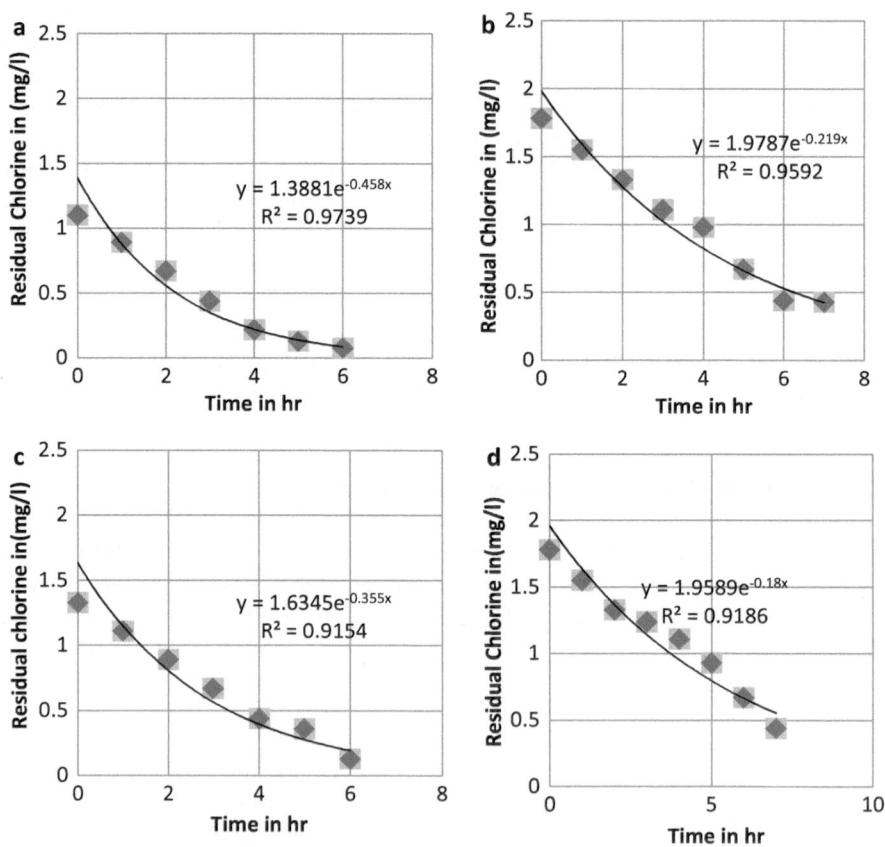

Fig. 17.5   a Bulk fluid chlorine decay and first order adjustment of raw water sample of PTW 505 Gulshan-5 (*Source* Authors). **b** Bulk fluid chlorine decay and respective first order adjustment of raw water sample of PTW 505 Gulshan-6. *Source* Authors

Rossman et al. (1993) found the value of k_b of -0.0239 h^{-1} for all pipes and tanks based on laboratory beaker tests of water taken from the service area. Vasconcelos et al. (1997) also found the first-order k_b values of -0.0345 h^{-1}, -0.0483 h^{-1}, -0.0095 h^{-1}, -0.0550 h^{-1}, -0.7375 h^{-1}, -0.0320 h^{-1} and -0.0033 h^{-1} for the site Bellingham, Fairfield, Harrisburg, North Marin (Aqueduct), North Marin (Stafford Lake), North Penn (Forest Park), North Penn (Keystone Tank), respectively. The value of k_b (-0.2975 h^{-1}), which is obtained from the current study, is remarkably higher than the value of k_b obtained by Rossman et al. (1993), but it falls within the range of values obtained by Vasconcelos et al. (1997). This shows that the k_b value can vary widely over several orders of magnitude depending on the type of samples. The higher value may be due to significant amounts of organic and inorganic reducing substances in the PTWs and moderately high temperatures over the DMA 505. Powell et al. (2000) studied the wall decay and bulk decay separately. They observed a significant variation in the bulk decay constant (k_b) with temperature, total organic carbon (TOC), and the initial chlorine concentration (C_0). As a relatively strong oxidizing agent, chlorine can react with inorganic and organic compounds in bulk water. As both organic and inorganic particles are present in varying concentrations and have different degrees of reactivity, as indicated in Table 17.1, the loss of chlorine over time is gradual. Clark et al. (1998) showed that the reactions of chlorine with organic matter make up most of the chlorine demand.

Table 17.1 Water quality parameter of the six PTWs of DMA 505

Parameters	Gulshan-4	Gulshan-5	Gulshan-6	Gulshan-8	Gulshan-9	Gulshan-10
pH	7.27	7.33	7.2	6.97	7.04	7.11
Alk, mg/l	48.0	151.0	128.0	96.0	105.0	79.0
Hardness, mg/l	98	104	90	88	78	84
Color, Hazen unit	2	4	2	2	2	2
NH$_3$, mg/l	0.00	0.00	0.00	0.00	0.00	0.00
Total coliform, N/ 100 ml	0	2	0	0	TNTC	0
Feacal coliform, N/ 100 ml	0	0	0	0	TNTC	0
TOC, mg/l	6.347	5.707	6.461	1.332	0.617	0.845
CO$_2$, mg/l	40.00	37.00	55.00	52.00	56.00	35.00
Fe, mg/l	0.04	0.18	0.04	0.02	0.16	0.04
As, mg/l	0.0048	0.0010	0.0022	0.00	0.0010	0.0026
Cl−, mg/l	3.00	8.00	11.00	28.00	18.00	23.00
Mn, mg/l	0.211	0.145	0.121	0.121	0.092	0.024

Source Authors

17.4.2   Calibration of Residual Chlorine Wall Decay Rate Constant (k_W)

For the measured k_b of -0.2975 h^{-1}, the residual chlorine for two representative intermediate points INP1/J505-172 and INP2/J505-198 was predicted for various values of k_w, ranging over several orders of magnitude ($k_w = -0.65$ ft/s–0.065 ft/s, -0.0065 ft/s and -0.00065 ft/s). The performance summary of the prediction to observed values is given in Table 17.2. The table shows that the prediction accuracy is higher when the value of k_w is -0.0065 ft/s. Therefore, the k_w value may be assumed to be -0.0065 ft/s for DMA 505. It is noted that the value of k_w is a sensitive input parameter for modeling residual chlorine in any specific water supply network (Castro and Neves 2003) The residual chlorine range lies between 0.29 mg/l and 0.34 mg/l for the same interval of observed time. Devarakonda et al. (2010) showed that chlorine disappears due to a combination of processes at the pipe wall. These processes include reactions between chlorine and biofilms attached to the distribution pipe wall, accumulated sediments, corrosion process, and mass transport process of chlorine and other reactants between the bulk flow and pipe wall.

Figures 17.6 and 17.7 compare the observed and predicted residual chlorine vs. time at the intermediate point INP1/J505-172 and INP2/J505-198 when k_w and k_b are fixed as -0.0065 ft/s and -0.2975 h^{-1}, respectively. It is evident from the figures that the predicted chlorine concentrations matched very well with the observed chlorine concentration for $k_w = -0.0065$ ft/s.

Therefore, k_w -0.0065 ft/s is a reasonable assumption for the EPANET residual chlorine model for DMA 505. Rossman et al. (1993) found k_w for a range of values between -0.0000057ft/s and -0.000017ft/s at the eight sampling locations in the network. Castro and Neves (2010) calibrated the pipe wall decay coefficient by trial–error method and found k_w as -0.002 ft/s and -0.007 ft/s in water distribution systems at the Lousada network. Vasconcelos et al. (1997) also found the k_w as -0.000028ft/s, -0.000000ft/s, -0.000010ft/s, -0.000057ft/s, -0.000057ft/s, 0.0000011ft/sand -0.0000011ft/s for the site at Bellingham, Fairfield, Harrisburg, North Marin (Aqueduct), North Marin (Stafford Lake), North Penn (Forest Park), North Penn (Keystone Tank), respectively. The k_w (-0.0065ft/s), obtained from the simulation results analysis under the current study, is remarkably higher than the value of k_w obtained by Rossman et al. (1993). Still, it falls within the range of values obtained by (Castro and Neves 2003). This shows that the k_w can vary widely over several orders of magnitude depending on several factors. Powell et al. (2000) investigated the factors that control wall chlorine decay through field surveys of several distribution pipes. The majority of the factors that have been shown to influence wall decay are pipe material and diameter (Shang et al. 2008), initial chlorine concentration (AWWARF 1996), corrosion, biofilm, and flow velocity. As a result, the chlorine will either decay due to reactions with compounds within the bulk water or at the pipe wall.

Residual chlorine concentration was measured at 21 selected points over the study area during the experimental period and compared with predicted values (Table 17.3).

Table 17.2 Agreement of predicted and observed time series value of residual chlorine for node INP1/J505-172 and INP2/J505-198

Time of residual chlorine prediction	Performance measure	When, $k_b = -0.2975\ h^{-1}$ and $k_w = -0.65$ ft/s	When, $k_b = -0.2975\ h^{-1}$ and $k_w = -0.065$ ft/s	When, $k_b = -0.2975\ h^{-1}$ and $k_w = -0.0065$ ft/s	When, $k_b = -0.2975\ h^{-1}$ and $k_w = -0.00065$ ft/s
INP1/J505-172	R^2	0.78	0.86	0.91	0.91
	RMSE	0.063	0.019	0.018	0.018
INP2/J505-198	R^2	0.75	0.93	0.88	0.88
	RMSE	0.082	0.028	0.025	0.018

Source Authors

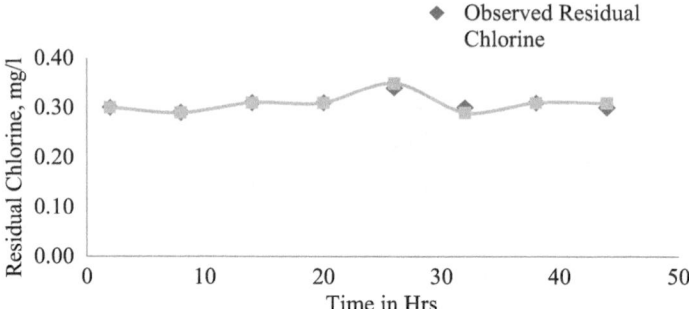

Fig. 17.6 Observed and predicted residual chlorine for the intermediate point INP1/J505-172 for $k_w = -0.0065$ ft/s and $k_b = -0.2975$ h^{-1}. *Source* Authors

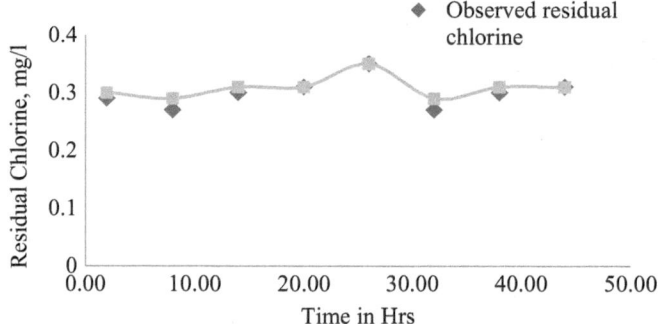

Fig. 17.7 Observed and predicted residual chlorine for the intermediate point INP2/J505-198 for $k_w = -0.0065$ ft/s and $k_b = -0.2975$ h^{-1}. *Source* Authors

It is seen from the table that the two concentrations are almost similar when calibrated k_w and measured k_b were used in the simulation.

17.4.3 Baseline Residual Chlorine Scenario of the Injected Chlorine Gas System

At the baseline, chlorine gas injection at a rate of 0.60 and 0.40 mg/l into two of the six injection points at R505-3 (PTW-Gulshan-6) and R505-3 (PTW-Gulshan-10), respectively, generates residual chlorine distribution pattern as shown in Fig. 17.8. The figure shows that more than 50% of the area of the network remains without chlorination. So, ensuring 100% chlorination over the network is recommended, maintaining a minimum 0.20 mg/l residual chlorine concentration.

Figure 17.9a–d show the histograms of the number of pipes with different residual chlorine concentrations under baseline scenarios at 2.00, 8.00, 14.00, and 20.00 h,

Table 17.3 Predicted and observed residual chlorine concentrations at the 21 selected points over the study area ($k_w = -0.0065$ ft/s and $k_b = -0.2975$ h^{-1})

Sl. No	Observed point	Observed data	Predicted data in residual chlorine model							
			Concentration of chlorine at 2:00 h	Concentration of chlorine at 8:00 h	Concentration of chlorine at 14:00 h	Concentration of chlorine at 20:00 h	Concentration of chlorine at 26:00 h	Concentration of chlorine at 32:00 h	Concentration of chlorine at 38:00 h	Concentration of chlorine at 44:00 h
		(mg/l)	(mg/l)	(mg/l)	(mg/l)	(mg/l)	(mg/l)	(mg/l)	(mg/l)	(mg/l)
1	P1 (J-155)	0.00	0.00	0.00	0.00	0.00	0.00	0.00	0.00	0.00
2	P2 (J-62)	0.00	0.00	0.00	0.00	0.00	0.00	0.00	0.00	0.00
3	P3 (J-54)	0.00	0.00	0.00	0.00	0.00	0.00	0.00	0.00	0.00
4	P4 (J-50)	0.00	0.00	0.00	0.00	0.00	0.00	0.00	0.00	0.00
5	P5 (J-48)	0.00	0.00	0.00	0.00	0.00	0.00	0.00	0.00	0.00
6	P6 (J-159)	0.00	0.00	0.00	0.00	0.00	0.00	0.00	0.00	0.00
7	P7 (J-161)	0.00	0.00	0.00	0.00	0.00	0.00	0.00	0.00	0.00
8	P8 (J-25)	0.00	0.00	0.00	0.00	0.00	0.00	0.00	0.00	0.00
9	P9 (J-26)	0.00	0.00	0.00	0.00	0.00	0.00	0.00	0.00	0.00
10	P10 (J-165)	0.00	0.00	0.00	0.00	0.00	0.00	0.00	0.00	0.00
11	P11 (J-10)	0.30	0.27	0.29	0.30	0.15	0.35	0.28	0.30	0.30
12	P12 (J-7)	0.45	0.37	0.45	0.45	0.46	0.37	0.45	0.45	0.46
13	P13 (J-1363A)	0.40	0.38	0.46	0.45	0.46	0.38	0.46	0.45	0.46
14	P16 (J-475)	0.00	0.00	0.00	0.00	0.00	0.00	0.00	0.00	0.00
15	P17 (J-82)	0.06	0.00	0.06	0.06	0.07	0.00	0.06	0.06	0.07

(continued)

Table 17.3 (continued)

Sl. No	Observed point	Observed data	Predicted data in residual chlorine model							
			Concentration of chlorine at 2:00 h	Concentration of chlorine at 8:00 h	Concentration of chlorine at 14:00 h	Concentration of chlorine at 20:00 h	Concentration of chlorine at 26:00 h	Concentration of chlorine at 32:00 h	Concentration of chlorine at 38:00 h	Concentration of chlorine at 44:00 h
		(mg/l)	(mg/l)	(mg/l)	(mg/l)	(mg/l)	(mg/l)	(mg/l)	(mg/l)	(mg/l)
16	P18 (J-174)	0.67	0.00	0.54	0.67	0.67	0.65	0.54	0.67	0.67
17	P19 (J-79)	0.63	0.62	0.51	0.63	0.63	0.62	0.51	0.63	0.63
18	P20 (J-202)	0.56	0.52	0.45	0.56	0.56	0.56	0.45	0.56	0.56
19	P21 (J-105)	0.00	0.00	0.00	0.00	0.00	0.00	0.00	0.00	0.00
20	P25 (J-129)	0.62	0.59	0.68	0.67	0.68	0.59	0.68	0.67	0.68
21	P27 (J-135)	0.47	0.32	0.34	0.44	0.47	0.35	0.34	0.44	0.47

Source Authors

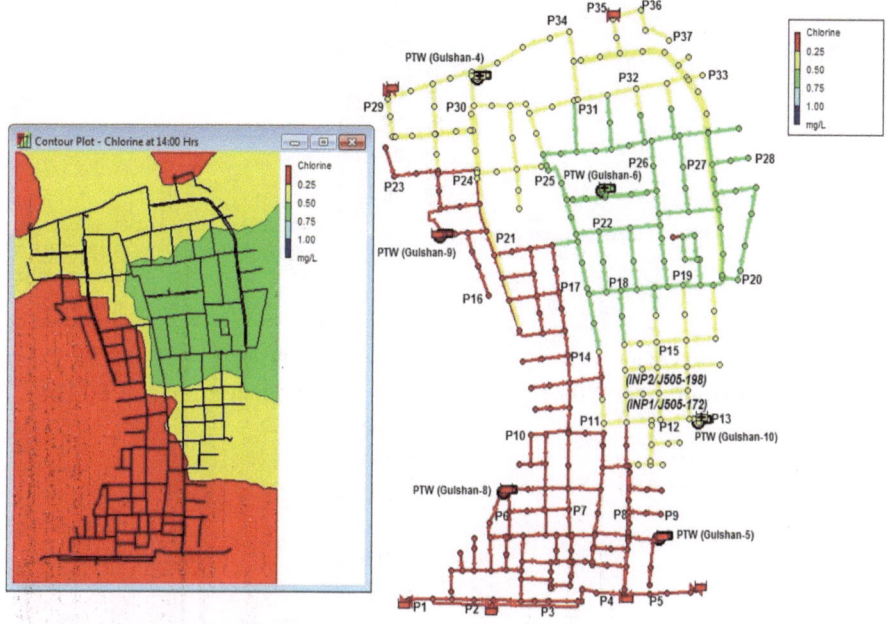

Fig. 17.8 Contour plot of baseline residual chlorine scenario at 14:00 h. *Source* Authors

respectively. Over 260 pipes were found with chlorine concentrations from 0.00 to 0.05 mg/l at four measuring times of the day.

The cumulative frequency distribution of the residual chlorine versus the percentage of pipes over 24 h at 2.00, 8.00, 14.00, and 20.00 h is illustrated in Fig. 17.10. Figure 17.10 shows no noteworthy difference between the Cl_2 coverage distance during the day, and many pipes (>63–66%) do not get Cl_2 from the current injection scheme. So, the injection locations need to be adjusted to increase coverage.

17.4.4   Improved Residual Chlorine Scenario of the Injected Chlorine Gas System

Baseline chlorine injection strategies were insufficient to maintain the minimum recommended residual chlorine concentrations (0.20 mg/l) throughout the distribution networks (Fig. 4.8). To ensure the availability of the required concentration in all parts of the network, we simulated the model for residual chlorine concentration by applying different chlorine injection strategies by trial and error method. In the simulation, we varied injection point locations and numbers, injecting at varying rates. After a few trials, we noticed a remarkable improvement in minimum residual chlorine distribution in the network.

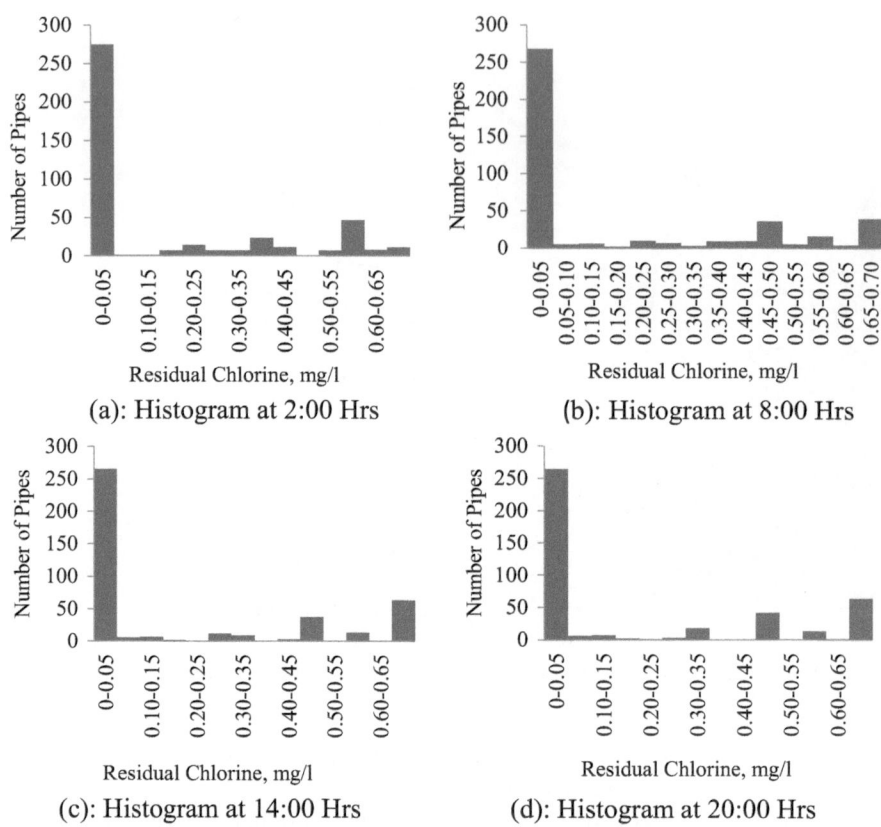

Fig. 17.9 Histogram of the number of pipes at different residual chlorine concentrations under baseline chlorination scenario at 2:00, 8:00, 14:00, and 20:00 h. *Source* Authors

In this strategy, the chlorine was injected in six points and was R505-3 (PTW-Gulshan-6), R505-4 (PTW-Gulshan-10), R505-5 (PTW-Gulshan-8), R505-2 (PTW-Gulshan-9), R505-1 (PTW-Gulshan-4) and R505-6 (PTW-Gulshan-5). However, the injected chlorine gas concentration was 0.20 mg/l for all six chlorination points. Figure 17.11 shows that the whole area is almost turquoise, meaning the residual chlorine concentration of 82% of the pipe is 0.22 mg/l over the 24-h observation period. It was noticed that the concentration of the residual chlorine changes spatially and temporally. The concentration of residual chlorine at the pipe network is below 0.25 mg/l.

Figure 17.12a–d show histograms of residual chlorine concentration and number of pipes at 2.00, 8.00, 14.00, and 20.00 h, respectively. The figures clearly indicate that the number of pipes is increasing gradually from 2 to 20 h at six-hour intervals. It is also observed that the total number of chlorinated pipes varies with time. The

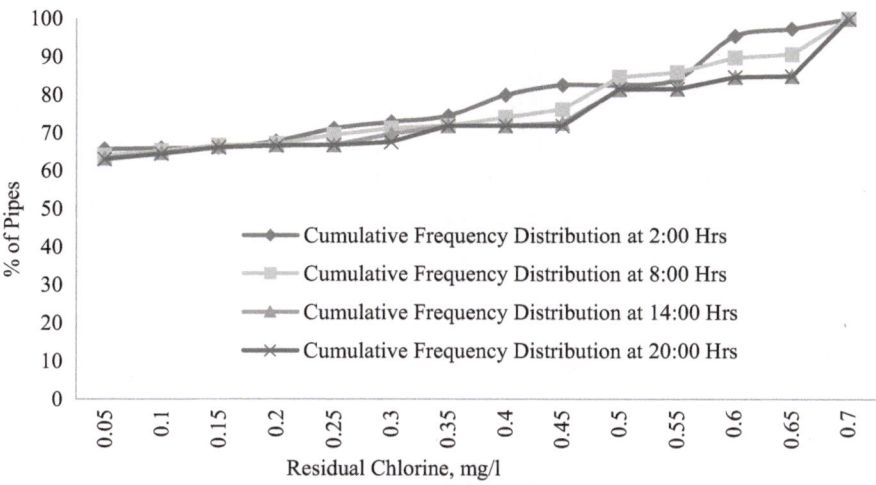

Fig. 17.10 Cumulative frequency distribution of residual chlorine with respect to percent of nonchlorinated pipes according to baseline chlorination scenario at 2:00, 8:00, 14:00, and 20:00 h. *Source* Authors

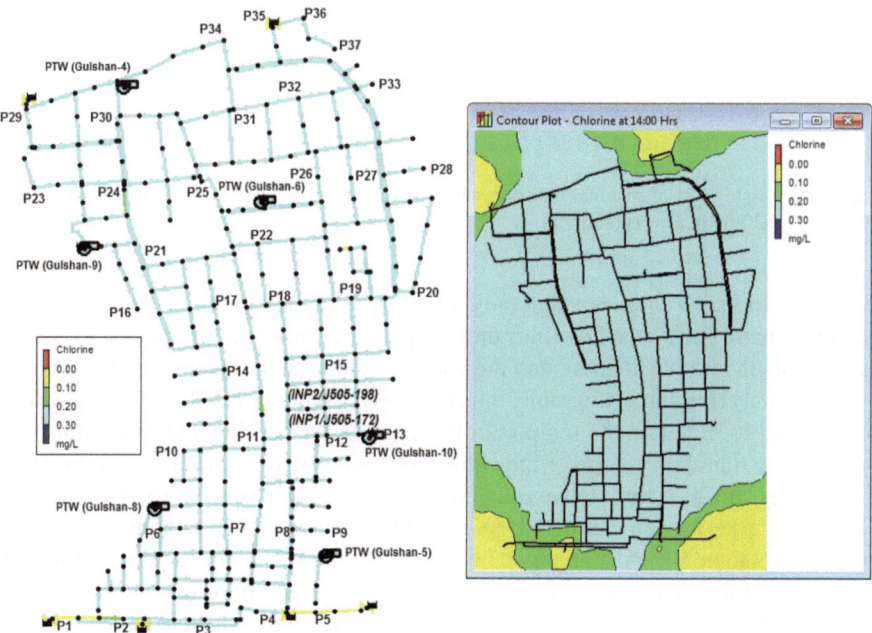

Fig. 17.11 Contour plot of residual chlorine from improved injection scenario at 14:00 h. *Source* Authors

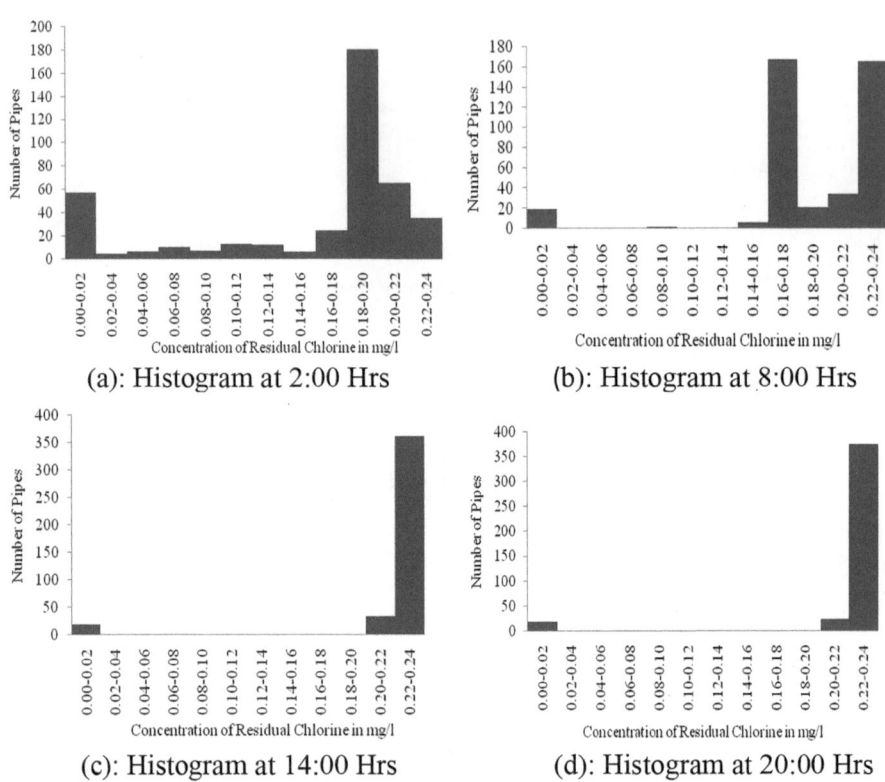

Fig. 17.12 Histogram of residual chlorine with respect to the number of pipes at 2:00 h, 8:00 h, 14:00 h, and 20:00 h. *Source* Authors

histogram in Fig. 4.11 shows that only a few pipes do not get chlorine due to insufficient demand for water flow into those pipes. The histogram indicates 400 pipes have come under chlorination, and the concentration range remains 0.20–0.22 mg/l.

After analyzing the histograms, the cumulative frequency distribution curves are constructed to illustrate the temporal difference in the residual chlorine concentration and the total percentage of the chlorinated pipes over a 24-h simulation period (Fig. 4.12). The 20-h cumulative frequency distribution curve remains up to 0.22 mg/l and then goes upward abruptly. At this stage, the percentage of nonchlorinated pipes is less than 5%. Also, the gap between the 2 and 20 h curves at concentration 0.22 mg/l is 82%. This indicates the maximum percentage of pipes within 0.22 mg/l residual chlorine concentration. So, this improvement is satisfactory to maintain the concentration below 0.25 mg/l.

Figure 17.13 suggests that the maximum residual chlorine concentration range was found to be 0.22–0.24 mg/l. This range is close to the standard residual chlorine concentration of 0.20–0.25 mg/l in the drinking water distribution network. The maximum allowable WHO standard free chlorine residual in drinking water is 5 mg/L. The minimum WHO recommended value for free chlorine residual in the

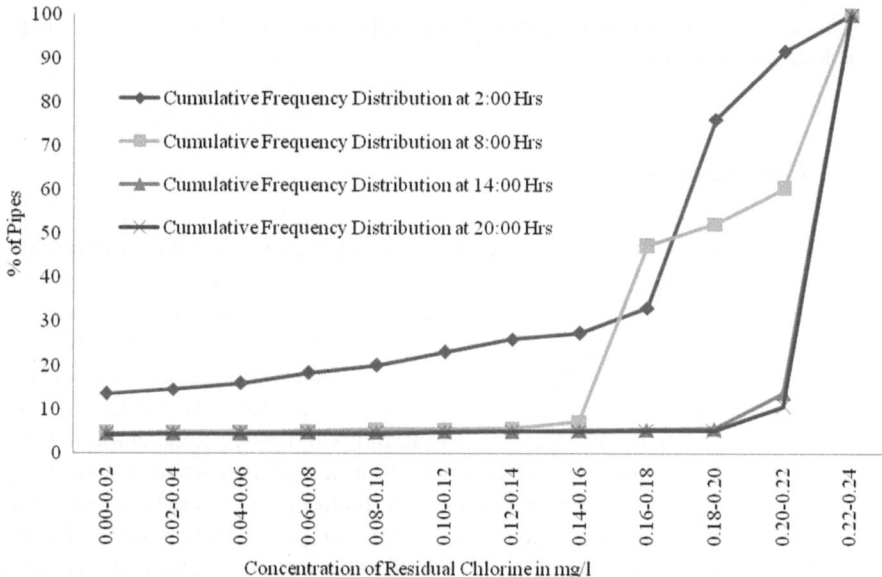

Fig. 17.13 Cumulative frequency distribution of residual chlorine with respect to percent of nonchlorinated pipes at 2:00, 8:00, 14:00, and 20:00 h. *Source* Authors

treated drinking water is 0.20 mg/L. It is recommended to not exceed 2.0 mg/L due to taste concerns, and chlorine residual decays over time in the stored water. So, in this case, the obtained concentration range is as nearly as the minimum WHO standard free residual chlorine concentration in the distribution network, which is relatively economical and reliable for practicing.

The residual chlorine improvement scenario has been applied to make a residual chlorine modeling in a specific DMA of DWASA. In this case, it was observed that the concentration of residual chlorine in the majority of pipes remains less than 0.25 mg/l concentration, which matches with 0.25 mg/l requirement. So, from the analysis, we found that six chlorine injection points are required to maintain the minimum WHO standard residual chlorine value uniformly throughout the whole water distribution network of DMA 505. It is also observed that it is impossible to cover the whole network under minimum residual chlorine concentration without missing any injection point from injecting chlorine at a certain dose. It is also examined that if chlorine concentration is increased at the injection points, it could not cover the whole network area with minimum uniform chlorine concentration. It is recommended to properly inject the chlorine gas at six chlorine injection points at a time for the effectiveness of the present chlorination system over the whole DMA.

There are 348 numbers of joints and 419 numbers of pipes in the water distribution network, but the concentration of residual chlorine always remains zero in nineteen pipes. This is due to insufficient water flow through those pipes. It is recommended that the areas served by those pipes should apply chlorine by other means (e.g., household application).

17.5   Conclusions

The objectives of this study were to determine the effectiveness of the baseline chlorination system adopted in a particular DMA of Dhaka city. Residual chlorine was measured and predicted at selected locations within the network over space and time under various operating conditions using EPANET. The average value of bulk residual chlorine decay rate constant k_b and wall residual chlorine decay rate constant k_w was determined as -0.2975 h^{-1} and -0.0065 ft/s, respectively. Under the baseline scenario, more than 50% of pipes were found with zero residual chlorine using EPANET analysis. A simulation of different chlorine injection strategies was made to ensure the full chlorine coverage of the DMA. Including additional chlorine injection points increased the coverage area; six injection points increased the chlorine coverage over 86% of the pipes. Even with all six injection points, 100% chlorine coverage could not be obtained over 24 h. In that case, it would be advised to have a separate chlorine addition to ensure water safety. This study was limited to the chlorine injection points fixed by DWASA. More sampling points within the distribution network are required for the validation of residual chlorine concentration. In the future, linear programming can be used to determine the most optimum locations of chlorine injection points in the system, potentially leading to better chlorine coverage over the entire DMA.

References

Al-Jasser AO (2007) Chlorine decay in drinking-water transmission and distribution systems: pipe service age effect. Water Res 41(2):387–396

APHA, AWWA, and WEF (2005) Standard methods for examination of water and wastewater, 21st edn. American Public Health Association, Washington, D.C.

American Water Works Association (1996) Internal corrosion of water distribution systems. American Water Works Association

Bello O, Abu-Mahfouz AM, Hamam Y, Page PR, Adedeji KB, Piller O (2019) Solving management problems in water distribution networks: a survey of approaches and mathematical models. Water 11(3):562

Brown D, Bridgeman J, West RJ (2011a) Predicting chlorine decay and THM formation in water supply systems. Rev Environ Sci Bio/technol 10(1):79–99

Brown S, Mcnabb R, Mchardy JP, Taylor K (2011b) Workplace performance, worker commitment, and loyalty. J Econ & Manag Strat 20(3):925–955

Castro FG, Kellison JG, Boyd S, Kopak AM (2010) A methodology for conducting integrative mixed methods research and data analyses. J Mixed Methods Res 4(4):342–360

Castro P, Neves M (2003) Chlorine decay in water distribution systems case study–lousada network. Elec J Env Agricult Food Chem Title 2(2):261–266

Castro P, Neves M (2010) Chlorine decay in water distribution system case study–lousada network. In: Environmental 2010: situation and perspectives for the European Union, 6–10 May 2003, Porto, Portugal (G11), pp 1–6

Chowdhury S, Champagne P, McLellan PJ (2009) Models for predicting disinfection byproduct (DBP) formation in drinking waters: a chronological review. Sci Total Environ 407(14):4189–4206

Clark RM (1998) Chlorine demand and TTHM formation kinetics: a second-order model. J Environ Eng 124(1):16–24

Clark DB, Kirisci L, Moss HB (1998) Early adolescent gateway drug use in sons of fathers with substance use disorders. Addict Behav 23(4):561–566

Deborde M, Gunten UV (2008) Reactions of chlorine with inorganic and organic compounds during water treatment—kinetics and mechanisms: a critical review. Water Res 42(1–2):13–51

Design Management Consultant (DMC) (2014) Commissioning report of DMA 505, DWASA

Devarakonda AK, Tummala P, Sandrala IP (2010) Security solutions to the phishing: transactions based on security questions and image. Inf Process Manag 70:565–567

Dongare P, Sharma KV, Kumar V, Mathew A (2023) Water distribution system modelling of GIS– remote sensing and EPANET for the integrated efficient design. J Hydroinformatics jh2023281

Edwards DK, Dumy VE, Mills AF (1979) Transfer processes. Hemisphere Publishing Corporation, Washington

Freese D and Nozaic J (2004) "Chlorine: is it really so bad and what are the alternatives. Water Institute of Southern Africa (WISA) biennial conference, pp 1212–1222. ISBN: 1-920-01728-3

Haas BK (1999) A multidisciplinary concept analysis of quality of life. West J Nurs Res 21:728–742

Harp DL (2002) Current technology of chlorine analysis for water and wastewater, technical information series, vol 17. Hach Co. Inc, USA, p 34

Hass CN, Karra SB (1984) Kinetics of wastewater chlorine demand exertion. J Water Pollut Control Fed 56:170–173

Kima H, Kima S, Koob J (2014) Prediction of the chlorine concentration in various hydraulic conditions in a pilot scale water distribution system. In: 12th international conference on computing and control for the water industry, CCWI2013, vol 70, pp 934–942

Levi Y (2009) Contraintes et enjeux dans l'évaluation et la gestion des risques sanitaires liés aux micropolluants émergents dans les eaux. Bull Acad Natl Med 193(6):1331–1344

Minear RA, Amy GL (1996) Water disinfection and natural organic matter: history and overview

Morin-Crini N, Lichtfouse E, Liu G, Balaram V, Ribeiro ARL, Lu Z, Stock F, Carmona E, Teixeira MR, Picos-Corrales LA, Moreno-Piraján JC (2022) Worldwide cases of water pollution by emerging contaminants: a review. Environ Chem Lett 20(4):2311–2338

Mostafa GN, Matta M, Halim AH (2013) Simulation of chlorine decay in water distribution networks using EPANET – case study. Civ Environ Res 3(13):100–116

Nagwan GM, Minerva EM, Hisham AH (2013) Simulation of chlorine decay in water distribution networks using EPANET – case study. Civ Environ Res 3(13):100–116

Neerugatti KRE, Veldurthi NK, Heo J (2022) Emerging pollutants in water bodies: a cause and effect analysis. In: Nano-enabled technologies for water remediation. Elsevier, pp 23–38

Powell RD, Krisskk LA, Mri JJMVD (2000) Preliminary depositional environmental analysis of CRP-2/2A, Victoria Land Basin, Antarctica: palaeoglaciological and palaeoclimatic inferences. Terra Antertica 7(3):313–322

Rossman AL, Clark MR, Grayman MW (1994) Modeling chlorine residuals in drinking-water distribution systems. J Environ Eng 120(4):803–820

Rossman L, Boulos P, Altman T (1993) Discrete volume-element method for network water-quality models. J Water Resour Plan Manag 119(5):505–517

Rossman LA (2000) EPANET 2 Users Manual. United States Environmental Protection Agency, EPA/600/R-00/057

Sadiq R, Rodriguez MJ (2004) Disinfection by-products (DBPs) in drinking water and predictive models for their occurrence: a review. Sci Total Environ 321(1–3):21–46

Shang F, Uber JG, Rossman LA (2008) Modeling reaction and transport of multiple species in water distribution systems. Environ Sci Technol 42:808–814

Stoianova I, Aisopou A (2014) Chlorine decay under steady and unsteady-state hydraulic conditions. In: 12th international conference on computing and control for the water industry, CCWI2013, vol 70, pp 1592–1601

Taleb MA, Mowafi S, El-Sayed H (2020) Utilization of keratin or sericin-based composite in detection of free chlorine in water. J Molecul Struct 1202:127379, https://doi.org/10.1016/j.molstruc.2019.127379

Vasconcelos JJ, Rossman LA, Mraman WM, Boulos PF (1997) Kinetics of chlorine decay. Am Water Works Assoc 89(7):54–65

Vieira P, Coelho ST, Loureiro D (2004) Accounting for the influence of initial chlorine concentration, TOC, iron and temperature when modelling chlorine decay in water supply. J Water Supply Res Technol-Aqua 53(7):453–467

Warton B, Heitz A, Joll C, Kagi R (2006a) A new method for calculation of the chlorine demand of natural and treated waters. Water Res 40(15):2877–2884

Warton DI, Wright IJ, Falster D, Westoby M (2006b) Bivariate line-fitting methods for allometry. Biol Rev Camb Philos Soc 81(2):259–291

Zhang SH, Reddick RL, Piedrahita JA, Maeda N (1992) Spontaneous hypercholesterolemia and arterial lesions in mice lacking apolipoprotein E. Science 258(5081):468–471

Part V
Priority Emerging Pollutants in the Hydrocycle

Chapter 18
Curbing the Environmental Implications of Emerging Nano-Pollutants: Current Developments in Preventing Environmental Exposure Potential and Adverse Effects

Mbuyiselwa Shadrack Moloi, Thabiso Mzinyati, Raisibe Florence Lehutso, Paul J. Oberholster, and Melusi Thwala

Abstract Commercialization of nano-enabled products (NEPs) being products that contain engineered nanomaterials (ENMs) is rapidly increasing. Most NEPs in markets exhibit high likelihood of ENMs release into the aquatic environment where they may induce undesirable effects; current data suggests rising nanopollution driven by rising commercialization of NEPs. Thus, measures to reduce the environmental exposure and impact of ENMs are required across all lifecycle phases. Herein, two strategies are proposed: safer- and sustainable-by-design (SSbD) strategy and policy development at the international level for ENMs/NEPs' environmental safety. The SSbD strategy seeks to balance the full exploitation of ENMs in NEPs while reducing their environmental exposure and impact at the design and manufacturing phase. This is achieved by integrating the knowledge of ENMs' physicochemical properties, exposure, and risk and designing out (reducing) the unfavourable properties. For both strategies, the current knowledge, shortcomings, and recommendations for successful implementation are discussed. Overall, SSbD and policy development can play a significant role in curbing the aquatic environmental risks associated with ENMs/PR-ENMs. However, both strategies are still in infancy and require comprehensive research for further development and implementation at country and international levels.

M. S. Moloi (✉) · P. J. Oberholster · M. Thwala
Centre for Environmental Management, University of the Free State, Bloemfontein, South Africa
e-mail: moloimbuyiselwa@gmail.com

T. Mzinyati
Department of Chemical Sciences, University of Johannesburg, Johannesburg, South Africa

M. S. Moloi · R. F. Lehutso
Water Research Centre, Council for Scientific and Industrial Research, Pretoria, South Africa

M. Thwala
Science Advisory and Strategic Partnerships, Academy of Science of South Africa, Pretoria, South Africa

© UNESCO 2025
S. Zandaryaa et al. (eds.), *Emerging Pollutants*, Advances in Water Security,
https://doi.org/10.1007/978-3-031-71758-1_18

405

Keywords Nano-enabled products · Exposure assessment · Nanotechnology market · Safer- and sustainable-by-design · Policy development · REACH · FDA

18.1 Introduction

Nanotechnology has revolutionized material design and product formulation due to the ability of engineered nanomaterials (ENMs) (1–100 nm) to improve the properties and functionalities of materials and products (Batchelor-McAuley et al. 2014). The application of ENMs is property-functionality-specific; meaning that ENMs are applied to improve specific properties and functionalities of products (Surette et al. 2019). Therefore, different ENMs are incorporated into products in a wide array of applications, for instance, nAg in textiles is used for antimicrobial properties and thus prevent odour (Gagnon et al. 2019), dietary supplements used in nanoscale to improve their bioavailability and maximize their bioactivity and safety (Jampilek et al. 2019), in the personal care industry various cosmetic products (e.g., UV filtration) have adopted nanotechnology (Kaur and Agrawal 2008; Mu and Sprando 2010). As a result, nano-enabled products (NEPs) commercialization has risen rapidly (Hansen et al. 2020). Various databases have been established to document NEPs' commercialization.

The earliest database published was the Consumer Product Inventory (CPI) by the Woodrow Wilson International Center for Scholars and the Project on Emerging Nanotechnologies with 54 products listed in 2005 (Vance et al. 2015). In 2016, the number of NEPs recorded was just over 1800 (Vance et al. 2015). However, this database has not been updated since. In 2012, the EU established The Nanodatabase with 1012 products reported in Europe by 2016 (Hansen et al. 2016). The total number of NEPs recorded is currently over 5,300 (Nanodatabase 2022); The Nanodatabase is an online inventory and is frequently updated. Others (Moeta et al. 2019; Zhang et al. 2015) have focused on country-level inventories and reported respectively for Singapore and South Africa; overall trends being somewhat similar to Consumer Product Inventory and The Nanodatabase (Moeta et al. 2019). Over an estimated 10,000 NEPs globally are currently in markets (StatNano 2023). Notably, the most used ENMs in NEPs globally include titanium dioxide ($nTiO_2$), silver (nAg), silicon dioxide (SiO_2), graphene, tungsten disulfide, and others (Fig. 18.1). These ENMs contain the beneficial properties for high consumption NEPs, such as high refractive index and broad UV spectrum ($nTiO_2$), chemical and photostability (nZnO), broad-spectrum antimicrobial and antifungal activity (nAg), and increased absorption, self-cleaning, transparency, and abrasion resistance (Moloi et al. 2021).

The rising commercialisation of NEPs has since triggered research into the environmental exposure and effects assessment of product-released ENMs (PR-ENMs) due to suspected and known effects of ENMs. Since around 2007, interest has grown focusing on categorising ENMs risk, concerning PR-ENMs this has been linked to the location of ENMs in the NEPs which influences ENMs' ease of release during the product lifecycle (Hansen et al. 2007, 2008). Accordingly, the products in the health

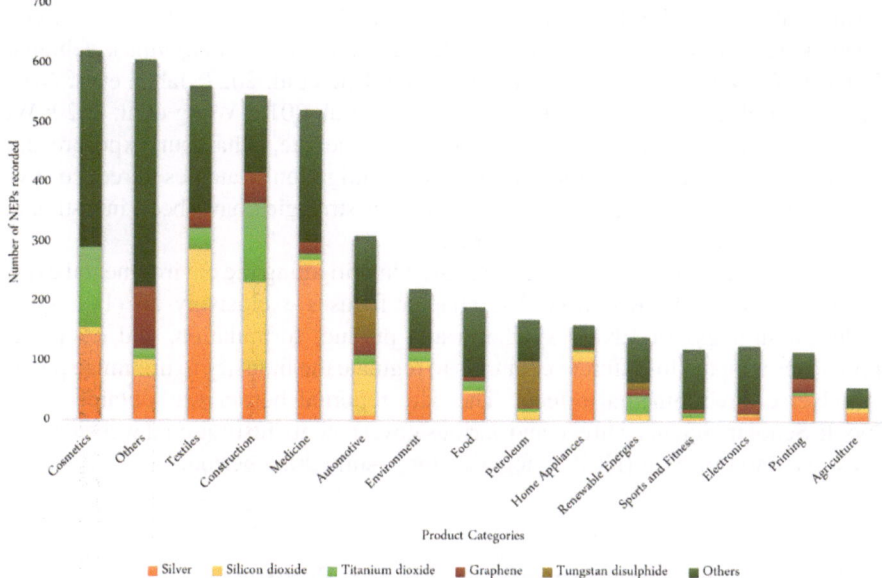

Fig. 18.1 Number of NEPs recorded on StatNano database by product category and ENMs types. *Source* Reproduced by the authors by using open-access data from StatNano 2023. *Note* StatNano is an open-access Nanotechnology Products Database. https://product.statnano.com/

and fitness category have the highest environmental exposure potential due to the ENMs' suspension within the liquid matrix (Hansen et al. 2016; Moeta et al. 2019). Thus, the PR-ENMs associated with this category have been reported in several environmental reports.

PR-ENMs which are either suspended in liquids or surface coated are most likely to be released in massive quantities to the aquatic environment during the product lifecycle which contributes towards escalating water nanopollution. Thus far, evidence shows considerable release of PR-ENMs from sunscreens, personal care products, paints, and textiles (Lehutso and Thwala 2021; Moloi et al. 2021). The release and evidence of nano pollution in aquatic environments have been reported (Lehutso and Thwala 2021; Moloi et al. 2021). A handful of studies have recovered PR-ENMs in surface water with concentrations of 10–15 μg/L (PR-nZnO), 0.0003–0.619 μg/L (PR-nAg), and 10–15 μg/L (PR-nTiO$_2$) (Gondikas et al. 2014; Labille et al. 2020; Markus et al. 2018; Reed et al. 2017). With these findings, it is evident that PR-ENMs occur in the aquatic environments which may lead to detrimental effects to the aquatic ecosystem as concentrations rise.

Once released, PR-ENMs enter the aquatic environment and interact with biotic and abiotic constituents in the water system (e.g., natural organic matter, electrolytes, and other pollutants) (Abbas et al. 2020). The interaction of PR-ENMs with surface water parameters/natural organic matter influences their transformation and subsequent bioavailability (Gagné et al. 2019). Ecotoxicity investigations have so far shown detrimental effects of PR-ENMs in aquatic bacteria (*Vibrio fischeri*), *Daphnia magna*, copepods (*Tigriopus japonicus*) microalgae (*Raphidocelis subcapitata*), zebrafish (*Danio rerio*), and plants (*Spirodela polyrhiza*) (Gao et al. 2022; Jahan et al. 2017; Künniger et al. 2014; Lehutso et al. 2021; Reed et al. 2016; Wong et al. 2020; Wu et al. 2019). The knowledge and understanding of the fate, behaviour, exposure, and risk of PR-ENMs provides a platform to devise mitigation strategies to reduce environmental exposure and risks. So far, mitigation strategies have been investigated using the knowledge from pristine ENMs.

Herein, this chapter reviews strategies used in mitigating the environmental exposure and toxicity of PR-ENMs. The chapter focuses exclusively on (1) a safer-by-design strategy for ENMs synthesis and product formulation, and (2) policy approaches adopted in different countries to regulate the industry to minimize potential adverse environmental effects. The data reported herein was obtained from Google Scholar, Science Direct, and various governments/institutional websites, and is restricted to the 2008 (NEPs categorization)—July 2023 period.

18.2 Safer- and Sustainable-By-Design Strategy for Engineered Nanomaterials' Safety

The increasing usage of NEPs and the rising ENMs' environmental exposure linked to NEP commercialisation have since triggered the investigation and implementation of a safety-by-design approach to ENMs synthesis and NEPs manufacturing. Safety and sustainable-by-design, also referred to as safe and sustainable-by-design, or safer and sustainable-by-design (herein SSbD for all versions) is a concept that integrates the knowledge of ENMs' potential adverse effects on the environment, and human health, in designing safer ENMs and NEPs to minimize the unfavourable aspects of the ENMs (Schwarz-Plaschg et al. 2017). This ensures that the environmental health and safety concerns of ENMs are minimized during the early stage of ENMs/NEPs development to control exposure, hazard, and risk (Rose et al. 2021). The simple schematic in Fig. 18.2 depicts measures that must be considered for successful SSbD implementation.

Although the SSbD research is still in the infancy stages, a few studies have investigated the implementation of SSbD in the various stages of the ENMs/NEPs' life cycle. The purpose of the reported studies was to establish the effectiveness of SSbD in limiting the potential nanopollution and toxicological impact on the aquatic ecosystem. In this section, investigations that assessed implementation of SSbD to decrease ENMs release from NEPs, reduction of ENMs' adverse effects, and reduction of ENMs persistence in the environment are discussed.

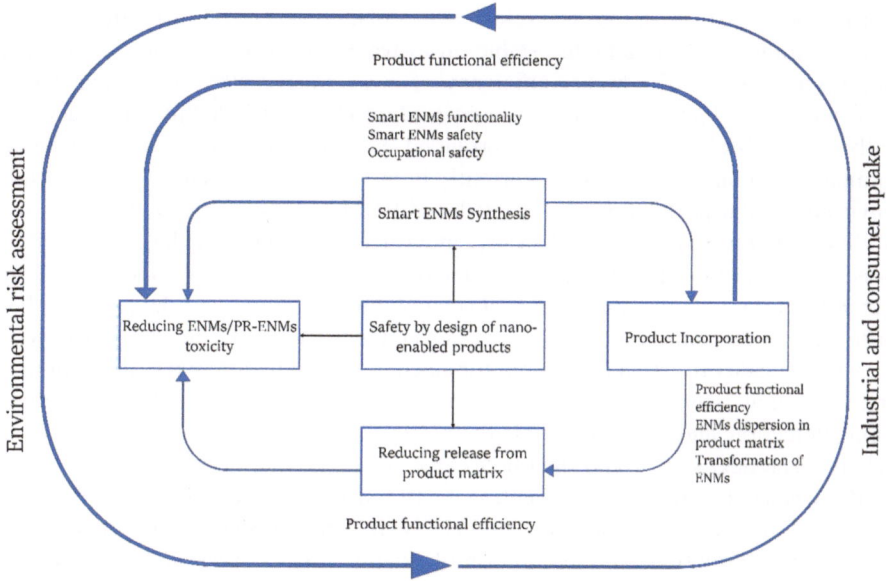

Fig. 18.2 Simple schematic indicating the considerations that must be made in successful SSbD implementation in product manufacturing. *Source* Authors

18.2.1 Synthesis of Smart Engineered Nanomaterials

The synthesis of smart ENMs is aimed at altering the specific physicochemical properties of the ENM to reduce their environmental exposure and toxicity (Kraegeloh et al. 2018). This is achieved by using methods such as doping (Aydın et al. 2019; Bayan et al. 2020; Pathak et al. 2018, 2020), surface coating or encapsulation (Marchioni et al. 2020; Moraes Silva et al. 2016), or reduction of ENM concentrations used (Shandilya and Capron 2017). Smart ENMs are therefore intended to be as efficient, but without the high release, bioavailability, and impact consequences. Currently, different 'smart' ENMs have been synthesized and investigated (Laisney et al. 2021; Marchioni et al. 2020; Shandilya and Capron 2017).

For nTiO$_2$, an ENM type that is frequently used in cosmetics (Hansen et al. 2016; Moeta et al. 2019; Vance et al. 2015; Zhang et al. 2015), a hybrid nanostructure was designed as an alternative to the currently used nTiO$_2$ (e.g. T-AVO, T-S) (Shandilya and Capron 2017; Slomberg et al. 2021). In this study, the smart nTiO$_2$, referred to as a hybrid, was synthesized by grafting nTiO$_2$ to cellulose nanocrystals (CNC). Compared to conventional nTiO$_2$ currently used in NEPs, the properties of the hybrid were intended to enhance the UV absorbance and stabilize the emulsion pickering with a lesser amount of nTiO$_2$ (Shandilya and Capron 2017). The smart nTiO$_2$ was synthesised using the sol–gel, a method that meets the requirements of safety by design (Shandilya and Capron 2017). Smart nTiO$_2$ (CNC-nTiO$_2$) exhibited a higher UV absorbance when compared to conventional nTiO$_2$ (Shandilya and Capron 2017).

Furthermore, the CNC-nTiO$_2$ was effective in replacing surfactants in product formulation and thus forming a highly stable oil-in-water pickering emulsion (Shandilya and Capron 2017). The higher efficiency of these smart ENMs meant that lower concentrations of nTiO$_2$ could be incorporated into NEPs. While this study successfully synthesized smart nTiO$_2$ and provided evidence of its improved efficiency compared to conventional nTiO$_2$ currently in use, the study failed to prove the full environmental aspects of SSbD (i.e., reduced ENMs release, toxicity, and bioavailability). Notwithstanding, the data presented here indicates that the hybridization was successful in improving the UV absorbance of nTiO$_2$. This implies that less quantity of the smart synthesized nTiO$_2$ can be incorporated into the products but still achieve the same efficiency as higher concentrations of conventional nTiO$_2$. However, further research is required for a better understanding of the interaction of the hybrid materials with the other components within the product emulsion or matrix, skin interaction, and SPF to ascertain their efficiency as alternatives for commercialization. The next step would then be to investigate the rate of release of the smart nTiO$_2$ when compared to the conventional nTiO$_2$ and subsequent ecotoxicity impact.

One of the challenges that conventional ENMs such as nAg present is their ability to undergo dissolution. The dissolution of pristine nAg (P-nAg) and the bioavailability of its ionic forms is well documented (Batchelor-McAuley et al. 2014; Beer et al. 2012; Kittler et al. 2010; Ma et al. 2012). Drawing from the P-nAg knowledge, the challenge of nAg dissolution used in biocidal products was addressed; nAg ions often cause the adverse effects induced by nAg (Van Aerle et al. 2013). Marchioni et al. (2020) synthesized 'smart' nAg by connecting nAg with a tri-thiol molecule to form larger and more stable nAg. The diameter of the smart nAg ranged between 40 and 200 nm, with the number of ENMs in each assembly ranging from 5 to 60 (Marchioni et al. 2020). Overall, the 'smart' nAg physicochemical properties were indicated to limit the release of nAg because of tight bondage with an organic coating that ensured minimal nAg dissolution (Marchioni et al. 2020). The 'smart' nAg design controlled the dissolution rate during NEPs use compared to conventional nAg (Marchioni et al. 2020). As evidenced in this study, the tri-thiol molecules protected the nAg from surface modification, thereby rendering the nAg assemblies more robust against bacteria (*Escherichia coli* and *Bacillus subtilis*). This means that this smart nAg efficiently delivered the results with a decreased risk of dissolution and thus the risk to the environment. Future studies should also investigate the period of degradation of these assemblies and the potential impact over time. Furthermore, this method needs to be assessed on another array of NEPs to establish the possible standardization of the NEPs created for biocidal activity.

Elsewhere, Laisney et al. (2021) controlled the reduction in ENM's environmental exposure by reducing the amount of ENMs incorporated in NEPs. In this study, the ENMs were coated with ligands [dopamine and polyacrylic acid (PAA)] during NEPs formulation (Laisney et al. 2021). The synthesized 'smart' TiO$_2$-PAA (nTiO$_2$-Dopa and n TiO$_2$-PAA) were found to adhere strongly to the NEPs matrix and thus prevent or limit the release and reduce environmental exposure and by extension, impacts (Laisney et al. 2021). To assess the SSbD concept, TiO$_2$-PAA was added to the paint at a concentration of 3.5% (w/w), and the paint aged for a year in the dark and interior light conditions. The photocatalytic activity was the main endpoint of this study and was investigated at 50 μg/mL (Laisney et al. 2021). Smart ENMs photocatalytic activity increased by 20% compared to the conventional nTiO$_2$ (Laisney et al. 2021). An increase in photocatalytic activity of 'smart' ENMs provided means to reduce the amount of ENMs added into NEPs; an exercise that maintains the ENMs' efficiency in NEPs and simultaneously limits environmental exposure (Laisney et al. 2021). While it is evidenced that the photocatalytic activity is significantly enhanced at low ENM concentrations due to the ligand coating, robust environmental investigations need to be conducted to ascertain the risk or non-existence thereof to the environment. This can be achieved through weathering studies using the paint incorporated with smart ENMs and subsequently, ecotoxicity studies of both the smart ENMs before incorporation into the product and the released smart ENMs from the weathering studies; the comparison of the latter and the former will shed the light into the durability of the ligand during the NEP's lifecycle.

It is worth appreciating that nanotechnology, and by extension nanotoxicology, is still developing, and thus, many data gaps still need to be filled. Therefore, the work reviewed herein provides initial evidence of ENMs' alteration to improve their safety. Specifically, evidence shows that if the concentration or volume of specific ENMs required for a specific activity can be established beforehand, it will enable manufacturers to produce NEPs with a more targeted amount of ENMs and thus a lower amount of ENMs per unit product. This will imply environmental exposure and risk reduction.

18.2.2 Reducing the Release of Engineered Nanomaterials from the Product Matrix

Although ENMs improve the efficiency of NEPs, one of the main challenges is that the ENMs are not permanently fixed to the NEPs matrix and are released throughout the NEPs life cycle (Azimzada et al. 2020a, b; Kaegi et al. 2010; Limpiteeprakan et al. 2016; Wong et al. 2020). Controlling the ENM concentrations, size, and rate of release may decrease the environmental exposure and subsequent ecotoxicity (Hwang et al. 2018; Lynch et al. 2014). Methods to limit environmental exposure across the life cycle have been assessed.

Reduction of the release of ENMs from NEP was investigated by depositing nAg onto the glass surface using a liquid flame spray (Brobbey et al. 2018). This was followed by depositing a thin layer of Al$_2$O$_3$ coating (2 and 15 nm) to improve the nanoparticle adhesion to the glass surface while immobilizing nAg to the glass surface. To assess the degree of reduction of leaching reduction by the surface coating, the rate of leaching of nAg from the surface was continuously measured over 48 h. The results indicated that the thin layer of Al$_2$O$_3$ was able to control the release of the nAg from the glass surface and reduce the initial leaching (Brobbey et al. 2018).

The thin layer (2 nm) of the coating material inhibited the release of nAg for 4 h while the thicker layer (15 nm) inhibited release for 48 h (Brobbey et al. 2018).

Although in this study, nAg was not coated before use but deposited on the NEP (glass surface) before being coated, the results show that nAg surface modification can help in reducing the leaching of ENMs into the environment. Surface coating of ENMs must be conducted with a material that will not be degraded by other materials within the product matrix. Furthermore, the durability of the specific coating material should be investigated over time. The thickness of the coating material has also been evidenced to play a role in the leaching rate and thus must be considered; the coating layer should be thick enough to decrease the release of ENMs but thin enough not to hinder the specific properties of the specific ENMs that enhance the NEP concerned.

18.2.3 Reducing Engineered Nanomaterials' Toxicity

Different types of ENMs are known to induce varying extent of adverse effects on aquatic organisms (Jahan et al. 2017; Künniger et al. 2014; Sendra et al. 2017; Spisni et al. 2016). Since ENMs' environmental release at varying amounts is invertible, efforts to lessen the effects and reduce the risk of the released ENMs have been explored. In the study of Xia et al. (2011), the toxicity effect studies showed that smart nZnO induced lesser effect compared to conventional nZnO. The smart nZnO (Fe-doped nZnO) was synthesized by flame spray pyrolysis doping and had a size range of 8.3–15 nm, while conventional ENMs were sized at 20.2 nm. The high crystallinity of nZnO was not affected by the smart nZnO synthesis procedure (Xia et al. 2011). The ecotoxicological assay was conducted by exposing the zebrafish embryo ($D.$ $rerio$) to conventional and Fe-nZnO at 5–120 h post-fertilization (hpf). Conventional nZnO interfered with embryo hatching but did not directly affect viability (Xia et al. 2011). Contrary, at maximum exposure concentration (5 $\mu g/mL$), the Fe-nZnO did not interfere with the hatching rate of the $D.$ $rerio$ embryo. Instead, the hatching rate increased with increasing amounts of Fe doping and reached near-normal hatching at 10 wt% doping level when compared to the control (without any particles) (Xia et al. 2011). This study showed that Fe–nZnO could significantly reduce the toxicity of nZnO to zebrafish. The toxicity reduction was owed to the Fe dopant's ability to reduce nZnO dissolution. In this study, the changes in ENMs' physicochemical properties reduced the potential toxicity. Future studies, therefore, need to investigate the extent to which doping can reduce the toxicity of nZnO at different organizational levels, i.e., microbial, invertebrate, and vertebrate levels. Doping has been proven to reduce dissolution which several studies attribute the toxicity to. Other studies therefore need to focus on other mechanisms of toxicity of various ENMs, especially the ENMs that do not undergo dissolution, and how doping can assist in reducing the toxicity of these ENMs.

Elsewhere, smart SiO_2 with lower toxicity potential, intended to be used for in $vitro$ biological testing, was synthesised (Jiménez et al. 2020). To reduce the toxicity of conventional quantum dots doped $nSiO_2$ ($CdSe-nSiO_2$) typically used in in $vitro$

biological testing as tracers, the toxic CdSe were substituted with organic pigment (dye) (Jiménez et al. 2020). Smart $nSiO_2$ induced reduced toxicity on the fish cell lines when compared to conventional $nSiO_2$; findings that indicates reduced ENMs risk (Jiménez et al. 2020). The life cycle assessment (LCA) analysis also showed that smart $nSiO_2$ posed negligible risk due to the reduced waste production because of the lesser amounts of water required during the $nSiO_2$ synthesis. The toxicity bioassays (*in vitro*, ROS) showed reduced toxicity in different freshwater organisms (Jiménez et al. 2020).

While the first study introduces the superiority and benefits of doping, the second study highlights the importance of the doping material. The more toxic dopant (CdSe) was substituted with the organic dye. The results show reduced toxicity and environmental risk throughout the lifecycle of the ENMs. Firstly, the doping process with more organic material produces less waste. Secondly, it induces comparably lower toxicity than CdSe doping. Therefore, Future studies need to not only dope materials but also assess the different dopants for benefit/risk analysis purposes. Secondly, there is also a need to use more organic and environmentally friendly materials for the doping process.

18.2.4   Challenges with Safer and Sustainable-by-Design Strategies Implementation at the Manufacturer Level

The use of ENMs and commercial penetration of NEPs is rapidly increasing worldwide, and proportional environmental exposure is expected. To reduce environmental exposure and subsequent effects, SSbD concepts must be explored and implemented in the early stages of production (i.e. manufacturing stage) (Cobaleda-Siles et al. 2017). SSbD in the initial stages enables the design of smart ENMs and incorporation into NEPs. Due to the infancy of the field (SSbD concept), fewer studies have investigated the concept; investigations have focused on the synthesis of smart ENMs as opposed to reducing adverse effects and reducing environmental exposure.

Although the synthesis of smart ENMs has been the main focus of the SSbD concept, one study (Laisney et al. 2021) incorporated the smart ENMs into the product to understand the holistic impact of such ENMs. In all other cases, the concepts of SSbD were assessed without incorporating smart ENMs into products, a limitation that can provide wrong prediction, as the matrix of NEPs is not accounted for; the matrix has previously been shown to contribute to the overall toxicity of the ENMs/NEPs (Reed et al. 2016; Schiavo et al. 2018; Wong et al. 2020). Therefore, smart ENMs must be analysed within the product matrix to understand the full exposure potential and impact of such ENMs.

Other challenges arising within the SSbD concepts are that manufacturers of NEPs depend on the readily available conventional ENMs. This significantly reduces the cost of production. However, it also means that the nanotoxicologists/risk assessors are not involved anywhere in the manufacturing processes of the NEPs. Only a few manufacturers have been reported to own the entire manufacturing chain and therefore can innovate to create smart ENMs that take into consideration human and environmental health (Jiménez et al. 2020). Although the current literature pool is small, it is apparent that there is an increasing interest in the concept of SSbD, specifically in synthesizing smart ENMs. However, most studies have not assessed the environmental exposure and subsequent toxicity of the released smart ENMs. Based on the successful synthesis of the smart ENMs, conclusions about environmental exposure and subsequent toxicity were assumed based on the physicochemical properties of the smart ENMs and not experimental data. While the physicochemical property data of ENMs can predict their behaviour, fate, and effects, in the case where NEP matrix is involved, the prediction is not clearcut as the ENMs interact with the NEPs matrix. The interaction of the ENMs with the NEPs matrix may alter ENMs' physicochemical properties.

The implementation of SSbD strategies depends solely on the information on the ENMs' detrimental physicochemical properties, the environmental release, and their toxicity. Currently, there is data paucity in all three areas, making it challenging to assess the SSbD competence. While information is available and a few studies have successfully implemented SSbD, the release, and ecotoxicity data are still crucial in informing the full implementation of SSbD.

Gaining extensive knowledge about ENMs released from NEPs and the toxicity of product released ENMs remains central to understanding the application of SSbD in any industry. However, this remains a great challenge due to the paucity of data that exists in ENM release and PR-ENMs toxicity. Thus, SSbD is still only a theoretical concept. Nevertheless, it holds great promise for the design of ENMs/NEPs with less potential detriment to the environment. The synthesis of smart ENMs and their incorporation into NEPs will ensure reduced environmental exposure. This will further lead to reduced toxicity. The case studies reviewed above show that, given the limited knowledge available, SSbD of NEPs should be prioritized while ENM release and toxicity data are concurrently being made available.

To ensure the exploration and uptake of SSbD, regulatory bodies may develop technical guidelines and policies for industries. These guidelines may include information on methods for SSbD implementation and the allowable concentration limits allowed for NEP formulations. The policies may further require the data for the environmental safety of the SSbD-based ENMs/NEPs.

18.3   Policies for Nanotechnology Environmental Safety

The current environmental exposure and risks of nanotechnology are uncertain. The development and implementation of policies as risk mitigation for nanotechnology relies heavily on the successful generation of robust data on environmental exposure and risk. Currently, there are huge gaps in the relevant scientific knowledge of the exposure and risks of ENMs and NEPs. Notwithstanding, several countries and/

or corporations have drafted and are implementing different policies to proactively address the imminent risks of nanotechnology.

In 2014, the U.S. Food and Drug Administration (FDA) published the guidance for industry, which was not meant as a regulatory policy document, to guide the industry in addressing the potential risks of nano-based products under FDA regulation (FDA 2014). This guidance would allow the FDA to assess the products' safety, effectiveness, public health impact, and regulatory status based on the materials' size (as per the provision of the U.S. National Nanotechnology Initiative) (NNI 2023), its associated physicochemical properties and behaviour (FDA 2014). Although this was just a guidance on the FDA's thinking process on regulating nano-based products under its jurisdiction, the official regulatory frameworks were subsequently published. The first was the regulation on the safety of nanomaterials in cosmetics which aimed to guide the use of nanomaterials in nano-enabled cosmetic products. According to this framework, manufacturers are expected to provide information on the safety of their products following the traditional safety assessment of other cosmetic products (FDA 2014). However, the FDA regulation does not require the pre-market registration of nano-enabled products. According to the regulation, nano-enabled cosmetic product manufacturers are just expected to not misbrand or adulterate their products (FDA 2014). Although the regulation seems flexible and unenforceable, the FDA does require that the manufacturers conduct full characterization of the products/nanomaterials used (physicochemical properties) and provide a complete toxicological profile (FDA 2014). Other regulatory frameworks include the use of nanomaterials in food for animals (FDA 2015), and most recently the guidance on drug and biological products that contain nanomaterials (FDA 2022). This work in the U.S.A is also supported by the U.S. Environmental Protection Agency (EPA)'s ongoing Research on Nanomaterials (U.S. Environmental Protection Agency (EPA) 2023a, b), which has also informed the Control of Nanoscale Materials under the Toxic Substances Control Act (U.S. Environmental Protection Agency (EPA) 2023a, b).

The European Commission published the regulatory framework (2018/1881, EC No 1907/2006 amendment) for 'nanoforms' materials through its Registration, Evaluation, Authorisation and Restriction of Chemicals (REACH) program (Juncker 2018). The purpose of the regulation was to provide guidelines on how manufacturers and importers and nano-based materials/products need to assess and document how the risks of the materials/products are controlled during manufacturing, usage, and throughout the supply chain (Juncker 2018). According to this regulation, the manufacturers/importers need to create a chemical safety document that provides complete information on characterization of the nanoform materials (size, shape, type, surface characteristics), toxicological and ecotoxicological profile, water solubility, and dustiness, among others (Juncker 2018). The regulation requires a full screening of the nano-based materials' health and environmental safety. Through the registration of these nano-based materials/products, the EU would be able to keep track of the market penetration and address safety concerns of nano-based products throughout manufacturing (occupational), usage (absorption/adsorption and environmental release), and disposal (end of life). The European Chemicals Agency

(ECHA) has also published a technical guideline that outlines the technical details of this regulation, to guide compliance (ECHA 2019).

The EU, through the Scientific Committee on Consumer Safety (SCCS), further introduced the Guidance on the Safety Assessment of Nanomaterials in Cosmetics which seeks to guide the safety of ENMs used in cosmetic products. In their framework, SCCS addresses the safety assessment of ENMs, their physicochemical characterization, exposure assessment, hazard identification, and dose–response characterization, and risk assessment (SCCS 2023). Through these pillars, SCCS guides the use of ENMs in cosmetic products; guidance is provided on the ENMs types, physicochemical properties, and safety measures to reduce environmental release and risk (SCCS 2023). Additionally, SCCS also guides the concentrations of ENMs that can be used in NEPs to ascertain their environmental and human health safety (SCCS 2019). The work done by the EU is currently underway, especially in the European region, to influence safer implementation of nanotechnology and nano-based products.

On the African continent, South Africa introduced the National Nanotechnology Strategy in 2007 which aimed to influence the uptake of nanotechnology to support industrial and economic development. Specifically, the strategy established R&D hubs with specific objectives to develop nanomaterials and nano-based devices for application purposes (Musee et al. 2010). Although the strategy does highlight that nanotechnology should be applied according to environmental safety standards (DST 2006), no provision was made on guiding safety and risk assessment, environmental exposure, and toxicity assessment. To this end, the Department of Science and Innovation established the Nanotechnology HSE Research Platform in 2015 to generate comprehensive data on safety and environmental risks of engineered nanomaterials in South Africa (Gulumian et al. 2023). Under this platform, the risk assessment of ENMs is conducted at production (ENMs synthesis), occupational (ENMs/NEPs manufacturing), usage, and end-of-life of ENMs and NEPs stages. This platform will inform and support regulation, decision-making, and successful implementation of the nanotechnology industry (Gulumian et al. 2023). Although this platform has generated some data (Project 0085/2015), experimental investigations are ongoing, and no policies or regulations have been effected; policies and regulations in other African countries are non-existent.

On the international network, the Organisation for Economic Cooperation and Development (OECD) has been working on generating data, compiling test guidelines and legislative guidelines for manufacturing (occupational), usage (exposure and hazard), and disposal (end-of-life) of ENMs for over a decade (OECD 2023). In one of their earliest reports, the OECD Working Party on Manufactured Nanomaterials (WPMN) indicated that there was a need for ENMs/NEPs regulation due to the increasing commercialisation and marketing of ENMs (OECD 2011). This was based on the Questionnaire on Regulated Nanomaterials: 2006–2009, the results of which were reported in 2010 (OECD 2011). Accordingly, legislative provisions such as consideration of ENMs safety, explicit labelling of 'nano' for products that incorporated ENMs, and pre-market notification (6 months) to the European Commission for nano-based cosmetic products (OECD 2011). In 2012, the OECD also published a framework for the risk assessment of manufactured nanomaterials (OECD 2012). Specifically for environmental risk assessment, this framework recommends the consideration of ENMs' behaviour (dissolution, agglomeration/aggregation, adsorption), persistence and degradation, distribution, transformation products, and bioaccumulation in the environmental matrices (OECD 2012). For water specifically, the water solubility of the ENMs is crucial (OECD 2012). Although this risk assessment framework has been adopted in different ways in various OECD countries, three components remain consistent for ENMs risk assessment: physicochemical properties of the ENMs, their toxicity and ecotoxicological profiles (OECD 2012). More than the ENMs in their pristine forms, this framework also highlights the research needs in the understanding of the concentrations of ENMs used in the NEPs, and their release into the environment (OECD 2012). Over the years, OECD has completed several projects and proposed several methodologies and test guidelines (OECD 2023) to address the objectives of these frameworks. Unsurprisingly, countries such as Japan, the USA, Chile, and Canada have developed their regulatory frameworks based on OECD recommendations (Allan et al. 2021).

The European Commission is deliberate in ensuring the regulation of nanotechnology, specifically for environmental safety. This includes the water resources, which are the major recipients of environmental contaminants. Although organizations such as the FDA (U.S.A.) and other countries are investing in the regulation of this industry, the EU is more deliberate in the implementation and ensuring compliance with the regulatory requirements. The FDA has so far only focused on nano-based products that are within their regulatory jurisdictions. Therefore, much more work is required to ensure that nanotechnology innovations do not compromise water resources. Data shows that even within the EU, the PR-ENMs in water resources are increasing; this could be biased due to a lack of data in other countries. The existing data in the EU can therefore be used as a baseline for influencing further research and policy development in OECD countries and beyond. It is worth noting that the current uptake of nanotechnology, globally, is increasing rapidly. While research data is necessary, policy development, implementation and compliance can be the fastest tool available to curb the potential risks of ENMs to the aquatic environment.

18.4 Conclusions and Recommendations

The environmental exposure and impact of ENMs/PR-ENMs are evident. The SSbD approach is therefore critical to ensuring that the increase in NEPs in the consumer market and their use do not compromise the environment. However, there is a paucity of data in critical areas needed to implement the SSbD strategies. The conducted studies indicate the possibility of a major shift to creating safer nanomaterials. Nonetheless, more data still need to be generated to validate these different methods. Moreover, more resources and access to research are needed at the industry/manufacturing level; manufacturers need to invest in research for the development of safer NEPs through safer ENM development. There is a need for the involvement of nanotoxicologists/risk assessors through the lifecycle of product development. Industries with limited access to such resources may depend on other partners or academic institutions for this research and development aspect. As a proposed concept and minute data pool currently available, only the advantages of SSbD can be presented with certainty. It can only be assumed that there will be reluctance on the uptake by the industry due to the potentially excessive costs of reconstructing their current processes.

At the regulatory level, there is some movement in research-based policy development. The EU has developed clear policies although the implementation still lags. This holds for other institutions such as the FDA. It is worth appreciating the work being done by the OECD in guidelines and test development for comprehensive risk assessment. However, this highlights the current challenge; there are only guidelines and recommendations put forward with the policy development and implementation at different countries' levels. Due to the import/export nature of nano-based products, there necessitate an international framework and/or treaties to guide policy implementation at an international level.

Overall, the aquatic environmental risk assessment of PR-ENMs and mitigation measures should be at the forefront of nanotechnology development. SSbD and policy development, as detailed herein, are the most effective and industrially relevant strategies to balance the uptake of nanotechnology and environmental safety.

References

Abbas Q, Yousaf B, Ullah H, Ali MU, Ok YS, Rinklebe J (2020) Environmental transformation and nano-toxicity of engineered nanoparticles (ENPs) in aquatic and terrestrial organisms. Crit Rev Environ Sci Technol 50(23):2523–2581. https://doi.org/10.1080/10643389.2019.1705721

Allan J, Belz S, Hoeveler A, Hugas M, Okuda H, Patri A, Rauscher H, Silva P, Slikker W, Sokull-Kluettgen B, Tong W, Anklam E (2021) Regulatory landscape of nanotechnology and nanoplastics from a global perspective. Regul Toxicol Pharmacol 122:104885. https://doi.org/10.1016/j.yrtph.2021.104885

Aydın H, Yakuphanoglu F, Aydın C (2019) Al-doped ZnO as a multifunctional nanomaterial: structural, morphological, optical, and low-temperature gas sensing properties. J Alloy Compd 773:802–811. https://doi.org/10.1016/j.jallcom.2018.09.327

Azimzada A, Farner JM, Hadioui M, Liu-Kang C, Jreije I, Tufenkji N, Wilkinson KJ (2020a) Release of TiO2 nanoparticles from painted surfaces in cold climates: characterization using a high sensitivity single-particle ICP-MS. Environ Sci Nano 7(1):139–148. https://doi.org/10.1039/c9en00951e

Azimzada A, Farner JM, Jreije I, Hadioui M, Liu-Kang C, Tufenkji N, Shaw P, Wilkinson KJ (2020b) Single- and multi-element quantification and characterization of TiO2 nanoparticles released from outdoor stains and paints. Front Environ Sci 8(June):1–13. https://doi.org/10.3389/fenvs.2020.00091

Batchelor-McAuley C, Tschulik K, Neumann CCM, Laborda E, Compton RG (2014) Why are silver nanoparticles more toxic than bulk silver? Towards understanding the dissolution and toxicity of silver nanoparticles. Int J Electrochem Sci 9(3):1132–1138

Bayan EM, Lupeiko TG, Pustovaya LE, Volkova MG, Butova VV, Guda AA (2020) Zn–F co-doped TiO2 nanomaterials: synthesis, structure, and photocatalytic activity. J Alloy Compd 822:153662. https://doi.org/10.1016/j.jallcom.2020.153662

Beer C, Foldbjerg R, Hayashi Y, Sutherland DS, Autrup H (2012) Toxicity of silver nanoparticles-nanoparticle or silver ion? Toxicol Lett 208(3):286–292. https://doi.org/10.1016/j.toxlet.2011.11.002

Brobbey KJ, Haapanen J, Gunell M, Toivakka M, Mäkelä JM, Eerola E, Ali R, Saleem MR, Honkanen S, Bobacka J, Saarinen JJ (2018) Controlled time release and leaching of silver nanoparticles using a thin immobilizing layer of aluminum oxide. Thin Solid Films 645:166–172. https://doi.org/10.1016/j.tsf.2017.09.060

Cobaleda-Siles M, Guillamon AP, Delpivo C, Vázquez-Campos S, Puntes VF (2017) Safer by design strategies. J Phys: Conf Ser 838(1). https://doi.org/10.1088/1742-6596/838/1/012016

DST (2006) The national nanotechnology strategy. Department of Science and Technology

ECHA (2019) Appendix for nanoforms applicable to the Guidance on Registration and Substance Identification (Draft (Public) Version 1.0). European Chemicals Agency

FDA (2014) Guidance for industry: safety of nanomaterials in cosmetic products, pp 1–16 [guidance document]. U.S. Department of Health and Human Services, Food and Drug Administration

FDA (2015) Guidance for industry: use of nanomaterials in food for animals (guidance document 220), pp 1–11. U.S. Department of Health and Human Services, Food and Drug Administration

FDA (2022) Guidance for industry: drug products, including biological products, that contain nano-materials (19640025 FNL), pp 1–33. U.S. Department of Health and Human Services, Food and Drug Administration

Gagné F, Auclair J, Turcotte P, Gagnon C, Peyrot C, Wilkinson K (2019) The influence of surface waters on the bioavailability and toxicity of zinc oxide nanoparticles in freshwater mussels. Comp Biochem Physiol Part - C: Toxicol Pharmacol 219:1–11. https://doi.org/10.1016/j.cbpc.2019.01.005

Gagnon V, Button M, Boparai HK, Nearing M, O'Carroll DM, Weber KP (2019) Influence of realistic wearing on the morphology and release of silver nanomaterials from textiles. Environ Sci Nano 6(2):411–424. https://doi.org/10.1039/C8EN00803E

Gao Z, Yu H, Li M, Li X, Lei J, He D, Wu G, Fu Y, Chen Q, Shi H (2022) A battery of baseline toxicity bioassays directed evaluation of plastic leachates—towards the establishment of bioanalytical monitoring tools for plastics. Sci Total Environ 828:154387. https://doi.org/10.1016/j.scitotenv.2022.154387

Gondikas AP, Von Der Kammer F, Reed RB, Wagner S, Ranville JF, Hofmann T (2014) Release of TiO2 nanoparticles from sunscreens into surface waters: a one-year survey at the old danube recreational lake. Environ Sci Technol 48(10):5415–5422. https://doi.org/10.1021/es405596y

Gulumian M, Thwala M, Makhoba X, Wepener V (2023) Current situation and future prognosis of health, safety, and environment risk assessment of nanomaterials in South Africa. South Afr J Sci 119(1/2). https://doi.org/10.17159/sajs.2023/11657

Hansen SF, Hansen OFH, Nielsen MB (2020) Advances and challenges towards consumerization of nanomaterials. Nat Nanotechnol 15(12):964–965. https://doi.org/10.1038/s41565-020-00819-7

Hansen SF, Heggelund LR, Revilla Besora P, Mackevica A, Boldrin A, Baun A (2016) Nanoprod-ucts—what is actually available to European consumers? Environ Sci Nano 3(1):169–180. https://doi.org/10.1039/c5en00182j

Hansen SF, Larsen BH, Olsen SI, Baun A (2007) Categorization framework to aid hazard identifi-cation of nanomaterials. Nanotoxicology 1(3):243–250. https://doi.org/10.1080/17435390701727509

Hansen SF, Michelson ES, Kamper A, Borling P, Stuer-Lauridsen F, Baun A (2008) Categorization framework to aid exposure assessment of nanomaterials in consumer products. Ecotoxicology 17(5):438–447. https://doi.org/10.1007/s10646-008-0210-4

Hwang R, Mirshafiee V, Zhu Y, Xia T (2018) Current approaches for safer design of engi-neered nanomaterials. Ecotoxicol Environ Saf 166:294–300. https://doi.org/10.1016/j.ecoenv.2018.09.077

Jahan S, Yusoff IB, Alias YB, Bakar AFBA (2017) Reviews of the toxicity behavior of five potential engineered nanomaterials (ENMs) into the aquatic ecosystem. Toxicol Rep 4:211–220. https://doi.org/10.1016/j.toxrep.2017.04.001

Jampilek J, Kos J, Kralova K (2019) Potential of nanomaterial applications in dietary supplements and foods for special medical purposes. Nanomaterials 9(2). https://doi.org/10.3390/nano90 20296

Jiménez AS, Puelles R, Pérez-Fernández M, Gómez-Fernández P, Barruetabeña L, Jacobsen NR, Suarez-Merino B, Micheletti C, Manier N, Trouiller B, Navas JM, Kalman J, Salieri B, Hischier R, Handzhiyski Y, Apostolova, MD, Hadrup N, Bouillard J, Oudart Y, Merino C, Garcia E, Liguori B, Sabella S, Rose J, Maison A, Galea KS, Kelly S, Štěpánková S, Mouneyrac C, Barrick A, Chatel A, Dusinska M, Rundén-Pran E, Mariussen E, Bressot C, Aguerre-Chariol O, Shandilya N, Goede H, Gomez-Cordon J, Simar S, Nesslany F, Jensen KA, van Tongeren M, Rodríguez Llopis I (2020) Safe(r) by design implementation in the nanotechnology industry. NanoImpact 20. https://doi.org/10.1016/j.impact.2020.100267

Juncker J-C (2018) Commission Regulation (EU) 2018/ 1881—of 3 December 2018—Amending Regulation (EC) No 1907 / 2006 of the European Parliament and of the Council on the Registration, Evaluation, Authorisation and Restriction of Chemicals (REACH) as regards Annexes I, III,VI, VII, VIII, IX, X, XI, and XII to address nanoforms of substances (L 308) European Union

Kaegi R, Sinnet B, Zuleeg S, Hagendorfer H, Mueller E, Vonbank R, Boller M, Burkhardt M (2010) Release of silver nanoparticles from outdoor facades. Environ Pollut 158(9):2900–2905. https://doi.org/10.1016/j.envpol.2010.06.009

Kaur I, Agrawal R (2008) Nanotechnology: a new paradigm in cosmeceuticals. Recent Pat Drug Delivery Formulation 1(2):171–182. https://doi.org/10.2174/187221107780831888

Kittler S, Greulich C, Diendorf J, Köller M, Epple M (2010) Toxicity of silver nanoparticles increases during storage because of slow dissolution under release of silver ions. Chem Mater 22(16):4548–4554. https://doi.org/10.1021/cm100023p

Kraegeloh A, Suarez-Merino B, Sluijters T, Micheletti C (2018) Implementation of safe-by-design for nanomaterial development and safe innovation: Why we need a comprehensive approach. Nanomaterials 8(4). https://doi.org/10.3390/nano8040239

Künniger T, Gerecke AC, Ulrich A, Huch A, Vonbank R, Heeb M, Wichser A, Haag R, Kunz P, Faller M (2014) Release and environmental impact of silver nanoparticles and conventional organic biocides from coated wooden façades. Environ Pollut 184:464–471. https://doi.org/10.1016/j.envpol.2013.09.030

Labille J, Slomberg D, Catalano R, Robert S, Apers-Tremelo ML, Boudenne JL, Manasfi T, Radakovitch O (2020) Assessing UV filter inputs into beach waters during recreational activity: a field study of three French Mediterranean beaches from consumer survey to water analysis. Sci Total Environ 706:136010. https://doi.org/10.1016/j.scitotenv.2019.136010

Laisney J, Rosset A, Bartolomei V, Predoi D, Truffier-Boutry D, Artous S, Bergé V, Brochard G, Michaud-Soret I (2021) TiO2nanoparticles coated with bio-inspired ligands for the safer-by-design development of photocatalytic paints. Environ Sci Nano 8(1):297–310. https://doi.org/10.1039/d0en00947d

Lehutso RF, Thwala M (2021) Assessment of nanopollution in water environments from commercial products. Nanomaterials 11(10):2537. https://doi.org/10.3390/nano11102537

Lehutso RF, Wesley-Smith J, Thwala M (2021) Aquatic toxicity effects and risk assessment of 'form specific' product-released engineered nanomaterials. Int J Mol Sci 22(22). https://doi.org/10.3390/ijms222212468

Limpiteeprakan P, Babel S, Lohwacharin J, Takizawa S (2016) Release of silver nanoparticles from fabrics during the course of sequential washing. Environ Sci Pollut Res 23(22):22810–22818. https://doi.org/10.1007/s11356-016-7486-3

Lynch I, Weiss C, Valsami-Jones E (2014) A strategy for grouping of nanomaterials based on key physico-chemical descriptors as a basis for safer-by-design NMs. Nano Today 9(3):266–270. https://doi.org/10.1016/j.nantod.2014.05.001

Ma R, Levard C, Marinakos SM, Cheng Y, Liu J, Michel FM, Brown GE, Lowry GV (2012) Size-controlled dissolution of organic-coated silver nanoparticles. Environ Sci Technol 46(2):752–759. https://doi.org/10.1021/es201686j

Marchioni M, Veronesi G, Worms I, Ling WL, Gallon T, Leonard D, Gateau C, Chevallet M, Jouneau PH, Carlini L, Battocchio C, Delangle P, Michaud-Soret I, Deniaud A (2020) Safer-by-design biocides made of tri-thiol bridged silver nanoparticle assemblies. Nanoscale Horiz 5(3):507–513. https://doi.org/10.1039/c9nh00286c

Markus AA, Krystek P, Tromp PC, Parsons JR, Roex EWM, de Voogt P, Laane RWPM (2018) Determination of metal-based nanoparticles in the river Dommel in the Netherlands via ultrafil-tration, HR-ICP-MS, and SEM. Sci Total Environ 631–632:485–495. https://doi.org/10.1016/j.scitotenv.2018.03.007

Moeta PJ, Wesley-Smith J, Maity A, Thwala M (2019) Nano-enabled products in South Africa and the assessment of environmental exposure potential for engineered nanomaterials. SN Appl Sci 1(6):1–13. https://doi.org/10.1007/s42452-019-0584-3

Moloi MS, Lehutso RF, Erasmus M, Oberholster PJ, Thwala M (2021) Aquatic environment expo-sure and toxicity of engineered nanomaterials released from nano-enabled products: current status and data needs. Nanomaterials 11(11):2868. https://doi.org/10.3390/nano11112868

Moraes Silva S, Tavallaie R, Sandiford L, Tilley RD, Gooding JJ (2016) Gold coated magnetic nanoparticles: from preparation to surface modification for analytical and biomedical applica-tions. Chem Commun 52(48):7528–7540. https://doi.org/10.1039/c6cc03225g

Mu L, Sprando RL (2010) Application of nanotechnology in cosmetics. Pharm Res 27(8):1746–1749. https://doi.org/10.1007/s11095-010-0139-1

Musee N, Brent AC, Ashton PJ (2010) A South African research agenda to investigate the potential environmental, health and safety risks of nanotechnology. South Afr J Sci 106(3/4), 6pp. https://doi.org/10.4102/sajs.v106i3/4.159

Nanodatabase (2022) The Nanodatabase. https://nanodb.dk/

NNI (2023) U.S. national nanotechnology initiative . https://www.nano.gov/

OECD (2011) Series on the safety of manufactured nanomaterials—regulated nanomaterials: 2006–2009 (No. 30 (ENV/JM/MONO(2011)52); OECD Environment, Health and Safety Publications, pp 1–53. Organisation for Economic Co-operation and Development

OECD (2012) Series on the safety of manufactured nanomaterials: important issues on risk assess-ment of manufactured nanomaterials (No. 33 (ENV/JM/MONO(2012)8); OECD Environ-ment, Health and Safety Publications, pp 1–57. Organisation for Economic Co-operation and Development

OECD (2023) Publications in the series on the safety of manufactured nanomaterials. https://www.oecd.org/env/ehs/nanosafety/publications-series-safety-manufactured-nanomaterials.htm

Pathak TK, Coetsee-Hugo E, Swart HC, Swart CW, Kroon RE (2020) Preparation and character-ization of Ce doped ZnO nanomaterial for photocatalytic and biological applications. Mater Sci Eng b: Solid-State Mater Adv Technol 261:114780. https://doi.org/10.1016/j.mseb.2020.114780

Pathak TK, Kroon RE, Swart HC (2018) Photocatalytic and biological applications of Ag and Au doped ZnO nanomaterial synthesized by combustion. Vacuum 157:508–513. https://doi.org/10.1016/j.vacuum.2018.09.020

Reed RB, Martin DP, Bednar AJ, Montaño MD, Westerhoff P, Ranville JF (2017) Multi-day diurnal measurements of Ti-containing nanoparticle and organic sunscreen chemical release during recreational use of a natural surface water. Environ Sci Nano 4(1):69–77. https://doi.org/10.1039/c6en00283h

Reed RB, Zaikova T, Barber A, Simonich M, Lankone R, Marco M, Hristovski K, Herckes P, Passantino L, Fairbrother DH, Tanguay R, Ranville JF, Hutchison JE, Westerhoff PK (2016) Potential environmental impacts and antimicrobial efficacy of silver- and nanosilver-containing textiles. Environ Sci Technol 50(7):4018–4026. https://doi.org/10.1021/acs.est.5b06043

Rose J, Auffan M, de Garidel-Thoron C, Artous S, Auplat C, Brochard G, Capron I, Carriere M, Cathala B, Charlet L, Clavaguera S, Heulin T, Labille J, Orsiere T, Peyron S, Rabilloud

T, Santaella C, Truffier-Boutry D, Wortham H, Masion A (2021) The SERENADE project; a step forward in the safe by design process of nanomaterials: the benefits of a diverse and interdisciplinary approach. Nano Today 37:101065. https://doi.org/10.1016/j.nantod.2020. 101065

SCCS (2019) Guidance on the safety assessment of nanomaterials in cosmetics. (Scientific Committee on Consumer Safety SCCS/1611/19). European Commission Directorate General for Health and Food Safety. https://data.europa.eu/doi/10.2875/40446

SCCS (2023) Guidance on the safety assessment of nanomaterials in cosmetics (2nd Revision) (SCCS/1655/23). European Commission Directorate General for Health and Food Safety. https:// data.europa.eu/

Schiavo S, Oliviero M, Philippe A, Manzo S (2018) Nanoparticles based sunscreens provoke adverse effects on marine microalgae *Dunaliella tertiolecta*. Environ Sci Nano 5(12):3011–3022. https:// doi.org/10.1039/c8en01182f

Schwarz-Plaschg C, Kallhoff A, Eisenberger I (2017) Making nanomaterials safer by design? NanoEthics 11(3):277–281. https://doi.org/10.1007/s11569-017-0307-4

Sendra M, Sánchez-Quiles D, Blasco J, Moreno-Garrido I, Lubián LM, Pérez-García S, Tovar-Sánchez A (2017) Effects of TiO2nanoparticles and sunscreens on coastal marine microalgae: Ultraviolet radiation is key variable for toxicity assessment. Environ Int 98:62–68. https://doi. org/10.1016/j.envint.2016.09.024

Shandilya N, Capron I (2017) Safer-by-design hybrid nanostructures: an alternative to conventional titanium dioxide UV filters in skin care products. RSC Adv 7(33):20430–20439. https://doi.org/ 10.1039/c7ra02506h

Slomberg DL, Catalano R, Bartolomei V, Labille J (2021) Release and fate of nanoparticulate TiO2 UV filters from sunscreen: effects of particle coating and formulation type. Environ Pollut 271:116263. https://doi.org/10.1016/j.envpol.2020.116263

Spisni E, Seo S, Joo SH, Su C (2016) Release and toxicity comparison between industrial- and sunscreen-derived nano-ZnO particles. Int J Environ Sci Technol 13(10):2485–2494. https:// doi.org/10.1007/s13762-016-1077-1

StatNano (2023) StatNano - Nanotechnology Products Database. https://statnano.com/

Surette MC, Nason JA, Kaegi R (2019) The influence of surface coating functionality on the aging of nanoparticles in wastewater. Environ Sci Nano 6(8):2470–2483. https://doi.org/10.1039/c9e n00376b

U.S. Environmental Protection Agency (EPA) (2023) Control of nanoscale materials under the toxic substances control act. https://www.epa.gov/reviewing-new-chemicals-under-toxic-substances-control-act-tsca/control-nanoscale-materials-under

U.S. Environmental Protection Agency (EPA) (2023) U.S. EPA's research on nanomaterials. https:// www.epa.gov/chemical-research/research-nanomaterials

Van Aerle R, Lange A, Moorhouse A, Paszkiewicz K, Ball K, Johnston BD, De-Bastos E, Booth T, Tyler CR, Santos EM (2013) Molecular mechanisms of toxicity of silver nanoparticles in zebrafish embryos. Environ Sci Technol 47(14):8005–8014. https://doi.org/10.1021/es401758d

Vance ME, Kuiken T, Vejerano EP, McGinnis SP, Hochella MF, Hull DR (2015) Nanotechnology in the real world: Redeveloping the nanomaterial consumer products inventory. Beilstein J Nanotechnol 6(1):1769–1780. https://doi.org/10.3762/bjnano.6.181

Wong SWY, Zhou GJ, Leung PTY, Han J, Lee JS, Kwok KWH, Leung KMY (2020) Sunscreens containing zinc oxide nanoparticles can trigger oxidative stress and toxicity to the marine copepod *Tigriopus japonicus*. Mar Pollut Bull 154:111078. https://doi.org/10.1016/j.marpol bul.2020.111078

Wu F, Harper BJ, Harper SL (2019) Comparative dissolution, uptake, and toxicity of zinc oxide particles in individual aquatic species and mixed populations. Environ Toxicol Chem 38(3):591–602. https://doi.org/10.1002/etc.4349

Xia T, Zhao Y, Sager T, George S, Pokhrel S, Li N, Schoenfeld D, Meng H, Lin S, Wang X, Wang M, Ji Z, Zink JI, Mädler L, Castranova V, Lin S, Nel AE (2011) Decreased dissolution of ZnO by iron doping yields nanoparticles with reduced toxicity in the rodent lung and zebrafish embryos. ACS Nano 5(2):1223–1235. https://doi.org/10.1021/nn1028482

Zhang Y, Leu YR, Aitken RJ, Riediker M (2015) Inventory of engineered nanoparticle-containing consumer products available in the Singapore retail market and likelihood of release into the aquatic environment. Int J Environ Res Public Health 12(8):8717–8743. https://doi.org/10.3390/ijerph120808717

Chapter 19
Direct Potable Reuse: A Prioritization of Emerging Contaminants for Monitoring Strategies and Pilot-Scale Advanced Treatment

Vinicius Diniz, Jarbas José Rodrigues Rohwedder, and Susanne Rath

Abstract Direct potable reuse (DPR) has emerged as a promising and practical solution to address the challenges of water scarcity (*e.g.*, climate dependency, limited space availability, water waste, and financial constraints) by treating wastewater to meet stringent drinking water quality standards given its potential to reduce water age within the distribution network, enhance water quality, and yield energy savings. Despite the potential benefits, the presence of emerging contaminants in treated wastewater has raised concerns among researchers and policymakers due to their unknown long-term toxicological risks, necessitating continuous research and rigorous monitoring to ensure the safety and sustainability of DPR initiatives. In the present work, ten emerging contaminants were assessed at a pilot-scale treatment plant in the city of Campinas, Brazil, previously selected based on consumption and previous monitoring data. The pilot-scale plant allowed for the evaluation of various treatment configurations, ultimately selecting reverse osmosis + photoperoxidation (UV/H_2O_2) + activated carbon as the most effective scheme for DPR. The effluent from the pilot plant met the guidelines for potable water set by Brazilian regulations. Among the monitored emerging contaminants, albendazole, carbamazepine, hydrochlorothiazide, sulfamethoxazole, and sucralose were identified as marker compounds for monitoring purposes, as they were detectable in both the influent and effluent of the pilot plant. The findings of this research contribute to the development of robust strategies for monitoring and mitigating the risks associated with emerging contaminants in DPR, ensuring the production of safe and sustainable drinking water sources.

Keywords Marker compounds · Potable reuse · Prioritization of emerging contaminants · Water purification · Water scarcity

V. Diniz (✉) · J. J. R. Rohwedder · S. Rath
Department of Analytical Chemistry, Institute of Chemistry, University of Campinas, Campinas, São Paulo, Brazil
e-mail: viniciusdiniz994@gmail.com

© UNESCO 2025
S. Zandaryaa et al. (eds.), *Emerging Pollutants*, Advances in Water Security,
https://doi.org/10.1007/978-3-031-71758-1_19

19.1   Introduction

Direct potable reuse (DPR) has emerged as an innovative and promising water management strategy, offering a potential solution to the pressing global water scarcity crisis (Soller et al. 2018; Liu et al. 2020).[1] By treating wastewater to a level that meets or exceeds potable water standards and subsequently distributing it directly into the drinking water supply, DPR closes the water loop and maximizes the utilization of water resources, providing a reliable and sustainable water supply for communities (Reddy et al. 2023).

The World Health Organization (WHO) has categorized potable reuse into three types: (i) unplanned potable reuse, which involves discharging treated or untreated wastewater into rivers, followed by downstream communities using the same water body as a drinking water source, without prior planning; (ii) indirect potable reuse, where treated wastewater is deliberately mixed with environmental buffers such as rivers, lakes, reservoirs, or aquifers, under controlled conditions, before being utilized as a drinking water source; and (iii) DPR, which entails directly introducing treated wastewater into a drinking water supply system, without prior discharge to an environmental buffer (WHO 2017).

While DPR holds great promise, its implementation poses challenges and costs due to the complex composition of municipal wastewaters, which contain recalcitrant compounds and diverse microbial communities (WHO 2017). In recent years, there have been increasing efforts to develop safe, robust, and cost-effective processes for DPR (Liu et al. 2020).

The WHO's Guidance for Producing Safe Drinking Water underscores the importance of reliability, redundancy, robustness, and resilience (the 4Rs criteria) in DPR projects (WHO 2017). Achieving these criteria requires defining quality indicators, including emerging contaminants, and employing multi-barrier treatment approaches (WHO 2017).

The multi-barrier approach has become the preffered configuration used in potable reuse treatment systems worldwide (Jeffrey et al. 2022; WHO 2017). The most commonly used multi-barrier process combines membrane filtration, oxidation, and media filtration (Table 19.1). In addition to ensuring the effective removal of a range of contaminants, the combination of these processes gives greater reliability to treatment systems by allowing for adaptability, security, and time to address potential process failures.

However, as the practice of DPR evolves, concerns have arisen regarding the presence and fate of emerging contaminants in treated wastewater (Wallmann et al. 2021; Villarin and Merel 2020). Emerging contaminants are a diverse range of chemical and biological substances that are not typically monitored or regulated in the drinking water industry, but have the potential to pose risks to human health and the environment (Puri et al. 2023; Mukhopadhyay et al. 2022). The analysis of sewage samples has revealed a staggering array of emerging contaminants, presenting a complex

---

[1] Brazil's Health Authority known as, Agência Nacional de Vigilância Sanitária (ANVISA): Portaria GM/MS N° 888, de 4 de maio de 2021.

Table 19.1 DPR installations worldwide

Project name and country	Start date and status	Capacity (m^3/day)	Unit process technology
Old Goreangab plant, Namibia	1969–2002/replaced	7,500	DAF, rapid sand filtration, GAC, Cl_2, blending (30–35%)
New Goreangab plant, Namibia	2002/operational	21,000	PAC, O_3, DAF, rapid sand filtration, O_3, BAC, GAC, UF and Cl_2
Beaufort West, South Africa	2011/built	1,000	Sand filtration, UF, RO, UV/H_2O_2, Cl_2
Big Spring, Texas, USA	2013/operational	7,000	MF, RO, UV/H_2O_2, DWTP
Wichita Falls, Texas, USA	2014–15/operational	26,000	MF, RO, UV, storage, DWTP
San Diego advanced water purification facility, California, USA	Under construction	68,000	MF, RO, UV/AOP
El Paso—advanced water purification facility, Texas, USA	Under design	38,000	MF, RO, UV/AOP, GAC, Cl_2
Morbylånga drinking water treatment plant, Sweden	2019/operational	4,000	Oxidation ($KMnO_4$), blending, UF, RO, UV, Cl_2

DAF: dissolved air flotation; GAC: granular activated carbon; Cl_2: chlorination; PAC: powder activated carbon; O_3: ozonation; BAC: biological activated carbon; UF: ultrafiltration; RO: reverse osmosis; UV/H_2O_2: photoperoxidation; DWTP: drinking water treatment plant; MF: microfiltration; AOP: advanced oxidation process
Source Authors

challenge for water resource management and treatment (Porto et al. 2019; Pivetta et al. 2020; Alves et al. 2021). The diverse range of emerging contaminants detected in sewage includes pharmaceuticals, hormones, microplastics, flame retardants, per- and polyfluoroalkyl substances, and a multitude of other chemical and biological compounds (Richardson and Kimura 2020). The detection of such a vast number of contaminants highlights the dynamic nature of our society and the constant intro- duction of new substances into the environment. This emphasizes the urgent need for continued research, monitoring, and proactive measures to mitigate the potential risks associated with these emerging contaminants and safeguard public health and the environment. Given the vast number of emerging contaminants, prioritization studies are urgently needed.

Prioritizing emerging contaminants in the context of DPR requires a multifaceted approach, considering human health risks, ecological impacts, and the technical feasi- bility of removal during treatment processes. The assessment of human health risks requires a thorough understanding of the toxicological properties of contaminants, their potential exposure pathways, and the potential for accumulation in the human body (Ritter et al. 2002). Additionally, the ecological risks associated with these contaminants must be evaluated to ensure the protection of sensitive ecosystems that may be impacted by the discharge of treated wastewater (Fawell and Ong 2012). Furthermore, the implementation of effective treatment technologies is crucial for the removal of emerging contaminants, necessitating consideration of their technical feasibility and cost-effectiveness (Diniz et al. 2023a, b), as well as the identification of appropriate marker compounds, in both raw and treated sewage, which can be quantified using current analytical methods. By integrating these various aspects, a comprehensive prioritization framework can guide decision-makers in identifying the most critical emerging contaminants that warrant attention and targeted mitigation strategies in the context of DPR, ultimately ensuring the production of safe and sustainable drinking water sources. Therefore, in the subsequent sections of this chapter, an in-depth discussion of the prioritization of emerging contaminants based on their recalcitrant properties is provided. This analysis facilitated the selection of specific contaminants for assessing the proper functioning of a pilot-scale DPR system.

The EPAR Capivari II wastewater treatment plant (WWTP), located in the city of Campinas (São Paulo state, Brazil), currently produces non-potable reuse water. This WWTP operates using one of the most advanced technologies in the world for sewage treatment, known as the membrane bioreactor system (MBR) and has installed a pilot-scale water treatment plant. Campinas is the third-largest city in São Paulo state, with 1,138,309 inhabitants. The region faced a significant water crisis in 2014 and 2015, resulting from a combination of political challenges, such as a lack of awareness of the root causes of the water crisis, which extends beyond the technical aspects, necessitating accounting for socio-ecological factors and the importance of prior planning, and an intense drought (Millington 2018). This water crisis prompted efforts to explore alternative water supply options that are not reliant on rainfall patterns. Since municipal wastewater systems are among the most reliable potential sources of water, even for drinking purposes (Liu et al. 2020), advances in wastewater treatment technologies have been made in the last few years. However, the presence of a wider variety of emerging contaminants in sewage can be a drawback for DPR. Therefore, special attention has been paid to the presence of emerging contaminants in Brazilian water matrices in the last few years (Marson et al. 2022).

In the last decade, our research group has worked with several emerging contaminant classes, including anesthetics, anthelmintics, antiacids, antiallergics, antidiabetics, antihypertensives, antilipidemics, antimicrobials, antipsychotics, artificial sweeteners, corticosteroids, cytotoxics, diuretics, steroidal anti-inflammatory drugs, and stimulants. Monitoring, degradation, leaching, and ecotoxicology studies were used to identify emerging contaminant markers suitable for monitoring purposes (Pivetta et al. 2020; Alves et al. 2021; Porto et al. 2019; Rodrigues-Silva et al. 2019; Caianelo et al. 2022; Diniz et al. 2020; Diniz et al. 2022; Diniz et al. 2021; Porto et al. 2021; Spina et al. 2021; Venancio et al. 2021; Caianelo et al. 2021; Diniz and Rath 2023; Diniz et al. 2023a, b). These studies have reported a wide array of emerging contaminants in WWTPs, with concentrations ranging from ng L^{-1} to μg L^{-1}. Bench-scale experiments have demonstrated that the majority of these emerging contaminants exhibit resistance to traditional treatment processes and pose toxicity risks to bacteria and algae. Therefore, the present work aimed to select emerging contaminants as potential marker compounds to assess the proper functioning of multi-barrier treatment in the pilot-scale treatment plant for DPR production installed at the EPAR Capivari II WWTP, which would facilitate further monitoring programs and speed up the decision-making process.

The pilot-scale water treatment plant allows the investigation of different treatment configurations consisting of reverse osmosis, photoperoxidation, and a fixed-bed column loaded with activated carbon. The emerging contaminants were quantified by bidimensional liquid chromatography coupled with tandem mass spectrometry (LC-UHPLC-MS/MS). As the primary outcome, the objective was to select emerging contaminants possessing the characteristics required for use as a marker compound.

19.2   EPAR Capivari II Wastewater Treatment Plant

The EPAR Capivari II WWTP in Campinas (-22.956785, -47.221532) can produce non-potable reuse water. This WWTP operates using an MBR system consisting of a bioreactor with three distinct anaerobic, anoxic, and aerated zones, together with a hollow fiber ultrafiltration membrane composed of polyvinylidene fluoride (LE-4040, DOW FILMTEC).

In the anaerobic zone, bacteria break down organic matter, such as wastewater biosolids and food wastes, in the absence of free and bound oxygen. In the anoxic zone, molecular or free oxygen (O_2) is also absent, but there may be the presence of nitrates (NO_3^-) or nitrites (NO_2^-). Denitrifying bacteria in this zone break down nitrogen products, releasing oxygen for their proliferation, while removing nitrogen compounds from the sewage. Subsequently, in the aerobic zone, a mechanical process introduces oxygen into the effluent. The oxygen is then exploited as a terminal electron acceptor for different reactions, such as the oxidation of organic material, which reduces the biochemical oxygen demand, oxidation of ammonia to nitrate, which reduces the ammonia concentration, and the uptake of phosphate, with the synthesis of polyphosphate, which reduces the effluent phosphate concentration. The hollow fiber ultrafiltration membrane, located at the end of the bioreactor, effectively removes protozoa, bacteria, and suspended solids from the treated water, resulting in values below the limit of detection of the methods.

Before entering the bioreactor and the membrane, the sewage undergoes a primary treatment process, which includes a mechanical medium bar screening system with a spacing of 15 mm, a rotary drum screen with 2 mm apertures, and a mechanical sand remover. This initial treatment helps to remove larger debris and solids from the sewage. As a result of these treatment processes, the final effluent of the EPAR Capivari II WWTP meets the most stringent criteria.

The treated effluent from the EPAR Capivari II WWTP has been successfully utilized for non-potable reuse purposes by the city of Campinas. This has led to

significant savings in potable drinking water, reaching up to 80% for the Campinas fire department, while nearby industries have achieved savings of up to 88%.
Additionally, the EPAR Capivari II WWTP features a pilot plant dedicated to further
research and development in DPR schemes, demonstrating the exploration of DPR
as a highly promising solution to address water scarcity challenges.

19.2.1   Pilot-Scale Treatment Plant

The pilot-scale treatment plant of the EPAR Capivari II WWTP was first designed by
Hespanhol et al. (2019) and comprised different treatment units employing activated
carbon, biological activated carbon, photoperoxidation, reverse osmosis, and ozonation. The current design of the plant (Diniz et al. 2023a, b) operates with a flow of 350
L h^{-1} and is composed of a multichannel system that allows the testing of different
treatment configurations for the effluent of the MBR system: I—photoperoxidation; II—activated carbon; III—reverse osmosis; IV—reverse osmosis + photoperoxidation; V—reverse osmosis + activated carbon; and VI—reverse osmosis +
photoperoxidation + activated carbon (Fig. 19.1).

The reverse osmosis unit is composed of a membrane (LE-4040, DOW FILMTEC)
with a feed spacer thickness of 34 mil and an active area of 7.2 m^2. The photoperoxidation operates using an ultraviolet reactor (17.5 L) equipped with 12 UV-C lamps

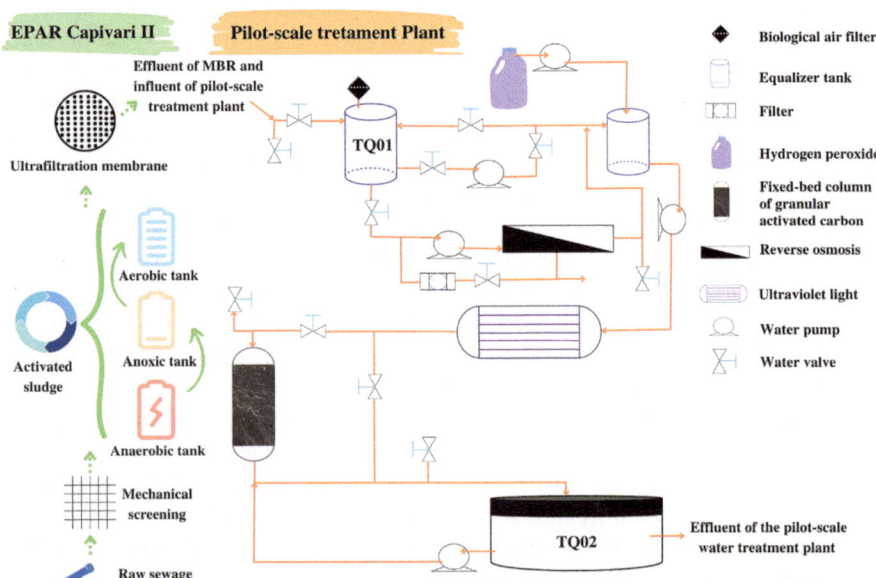

Fig. 19.1 Schematic description of the EPAR Capivari II wastewater treatment plant and the pilot-
scale water treatment plant. TQ01 and TQ02 are equalizer tanks. *Source* Authors

(HNS 55W G13 HO, Osram). The total UV-C dose applied is 1590 mJ cm^{-2}. Before the ultraviolet reactor, there is an equalizer tank that receives a hydrogen peroxide dose of 6 mg L^{-1}. The adsorption unit consists of a fixed-bed column (2.2 m × 0.2 m and 0.5 m of freeboard) loaded with 27 kg of activated carbon (Charbon500, 12 × 24 mesh, Carbonado, Brazil). The activated carbon loaded in the fixed column was characterized in our previous study (Diniz et al. 2023a, b). Briefly, The activated carbon had an average pore diameter of 2.6 nm and a total pore volume of 0.36 cm^3 g^{-1}, of which 75% consisted of micropores ($\leq$2 nm). The N$_2$ adsorption isotherm for the activated carbon (specific surface area of 536.5 m^2 g^{-1}) was type IV with H4 hysteresis, indicating cylindrical pore channels, which was confirmed by scanning electron microscopy. Thermogravimetric analysis showed a mass loss of up to 6% at 105 °C, with high thermal stability up to 500 °C.

The activated carbon consisted mainly of carbon (48.9%), oxygen (19.8%), and silicon (12.7%), with other elements present. Fourier transform infrared spectroscopy revealed various functional groups, including -OH, C–H, C=C, carbonyl C=O, and C–O of phenols and alcohols, as well as Si–O bonds of silicates. Boehm titration experiments indicated the presence of carboxylic, phenol, and lactone groups on the activated carbon surface and also the presence of basic groups on the surface of the activated carbon. Further point of zero charge (PZC) determinations confirmed the basic character of the activated carbon with a pH$_{PZC}$ of 9.2. Raman and X-ray diffraction spectra confirmed the expected graphitic structure and the disordered nature of the activated carbon.

19.2.2 Prioritization of Emerging Contaminants to Be Monitored at the EPAR Capivari II WWTP and the Pilot-Scale Treatment Plant

Understanding the dynamics and occurrence of emerging contaminants in WWTPs is crucial for identifying compounds that can serve as chemical markers to assess treatment efficiency. In this study, particular focus was placed on selecting pharmaceutical active ingredients and artificial sweeteners as marker compounds to address the quality of water produced by the pilot-scale treatment plant installed at the EPAR Capivari II WWTP. Key characteristics of a marker compound are its presence in both raw and treated sewage, as well as its easy quantification using available analytical methods. This implies that the compound should be resistant to removal or be moderately removed during sewage treatment. Therefore, pharmaceutical active ingredients compounds that are widely consumed, minimally metabolized in the human body, and excreted as the parent compound were preferred for prioritization. However, predicting pharmaceutical active ingredients concentrations in Brazilian wastewater is challenging, due to the lack of official consumption data in governmental databases. Furthermore, it needs to be considered that the consumption of pharmaceutical active ingredients may vary according to region and country (Marson et al. 2022). For the prioritization of pharmaceutical active ingredients, sales information for the year 2018 (number of boxes and market size) was obtained from IQVIA Soluções de Tecnologia do Brasil Ltda., a non-governmental company, enabling estimation of the daily dose and the pharmaceutical active ingredients consumption (kg/year/inhab). Estimation of consumption was based on a worst-case scenario, considering the highest quantity of units per package and the highest pharmaceutical active ingredients amount (mg) in each unit. The predicted concentrations of the pharmaceutical active ingredients in the influents of the WWTP were calculated based on the European Medicine Agency guideline (EMEA 2006). The calculations assumed a linear distribution (Eqs. 19.1 and 19.2) of consumption during the year and throughout the local population (Pivetta et al. 2020). In this first prioritization work, 45 pharmaceutical active ingredients were initially selected: sulfamethoxazole, ciprofloxacin, azithromycin, cefepime, ampicillin, trimethoprim, penicillin, piperacillin, tetracycline, cloxacillin, levofloxacin, imipenem, amoxicillin, chloramphenicol, cefazolin, fluoxetine, bupropion, escitalopram, clonazepam, carbamazepine, nortriptyline, amitriptyline, sertraline, trazodone, alprazolam, diazepam, hydrochlorothiazide, albendazole, ricobendazole, prednisolone, lidocaine, diclofenac, ibuprofen, acetaminophen, piroxicam, capecitabine, simvastatin, metformin, atenolol, propranolol, captopril, fexofenadine, dexamethasone, ranitidine, and caffeine.

$$PhAI_{consumption} = \frac{Pck * N_{units} * Strength * 10^{-6}}{Inhab_{Brazil}} \tag{19.1}$$

where $PhAI_{consumption}$ is the annual sales in Brazil in 2018 (kg y^{-1} inhab^{-1}), Pck is the number of packages sold in one year, N_{units} is the maximum number of units in one package, Strength is the amount (mg) of the active substance in one unit of the package, and $Inhab_{Brazil}$ is the estimated population of Brazil in 2018 (210 million).

$$PEC_{WWTPin} = \frac{PhAI_{consumption} * F_{excretion} * 10^{12}}{365 * Inhab_{wwtp} * V_{wwtp}} \tag{19.2}$$

where PEC_{WWTPin} is the predicted concentration in the raw sewage (ng L^{-1}), $F_{excretion}$ is the fraction of the pharmaceutical active ingredients excreted, $inhab_{WWTP}$ is the number of inhabitants served by the WWTP, and V_{WWTP} is the volume of wastewater produced per inhabitant per day (L inhab^{-1} d^{-1}). The volume of wastewater produced per inhabitant can vary significantly depending on several factors, including the country, region, urbanization level, water usage patterns, and industrial activities. In this work, the volume of wastewater produced per inhabitant was estimated to be around 136 L per day, and it was considered that 182,000 inhabitants were served by this WWTP.

In addition, permitted artificial sweeteners in Brazil were also considered for testing, including aspartame, saccharin, sucralose, acesulfame, neotame, and cyclamate. Low-calorie sweeteners have gained considerable attention, due to their widespread consumption worldwide. It is estimated that around 28% of the global population consumes artificial sweeteners daily (Shankar et al. 2013). Most artificial sweeteners are minimally metabolized in vivo and are primarily excreted unchanged, except for aspartame, which is metabolized to phenylalanine, aspartic acid, and methanol (Li et al. 2020). Despite the high global consumption of artificial sweeteners (Luo et al. 2019; Li et al. 2021), these compounds are poorly removed by WWTPs (Alves et al. 2021) and specific data on Brazilian consumption are lacking.

For analysis of the emerging contaminants in raw and treated wastewater, analytical methods using LC-UHPLC-MS/MS were developed and validated. The strategy was to avoid laborious sample preparation procedures before quantitation. The samples received the addition of surrogates, followed by filtration and direct injection into the LC-UHPLC-MS/MS system. Table 19.2 shows an example of the results for 12 selected compounds measured in the raw and treated sewage, together with the calculated PEC values (Table 19.2).

The predicted concentrations (PEC_{WWTP}) of the selected compounds were compared with the measured concentrations (MEC_{WWTP}), to determine the accuracy of the prediction model. Some compounds are efficiently removed during wastewater treatment, such as acetaminophen, metformin, acesulfame, and cyclamate (Table 19.2) (Alves et al. 2021; Pivetta et al. 2020; Porto et al. 2019; Diniz et al. 2023a, b). Therefore, it would not be appropriate to use these compounds as markers to evaluate the processes in the pilot-scale treatment plant.

Considering these aspects, the following compounds were selected for further monitoring: albendazole and its metabolite ricobendazole, carbamazepine, diclofenac, hydrochlorothiazide, propranolol, sulfamethoxazole, sucralose, and saccharin. Although propranolol was not monitored in previous studies, subsequent campaigns demonstrated its suitability as a marker residue, so it was included for further monitoring. Caffeine was added as a target analyte since it has been widely used for this purpose (Buerge et al. 2003).

Monitoring these emerging contaminants during several campaigns led to the selection of ten out of the 51 (pharmaceutical active ingredients + artificial sweeteners) potential marker compounds initially considered: albendazole, caffeine, carbamazepine, diclofenac, hydrochlorothiazide, propranolol, ricobendazole, saccharin, sulfamethoxazole, and sucralose. These compounds were used to evaluate the treatment process in the pilot-scale treatment plant.

19.3   Emerging Contaminants in Raw and Treated Effluent of the EPAR Capivari II WWTP

In this study, the ten previouly selected emerging contaminants (see sect. 19.2) were monitored in the effluent of the MBR system (sampled at TQ-01, Fig. 19.1), during four different sampling campaigns conducted between 2021 and 2022. Diclofenac

Table 19.2 Data used to calculate the predicted concentrations of the 12 selected emerging contaminants in the raw sewage at the EPAR Capivari II WWTP and measured median concentrations of three campaigns

Pharmaceutical active ingredients (PhAI)	Annual sales consumption (kg/year)	$PhAI_{consumption}$ (kg/year/inhab)	Fraction excreted in urine (%)	Estimated annual discharge into the effluent (kg)	PEC_{WWTP} (raw) (ng/L)	MEC_{WWTP} (raw) (ng/L), n = 3	MEC_{WWTP} post-MBR (ng/L), n = 3	Removal (%)
Acetaminophen	2,494,949.7	0.011881	5	108.1	11,945	4900	< 500	> 90
Albendazole	4581.7	0.000022	1	0.040	4	250	250	ND
Carbamazepine	37,100	0.000177	3	1.0	107	517	585	−13.1
Diclofenac	116,101.4	0.000553	35	35.2	3891	800	<500	>63
Hydrochlorothiazide	83,774.6	0.000399	100	114.9	8022	1800	3900	−117
Propranolol	62,307.5	0.000297	25	13.5	1492	NE	NE	NE
Sulfamethoxazole	29,671.8	0.000141	54	13.9	1534	250	250	ND
Metformin	6,794,547.1	0.032355	90	5299.7	585,554	20,000	700	96.5
Sucralose	NA	NA	100	NA	NA	23,000	23,000	ND
Saccharin	NA	NA	>92	NA	NA	46,000	3000	94
Acesulfame	NA	NA	100	NA	NA	42,000	ND	100
Cyclamate	NA	NA	100	NA	NA	138,000	ND	100

NA: not available; NE: not evaluated; ND: not detected; PEC_{WWTP} is the predicted concentration in the raw sewage; MEC_{WWTP} is the measured concentration in the WWTP.
Source Authors

and ricobendazole were not detected in any of the campaigns, while sulfamethoxazole was absent in the third campaign (Table 19.3). On the other hand, sucralose consistently appeared at the highest concentration in all four campaigns, with values ranging from 21,607 to 98,402 ng L^{-1}. Hydrochlorothiazide was also observed at high concentrations in all the campaigns, with concentrations ranging from 5,554 to 19,958 ng L^{-1}. Sulfamethoxazole, in contrast, was quantified at the lowest concentrations, with values ranging from 103 to 707 ng L^{-1}. It is pertinent to highlight that the minimum and maximum concentrations presented in Table 19.3 pertain to individual samples within each respective campaign.

These findings highlight the effectiveness of the MBR system in reducing the concentrations of certain emerging contaminants, such as acetaminophen, acesulfame, and diclofenac, to below detection limits. However, the persistence of sucralose and hydrochlorothiazide at relatively high concentrations in the effluent shows the importance of continued monitoring programs and advanced treatment strategies to ensure the effective removal of emerging contaminants in DPR schemes. These findings emphasize the importance of ongoing research efforts to optimize treatment processes and ensure the safety and sustainability of DPR initiatives.

19.4 Studies in the Pilot-Scale Treatment Plant

The pilot-scale treatment plant of the EPAR Capivari II served as a valuable testing ground for the processes most widely implemented in DPR schemes, namely membrane filtration, oxidation, and media filtration. The goal of these studies at the pilot-scale plant was to validate the selected treatment process, optimize system performance, and ensure that the treated water meets the necessary regulatory standards for potable reuse. This is an essential step in the development and implementation of water reuse projects, providing valuable data and insights, before moving to larger-scale operations.

The studies were conducted with the real treated MBR effluent of the EPAR Capivari II WWTP. It became evident that none of the individual processes in the pilot-scale plant when applied, could achieve complete removal of all the target emerging contaminants to below the limit of quantification obtained for the analytical method (Fig. 19.2). Overall, the removal efficiency was in the following order: reverse osmosis > activated carbon > photoperoxidation. However, it was noteworthy that regardless of the process employed, sulfamethoxazole, a well-known antimicrobial agent known for inducing the development of antimicrobial resistance genes in bacteria (Larcher and Yargeau 2012), was consistently eliminated to concentrations below the limit of quantification. Hence, the removal of sulfamethoxazole, as well as other antimicrobials, appears as a critical prerequisite for all DPR processes.

To address the challenges encountered in the previous experiments, further investigations were conducted to assess the efficacy of combining reverse osmosis with either activated carbon or photoperoxidation. In both configurations, all the emerging

Table 19.3 Concentrations of emerging contaminants in the effluent of the MBR system during campaigns in 2021 and 2022 (n = 6 for each campaign)

Emerging contaminant	Campaign average (s) (ng L^{-1})				Minimum concentration (ng L^{-1})	Maximum concentration (ng L^{-1})	Frequency (%)
	1st	2nd	3rd	4th			
Albendazole[1]	1,779 (203)	1,795 (268)	6,431 (1253)	1,757 (313)	1,217	8,014	100.0
Caffeine[1]	2,153 (141)	2,008 (73)	774 (319)	1,569 (126)	324	2,311	100.0
Carbamazepine[1]	2,361 (567)	3,083 (1,497)	2,900 (1,466)	1,969 (123)	1,788	5,437	100.0
Hydrochlorothiazide[3]	9,241 (917)	8,720 (2,353)	1,1741 (5,395)	8,715 (316)	5,554	19,658	100.0
Propranolol[2]	2,487 (2,668)	2,101 (1,012)	865 (354)	NA	510	6,012	75.0
Saccharin[3]	3,126 (259)	2,929 (190)	3,172 (290)	2,685 (116)	2503	3,642	100.0
Sulfamethoxazole[1]	464 (142)	318 (163)	<LOQ	601 (46)	130	707	75.0
Sucralose[3]	42,186 (5,499)	46,999 (34,848)	27,438 (4,129)	30,133 (909)	21,607	98,402	100.0

Limit of quantification: [1]100 ng L^{-1}; [2]500 ng L^{-1}; [3]1000 ng L^{-1}. NA: not available; s: standard deviation
Source Authors

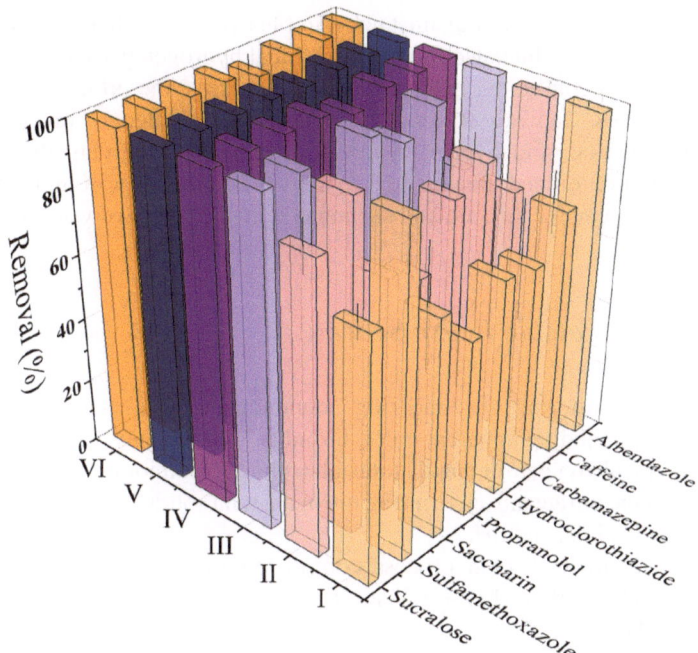

Fig. 19.2 Removal of the emerging contaminants by the different processes in the pilot-scale treatment plant: I—photoperoxidation (UV/H$_2$O$_2$); II—activated carbon; III—reverse osmosis (RO); IV—reverse osmosis + photoperoxidation; V—reverse osmosis + activated carbon; VI—reverse osmosis + photoperoxidation + activated carbon. *Source* Authors

contaminants, except hydrochlorothiazide, were successfully reduced to concentrations below the limit of quantification of the analytical method. However, despite achieving the desired removal of emerging contaminants, the implementation of a two-step treatment scheme raised concerns about compromising the four essential pillars of the treatment plant: reliability, resilience, robustness, and redundancy (the 4Rs). This concern arises from the possibility that if one of the treatment steps were to malfunction, the other might be unable to maintain the high quality of the effluent.

The utilization of photoperoxidation, an alternative to primary disinfection (Kruithof et al. 2007), may bring a new challenge related to the formation of undesirable byproducts (Spina et al. 2021; Venancio et al. 2021). Due to this issue, the introduction of activated carbon, capable of adsorbing and removing such undesirable byproducts, became essential. However, activated carbon alone lacks disinfection capabilities. Consequently, a novel configuration was tested, consisting of reverse osmosis, photoperoxidation, and activated carbon. This configuration not only allowed the reverse osmosis membrane to effectively remove particles larger than 7 Å and a portion of the emerging contaminants (Franke et al. 2021), but also enabled the residual amounts of the emerging contaminants to be oxidized by photoperoxidation, serving as a primary disinfection unit. In addition, the activated carbon acted as a robust barrier against undesirable byproducts that might arise from the oxidation process, while also removing residual hydrogen peroxide.

By strategically combining these treatment processes, the pilot-scale treatment plant of the EPAR Capivari II aimed to address the challenges of contaminant removal, disinfection, and byproduct mitigation, while preserving the 4Rs. The proposed configuration had the potential to deliver treated water of high quality for DPR, paving the way for the safe and sustainable management of water resources.

19.4.1 Direct Potable Reuse Scheme

In addition to monitoring the emerging contaminants, various parameters were also monitored, including apparent color, turbidity, conductivity, biological oxygen demand, chemical oxygen demand, total phosphorus, ammoniacal nitrogen, total Kjeldahl nitrogen, nitrate nitrogen, nitrite nitrogen, total solids, total dissolved solids, total suspended solids, and pH.

The selected pilot-scale plant process, comprising reverse osmosis + photoperoxidation + activated carbon ($RO + UV/H_2O_2 + AC$), was evaluated during different campaigns in the year 2022.

Initially, sampling of the influent and effluent at the pilot plant was conducted in a manner that allowed the plant to operate only during the sample collection period. During one of these campaigns (C1), it was observed that the biological oxygen demand and chemical oxygen demand values for the effluent from the pilot plant (using $RO + UV/H_2O_2 + AC$) were higher than the values for the raw sewage and the MBR effluent. Additionally, an exceptionally high conductivity value was measured for the sample collected at TQ02 (Fig. 19.3). This discrepancy indicated a malfunction in the RO membrane and the formation of a biofilm on the surface of the activated carbon. To address this issue, both units were replaced, with subsequent results returning to the expected values (as shown for the samples from C3 at TQ02).

Subsequently, the pilot-scale water treatment plant remained in continuous operation from March to December 2022. During this period, four sample collection campaigns (C2, C3, C4, and C5) were conducted. Figure 19.3 presents a comparison between the results of campaigns C1 and C3 for biological oxygen demand, chemical oxygen demand, and conductivity, obtained from samples collected at the WWTP inlet (raw sewage) and the TQ01 inlet (post-MBR) and TQ02 outlet (pilot plant effluent).

These results highlight the importance of regular monitoring and maintenance of the treatment plant to ensure optimal performance and the desired water quality. The replacement of malfunctioning units successfully resolved the issues observed in the first campaign, enabling the achievement of consistent and satisfactory results in the subsequent sampling campaigns.

It was also observed that the apparent color, total solids, nitrate nitrogen, and water hardness levels decreased significantly after treating the effluent using the RO + UV/

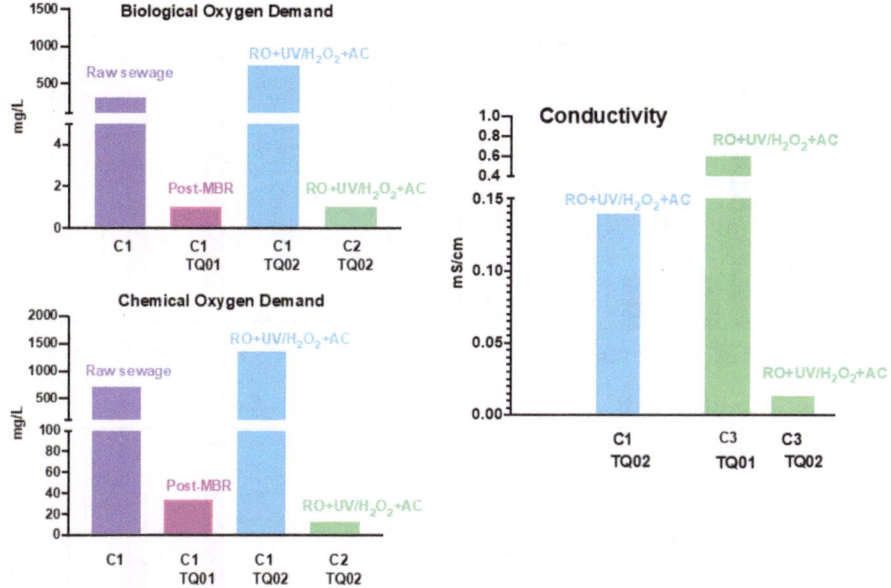

Fig. 19.3 Biological oxygen demand, chemical oxygen demand, and conductivity of the raw sewage and samples obtained post-MBR and after the RO + UV/H₂O₂ + AC treatment, in two campaigns (C1 and C3). *Source* Authors

H_2O_2 + AC process (Fig. 19.4). Furthermore, the treatment process contributed to decreases of sulfate, sodium, and phosphorus.

In all the campaigns, the selected emerging contaminants were quantified, with the results showing that the pilot-scale treatment plant successfully removed these contaminants to below the limit of quantification of the analytical method. The possibility of quantifying albendazole, carbamazepine, hydrochlorothiazide, and sucralose by the developed and validated analytical methods, coupled with the effective removal of the compounds using the reverse osmosis + photoperoxidation + activated carbon system (Fig. 19.5), established them as potential markers for monitoring removal efficiency and the quality of the treated effluent at the pilot plant.

The online LC-UHPLC-MS/MS technique was shown to be highly effective, as evidenced by the limits of quantification achieved for various compounds. The detection limit for sucralose was 1000 ng L^{-1}, while very low values of 10 ng L^{-1} were obtained for carbamazepine and albendazole. In turn, a detection limit of 100 ng L^{-1} was obtained for hydrochlorothiazide. It is noteworthy that sulfamethoxazole, despite being eliminated by all the tested processes, warrants inclusion in monitoring programs due to its easy quantification (quantification limit of 10 ng L^{-1}) and the potential risk it poses by stimulating resistance genes in bacteria.

A major advantage of employing this method is its streamlined process, which facilitates routine analysis, as samples only require filtration and the addition of surrogate deuterated internal standards, before injection into the instrument. The limits

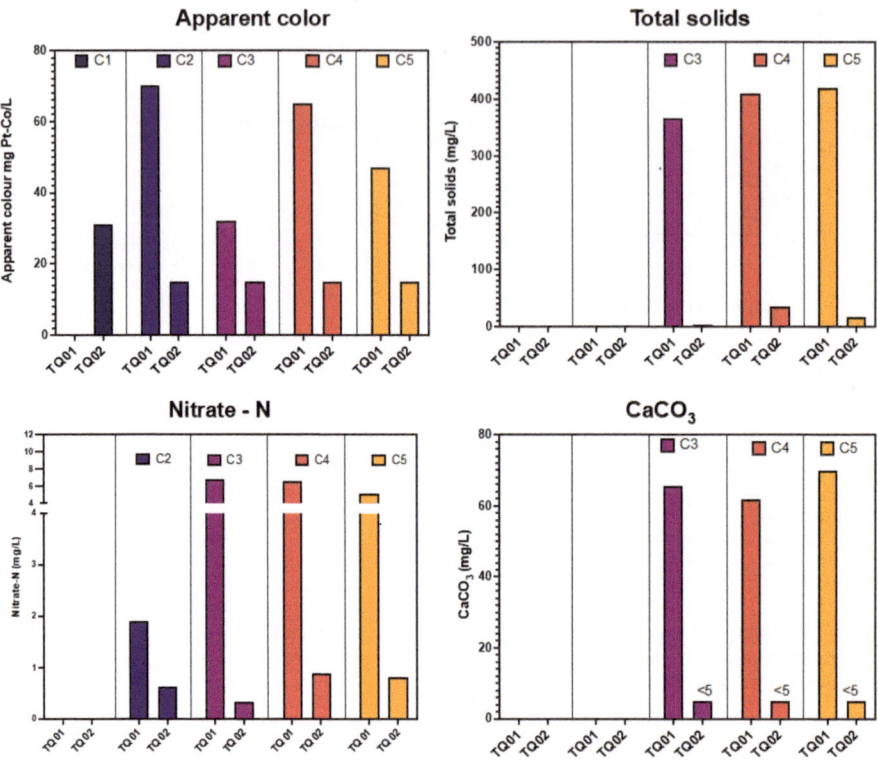

Fig. 19.4 Nitrate-N, total solids, apparent color, and CaCO₃ (water hardness) values for the samples collected at TQ01 (post-MBR) and TQ02 (effluent treated by RO + UV/H₂O₂ + AC). *Source* Authors

of quantification could be further improved by implementing a pre-concentration step utilizing a solid-phase extraction cartridge, which could achieve values approximately 100-fold lower. However, it is important to note that this enhanced procedure would demand more time and incur additional expenses. For monitoring purposes at the EPAR II Capivari II pilot-scale treatment plant, offline preconcentration steps to reduce the limits of quantification would not seem to be necessary. In this case, the simple addition of surrogate deuterated internal standards, followed by filtration and direct injection into the LC-UHPLC-MS/MS, provided lower cost and easier handling.

Furthermore, it is extremely important to emphasize that the effluent from the pilot-scale treatment plant fully complied with all the potability criteria stipulated by Brazilian legislation, which include consideration of organoleptic properties, inorganic and organic substances posing health risks (including pesticides and metabolites), protozoa, and bacteria. This accomplishment demonstrates the excellent potential of the MBR + reverse osmosis + photoperoxidation + activated carbon scheme for use in DPR. Nevertheless, confirmation of the safety of the treated effluent requires

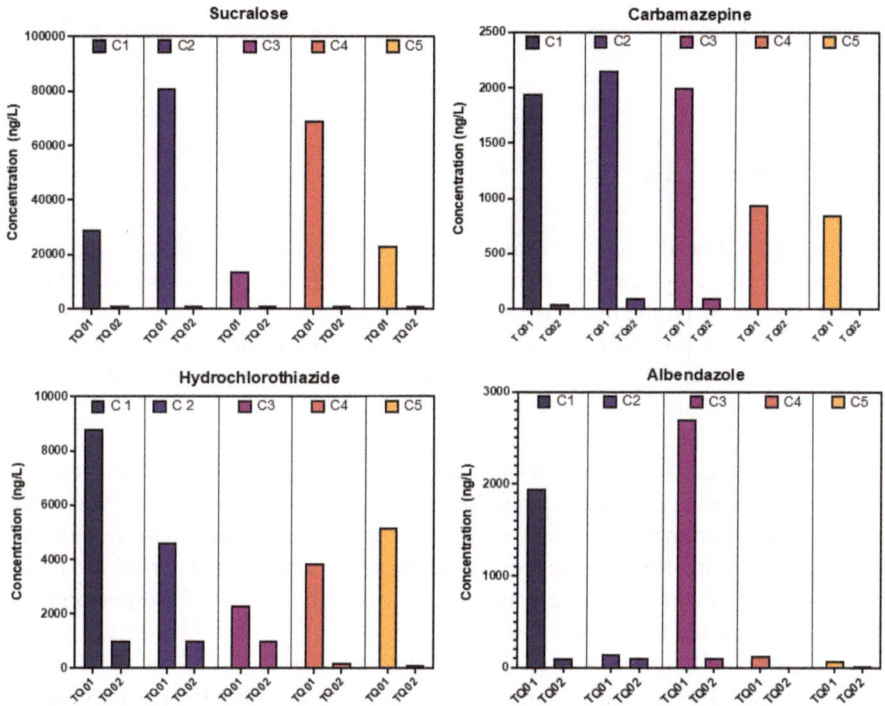

Fig. 19.5 Concentrations of the selected marker compounds in the samples collected at TQ01 (post-MBR) and TQ02 (effluent treated by RO + UV/H$_2$O$_2$ + AC). *Source* Authors

further comprehensive studies to investigate the ability of the system to deal with different viruses and bacteria.

19.5 Future Work

Although the findings of the present study highlighted the quality of the effluent of the pilot-scale treatment plant, it is essential to recognize that no disinfection/ chlorination studies had been conducted. Disinfection units are mandatory for all Brazilian drinking water treatment plants and, depending on the process used, they may result in the formation of disinfection byproducts (Pandian et al. 2022). In addition, as future steps, it would be beneficial to consider the continuation of studies involving the evaluation of bacteria, protozoa, and viruses, as well as toxicity assays (Zhu et al. 2022). These additional investigations should provide a comprehensive understanding of the potential risks associated with the treated effluent. Assessment of microbial pathogens and toxicity levels is essential to ensure the safety of the water and its suitability for the intended reuse purposes, reinforcing the need for comprehensive and rigorous research to guarantee the efficacy and safety of DPR practices.

19.6 Conclusions

The findings obtained from the work at the pilot-scale treatment plant at the EPAR Capivari II WWTP demonstrated the viability of producing DPR water from MBR-treated sewage, using the integrated multibarrier process of RO + UV/H_2O_2 + AC. This combination proved to be effective in removing a wide range of emerging contaminants, ensuring the production of high-quality treated effluent that meets the potability criteria stipulated by Brazilian legislation.

Key indicators including conductivity, apparent color, water hardness, and chemical oxygen demand, together with specific emerging contaminants such as sucralose, hydrochlorothiazide, carbamazepine, sulfamethoxazole, and albendazole, were identified as crucial parameters for assessment of the proper functioning of the treatment at the pilot-scale plant. Monitoring these indicators over an extended period and across different seasons can provide a comprehensive understanding of the performance and efficiency of the treatment.

While the results are promising, it is essential to acknowledge the need for further studies and continuous monitoring to ensure the safety of the treated effluent and its suitability for direct potable reuse. Future investigations should include disinfection/chlorination studies to assess the potential formation of disinfection byproducts and evaluate the presence of microbial pathogens. Additionally, comprehensive toxicity assays will help to confirm the overall safety of the water for its intended reuse purposes. It is also important to recognize that the selection of markers for emerging contaminants may require adjustments over time, as it depends on the dynamic consumption patterns of the population. Therefore, ongoing research and monitoring are critical to stay abreast of emerging contaminants and continuously optimize the treatment process.

In conclusion, the successful operation of the pilot-scale treatment plant and the consistent removal of emerging contaminants using the RO + UV/H_2O_2 + AC scheme demonstrate the promising potential of DPR as a sustainable solution for water scarcity challenges. Nevertheless, the pursuit of safe and reliable DPR requires a multidisciplinary and proactive approach, encompassing continuous monitoring, advanced treatment strategies, and comprehensive risk assessment, to ensure the long-term sustainability and safety of water resources. The collaborative efforts of researchers, policymakers, and stakeholders are paramount to achieving the vision of a future where DPR plays a vital role in securing a reliable and sustainable water supply for communities worldwide.

Acknowledgements The authors are grateful for the financial support provided by the Brazilian public agencies São Paulo State Research Foundation (INCTAA, FAPESP #2014/50951-4, FAPESP #2021/03239-0) and CNPq (#465768/2014-8 and #301737/2017-7). Scholarships were awarded to

V.D. (FAPESP #2021/08123-0 and #2022/11350-1). The authors also would like to thank SANASA for their support during the development of this study.

References

Alves PDC, Rodrigues-Silva C, Ribeiro AR, Rath S (2021) Removal of low-calorie sweeteners at five Brazilian wastewater treatment plants and their occurrence in surface water. J Environ Manag 289

Buerge IJ, Poiger T, Muller MD, Buser HR (2003) Caffeine, an anthropogenic marker for wastewater contamination of surface waters. Environ Sci Technol 37:691–700

Caianelo M, Espíndola JC, Diniz V, Spina M, Rodrigues-Silva C, Guimaraes JR (2022) Gatifloxacin photocatalytic degradation in different water matrices: antimicrobial activity and acute toxicity reduction. J Photochem Photobiol A-Chem 430

Caianelo M, Rodrigues-Silva C, Maniero MG, Diniz V, Spina M, Guimaraes JR (2021) Evaluation of residual antimicrobial activity and acute toxicity during the degradation of gatifloxacin by ozonation. Water Sci Technol 84:225–236

Diniz V, Cunha DGF, Rath S (2023) Adsorption of recalcitrant contaminants of emerging concern onto activated carbon: a laboratory and pilot-scale study. J Environ Manag 325

Diniz V, Rath G, Rath S, Araujo LS, Cunha DGF (2022) Competitive kinetics of adsorption onto activated carbon for emerging contaminants with contrasting physicochemical properties. Environ Sci Pollut Res 29:42185–42200

Diniz V, Rath G, Rath S, Rodrigues-Silva C, Guimaraes JR, Cunha DGF (2021) Long-term ecotoxicological effects of ciprofloxacin in combination with caffeine on the microalga Raphidocelis subcapitata. Toxicol Rep 8:429–435

Diniz V, Rath S (2023) Adsorption of aqueous phase contaminants of emerging concern by activated carbon: comparative fixed-bed column study and in situ regeneration methods. J Hazard Mater

Diniz V, Rath S, Crick CR (2023) Synthesis and characterization of TiO2-carbon filter materials for water decontamination by adsorption-degradation processes. J Environ Manag 346

Diniz V, Reyes GM, Rath S, Cunha DGF (2020) Caffeine reduces the toxicity of albendazole and carbamazepine to the microalgae Raphidocelis subcapitata (Sphaeropleales, Chlorophyta). Int Rev Hydrobiol 105:151–161

EMEA (2006) European medicine agengy guideline on the environmental risk assessment of medicinal products for human use: (EMEA/CHMP/SWP/4447/00)

Fawell J, Ong CN (2012) Emerging contaminants and the implications for drinking water. Int J Water Resour Dev 28:247–263

Franke V, Ullberg M, McCleaf P, Walinder M, Kohler SJ, Ahrens L (2021) The price of really clean water: combining nanofiltration with granular activated carbon and anion exchange resins for the removal of per- and polyfluoralkyl substances (PFASs) in drinking water production. ACS ES&T Water 1:782–795

Hespanhol I, Rodrigues R, Mierzwa JC (2019) Direct potable water reuse – technical feasibility study using a pilot plant. Rev DAE 67

Jeffrey P, Yang Z, Judd SJ (2022) The status of potable water reuse implementation. Water Res 214

Kruithof JC, Kamp PC, Martijn BJ (2007) UV/H2O2 treatment: a practical solution for organic contaminant control and primary disinfection. Ozone-Sci & Eng 29:273–280

Larcher S, Yargeau V (2012) Biodegradation of sulfamethoxazole: current knowledge and perspectives. Appl Microbiol Biotechnol 96:309–318

Li DD, O'Brien JW, Tscharke BJ, Choi PM, Ahmed F, Thompson J, Mueller JF, Sun HW, Thomas KV (2021) Trends in artificial sweetener consumption: a 7-year wastewater-based epidemiology study in Queensland, Australia. Sci Total Environ 754

Li D, O'Brien JW, Tscharke BJ, Choi PM, Zheng Q, Ahmed F, Thompson J, Li J, Mueller JF, Sun H, Thomas KV (2020) National wastewater reconnaissance of artificial sweetener consumption and emission in Australia. Environ Int 143:105963

Liu L, Lopez E, Duenas-Osorio L, Stadler L, Xie YF, Alvarez PJJ, Li QL (2020) The importance of system configuration for distributed direct potable water reuse. Nat Sustain 3:548–555

Luo J, Zhang Q, Cao M, Wu L, Cao J, Fang F, Li C, Xue Z, Feng Q (2019) Ecotoxicity and environmental fates of newly recognized contaminants-artificial sweeteners: a review. Sci Total Environ 653:1149–1160

Marson EO, Paniagua CES, Gomes O, Goncalves BR, Silva VM, Ricardo IA, Starling MCVM, Amorim CC, Trovo AG (2022) A review toward contaminants of emerging concern in Brazil: Occurrence, impact and their degradation by advanced oxidation process in aquatic matrices. Sci Total Environ 836

Millington N (2018) Producing water scarcity in Sao Paulo, Brazil: the 2014–2015 water crisis and the binding politics of infrastructure. Polit Geogr 65:26–34

Mukhopadhyay A, Duttagupta S, Mukherjee A (2022) Emerging organic contaminants in global community drinking water sources and supply: a review of occurrence, processes and remediation. J Environ Chem Eng 10

Pandian AMK, Rajamehala M, Singh MVP, Sarojini G, Rajamohan N (2022) Potential risks and approaches to reduce the toxicity of disinfection by- product - a review. Science Total Environ 822

Pivetta RC, Rodrigues-Silva C, Ribeiro AR, Rath S (2020) Tracking the occurrence of psychotropic pharmaceuticals in Brazilian wastewater treatment plants and surface water, with assessment of environmental risks. Sci Total Environ 727

Porto RS, Pinheiro RSB, Rath S (2021) Leaching of benzimidazole antiparasitics in soil columns and in soil columns amended with sheep excreta. Environ Sci Pollut Res 28:59040–59049

Porto RS, Rodrigues-Silva C, Schneider J, Rath S (2019) Benzimidazoles in wastewater: analytical method development, monitoring and degradation by photolysis and ozonation. J Environ Manage 232:729–737

Puri M, Gandhi K, Kumar MS (2023) Emerging environmental contaminants: a global perspective on policies and regulations. J Environ Manag 332

Reddy KR, Kandou V, Havrelock R, El-Khattabi AR, Cordova T, Wilson MD, Nelson B, Trujillo C (2023) Reuse of treated wastewater: drivers, regulations, technologies, case studies, and Greater Chicago area experiences. Sustainability 15

Richardson SD, Kimura SY (2020) Water analysis: emerging contaminants and current issues. Anal Chem 92:473–505

Ritter L, Solomon K, Sibley P, Hall K, Keen P, Mattu G, Linton B (2002) Sources, pathways, and relative risks of contaminants in surface water and groundwater: a perspective prepared for the Walkerton inquiry. J Toxicol Environ Health-Part A-Curr Issues 65:1–142

Rodrigues-Silva C, Porto RS, dos Santos SG, Schneider J, Rath S (2019) Fluoroquinolones in hospital wastewater: analytical method, occurrence, treatment with ozone and residual antimicrobial activity evaluation. J Braz Chem Soc 30:1447–1457

Shankar P, Ahuja S, Sriram K (2013) Non-nutritive sweeteners: review and update. Nutrition 29:1293–1299

Soller JA, Eftim SE, Nappier SP (2018) Direct potable reuse microbial risk assessment methodology: sensitivity analysis and application to State log credit allocations. Water Res 128:286–292

Spina M, Venancio W, Rodrigues-Silva C, Pivetta RC, Diniz V, Rath S, Guimaraes JR (2021) Degradation of antidepressant pharmaceuticals by photoperoxidation in diverse water matrices: a highlight in the evaluation of acute and chronic toxicity. Environ Sci Pollut Res 28:24034–24045

Venancio WAL, Rodrigues-Silva C, Spina M, Diniz V, Guimaraes JR (2021) Degradation of benzimidazoles by photoperoxidation: metabolites detection and ecotoxicity assessment using Raphidocelis subcapitata microalgae and Vibrio fischeri. Environ Sci Pollut Res 28:23742–23752

Villarin MC, Merel S (2020) Paradigm shifts and current challenges in wastewater management. J Hazard Mater 390

Wallmann L, Krampe J, Lahnsteiner J, Radu E, van Rensburg P, Slipko K, Wogerbauer M, Kreuzinger N (2021) Fate and persistence of antibiotic-resistant bacteria and genes through a multi-barrier treatment facility for direct potable reuse. Water Reuse 11:373–390
WHO (2017) Potable reuse: guidance for producing safe drinking-water
Zhu KH, Ren HW, Lu Y (2022) Potential Biorisks of Cryptosporidium spp. and Giardia spp. from reclaimed water and countermeasures. Curr Pollut Rep 8:456–476

Chapter 20
Prioritization of Emerging Pollutants Used for Fingerprinting Specific Water Sources

Olutobi Daniel Ogunbiyi, Maria Guerra de Navarro, Carolina Cuchimaque Lugo, Courtney Heath, Joshua Omaojo Ocheje, Luciana Teresa Dias Cappelini, and Natalia Quinete

Abstract Emerging Pollutants (EPs) are synthetic organic compounds identified in the environment that may potentially threaten ecological systems and human health. They can enter water systems from a variety of sources and are of particular concern due to their potential for bioaccumulation and toxicological effects. Their origin and environmental impact are not yet fully understood, so the monitoring of these compounds by regulatory bodies is not yet completely established. EPs cover a broad spectrum of substances, including pharmaceuticals, personal care products, endocrine disruptors, pesticides, and microplastics, that are not always detected by traditional monitoring targeted analysis. Non-targeted analysis (NTA) or suspect screening (SS) employing high-resolution mass spectrometry has emerged as a promising tool for environmental monitoring to identify and prioritize a broader spectrum of emerging contaminants that are not routinely monitored and their transformation products. Here we present a comprehensive review of NTA strategies for screening EPs, covering major insights on the current methodologies and limitations, workflows used for NTA and considerations during sampling, sample preparation, instrumental data acquisition, analysis, and approaches for identification and prioritization of EP in water sources. We have ultimately prioritized predominant and toxic compounds in groundwater, drinking water, surface water, wastewater, and leachates.

Keywords Non-targeted analysis · High-resolution mass spectrometry · Endocrine disruptors · Groundwater · Surface water · Wastewater

O. D. Ogunbiyi · M. Guerra de Navarro · C. Cuchimaque Lugo · C. Heath · J. O. Ocheje · L. T. Dias Cappelini · N. Quinete (✉)
Institute of Environment, Florida International University, Miami, FL, USA
e-mail: nsoaresq@fiu.edu

Department of Chemistry and Biochemistry, Florida International University, North Miami, FL, USA

© UNESCO 2025
S. Zandaryaa et al. (eds.), *Emerging Pollutants*, Advances in Water Security,
https://doi.org/10.1007/978-3-031-71758-1_20

447

20.1 Introduction: Emerging Contaminants

Think of a drop of water on its course, a traveler through the pathways of our modern world, and hidden secrets within it. But what if we could decipher these secrets, discovering their origin and the stories they carry from the pollutants hidden in them? Water becomes a storyteller of pollution, and every droplet holds the fingerprint to safeguard our most vital resource. Let's immerse ourselves in the scientific journey to unscramble the enigmatic field of 'Emerging Pollutants (EP).'

The EPs, also known as 'contaminants of emerging concern-CEC' encompassing a collection of mostly naturally occurring, manufactured, or manmade chemicals or materials recently identified in the environment, originating from everyday human activities, including residential, agricultural, and industrial activities. They are primarily organic compounds, characterized by their polar nature, limited longevity, and low-level environmental presence, leading to their classification as pseudo-persistent pollutants, including over 3,000 chemicals and their derivatives, such as pharmaceuticals and personal-care products (PPCPs), steroids, natural and synthetic hormones, endocrine-disrupting compounds (EDCs), cosmetics, pesticides, heavy metals, coatings, nanomaterials, per- and polyfluoroalkyl substances (PFAS), phosphoric ester flame retardants, drugs of abuse (DoAs) and their metabolites, microplastics, food additives, surfactants, and industrial additives and agents, siloxanes, sweeteners, and gasoline additives (Li et al. 2023; Bletsou et al. 2015; Puri et al. 2023; Rosenfeld et al. 2011). Their sources, fate, behavior, and potential threats to the environmental ecosystems and human health and safety are poorly recognized or understood. Because of this lack of knowledge, monitoring routine programs for EPs is not established by regulatory agencies. With technological improvements, research, and monitoring data, EPs can be detected, identified, and quantified, providing data to support the development of guidelines or decision frameworks. Research continuously arises reporting detection of known and novel EPs. In fact, while findings in drinking water and groundwater have reported some compounds that are regulated, but the majority are not (Kaserzon et al. 2017; Kiefer et al. 2021; Li et al. 2023; Newton et al. 2018; Pinasseau et al. 2019; Postigo et al. 2021; Segura et al. 2011; Soulier et al. 2016). Studies in surface water and wastewater also shows increasing discovery of new EPs (Abafe et al. 2023; Angeles et al. 2020; Ccanccapa-Cartagena et al. 2019; Ebele et al. 2020; Eysseric et al. 2021a; Golovko et al. 2021; Jacob et al. 2021; Menger et al. 2021; Segura et al. 2011; Troxell et al. 2022; Venkatesan and Halden 2014; Wang et al. 2020). Therefore, the list of these contaminants is constantly being updated and might be considered a priority in future regulations regarding their adverse effects and persistence (Atugoda et al. 2021).

A large amount of new and even unknown organic pollutants considered as organic micropollutants (OMPs), in a scale of μg/L or even ng/L, are introduced annually in the environment, including water systems. Regular use of those novel chemical compounds leads to public exposure of these substances via inhalation, consumption, and skin contact, raising alarms about aquatic ecotoxicity, human toxicity via aquatic exposure routes (e.g., water consumption and recreational use), bioaccumulation, biomagnification, and potential health repercussions because of the limited toxicological data (Paszkiewicz et al. 2022). For example, ecological effects have been demonstrated at certain levels of pharmaceuticals in fishes, algae, and mussels, leading to alteration in hepatic gene expression in fishes and effects in the photosynthetic apparatus in algae, among others. Also, the presence of hormone compounds in the environment could result in endocrine disruption affecting many animals, including mammals, birds, and fish (Li 2014). In humans, microbiota imbalance, metabolic alterations, pseudo-membranous colitis, certain types of cancer, birth defects, developmental delay in reproductive organs, endometriosis, infertility, immune system toxicity, bioaccumulation, diseases related to ulcers, thyroid disease, antibiotic resistance effects have been associated with the presence of antibiotic residues, PFAS, and disinfection by-products (DBPs) in environmental matrices (González-González et al. 2022; Pradhan et al. 2023).

The importance of emerging pollutants and their transformation products cannot be underestimated. Transformation products (TPs) are primarily generated via oxidation, hydroxylation, hydrolysis, photodegradation, conjugation, cleavage, dealkylation, as well as methylation and demethylation (Bletsou et al. 2015). Although TPs are commonly found in low concentrations and tend to be less toxic and possess greater polarity than the original parent compounds, occasionally, they might exhibit higher toxicity and persistency than their precursors and can also be detected in higher concentrations. The EPs and their TPs enter, directly or indirectly, into water systems from diverse origins, like as atmospheric deposition, agriculture and urban run-off, industrial discharges, landfill leachates, and even incidental spills and improper disposal of medications. The indirect pathways might involve vertical migration through soils, allowing them to reach groundwater. Also, they can potentially get into surface waters through lateral movement of surface run-off or permeate through subsurface drainage systems, eventually leading these pollutants to enter streams, major rivers, reservoirs, estuaries, and oceans (Bletsou et al. 2015). There is no universal approach that effectively removes all these contaminants from the wastewater treatment plants (WWTPs), resulting in the identification of these compounds in effluent wastewaters, urban and landfill leachates, and subsequently in ground and surface waters, which are used to produce drinking water (Bade et al. 2016). As a result, a comprehensive understanding of their occurrence, distribution, fate, and potential risks is essential to formulate effective strategies for pollution management, ecological preservation, and safeguarding human health.

In the United States of America, the Environmental Protection Agency (U.S. EPA) regularly develops the Contaminant Candidate List (CCL) as a first step in assessing water pollutants. It is followed by an evaluation and regulative process-data collection, further research to sufficiently comprehend concentrations at what the listed contaminants occur in drinking water as well as evaluate potential human health effects- to periodically prepare regulatory determinations for CCL (https://www.epa.gov/ccl). Likewise, organizations in the European Union as the European Chemical Agency-ECHA (https://echa.europa.eu/) and the European Environmental Agency- EEA (https://www.eea.europa.eu) cooperate to understand and address practical approaches to manage emerging contaminants, engaging with industry and stakeholders, and communicating information about emerging pollutants to policy-makers, community, and stakeholders, respectively, promoting informed decisions. The ECHA identifies and designates under scientific principles a candidate list of Substances of Very High Concern (SVHCs) through the Registration, Evaluation, Authorization, and Restriction of Chemicals (REACH) regulation. Emerging pollutants posing potential environmental and human health risks are required to be monitored.

Identification and assessment of EPs within water systems have been traditionally approached through targeted analysis methods, which focus on a limited set of known compounds. A broad and commonly used instrumental technique is liquid chromatography (LC) coupled with mass spectrometry (LC-MS), employing diverse types of mass analyzers, for the assessment of more polar, higher soluble in water, and larger organic molecules, that are sensitive to heat, making it well-suited not only for EPs but also TPs, which usually possess higher polarity than their precursor molecules (Aceña et al. 2015; Bader et al. 2016, Eysseric et al. 2021a; Sjerps et al. 2016). Common mass analyzers available nowadays are triple quadrupole (QQQ), time-of-flight (TOF), ion-trap (IT), Orbitrap and hybrid systems (e.g., quadrupole time-of-flight (Q-TOF), quadrupole-linear ion trap (Q-LIT), linear ion trap-Orbitrap or quadrupole-Orbitrap), being the preferred methods for assessing EPs and their TPs in aqueous samples (Aceña et al. 2015; Bletsou et al. 2015; Hollender et al. 2017). However, the dynamic nature of emerging pollutants suggests a significant challenge for conventional methods, as they may overlook newly introduced contaminants or transformation and degradation products. Here, suspect screening (SS) or non-targeted analysis (NTA) approaches are favored by advancements in the use of high-resolution mass spectrometry (HRMS), coupled with either gas or liquid chromatography (Bade et al. 2016; González-Gaya et al. 2021; Hollender et al. 2017; Paszkiewicz et al. 2022; Place et al. 2021; Troxell et al. 2022). NTA is used to comprehensively detect various compounds without prior knowledge of their presence or pollution sources. HRMS, with its unparalleled resolving power, enables the precise and accurate determination of molecular masses and structural information, thus facilitating the identification of emerging pollutants even in complex matrices, such as wastewater, leachates, surface water, and groundwater. In this chapter, we will focus on the comprehensive review of available literature to identify NTA strategies and workflows for screening, identification, and prioritization of emerging pollutants by fingerprinting specific water sources using LC-HRMS, and exploring common and appropriate NTA sampling techniques, sample preparation, instrumental analysis, data processing and visualization tools, to identify a wide array of EPs, TPs, and their metabolites. A generic summary is shown in Fig. 20.1.

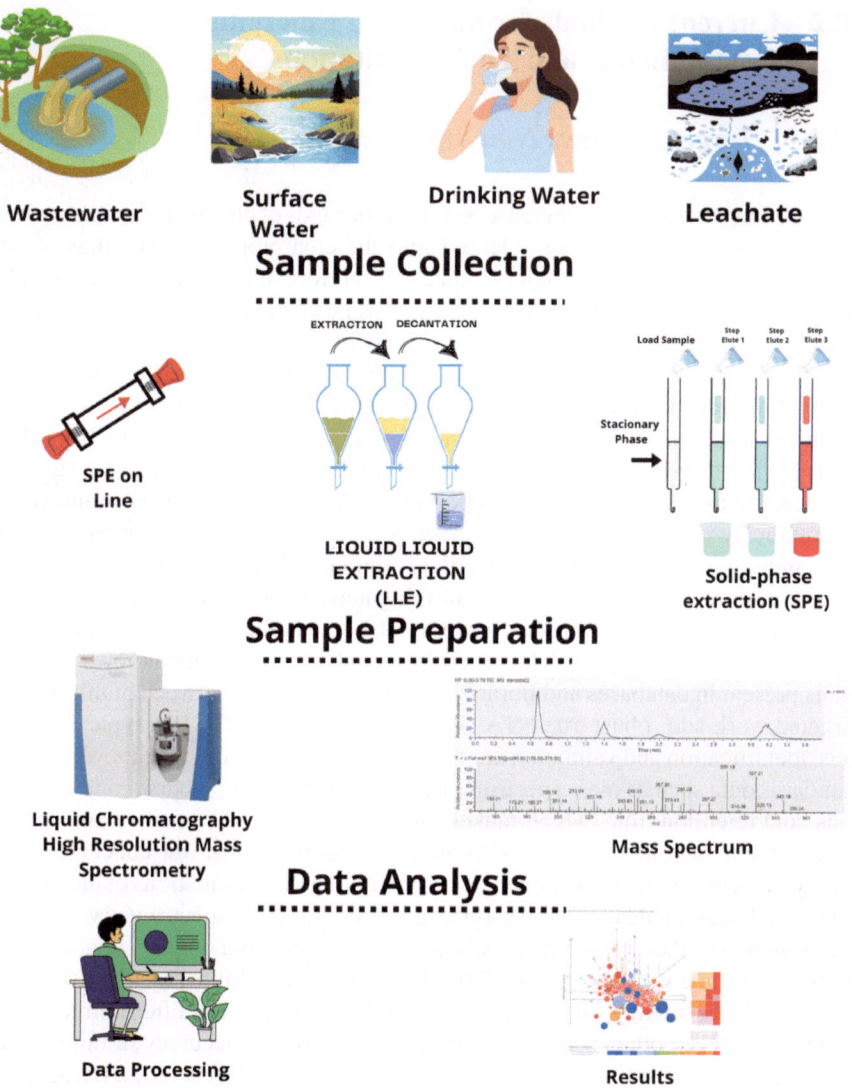

Fig. 20.1 A generic summary of the NTA workflow. *Source* Authors, drawn by Luciana Cappelini

20.2   Current Methods for Screening Emerging Contaminants and Their Limitations

20.2.1   Targeted Versus Non-Targeted Analysis

As anthropogenic activities increase, several thousands of chemicals, including transformation products, are being released into the environment yearly, thus creating concerns for human and ecological impacts. Therefore, it is crucial to develop robust and highly selective instruments that can screen and elucidate these wide spectrums of environmental contaminants. Common analytical techniques used for this approach include- targeted, suspect screening and non-targeted analysis. Targeted analysis (TA) allows for the precise detection and quantification of known chemicals whose chemical structure and names are already predefined. Non-targeted analysis (NTA) is an exploratory-based approach used for comprehensive analysis of a sample to characterize as many compounds as possible, allowing the discovery of unexpected or previously unknown compounds, elucidating their structural components as well as unambiguous assignment of molecular formulas to ions (Guo et al. 2020). Hence, it is a vital analytical technique for identifying new markers and transformation products (Badea et al. 2020; Paszkiewicz et al. 2022). Another approach within NTA is called suspect screening, which is useful for analyzing "known unknown" contaminants present in databases and libraries and acts as a bridge between highly sensitive targeted work and robust true NTA (Paszkiewicz et al. 2022). In a typical TA analysis, identification and validation are essentially based on the use of known reference standards data such as mass spectrum, MS/MS spectrum of precursor and fragments ions, and retention time, which makes this technique highly sensitive and selective for detection and quantification of known analytes that are at low concentrations in the aquatic environment. However, it is worth noting that in situations of data acquisition in full-scan mode using triple-quadrupole (QqQ) mass spectrometry, sensitivity drops impeding their screening capabilities for analytes found in low concentrations in the environment (Krauss et al. 2010). Furthermore, multiple reaction monitoring, used in TA, depending on the chromatographic separation is often limited to few targeted analytes, otherwise, instrumental sensitivity and accuracy diminish due to too short acquisition time and temporally insufficient peak resolution (Krauss et al. 2010). Since there is also a limited number of authentic standards available for the quantification of a wider range of analytes, this makes NTA a powerful and holistic tool for screening, characterization, identification, and prioritization of chemicals from water samples (Koronaiou et al. 2022), due to their high-resolution capabilities and wide applicability to a variety of environmental, product, and biological matrices (Fig. 20.2). Furthermore, several chemical compounds are being released into the environment, hence it is crucial to develop a monitoring and regulatory framework for these chemicals through prioritization. Prioritization aims to evaluate risk associated to chemical exposure by categorizing these contaminants as low or high priority. This underscore NTA as a resourceful approach since it captures a wider spectrum of toxic chemicals in environmental samples (Hong et al. 2016; Lee et al. 2019; Simon

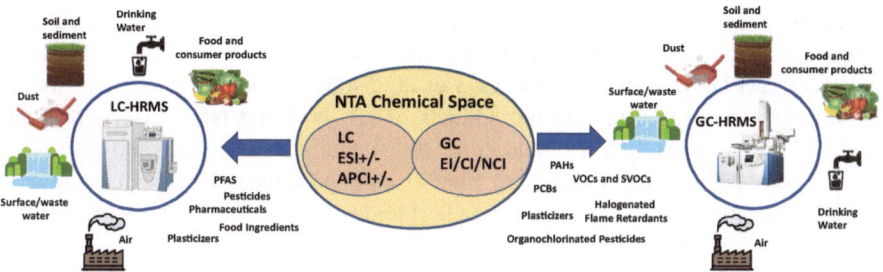

Fig. 20.2* Schematic diagram of application of non-targeted analysis for screening wide spectrum of chemicals. *Source* adapted from Manz et al. 2023

et al. 2013; Yue et al. 2015). Unlike TA where extraction and rigorous purification are needed for the isolation and fortification of targeted analytes, NTA requires that the extraction, including the clean-up steps be non-selective and minimal as much as possible to cover the wide range of various chemicals present in water samples, in this way avoiding conditions that could lead to the removal of specific classes of chemicals, which would restrict the NTA chemical space. At the same time, the lack of minimal clean-up will likely cause matrix build-up and affect the sensitivity of this method. In NTA, ultrafiltration or "dilute-and-shoot" approach are recommended; although, these procedures can suffer from analyte loss due to the matrix effect, thus affecting their sensitivity (El-Deen and Shimizu 2022; Guo et al. 2020; Paszkiewicz et al. 2022). Furthermore, there could also be potential loss of analyte if dilution is below the detection limit, which simultaneously leads to low sensitivity for the identification and detection of analytes (Guo et al. 2020). This would invariably compromise the reliability and identification accuracy of NTA. However, besides the challenges associated with data analysis complexity, limited identification (dependable on presence in databases or confirmation with authentic standards), false positive and negative rates, and quantification issues, NTA serves as a complementary technique to TA, bringing new insights and knowledge on environmental transformation and degradation products and potential metabolites.

20.2.2 NTA Data Acquisition and Processing

Most common soft ionization methods used in liquid chromatography tandem MS/MS analysis include electrospray ionization (ESI) and atmospheric pressure chemical ionization (APCI). ESI is best used for the ionization of high molecular weight and polar compounds, while APCI is more advantageous in the screening of nonpolar to thermal stable polar compounds (Cheng et al. 2015; Ng et al. 2022). High-resolution mass spectrometry (HRMS) coupled with an ESI source is often the most used ionization mode in NTA. Nevertheless, the advantages of using APCI alongside ESI, as complementary techniques, for wider coverage of chemical compounds in the NTA chemical space have been previously reported (Ng et al. 2022; Singh et al. 2020), especially since APCI is found to be less prone to signal loss/suppression due to matrix effect (Duncan et al. 2014; Gao et al. 2005), reduce false positive identification (less formation of cationic adducts of K^+, Na^+ and NH_4^+) (Byrdwell and Neff 2002; Lee et al. 2015; Liigand et al. 2020) and more energetic and can ionize some chemical compounds which are otherwise not possible in ESI, such as polyaromatic hydrocarbons (PAHs), phenylurea herbicides, bisphenols, and organochlorine pesticides (OCPs) (Eichman et al. 2017; Straube et al. 2004).

Data-dependent acquisition (DDA) and data-independent acquisition (DIA) are two common data acquisition methods used for the tentative identification of potential candidates in NTA. In DIA, ions selected within a specified m/z range are further fragmented and analyzed in the MS/MS stage (2nd stage) to detect as many unknown compounds as possible, since the MS and MS/MS data of unknown compounds are obtained within one chromatographic run, it allows for the comprehensive screening and detection of all peaks in samples without the establishment of any intensity threshold or mass filter before injection. Whereas, in DDA mode, the most intense peaks of precursor ions are selected based on predesigned criteria for dissociation. This criterion is streamlined to duty cycle and scan speed and other criteria such as charge state, isotopic patterns, and specific m/z values (Martínez-Bueno et al. 2019). The most intense ions are then subsequently subjected to fragmentation in the MS^2 mode. The MS/MS spectra of the resulting scan can be directly compared to library spectra or used for structural elucidation of prioritized compounds (Fisher et al. 2021; Ogunbiyi et al. 2023). However, this implies that many lower abundances in DDA mode will not be dissociated. Nonetheless, recent advances in automated, recurrent DDA allow for increased sample injection and more MS/MS screening events per MS scan for comprehensive coverage of ions (Broeckling et al. 2018). Hybrid mass analyzers used in NTA include time-of-flight (TOF) analyzers, Orbitrap (Q-Exactive) analyzers, and Ion mobility mass spectrometry (IMS). Most DIA modes are amenable for use on a TOF analyzer due to their high data acquisition speed. Whereas Orbitrap mass analyzers are suitable for both DIA and DDA due to their increased scanning speed (Kaserzon et al. 2017; Schymanski et al. 2015).

A very complex raw data is generated using tandem mass analyzers, emanating from several thousands of natural and anthropogenic chemical compounds present in environmental samples. Hence, these enormous datasets need to be reduced through data *processing*. Data processing steps normally involve mass correction, retention time alignment, peak picking, and grouping of signals that are homologous. i.e., having the same molecular structure. A typical workflow follows (Fig. 20.3): (a) peak picking and detection using mass filtering from an instrumental run, (b) retention time alignment and blank subtraction (c) molecular formula assignment to the exact mass of study, and (d) database search using several creditable online platforms (e.g., Chemspider, PubMed, CompTox etc.) for compound identification based on fragmentation and isotopic patterns (Bader et al. 2016; Fisher et al. 2021; Krauss et al. 2010; Ogunbiyi et al. 2023). Signal intensity, signal-to-noise ratio (S/N), mass-to-charge ratio (m/z), or retention time (RT) are common criteria used for deconvoluting complex candidate lists during data processing, thus, ensuring high data prioritization (Bletsou et al. 2015; Schymanski et al. 2015). Data processing can be somewhat challenging especially when wrong peaks are selected as this can result in the omission of relevant signals (false negatives). In contrast, if too many peaks are integrated, component lists will comprise too much "background noise" as the prioritized candidate, leading to an increase in false positives. Apart from the commercially available data processing software such as Compound Discover, MassHunter Profinder, Mass Profiler, SCIEX OS, ChromaTOF, and other free, open-source software e.g., XCMS (Hohrenk et al. 2020; Smith et al. 2006), SIRIUS + CSI:FingerID (Dührkop et al. 2019), RMassBank (Stravs et al. 2013), MetFrag (Ruttkies et al. 2019), Fluoromatch (for PFAS) (Koelmel et al. 2021) and enviMass (Hohrenk et al. 2020) are often used, however, problems associated with data handling and advanced programming skills remains a limitation. Most of the software utilized have different inbuilt algorithms, which can potentially affect the data processing steps. Hence, it is suggested that a strategic data processing approach be implemented to prevent the presence of false positives in the preferred candidate list (Feng et al. 2021; Hohrenk et al. 2020). The process of eliminating irrelevant/unwanted data while preserving compounds of high preference from complex data mixtures is known as *data prioritization*. Common prioritization methods recently employed for NTA include trend analysis, model-based prioritization, binary sample comparison, Schymanski scale system, and toxicity levels using e.g., EPA Comptox, DSSTox, and ExpoCast (Leendert et al. 2015; Paszkiewicz et al. 2022; Schymanski et al. 2014).

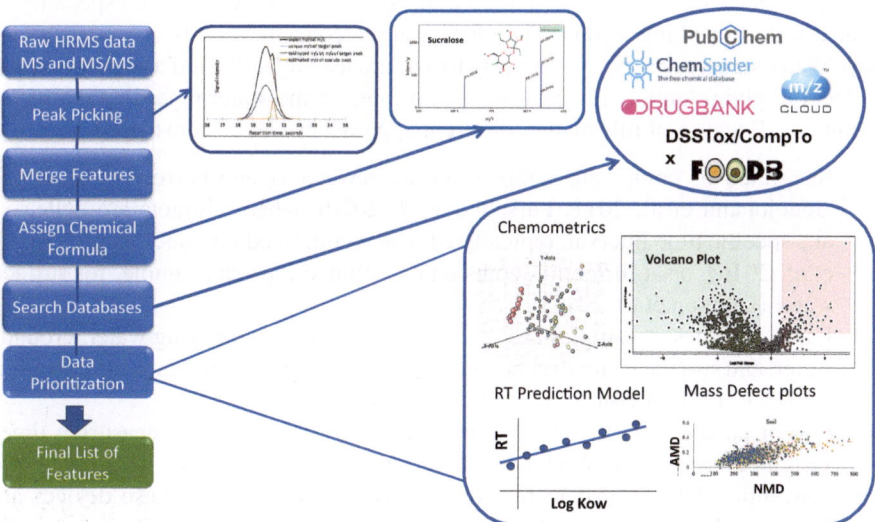

Fig. 20.3* Schematic diagram showing the general flowchart of data processing for non-targeted analysis. *Source* adapted from Fisher et al. 2021

20.3 NTA Approaches for Identification and Prioritization of Emerging Contaminants in Water Sources

20.3.1 Sampling Considerations

The chemical space assessed in non-targeted analysis starts from the sampling collection plan. The container chosen to collect the samples impacts the type of compounds that could be detected during the analysis. Glass containers cleaned and previously combusted have been used for NTA in drinking water and groundwater in studies assessing pesticides and pharmaceuticals. Polypropylene (PP) or high-density polyethylene (HDPE) are also part of the options to be considered, especially if one of the goals is to include NTA of PFAS as part of the possible group of anthropogenic compounds. When volatile compounds are a matter of concern, well-cap containers are required to sample; this type of container commonly includes Teflon.

Besides the container, it has been proven that repeatability is a key factor in NTA. Hence, a sampling plan must include sample replicates (at least 2 or 3). Combining replicates with rigorous filter criteria for peak recognition improves NTA data quality. Regular quality controls regarding trip, field, lab, or any other blanks that would normally be considered in a target analysis must also be considered in NTA. This will help with the background subtraction necessary for eliminating false positives features (Bader et al. 2016; Ng et al. 2020).

As a technique to widely explore different groups of pollutants in a sample simultaneously, NTA faces a challenge regarding sensitivity and detection, especially in treated water (drinking water). Therefore, sample concentration is needed as a first step. Water samples for NTA, similarly to TA, are usually collected as a grab sample collection, giving only a glance at what is present in the water regarding emerging pollutants. Because of this matter, different approaches can be considered:

(a) Integrated or composite samples for wastewater (Gago-Ferrero et al. 2015; Godejohann et al. 2011; Purschke et al. 2020), where aliquots are collected at a specific time interval, typically 24 h, and combined into one sample (Sjerps et al. 2016), or at different depths and combined into one sample, for surface water (Dimzon et al. 2018).

(b) Large volumes of samples (12–24 L) are collected for drinking water, groundwater, and surface water to concentrate the sample during the sample preparation step (5 L used for concentration) (Postigo et al. 2021).

(c) Use of passive samplers for groundwater and surface water, measuring time-weighted average (TWA) concentrations of pollutants and decreasing the conventional level of detection of a particular compound. These devices are known as Polar Organic Chemical Integrative Samplers (POCIS); they can be deployed over a couple of months in an aquatic environment and have revealed the presence of ECC, e.g., pharmaceuticals (Soulier et al. 2016), pesticides (Sultana et al. 2018), and more specifically, PFAS with modifications of the originals POCIS, such as WAX (weak-anion exchange) and HLB (hydrophilic-lipophilic balance) to favor short-chain or long-chain adsorption rates into the device (Gobelius et al. 2019; Kaserzon et al. 2012)

(d) Sampling points of use (home filters) for drinking water. In North Carolina, it was designed a campaign to collect home filters for drinking water, which contained granular activated carbon (GAC), and extract the pollutants retained in the filters (Newton et al. 2018).

Although grab and composite sampling methods can be used for groundwater, drinking water, leachates, surface water, and wastewater, composite sampling is gaining more interest in wastewater research. For wastewater, another option is targeted surveillance that involves sampling at specific locations in the wastewater network, such as culverts outside of buildings, which receive wastewater inputs from the target population only (Augusto et al. 2022; Ort et al. 2010; Saravanan et al. 2021).

The advantages of composite sampling are that it provides representativeness of the water flow over time, considering variations in flow rate and contaminant concentrations. It can reduce analytical costs by grouping several random samples into a composite sample, however individual sample information may be lost and the ability to assess variability may also be reduced (Blum et al. 2017; Lai et al. 2021). These sampling methods can be used individually or in combination to obtain representative water samples for analysis and monitoring. Prior to the extraction of analytes from the sample matrix, water samples are also subjected to ultrafiltration to remove any suspended particulate matter (Wang et al. 2020b); this is done to endure the removal of particulate-bound pollutants.

20.3.2 Sample Preparation Techniques for Water NTA

Solid phase extraction (SPE) is often used in sample preparation for the extraction and simultaneous clean-up to isolate contaminants from water samples. This extraction technique is used in both TA and NTA. One of the most used solid phase materials is the Oasis hydrophilic–lipophilic balance (HLB) sorbent, which consists of a reverse-phase sorbent that can retain a wide range of acid, base, and neutral compounds, making it ideal for NTA. In fact, it has been commonly reported for extraction of EPs by NTA in wastewater (Challis et al. 2020; Wang et al. 2020a; Tadić et al. 2022), surface water (Angeles et al. 2020; Tadić et al. 2022; Sjerps et al. 2016) and groundwater/drinking water (Soulier et al. 2016; Tröger et al. 2018). Commonly, an SPE procedure in an HLB cartridge involves a first step of conditioning with an organic solvent like methanol, in some cases, it also includes an acidic buffer, followed by the loading of the sample. After the loading step, a wash procedure is done to remove interferences, and finally, the extracted and concentrated sample is eluted from the cartridge. When a large volume of sample is concentrated, a solid phase extraction (SPE) disk is used to concentrate the sample instead of a cartridge

(Tröger et al. 2021; Postigo et al. 2021). Other commercially sold sorbent used for NTA includes HR-X, C18, and Oasis WAX. Organic compounds such as PFAS have exhibited better recoveries with the weak anion exchange (WAX) sorbent in the target analysis, hence, it has been also used in NTA protocols where PFAS are the monitored goal (Li et al. 2023). In order to expand the screening of a wide spectrum of organic contaminants, commercial sorbents have been combined e.g., anionic and cationic exchangers (Strata-X-AW and Strata-X-CW) and non-polar phases (e.g., Isolute ENV+) (Wang et al. 2020). Meng et al. (2020) utilized a polymeric-packed SPE cartridge comprising MCX, Oasis WAX, and HLB to comprehensively extract organic contaminants from surface water. Alternatively, to improve the extraction efficiency of sample analytes, self-designed SPE columns are also used. For example, Schemeth et al. (2019) developed and fabricated a q-p(NVI/EGDMA) sorbent and used it in conjunction to available, commercially sold sorbents such as Oasis MAX, Oasis HLB, and Strata X-A to extract polar transformation products of polycyclic aromatic hydrocarbon (PAHs) in water samples (Schemeth et al. 2019; Schlabach et al. 2013). It is, however, important to note that potential analyte loss due to filtration of samples before extraction is of concern, however particulate presence can, in turn, lead to clogging of the SPE cartridge. Previous research conducted by Du et al. (2017) revealed that Infinity SPE wasn't easily clogged by particulates from surface runoffs (leachates) compared to Oasis HLB, being recommended for the extraction of organic contaminants from stormwater and other particulate-bound contaminants in water samples (Du et al. 2017; Hajeb et al. 2022).

SPE concentration step could also be done online, having the same operating principle as offline SPE, but the system is completely automated, using a combination of pumps and valve systems that automatically will precondition an online SPE column, load the sample, wash, elute, and immediately inject to a separation column while the next sample is being extracted (Segura et al. 201, Hajeb et al. 2022). Recent research has utilized an online-SPE-based NTA method to extract analytes from surface water and drinking water samples (Li et al. 2023). Online SPE is promising since it reduces the extraction time, requires a small volume of samples, and has little personnel interference. Furthermore, it is regarded as a faster alternative for the isolation and intermittent concentration of analytes from samples matrices than other traditional extraction methods (Lacina et al. 2011; Ogunbiyi et al. 2023). Some researchers have also used liquid–liquid Extraction (LLE) (Belay et al. 2022; Bergfors et al. 2021), and vacuum-assisted evaporation to avoid losing polar compounds after adding a mixture of the isotopically labeled internal standard (Kiefer et al. 2021), and others had only done a filtration through a 0.45 μm membrane and direct injection into the chromatography system.

20.3.3 Instrumentation and Data Processing Tools

After the extraction and preconcentration step is complete, samples are analyzed on an LC-HRMS and separated on an analytical column (e.g., C18). The implementation of NTA combined with HRMS for fingerprinting potential contaminants in water has the ability to detect and identify a wide range of emerging contaminants (Hollender et al. 2019). Typical examples of instruments used include LC-orbitrap system, LC-QTOF, GC x GC-ToF-MS, among others. Liquid chromatography coupled to ion mobility quadrupole time-of-flight mass spectrometry (LC-IM-QTOF-MS), a technique that combines the separating power of liquid chromatography with high-resolution mass spectrometry and ion mobility measurements, can be useful in providing additional identification criteria, such as collision cross-section values especially for isobaric chemical compounds with indistinguishable m/z (Hinnenkamp et al. 2022). In addition, gas chromatography (GC)-MS spectrometry, gas chromatography-tandem mass spectrometry (GC-MS-MS), and GC-Orbitrap-MS are powerful analytical techniques used to determine volatile contaminants e.g., brominated diphenyl ethers (BDEs), pesticides and polychlorinated biphenyls (PCBs) in water matrices by NTA. These techniques provide high sensitivity, selectivity, and the ability to identify a wide range of organic compounds in complex matrices such as wastewater, and surface water, among others (Domínguez et al. 2020; Eysseric et al. 2021b).

Data processing stage is one of the rigorous and time-consuming aspects of NTA. There are different developed software to help with peak detection (finding-alignment-integration), blank background correction, mass error limits, elemental composition, and isotopic pattern evaluation; the most widely used is the commercial software Compound Discoverer and other free open-sourced software such as XCMS, enviMass, Fluoromatch and RMassBank, for the identification and prioritization of emerging contaminants in water. The definition of tolerance levels for peak picking, mass accuracy tolerance, and the strict application of the specified threshold (retention time, peak area/ intensity, number of fragmentation patterns, isotopic ratio, and signal-to-noise ratio) improves the data quality and reduces false positive and negative results. Applying these criteria allows the merging and grouping of features, which after the generated molecular formula is searched in the database to identify the feature with a certain amount of probability and confidence (Ng et al. 2020). Additionally, a combination of several software processing tools can come in handy as they help to boost the confidence level and reliability of screened compounds as well as reduce the number of false positives in candidate lists (Guo et al. 2020). Although different approaches have been utilized for confirming and characterizing suspected and identified chemical compounds, most published literature on the prioritization of chemicals has been unequivocally based on the Schymanski scheme which uses a level system for communicating confidence levels of identified and prioritized chemicals (Schymanski et al. 2014).

20.3.4   Prioritization of Emerging Contaminants

Considering the development and availability of analytical techniques, instrumentation, and computational tools for continuous monitoring and assessment of EPs in relevant water sources, such as groundwater, drinking water, surface water, wastewater, and leachate, it is essential to improve our understanding of how these pollutants are being prioritized and their potential impact on aquatic life and human health. In general, the criteria for identifying and prioritizing chemical tracers in water sources should be based on (i) Prevalence: chemicals should be commonly found in waterways (Golovko et al. 2021; Venkatesan and Halden 2014), (ii) Persistence: chemicals that are resistant to degradation and have a long environmental half-life should be prioritized (Golovko et al. 2021). Deere et al. in (2021) adopted the EPAs Estimation Programs Interface (EPI) suite to interpret the persistence and degradation potentials of organic contaminants in environmental samples. Any contaminants which surpass half-life thresholds of >2 months in water were categorized as persistent and assigned a score of 1 (Deere et al. 2021), (iii) Bioaccumulation potential: chemicals that have a high potential to accumulate in living organisms, such as those with high bioconcentration factors, should be considered (Golovko et al. 2021). Organic carbon–water partition (K_{OC}) and octanol–water (K_{OW}) coefficient are used for determining accumulation potential of organic compounds. Basically, any chemical that exceeded the guideline of $\log 10\ K_{OC} \geq 3$ and $\log 10\ K_{OW} \geq 4$ are considered to be highly bioaccumulative in biota or sediments (Deere et al. 2021; OECD 2001), (iv) Toxicity: chemicals that have the potential to pose significant risks to human health and the environment should be prioritized (Golovko et al. 2021; Venkatesan and Halden 2014). Various toxicological tools and models used includes EPAs ecotoxicology knowledgebase (ECOTOX), ecological structure activity relationships (ECOSAR), CompTox and ToxCast. Toxicity of chemicals in aqueous phase is classified as acute when values <10,000 µg/L and chronic when values are <100 µg/L according to the ECOSAR-based model (Deere et al. 2021), (v) Production volume: chemicals with high production volumes and widespread use should be considered as they are more likely to be present in the environment at high concentrations (Golovko et al. 2021). The summary of chemical features identified and prioritized in groundwater, drinking water, surface water, wastewater, and leachates in different locations using NTA is shown in Table 20.1.

20.3.4.1   Groundwater and Drinking Water

Groundwater is a very common water source for drinking water purposes. It tends to be preferable to surface water due to less environmental exposure and interaction with different sources of contamination. Nevertheless, the aquifer constituents and structure define what percolates into the basin and its rate. The EPs tend to be in very low concentrations in groundwater and drinking water, below ppb levels. Yet, that low concentration could greatly impact the environment and biota, including

Table 20.1 Summary of chemical features tentatively identified in groundwater, drinking water, surface water, wastewater and leachates from different locations using non-targeted analysis

Sample type	Sample preparation	Location	Analytical techniques	Class of analytes detected	Selected list of top prioritized chemicals	References
Surface water	Oasis HLB bulk cartridge (POCIS)	Pyris River catchment located in Uppsala, Sweden	UPLC-Q-TOF; C18 analytical column (50 mm × 2.1 mm,1.7 μm)	Halogenated (polar) micropollutants	Chlorzoxazone, 3-(3,4-dichlorophenyl)-1,1-dimethylurea, Diflufenican, 2,4-disulfamyl-5-trifluoromethylaniline, perfluoropentanesufonic acid (PFPeA), 5-amino-2-chlorotoluene-4-sulfonic acid, (2-chlorophenyl)(hydroxy)methanesulfonic acid	Menger et al. (2021)
Melted iceberg, seawater	SPE (Oasis HLB)	Arctic environment	GC-TOF-MS	Persistent organic pollutants (POPs)	Polychlorinated bi phenyls (PCBs), siloxanes, Polyaromatic hydrocarbons (PAHs), phthalates, novel brominated fire retardants (OPFRs), and synthetic musk compounds (SMcs)	Lee et al. (2019)
Surface water (river waters)	SPE (Oasis HLB)	Jiangsu (China)	LC-Q-TOF-MS, XBD-C18 (2.1 × 150 mm, 3.5 μm)	Chemicals present in wastewater treatment plants (WWTPs)	Eleven compounds detected at the highest level—tricyclazole, rivastigmine, atrazine, metalaxyl, azoxystrobin, carbendazim, clomazone, phenacetin, prometryn, amantadine, and diuron	Liu et al. (2020)
Surface water (lake water)	Mix mode polymeric sorbent comprising of Oasis WAX, Oasis HLB and Oasis MCX (1.5:2:1.5)	Dianshan lake, Shanghai (China)	LC-Q-Orbitrap, Hypersil Gold C18 analytical column (2.1 × 50 mm, 1.9 μm)	Herbicides, insecticides, flame retardants, surfactants, fungicides, plasticizers, and pharmaceuticals (antibiotics and antihypertensive drugs)	Diisobutylphthalate, triphenyl phosphate, Methoxyfenozide, tebuconazole, perfluorooctanoic acid (PFOA), perfluorooctane sulfonic acid (PFOS), Valsartan, Telmisartan Clindamycin, Erythromycin	Meng et al. (2020)
Surface water (river water) and effluent wastewaters	SPE (Oasis HLB sorbent)	Turia River (Spain)	LC-Q-TOF MS, C18 column (2.1 × 150 mm, 3 μm)	Mycotoxins, pesticides, and pharmaceuticals	Tanacaine, paroxypropione, ampyrone, rolipram, cyclopent, ibuverine, 5-CT, eprosartan, safingol, alprenolol, tinabinol and crotetamide	Ccanccapa-Cartagena et al. (2019)

(continued)

Table 20.1 (continued)

Sample type	Sample preparation	Location	Analytical techniques	Class of analytes detected	Selected list of top prioritized chemicals	References
Surface water, ground water, coastal waters	Filtration followed by online SPE	Miami, FL	HPLC-Q-Exactive Orbitrap, Hypersil GOLD aQ C18 encapped column	Insect repellants, personal care products, industrial compounds, metabolites, herbicides, insecticides	Diethyltoluamide (DEET), azelaic acid, stearic acid, 1-stearoylglycerol, oleic acid, icaridin, monoolein, ethyl myristate, oxybenzone, sorbitan monopalmitate,	Troxell et al. (2022)
Surface water (river water)	Filtration followed by application of regenerative (cellulose) filters and direct injection	Nidda (Germany)	LC-Q-TOF MS, C18 column (2.1 × 150 mm, 3 µm), guard column (2.0 × 4.0 mm)	Hypertoxic microcystins, Nylostab S-EED®	H-benzimidazole-sulfonic acid, N1,N3-Bis(2,2,6,6-tetramethylpiperidin-4-yl)-isophtalamide (Nylostab S-EED), microcystin YR, microcystin RR	Köppe et al. (2020)
Surface water (river water)	Filtration, SPE with mixture of different sorbents (Strata-X-AW, Strata-X-CX, Oasis HLB, Isolute ENV + (1:1:2:1.5))	Rhine (Germany)	LC-Q-TOF MS, C18 column (2.1 × 50 mm, 3.5 µm)	Biocides, artificial sweeteners, industrial chemicals, illicit pesticides, drugs, pharmaceuticals, corrosion inhibitors and others	Metformin, Gabapentin, Tetraglyme, Triethyl phosphate, Melamine, 1,3-Dimethyl-2-imidazolidone, 2,4-Dichlorobenzoic acid, 3-(2-chloroanilino)-propanoic acid, Tizanidine and 1,3-Dimethyl-2-imidazolidinone	Ruff et al. (2015)
Surface water	Filtration followed by offline SPE (Oasis HLB)	Africa	UHPLC-Q-TOF; HPH C18 column (2.1 × 100 mm, 2.7 µm)	Antiretroviral drugs, pharmaceuticals	Prednisolone, ritonavir, lamotrigine, nevirapine	Wood et al. (2017)

(continued)

Table 20.1 (continued)

Sample type	Sample preparation	Location	Analytical techniques	Class of analytes detected	Selected list of top prioritized chemicals	References
Surface water	Filtration followed by use of multiple mixture/layer of sorbents (Strata-X-AW, Strata-X-CX, Oasis HLB, Isolute ENV + (1:1:2:1.5))	Yeongsan River (Korea)	LC-Q-Orbitrap MS, C18 column (2.1 × 50 mm, 3.5 μm)	Personal care products (PPCPs), 58 pharmaceuticals	Carbamazepine, paraxanthine, cimetidine, naproxen, metformin, cetirizine, caffeine, climbazole, fexofenadine, fluconazole, tramadol, and lidocaine	Park et al. (2018)
Surface water and drinking water	Online SPE Hypersil GOLD aQ (20 × 2.1 mm, 12 μm, Thermo Scientific, USA)	Biscayne Bay, Everglades National park, Miami-Dade (Florida)	LC-Q-Orbitrap MS, Hypersil GOLD aQ C18 polar end capped (100 × 2.1, 1.9 μm, Thermo Scientific, USA)	Chemical compounds were categorized as pesticides, phthalates, bisphenols, surfactants, polyethylene glycol, PAHs, PBDEs and pharmaceuticals	Fenuron, Tetrahydropteridine, 2-Hydroxyatrazine, 3-Methyldioxyindole, Dodecylsuccinic Anhydride, Phenacetin, Geranyl isopentanoate, N-benzyl-N,N'-dimethylamine, Benzoyleneurea, 3,5-Di-tert-butyl-4-hydroxybenzaldehyde, Stearic acid, 4-Trimethylammoniobutanal etc	Ng et al. (2022)
Surface water (streams of agricultural watershed)	Glass fiber beads to prevent clogging were added to SPE (Oasis HLB, 200 mg, Waters, USA)	Wapato Creek, Boise Creek, Newaukum, Joe Leary Slough (USA)	UHPLC-Q-TOF, C18 column (2.1 × 100 mm, 1.8 μm)	Pesticides	Five pesticides were prioritized based on their risk quotient (imidacloprid, azoxystrobin, 4-hydroxy-chlorothalonil, fludioxonil and 2-methyl-4-chlorophenoxyacetic acid	Tian et al. (2021)

(continued)

Table 20.1 (continued)

Sample type	Sample preparation	Location	Analytical techniques	Class of analytes detected	Selected list of top prioritized chemicals	References
Drinking water	Activated carbon filters were lyophilized, Soxhlet extraction	North Carolina, USA	LC-TOF HRMS, C8 column (2.1 × 50 mm, 3.5 μm)	Pesticides, personal care products, consumer and industrial process compounds, food additives, colorants, herbicides, fragrances, and antimicrobial. 12 of the compounds were identified at level 1	Prioritization criteria: Human health, toxicity data from Tox21 US Federal collaborative, and EPA ToxCast program. 1,2-Benzisothiazolin-3-one*, diethylene glycol, N-[3-(Dimethylamino)propyl] methacrylamide, Nonylparaben, Dipentylphtalate, 2-[2-(2-Butoxyethoxy)ethoxy]ethanol* N, N-Dimethyldodecan-1-amine*, sucralose, PFOS*, 2-(2-Ethoxyethoxy)ethyl acetate*, TDCPP*, Zearalanol, PFOA* *Confirmed with standards	Newton et al. (2018)
Effluent, surface water, groundwater, and drinking water	Effluents: 24 h flow—corrected samples. Other samples grab samples. Acidified to pH2.3 and then OASIS HLB columns (Waters, USA)	Netherlands	LC-LTQ (Linear Ion Trap, Orbitrap HRMS) C18 column (150 mm × 2.0 mm, 3 μm)	Pharmaceuticals, industrial compounds, plant protection products, pesticides. 24 of the compounds were identified at Level 1	Prioritization criteria: 76% chemicals included in REACH with a production rate above 100 tonnes/year, 11% pesticides and biocides, and 6% pharmaceuticals. for the suspect list Caffeine, carbamazepine 10,11-epoxide, metoprolol, 1,2-benzisothiazol-3(2H)-on, 4-Methyl-1H-benzotriazole, tributyl phosphate, chloridazon, dimetheanime O, dimethomorph	Sjerps et al. (2016)

(continued)

Table 20.1 (continued)

Sample type	Sample preparation	Location	Analytical techniques	Class of analytes detected	Selected list of top prioritized chemicals	References
Groundwater and stormwater runoff	Passive sampler: SDB-RPS and SDB-XC Empore™ disks, 47 mm diameter, 12 µm particle size, 0.5 mm thick. Extraction: Liquid extraction MeOH and Acetone	Lyon, France	LC-TOF HRMS, C18 column (2.1 × 100 mm, 2.2 µm)	Fungicides, herbicides, insecticides, antibiotics, antiepileptics, antihypertensive and non-steroidal anti-inflammatory drugs, and metabolites. 40 of the compounds were identified at level 1	Atrazine, atrazine-desethyl, DEET, Dichlorobenzamide, Diuron, metolachlor, adenine, adenosine, caffeine, and carbamazepine	Pinasseau et al. (2019)
Wastewater treatment plant effluent, surface water, and drinking water	Direct injection, only samples that exhibit high particulate were centrifuged before injection	Germany	UPLC-IM-Q-TOF-MS. BEH amide (2.1 × 5 mm) 1.8 µm precolumn connected to an HSS T3 (reverse phase) (2.1 × 100 mm) 1.8 µm main column	Pharmaceuticals and transformation products	Prioritize criteria: health related issues. 10,11-dihydroxy-10,11 dihydrocarbamazepine, candesartan, carbamazepine and valsartan acid, gabapentin, gabapentin-lactam and iopamidol, and for 1H-benzotriazole, chlorothalonil M-12 and methyl-desphenyl-chloridazon	Hinnenkamp et al. (2022)

(continued)

Table 20.1 (continued)

Sample type	Sample preparation	Location	Analytical techniques	Class of analytes detected	Selected list of top prioritized chemicals	References
Groundwater	Vacuum-assisted evaporation on a Syncore® Analyst (BÜCHI, Switzerland)	Switzerland	LC-ESI-Orbitrap. column (Atlantis T3, 3 μm, 3 × 150 mm; waters, Ireland)	Pharmaceutical, industrial chemicals, pesticides, and transformation products. 12 compounds were detected in level 1, and 5 in level 2	Novel compounds detected for the first time in environmental samples: 2,5-dichlorobenzenesulfonic acid, phenylphosphonic acid, and O-des[2-aminoethyl]-O-carboxymethyl dehydroamlodipine, (TPs of amlodipine), pyroxsulam TP PSA, and metolachlor TPs: SYN542490, CGA357704 SYN542607, SYN547977, SYN542489, SYN542491, SYN547969 and SYN542488	Kiefer et al. (2021)
Groundwater	SPE (Oasis HLB)	France	LC-ESI-QTOF MS, ACQUITY BEH C18 1.7 μm, 150 mm × 2.1 mm column (waters)	Using NTA and suspect screening approach pesticides, pharmaceuticals, personal care products, and industrial compounds were detected, some transformation products were detected for the first time in environmental samples	5-methyl-1 H-benzotriazole, atrazine, pirimicarb, 2-aminobenzimidazole (TP of carbendazim), guaifenesin and desmethyl-dextrophan (TP of dextromethorphan), tramadol, ketorolac, des-venlafaxine, didesmethylvenlafaxine (TPs of venlafaxine), primidone, 2-aminobenzimidazole, carbendazim	Soulier et al. (2016)

(continued)

Table 20.1 (continued)

Sample type	Sample preparation	Location	Analytical techniques	Class of analytes detected	Selected list of top prioritized chemicals	References
Drinking water	HLB-H disk (Horizon Technology)	Sweden	LC-ESI(-)-Orbitrap MS, Purospher® STAR RP-18 end-capped column (2 μm particle size, 150 × 2.1 mm)	Presumed halogenated polyphenolic and highly unsaturated compounds. 86 DBPs were detected at level 3. 4 compounds were identified level 1	2-chloroacetic acid, 2,2-dichloroacetic acid, 2 -bromo,2 -chloroacetic acid, and 2,2 -dibromoacetic acid, Level 1	Postigo et al. (2021)
Drinking water	SPE-DEX semi-automated extraction system (4790 SPE-DEX®; Horizon Technology, Salem, New Hampshire, USA) with HLB-M SPE disks (47 mm, Atlantic, Horizon Technology)	Belgium China, Czech Republic, Germany, Italy, Japan, Spain, Sweden, Switzerland, the Netherlands, and Vietnam	Two columns were used: an Acquity UPLC BEH-C18 (Waters, 2.1 × 100 mm, 1.7 μm particle size) for the negative ionization mode and a Acquity UPLC HSS T3-C18 (Waters, 2.1 × 100 mm, 1.8 μm particle size) for the positive mode analysis	Suspect screening: 115 CECs were detected among PFAS, pharmaceuticals, pesticides and other compounds	Prioritization criteria: log D (pH adjusted octanol/water partition coefficient), log Koc (organic carbon/water partition coefficient), log Sw (water solubility), log BCF (bioconcentration factor), biodegradation, QI (quantity index), EI (exposure index) for water, EI for sewage treatment and EI for consumers D-(−)-salicin, 4-hydroxyphenylpyruvic acid, serotonin, salidroside, ginkgolide A, ginkgolide J, ginkgolide C, helicin, chlorogenic acid, 5-amino-2-hydroxy-3-sulfobenzoic acid, 7H- dodecafluoroheptanoic acid, DL-vanillactic acid lithium salt hydrate, dimidium bromide, dhurrin, asperuloside, γ-glu-cys and 4,4´ -disulfanediylbis (2-aminobutanoic acid)	Tröger et al. (2021)

(continued)

Table 20.1 (continued)

Sample type	Sample preparation	Location	Analytical techniques	Class of analytes detected	Selected list of top prioritized chemicals	References
Wastewater	Strata X polymeric reversed phase cartridge	South Africa	ABSciex 6600 series TripleTOF coupled with an ABSciex ExionLC™ ultrahigh performance liquid chromatography system (AB Sciex, Framingham, USA)	Compounds detected from pharmaceuticals, pesticides, personal care products, biological compounds, food additives, industrial chemicals to microplastics additives	Abacavir, tenofovir, efavirenz at Level 1 (antiretroviral drugs), tramadol and its metabolite o-desmethyl-cis-tramadol (opioid), jasmonic acid (plant hormone), N, N-dimethylaniline and 2,4,5-trimethylaniline (intermediates in manufacture of dyes, pigments, etc.), tetraethylene glycol (TEG) (industry and domestic consumer products), fenpropidin (fungicide), sabinene and piperitone (essential oils in food and cosmetic industries)	Abafe et al. (2023)
Wastewater	SPE (HRX)	P Bitterfeld-Wolfen, Saxony-Anhalt, Germany	LC-HRMS (ORBITRP)	Herbicides, UV filters, dyes, pharmaceuticals, transformation products	Clomazone (herbicide), benzophenone-4 and benzothiazole (UV filters), 2-(2-benzothiazolylthio) ethanol (rubber, industrial use), 5-chloro-2,6-dimethyl-4-pyrimidinamine (industrial), pentobarbital and secobarbital (barbiturates)	Hug et al. (2014)
Wastewater	Liquid liquid and Soxhlet Extraction	Sweden	GC-ToF-MS	Pharmaceuticals, plastic additives, PCPs, UV filters, organophosphorus flame retardants	Galaxolide, (synthetic fragrance) α- tocopheryl acetate, (cosmetics) octocrylene, (UV blocker) 2,4,7,9-tetramethyl-5-decyn-4,7-diol, (personal care products) benzophenone, (UV blocker) 2-(methylthio)benzothiazole, (paper and pupl industries) tris(2-butoxyethyl) phosphate, (flame retardant) acetylsalicylic acid (pain reliever), caffeine, carbazepine (antiepileptic), MTBT and n-BBSA (plastic additives, 2,3-dichlorobenzonitrile and DEET (pesticides)	Blum et al (2017)
Wastewater	SPE (Oasis HLB)	China	LC-QToF-MS	Pesticides, pharmaceuticals, intermediates, plasticizers, dyes, fertilizers	Prometryn, clomazone, metholachlor, atrazine and diuron(herbicides), azoxystrobin, carbenzaim, metalaxyl, and tricyclazole (fungicide), rivastigmine, (cholinesterase inhibitor medication), thiamethoxam, (insecticide) telmisartan, (blood vessel relaxer medication) (all at Level 1)	Liu et al. (2020)

(continued)

Table 20.1 (continued)

Sample type	Sample preparation	Location	Analytical techniques	Class of analytes detected	Selected list of top prioritized chemicals	References
Urban Runoff	SPE (Oasis HLB)	France	UPLC coupled with IMS-QToF—Vion, Waters	Rubber vulcanization accelerator DPG Additives, Polycyclic aromatic hydrocarbons (PAHs), Alkylphenols and Phthalates	Ca, Fe, Mg, K, Na, among other metals, dibuytl phthalate (DiBP), naphthalene, (polyaromatic hydrocarbons), bisphenol A (alkylphenol), BDE 100 (polybrominated biphenyl ether), perfluorooctanoic acid (perfluroalkyl carboxylic acid, 1H-benzotriazole (benzotriazoles)	Gasperi et al. (2022)
Landfill Leachate	Ultrasonic bath, SPE with Oasis HLB cartridges	Guangzhou, South China	LC-QTOF-MS	Pharmaceutical Intermediates, Personal care products, Food additives, Industrial chemicals, Pesticides and Transformation products	4-(1,1,3,3-Tetramethylbutyl) phenol. 1,2-Benzisothiazolin-3-one 1,3-Diphenylguanidine, Benzotriazole, 1,3-Di-o-tolyl guanidine, 6PPD-quinone, 2-hydroxybenzothiazole, amino benzothiazole and t-butyl hydroquinone	Han et al (2022)
Landfill leachate	LLE	Lahti, Southern Finland	GC-TOF-MS equipped with a GC Pal injection system	Industrial chemicals, Pharmaceuticals and Additives pesticides	Bisphenol A (BPA), cotinine. N, N-diethyltoluamide (DEET); followed by lidocaine (89%), camphor, benzophenone, Naphthalene, Amphetamine Dicyclopentyldimethoxysilane; (2) hydroxypropylstearate/ (3)-hydroxypropylstearate/1-hydroxypropan-2-ylstearate; laurolactam monomer (plastic monomer); p-t-Octylphenol; Diisopropyl xanthate	Hajeb et al. (2022)

Source Authors

humans. Drinking water treatment plants (DWTPs) include a disinfection step, which could be accomplished by physical or chemical procedures, adding a diversity of chemicals to the final product. These are named disinfection byproducts (DBPs), formed when disinfecting products, such as chlorine and/or ozone interact with the organic materials in water, whereas some of them have been proven to be toxic and carcinogenic to humans. Still, a major concern is that the DBPs are present in drinking water as a mixture that depends on the disinfection conditions applied and the water source's characteristics, making it a complex mix to monitor based on only targeted analysis.

Non-target analysis has been used as a tool to identify DBPs and their precursors; the first studies were conducted using gas chromatography and high-resolution mass spectrometry (GC- ESI(-)-Orbitrap MS) (Postigo et al. 2016), and more recently using Fourier transform ion cyclotron resonance (FT-ICR MS) and liquid chromatography and high-resolution mass spectrometry (LC-ESI(-)-Orbitrap MS) (Postigo et al. 2021). It is presumed that the mixture of DBPs is comprised of halogenated polyphenolic and highly unsaturated compounds, and their identification brought attention to the monitoring needs of a chlorine-based disinfection facility.

A study in Sweden (Tröger et al. 2018), including surface water before and after a WWTP effluent discharge and DWTP, identified pesticides, pharmaceuticals, personal care products, drug-related compounds, food-additives, and PFAS as part of the EPs detected through NTA. Tröger et al. (2021), continue their research path in drinking water assessment and different raw and drinking water samples collected in different countries from Europe (Belgium, Czech Republic, Germany, Italy, Spain, Sweden, Switzerland, and The Netherlands) and Asia (China, Japan, and Vietnam). Compounds such as pesticides, pharmaceuticals, PFAS, and other EPs, including Sucralose (a tracer indicator of wastewater), were detected. The United States has also detected pharmaceuticals in their drinking water treatment plant exits (Furlong et al. 2017). In France, Soulier et al. (2016) applied a passive sampler to alluvial aquifers impacted by agricultural pollution and urban effluents over several months. Using the passive sampler allowed the detection of compounds not detected by grab sampling only. This study identified pesticides, pharmaceuticals, and their transformation products for the first time on the sites sampled. A molecular fingerprint was created with NTA results and could be improved by identifying the detected unknown compounds. Using the same NTA approach of passive samplers in groundwater but having access to the proper standards, Pinasseau et al. 2019, reported for the first time the detection of ethidimuron and fluopyram (pesticides), adenine, adenosine, and nicotinamide (pharmaceuticals), in European groundwaters. Among the predominant compounds detected in groundwater, the following chemicals,11-dihydrocarbamazepine, carbamazepine, candesartan, valsartan acid, iopamidol, gabapentin-lactam, methyl-desphenyl-chloridazon were prioritized based on potential health risk.

20.3.4.2 Wastewater

Wastewater is water used in various applications or processes, such as domestic, industrial, commercial, or agricultural activities, and often contains contaminants due to these uses (Ševčík 2005), such as pharmaceuticals, personal care products, PFAS, pesticides, x-ray contrast media, endocrine disruptors, and medicines (Kumar et al. 2022). By comprehensively screening emerging contaminants in wastewater, scientists can develop sustainable removal technologies and strategies to mitigate their presence in the environment (Kumar et al. 2022). Removing contaminants from wastewater is essential for protecting public health and the environment by eliminating these EPs before the water is returned to the environment or reused. Currently, there are several methods and technologies used to treat wastewater, such as physical treatment that removes solid particles and some contaminants using techniques such as sedimentation, filtration, and aeration (Saravanan et al. 2021); biological treatment, which takes advantage of bacteria and other microorganisms action to decompose organic substances, this treatment is often used in combination with other treatment steps such as chlorination and UV treatment (Hussain et al. 2021; Samer 2015); chemical treatments that are used to speed up disinfection and induce chemical reactions that help remove contaminants from wastewater (Samer 2015); and adsorption that involve the use of materials, such as activated carbon, to remove contaminants by binding them to the adsorbent surface (Morin-Crini et al. 2022). In addition to these methods, other more advanced wastewater treatments have been used, which include ozone treatment, membrane bioreactors (MBR), advanced oxidation processes (AOP), UV in conjunction with advanced oxidation, and nanotechnology, among others (Al-Asheh et al. 2021; Chuang et al. 2017; Wert et al. 2009). Although these improved systems using AOP have shown efficient removal of some EPs, many have not been applied on an industrial scale (Arzate et al. 2017; Gomes et al. 2017; Rizzo et al. 2019; Rout et al. 2021; Sheng et al. 2016). Nevertheless, conventional wastewater treatment and even some existing advanced treatments are not able to properly eliminate most EPs due to factors such as non-biodegradability, high polarity, and water solubility, structural complexity of the compounds, and can in some cases contribute to the breakdown of precursor compounds increasing the concentration of some EPs (Alvarino et al. 2018; Chen et al. 2021; Sheng et al. 2016).

Several studies have applied NTA to discover known and unknown emerging pollutants in wastewater (Golovko et al. 2021; Venkatesan and Halden 2014; Wang et al. 2021); among them, one has identified persistent and mobile organic compounds (PMOC) as contaminants in wastewater for the first time using NTA when analyzing samples from 22 wastewater treatment plants. In addition, a study done in wastewater in Greece found 284 compounds, where thirteen of them were confirmed with reference standards as being pharmaceuticals and surfactants (Gago-Ferrero et al. 2015). Another very relevant study has applied SS to identify potentially persistent, bioaccumulative, and toxic domestic wastewater contaminants, detecting the occurrence of galaxolide, α-tocopherol acetate, octocrylene, 2,4,7,9-tetramethyl-5-decyn-4,7-diol, and various retardant chlorinated organophosphates (Blum et al. 2017).

Among the predominant compounds detected in wastewater, the following chemicals such as diethyltoluamide (DEET), lauryl guanidine, fenpropidin, strychnine, N, N-dimethylaniline, 1H-indole-3- popanoic acid, epoxyfarnesenic acide methyl ester were prioritized based on their detection frequency and confirmed using available authentic reference standard to improve confidence (Abafe et al. 2023).

20.3.4.3   Surface Water

Surface water is easily accessible and used for drinking water, livestock, hydropower, industry, recreation, and irrigation. It is present in rivers, lakes, reservoirs, springs, wetlands and can be marine, brackish, or freshwater. Because surface water is readily available, it participates in many hydrological processes and is also more exposed to anthropogenic contaminants compared to deeper source water like groundwater (Gleeson et al. 2020). Agricultural and stormwater runoff, municipal and industrial waste, hospital and wastewater effluent, wet and dry atmospheric deposition, and sewage overflow are sources of herbicides, insecticides, flame retardants, surfactants, plasticizers, pharmaceuticals, industrial chemicals, illicit drugs, phthalates, fluorinated chemicals, hormones, and personal care products into surface water bodies (Katsanou and Karapanagioti 2017) and have been monitored by target and non-target analysis. Decreasing water quality can lead to eutrophication, lower dissolved oxygen, toxicity, and reproductive and behavioral issues in aquatic animals. In humans, these contaminants can lead to skin diseases, gastrointestinal and neurological issues, kidney, blood, liver, and lung cancers (Lin et al. 2022). A suspect screening study in Sweden found seven novel organic halogenated molecules in surface water affected by wastewater using UPLC-QTOF-MS, and among the identified compound classes were pharmaceuticals, pesticides, industrial, and surfactants (Menger et al. 2021). Another study incorporated workflows for targeted analysis, suspect screening and NTA using an LC-QTOF-MS in which 68 contaminants were tentatively identified, which included 51 pesticides, 15 pharmaceuticals, and two mycotoxins in river water of the Mediterranean River Basin. In 10 water samples, imazalil, tebucanzole, nytenpriram, metalaxyl, thiabendazole, and oxytetracycline (herbicides and pesticides) were confirmed by NTA with standards (Level 1 at the Schymanski scheme) (Ccanccapa-Cartagena et al. 2019). In Quebec, Canada, NTA has been used to identify novel transformation products in surface water, in which they have reported 106 transformation products and 176 congeners, whereas eight transformation products and 20 congeners were reported for the first time (not present yet in databases at the time of publication) (Eysseric et al. 2021b). Insect repellent, personal care products, industry-related compounds, and a food additive metabolite (top 10 compounds) were tentatively identified in South Florida's freshwater, septic-tank influenced groundwater, and coastal waters (Troxell et al. 2022). Furthermore, chemicals can be prioritized based on criteria such as KOC, BCF, quality index (QI), exposure index (EI). For example, Tröger et al. in (2021) prioritized chemicals such as lamotrigine, caffeine, oxazepam, diclofenac, and PFAS such as perfluorooctanesulfonamide (FOSA), perfluorooctanoic acid (PFOA), perfluorooctanoic acid (PFOS), perfluorohexanesulfonic acid (PFHxS), perfluorobutylsulfonic acid (PFBS), perfluorodecanoic acid (PFDA) and perfluorohexanoic acid (PFHxA) in surface and tap water using QI, BCF and KOC (Tröger et al. 2021).

20.3.5 Urban and Landfill Leachates

Leachates from landfills and urban run-off consist of complex organic and inorganic chemical mixtures considered a major source of environmental pollution (Vaverková 2019). They are formed due to the gradual degradation of waste and the action of washing out by rainwater percolating the waste. The chemical composition of landfill leachate and runoffs varies per time and across sites due to changes in seasonal precipitation, landfill age, and waste composition (Masoner et al. 2014). Even though the contaminants and their respective toxicity are not consistent across landfills, the generated leachate is potentially toxic and can contaminate nearby underground and surface water (Mojiri et al. 2020). Xenobiotic organic compounds with endocrine disrupting properties such as PFAS, heavy metals, pharmaceuticals, and other EPs have been detected in landfill leachates and runoffs at concentrations ranging from ng/L to µg/L (Nika et al. 2023). Different technologies along with biological and physicochemical processes have been employed in treating leachates before final discharge (Mojiri et al. 2020), however, due to a lack of engineered structure or inadequate infrastructure in some places, landfills produce leachates that are discharged into the environment without proper treatment. This poses a significant threat to the health of the receiving ecosystem and requires continual monitoring to protect the environment (Nika et al. 2023; Vaverková 2019). Recently, NTA methods have been applied in several studies to identify EPs in landfill leachates and urban runoffs (Hajeb et al. 2022; Han et al. 2022; Jernberg et al. 2013; Nika et al. 2023; Sibiya et al. 2019). For instance, 1,3-Diphenylguanidine, Benzotriazole, 1,3-Di-o-tolylguanidine, 6PPD-quinone, 2-hydroxybenzothiazole, amino benzothiazole and t-butylhydroquinone were identified by NTA in urban runoff areas (Gasperi et al. 2022), in which only two compounds were confirmed with reference standard out of seven compounds identified in this study. Similarly, in another study by Han et al. (2022) on screening of EPs in landfill leachates and groundwater, 242 chemicals were tentatively identified and 26 of them were confirmed with reliable reference standards while 142 were quantitated with standards using target analysis which confirmed the presence of the most abundant compounds (Acesulfame, bisphenol F and ketoprofen) in the raw leachates at concentrations of 272–1780 µg/L. According to Masoner et al. (2014), among the most frequently detected EPs in landfill leachates across the United States are Bisphenol A (BPA), cotinine, and N, N-diethyltoluamide (DEET); followed by lidocaine (89%), camphor, benzophenone, naphthalene, and amphetamine with concentration in six orders of magnitude ranging from ng/L to mg/L. Other studies also reported micropollutants such as Hexabromocyclododecane (HBCDD), tetrabromobisphenol A (TBBPA), PFAS such as PFOA, PFOS, PFDS, among others, as the most prioritized in urban runoffs due to prevalence/ higher detection frequency, bioaccumulative properties and higher concentration (Gasperi et al. 2022).

20.4   Conclusions and Future Research Needs

Non-targeted analysis (NTA) using high-resolution mass spectrometry is a broad and holistic tool to monitor, identify, characterize, and prioritize emerging pollutants that are not routinely monitored or regulated in targeted approaches and the degradation and transformation products that may occur in the environment or be produced during drinking and wastewater treatment processes. NTA is helping to identify a broad spectrum of emerging contaminants in surface water, wastewater, leachate, groundwater, and drinking water. In addition to known and regulated compounds, emerging pollutants and transformation products can have different toxicity and behavior compared to the original compounds, and yet these contaminants may be interacting with each other, which may lead to synergistic or antagonistic effects. Chemical monitoring of contaminants can help to understand their spatial and temporal variability, behavior, fate, and toxicity in the environment, which will assist in the evaluation and mitigation risks and development of regulatory standards and guidelines related to emerging contaminants in different water sources. Advances and improvements in NTA tools and software to overcome lengthy data processing, time-consuming, inconsistencies among different platforms (e.g., algorithms), and/or to make the process more transparent (especially for commercial software) is a major need in the NTA research area, which would facilitate the acceptability and application of data generated by NTA by stakeholders and governmental agencies for regulatory purposes.

Acknowledgements This material is based upon work supported by the National Science Foundation under Grant No. HRD-1547798 and Grant No. HRD-2111661. These NSF Grants were awarded to Florida International University as part of the Centers of Research Excellence in Science and Technology (CREST) Program. This is contribution number #1784 from the Institute of Environment, a Preeminent Program at Florida International University.

References

Abafe OA, Lawal MA, Chokwe TB (2023) Non-targeted screening of emerging contaminants in South African surface and wastewater. Emerg Contam 9:100246. https://doi.org/10.1016/j.emcon.2023.100246

Aceña J, Stampachiacchiere S, Pérez S, Barceló D (2015) Advances in liquid chromatography - high-resolution mass spectrometry for quantitative and qualitative environmental analysis. Anal Bioanal Chem 407:6289–6299. https://doi.org/10.1007/s00216-015-8852-6

Al-Asheh S, Bagheri M, Aidan A (2021) Membrane bioreactor for wastewater treatment: a review. Case Stud Chem Environ Eng 4:100109. https://doi.org/10.1016/j.cscee.2021.100109

Alvarino T, Suarez S, Lema J, Omil F (2018) Understanding the sorption and biotransformation of organic micropollutants in innovative biological wastewater treatment technologies. Sci Total Environ 615:297–306. https://doi.org/10.1016/j.scitotenv.2017.09.278

Angeles LF, Islam S, Aldstadt J, Saqeeb KN, Alam M, Khan MA et al (2020) Retrospective suspect screening reveals previously ignored antibiotics, antifungal compounds, and metabolites in Bangladesh surface waters. Sci Total Environ 712. https://doi.org/10.1016/j.scitotenv.2019. 136285

Arzate S, García Sánchez JL, Soriano-Molina P, Casas López JL, Campos-Mañas MC, Agüera A et al (2017) Effect of residence time on micropollutant removal in WWTP secondary effluents by continuous solar photo-fenton process in raceway pond reactors. Chem Eng J 316:1114–1121. https://doi.org/10.1016/j.cej.2017.01.089

Atugoda T, Vithanage M, Wijesekara H, Bolan N, Sarmah AK, Bank MS et al (2021) Interactions between microplastics, pharmaceuticals and personal care products: implications for vector transport. Environ Int 149. https://doi.org/10.1016/j.envint.2020.106367

Augusto MR, Claro ICM, Siqueira AK, Sousa GS, Caldereiro CR, Duran AFA et al (2022) Sampling strategies for wastewater surveillance: evaluating the variability of SARS-COV-2 RNA concentration in composite and grab samples. J Environ Chem Eng 10:107478. https://doi.org/10.1016/ j.jece.2022.107478

Bade R, Causanilles A, Emke E, Bijlsma L, Sancho JV, Hernandez F et al (2016) Facilitating high resolution mass spectrometry data processing for screening of environmental water samples: an evaluation of two deconvolution tools. Sci Total Environ 569–570:434–441. https://doi.org/10. 1016/j.scitotenv.2016.06.162

Badea SL, Geana EI, Niculescu VC, Ionete RE (2020) Recent progresses in analytical GC and LC mass spectrometric based-methods for the detection of emerging chlorinated and brominated contaminants and their transformation products in aquatic environment. Sci Total Environ 722. https://doi.org/10.1016/j.scitotenv.2020.137914

Bader T, Schulz W, Kümmerer K, Winzenbacher R (2016) General strategies to increase the repeatability in non-target screening by liquid chromatography-high resolution mass spectrometry. Anal Chim Acta 935:173–186. https://doi.org/10.1016/j.aca.2016.06.030

Belay MH, Precht U, Mortensen P et al (2022) A fully automated online SPE-LC-MS/MS method for the determination of 10 pharmaceuticals in wastewater samples. Toxics 10. https://doi.org/ 10.3390/toxics10030103

Bergfors SN, Huynh K, Jensen AE, Sundberg J (2021) Non-target screening of organic compounds in offshore produced water by GC×GC-MS. PeerJ Anal Chem 3:e11. https://doi.org/10.7717/ peerj-achem.11

Bletsou AA, Jeon J, Hollender J, Archontaki E, Thomaidis NS (2015) Targeted and non-targeted liquid chromatography-mass spectrometric workflows for identification of transformation products of emerging pollutants in the aquatic environment. TrAC Trends Anal Chem 66:32–44. https://doi.org/10.1016/J.TRAC.2014.11.009

Blum KM, Andersson PL, Renman G, Ahrens L, Gros M, Wiberg K et al (2017) Non-target screening and prioritization of potentially persistent, bioaccumulating and toxic domestic wastewater contaminants and their removal in on-site and large-scale sewage treatment plants. Sci Total Environ 575:265–275. https://doi.org/10.1016/j.scitotenv.2016.09.135

Broeckling CD, Hoyes E, Richardson K, Brown JM, Prenni JE (2018) Comprehensive tandem-mass-spectrometry coverage of complex samples enabled by data-set-dependent acquisition. Anal Chem 90:8020–8027. https://doi.org/10.1021/acs.analchem.8b00929

Byrdwell WC, Neff WE (2002) Dual parallel electrospray ionization and atmospheric pressure chemical ionization mass spectrometry (MS), MS/MS and MS/MS/MS for the analysis of triacylglycerols and triacylglycerol oxidation products. Rapid Commun Mass Spectrom 16:300–319. https://doi.org/10.1002/rcm.581

Ccanccapa-Cartagena A, Pico Y, Ortiz X, Reiner EJ (2019) Suspect, non-target and target screening of emerging pollutants using data independent acquisition: assessment of a Mediterranean River basin. Sci Total Environ 687:355–368. https://doi.org/10.1016/j.scitotenv.2019.06.057

Challis JK, Almirall XO, Helm PA, Wong CS (2020) Performance of the organic-diffusive gradients in thin-films passive sampler for measurement of target and suspect wastewater contaminants. Environ Pollut 261. https://doi.org/10.1016/j.envpol.2020.114092

Cheng SC, Jhang SS, Huang MZ, Shiea J (2015) Simultaneous detection of polar and nonpolar compounds by ambient mass spectrometry with a dual electrospray and atmospheric pressure chemical ionization source. Anal Chem 87:1743–1748. https://doi.org/10.1021/ac503625m

Chuang Y-H, Chen S, Chinn CJ, Mitch WA (2017) Comparing the UV/monochloramine and UV/free chlorine Advanced Oxidation Processes (AOPs) to the UV/hydrogen peroxide AOP under scenarios relevant to potable reuse. Environ Sci Technol 51:13859–13868. https://doi.org/10.1021/acs.est.7b03570

Deere JR, Streets S, Jankowski MD, Ferrey M, Chenaux-Ibrahim Y, Convertino M, Isaac EJ, Phelps NBD, Primus A, Servadio JL, Singer RS, Travis DA, Moore S, Wolf TM (2021) A chemical prioritization process: applications to contaminants of emerging concern in freshwater ecosystems (phase I). Sci Total Environ 772:146030. https://doi.org/10.1016/j.scitotenv.2021.146030

Dimzon IKD, Morata AS, Müller Janine, Yanela RK, Lebertz S, Weil H et al (2018) Trace organic chemical pollutants from the lake waters of San Pablo City, Philippines by targeted and non-targeted analysis. Sci Total Environ 639:588–595. https://doi.org/10.1016/j.scitotenv.2018.05.217

Domínguez I, Arrebola FJ, Martínez Vidal JL, Garrido FA (2020) Assessment of wastewater pollution by gas chromatography and high resolution Orbitrap mass spectrometry. J Chromatogr A 1619:460964. https://doi.org/10.1016/j.chroma.2020.460964

Dührkop K, Fleischauer M, Ludwig M et al (2019) SIRIUS 4: a rapid tool for turning tandem mass spectra into metabolite structure information. Nat Methods 16:299–302. https://doi.org/10.1038/s41592-019-0344-8

Duncan KD, Vandergrift GW, Krogh ET, Gill CG (2014) Ionization suppression effects with condensed phase membrane introduction mass spectrometry: methods to increase the linear dynamic range and sensitivity. J Mass Spectrom 50:437–443. https://doi.org/10.1002/jms.3544

Ebele AJ, Oluseyi T, Drage DS, Harrad S, Abou-Elwafa Abdallah M (2020) Occurrence, seasonal variation and human exposure to pharmaceuticals and personal care products in surface water, groundwater and drinking water in Lagos State, Nigeria. Emerging Contam 6:124–132. https://doi.org/10.1016/j.emcon.2020.02.004

Eichman HJ, Eck BJ, Lagalante AF (2017) A comparison of electrospray ionization, atmospheric pressure chemical ionization, and atmospheric pressure photoionization for the liquid chromatography/tandem mass spectrometric analysis of bisphenols. Application to bisphenols in thermal paper receipts and U.S. currency notes. Rapid Commun Mass Spectr 31:1773–1778. https://doi.org/10.1002/rcm.7950

El-Deen AK, Shimizu K (2022) Suspect and non-target screening workflow for studying the occurrence, fate, and environmental risk of contaminants in wastewater using data-independent acquisition. J Chromatogr A 1667. https://doi.org/10.1016/j.chroma.2022.462905

Eysseric E, Beaudry F, Gagnon C, Segura PA (2021a) Non-targeted screening of trace organic contaminants in surface waters by a multi-tool approach based on combinatorial analysis of tandem mass spectra and open access databases. Talanta 230:122293. https://doi.org/10.1016/j.talanta.2021.122293

Eysseric E, Gagnon C, Segura P.A (2021b) Identifying congeners and transformation products of organic contaminants within complex chemical mixtures in impacted surface waters with a top-down non-targeted screening workflow. Sci Total Environ 822. https://doi.org/10.1016/j.scitotenv.2022.153540

Feng X, Li D, Liang W, Ruan T, Jiang G (2021) Recognition and prioritization of chemical mixtures and transformation products in Chinese estuarine waters by suspect screening analysis. Environ Sci Technol 55:9508–9517. https://doi.org/10.1021/acs.est.0c06773

Fisher CM, Croley TR, Knolhoff AM (2021) Data processing strategies for non-targeted analysis of foods using liquid chromatography/high-resolution mass spectrometry. TrAC - Trends Anal Chem 136. https://doi.org/10.1016/j.trac.2021.116188

Furlong ET, Batt AL, Glassmeyer ST, Noriega MC, Kolpin DW, Mash H et al (2017) Nationwide reconnaissance of contaminants of emerging concern in source and treated drinking waters of the United States: pharmaceuticals. Sci Total Environ 579:1629–1642. https://doi.org/10.1016/j.scitotenv.2016.03.128

Gago-Ferrero P, Schymanski EL, Bletsou AA, Aalizadeh R, Hollender J, Thomaidis NS (2015) Extended suspect and non-target strategies to characterize emerging polar organic contaminants in raw wastewater with LC-HRMS/MS. Environ Sci Technol 49:12333–12341. https://doi.org/10.1021/acs.est.5b03454

Gao S, Zhang ZP, Karnes HT (2005) Sensitivity enhancement in liquid chromatography/atmospheric pressure ionization mass spectrometry using derivatization and mobile phase additives. J Chromatogr B Anal Technol Biomed Life Sci 825:98–110. https://doi.org/10.1016/j.jchromb.2005.04.021

Gasperi J, Le Roux J, Deshayes S, Ayrault S, Bordier L, Boudahmane L, Budzinski H, Caupos E, Caubriere N, Flanagan K, Guillon M, Huynh N, Labadie P, Meffray L, Neveu P, Partibane C, Paupardin J, Saad M, Varnede L, Gromaire M-C (2022) Micropollutants in urban runoff from traffic areas: target and non-target screening on four contrasted sites. Water 14(3):394. https://www.mdpi.com/2073-4441/14/3/394

Gleeson T, Wang-Erlandsson L, Porkka M, Zipper SC, Jaramillo F, Gerten D et al (2020) Illuminating water cycle modifications and Earth system resilience in the Anthropocene. Water Resour Res 56. https://doi.org/10.1029/2019WR024957

Gobelius L, Persson C, Wiberg K, Ahrens L (2019) Calibration and application of passive sampling for per- and polyfluoroalkyl substances in a drinking water treatment plant. J Hazard Mater 362:230–237. https://doi.org/10.1016/j.jhazmat.2018.09.005

Godejohann M, Berset JD, Muff D (2011) Non-targeted analysis of wastewater treatment plant effluents by high performance liquid chromatography-time slice-solid phase extraction-nuclear magnetic resonance/time-of-flight-mass spectrometry. J Chromatogr A 1218:9202–9209. https://doi.org/10.1016/j.chroma.2011.10.051

Golovko O, Örn S, Sörengård M, Frieberg K, Nassazzi W, Lai FY et al (2021) Occurrence and removal of chemicals of emerging concern in wastewater treatment plants and their impact on receiving water systems. Sci Total Environ 754:142122. https://doi.org/10.1016/j.scitotenv.2020.142122

Gomes J, Costa R, Quinta-Ferreira RM, Martins RC (2017) Application of ozonation for pharmaceuticals and personal care products removal from water. Sci Total Environ 586:265–283. https://doi.org/10.1016/j.scitotenv.2017.01.216

González-Gaya B, Lopez-Herguedas N, Bilbao D, Mijangos L, Iker AM, Etxebarria N, Irazola M, Prieto A, Olivares M, Zuloaga O (2021) Suspect and non-target screening: the last frontier in environmental analysis. Anal Methods 13(16):1876–1904. https://doi.org/10.1039/d1ay00111f. Royal Society of Chemistry

González-González RB, Sharma P, Singh SP, Américo-Pinheiro JHP, Parra-Saldívar R, Bilal M, Iqbal HMN (2022) Persistence, environmental hazards, and mitigation of pharmaceutically active residual contaminants from water matrices. Sci Total Environ 821:153329. https://doi.org/10.1016/j.scitotenv.2022.153329

Guo Z, Huang S, Wang J, Feng YL (2020) Recent advances in non-targeted screening analysis using liquid chromatography - high resolution mass spectrometry to explore new biomarkers for human exposure. Talanta 219. https://doi.org/10.1016/j.talanta.2020.121339

Hajeb P, Zhu L, Bossi R, Vorkamp K (2022) Sample preparation techniques for suspect and non-target screening of emerging contaminants. Chemosphere 287(Pt 3):132306. https://doi.org/10.1016/j.chemosphere.2021.132306

Han Y, Hu L-X, Liu T, Liu J, Wang Y-Q, Zhao J-H, Liu Y-S, Zhao J-L, Ying G-G (2022) Non-target, suspect and target screening of chemicals of emerging concern in landfill leachates and groundwater in Guangzhou, South China. Sci Total Environ 837:155705. https://doi.org/10.1016/j.scitotenv.2022.155705

Hinnenkamp V, Balsaa P, Schmidt TC (2022) Target, suspect and non-target screening analysis from wastewater treatment plant effluents to drinking water using collision cross section values as additional identification criterion. Anal Bioanal Chem 414:425–438. https://doi.org/10.1007/s00216-021-03263-1

Hohrenk LL, Itzel F, Baetz N, Tuerk J, Vosough M, Schmidt TC (2020) Comparison of software tools for liquid chromatography-high-resolution mass spectrometry data processing in nontarget screening of environmental samples. Anal Chem 92:1898–1907. https://doi.org/10.1021/acs.analchem.9b04095

Hollender J, van Bavel B, Dulio V, Farmen E, Furtmann K, Koschorreck J et al (2019) High resolution mass spectrometry-based non-target screening can support regulatory environmental monitoring and chemicals management. Environ Sci Eur 31:42. https://doi.org/10.1186/s12302-019-0225-x

Hollender J, Schymanski EL, Singer HP, Ferguson PL (2017) Nontarget screening with high resolution mass spectrometry in the environment: ready to go? Environ Sci Technol 51:11505–11512. https://doi.org/10.1021/acs.est.7b02184

Hong S, Giesy JP, Lee JS, Lee JH, Khim JS (2016) Effect-directed analysis: current status and future challenges. Ocean Sci J 51:413–433. https://doi.org/10.1007/s12601-016-0038-4

Hug C, Ulrich N, Schulze T, Brack W, Krauss M (2014) Identification of novel micropollutants in wastewater by a combination of suspect and nontarget screening. Environ Pollut 184:25–32. https://doi.org/10.1016/j.envpol.2013.07.048

Hussain A, Kumari R, Sachan SG, Sachan A (2021) Biological wastewater treatment technology: advancement and drawbacks. In: Microbial ecology of wastewater treatment plants, pp 175–192. Elsevier. https://doi.org/10.1016/B978-0-12-822503-5.00002-3

Jernberg J, Pellinen J, Rantalainen A-L (2013) Qualitative nontarget analysis of landfill leachate using gas chromatography time-of-flight mass spectrometry. Talanta 103:384–391. https://doi.org/10.1016/j.talanta.2012.10.084

Jacob P, Barzen-Hanson KA, Helbling DE (2021) Target and nontarget analysis of per- and polyfluoroalkyl substances in wastewater from electronics fabrication facilities. Environ Sci Technol 55(4):2346–2356. https://doi.org/10.1021/acs.est.0c06690

Kaserzon SL, Kennedy K, Hawker DW, Thompson J, Carter S, Roach AC et al (2012) Development and calibration of a passive sampler for perfluorinated alkyl carboxylates and sulfonates in water. Environ Sci Technol 46:4985–4993. https://doi.org/10.1021/es300593a

Kaserzon SL, Heffernan AL, Thompson K, Mueller JF, Gomez Ramos MJ (2017) Rapid screening and identification of chemical hazards in surface and drinking water using high resolution mass spectrometry and a case-control filter. Chemosphere 182:656–664. https://doi.org/10.1016/j.chemosphere.2017.05.071

Katsanou, K, Karapanagioti, H (2017) Surface water and groundwater sources for drinking water in Applications of Advanced Oxidation Processes (AOPs) in drinking water treatment. Hdb Env Chem 67. https://doi.org/10.1007/698_2017_140

Kiefer K, Du L, Singer H, Hollender J (2021) Identification of LC-HRMS nontarget signals in groundwater after source related prioritization. Water Res 196. https://doi.org/10.1016/j.watres.2021.116994.

Koronaiou LA, Nannou C, Xanthopoulou N et al (2022) High-resolution mass spectrometry-based strategies for the target analysis and suspect screening of per- and polyfluoroalkyl substances in aqueous matrices. Microchem J 179. https://doi.org/10.1016/j.microc.2022.107457

Köppe T, Jewell KS, Dietrich C, Wick A, Ternes TA (2020) Application of a non-target workflow for the identification of specific contaminants using the example of the Nidda river basin. Water Res 178. https://doi.org/10.1016/j.watres.2020.115703

Krauss M, Singer H, Hollender J (2010) LC-high resolution MS in environmental analysis: from target screening to the identification of unknowns. Anal Bioanal Chem 397:943–951. https://doi.org/10.1007/s00216-010-3608-9

Kumar R, Qureshi M, Vishwakarma DK, Al-Ansari N, Kuriqi A, Elbeltagi A et al (2022) A review on emerging water contaminants and the application of sustainable removal technologies. Case Stud Chem Environ Eng 6:100219. https://doi.org/10.1016/j.cscee.2022.100219

Lacina O, Hradkova P, Pulkrabova J, Hajslova J (2011) Simple, high throughput ultra-high performance liquid chromatography/tandem mass spectrometry trace analysis of perfluorinated alkylated substances in food of animal origin: milk and fish. J Chromatogr A 1218:4312–4321. https://doi.org/10.1016/j.chroma.2011.04.061

Lai A, Singh RR, Kovalova L, Jaeggi O, Kondić T, Schymanski EL (2021) Retrospective non-target analysis to support regulatory water monitoring: from masses of interest to recommendations via in silico workflows. Environ Sci Eur 33:43. https://doi.org/10.1186/s12302-021-00475-1

Lee HR, Kochhar S, Shim SM (2015) Comparison of electrospray ionization and atmospheric chemical ionization coupled with the liquid chromatography-tandem mass spectrometry for the analysis of cholesteryl esters. Int J Anal Chem 2015. https://doi.org/10.1155/2015/650927

Lee S, Kim K, Jeon J, Moon HB (2019) Optimization of suspect and non-target analytical methods using GC/TOF for prioritization of emerging contaminants in the Arctic environment. Ecotoxicol Environ Saf 181:11–17. https://doi.org/10.1016/j.ecoenv.2019.05.070

Leendert V, Van Langenhove H, Demeestere K (2015) Trends in liquid chromatography coupled to high-resolution mass spectrometry for multi-residue analysis of organic micropollutants in aquatic environments. TrAC - Trends Anal Chem 67:192–208. https://doi.org/10.1016/j.trac.2015.01.010

Li X, Cui D, Ng B, Ogunbiyi OD, Guerra de Navarro M, Gardinali P et al (2023) Non-targeted analysis for the screening and semi-quantitative estimates of per-and polyfluoroalkyl substances in water samples from South Florida environments. J Hazard Mater 452. https://doi.org/10.1016/j.jhazmat.2023.131224

Liigand J, Wang T, Kellogg J, Smedsgaard J, Cech N, Kruve A (2020) Quantification for non-targeted LC/MS screening without standard substances. Sci Rep 10. https://doi.org/10.1038/s41598-020-62573-z

Lin L, Yang H, Xu X (2022) Effects of water pollution on human health and disease heterogeneity: a review. Frontiers 10. https://doi.org/10.3389/fenvs.2022.880246

Liu W, Yao H, Xu W, Liu G, Wang X, Tu Y et al (2020) Suspect screening and risk assessment of pollutants in the wastewater from a chemical industry park in China. Environ Pollut 263:114493. https://doi.org/10.1016/j.envpol.2020.114493

Manz KE, Feerick A, Braun JM, Feng Y-L, Hall A, Koelmel J et al (2023) Non-targeted analysis (NTA) and suspect screening analysis (SSA): a review of examining the chemical exposome. J Expo Sci Environ Epidemiol 33:524–536. https://doi.org/10.1038/s41370-023-00574-6

Martínez-Bueno MJ, Gómez Ramos MJ, Bauer A, Fernández-Alba AR (2019) An overview of non-targeted screening strategies based on high resolution accurate mass spectrometry for the identification of migrants coming from plastic food packaging materials. TrAC - Trends Anal Chem 110:191–203. https://doi.org/10.1016/j.trac.2018.10.035

Masoner JR, Kolpin DW, Furlong ET, Cozzarelli IM, Gray JL, Schwab EA (2014) Contaminants of emerging concern in fresh leachate from landfills in the conterminous United States. Environ Sci Process Impacts 16(10):2335–2354. https://doi.org/10.1039/c4em00124a

Meng D, Fan DL, Gu W, Wang Z, Chen YJ, Bu HZ et al (2020) Development of an integral strategy for non-target and target analysis of site-specific potential contaminants in surface water: a case study of Dianshan Lake, China. Chemosphere 243:125367. https://doi.org/10.1016/j.chemosphere.2019.125367

Menger F, Ahrens L, Wiberg K, Gago-Ferrero P (2021) Suspect screening based on market data of polar halogenated micropollutants in river water affected by wastewater. J Hazard Mater 401. https://doi.org/10.1016/j.jhazmat.2020.123377

Mojiri A, Zhou JL, Ratnaweera H, Ohashi A, Ozaki N, Kindaichi T, Asakura H (2020) Treatment of landfill leachate with different techniques: an overview. Water Reuse 11(1):66–96. https://doi.org/10.2166/wrd.2020.079

Morin-Crini N, Lichtfouse E, Fourmentin M, Ribeiro ARL, Noutsopoulos C, Mapelli F et al (2022) Removal of emerging contaminants from wastewater using advanced treatments. A review. Environ Chem Lett 20:1333–1375. https://doi.org/10.1007/s10311-021-01379-5

Newton SR, McMahen RL, Sobus JR, Mansouri K, Williams AJ, McEachran AD et al (2018) Suspect screening and non-targeted analysis of drinking water using point-of-use filters. Environ Pollut 234:297–306. https://doi.org/10.1016/j.envpol.2017.11.033

Ng B, Quinete N, Gardinali P (2022) Differential organic contaminant ionization source detection and identification in environmental waters by nontargeted analysis. Environ Toxicol Chem 41:1154–1164. https://doi.org/10.1002/etc.5268

Ng B, Quinete N, Gardinali PR (2020) Assessing accuracy, precision and selectivity using quality controls for non-targeted analysis. Sci Total Environ 713. https://doi.org/10.1016/j.scitotenv.2020.136568

Nika M-C, Alygizakis N, Arvaniti OS, Thomaidis NS (2023) Non-target screening of emerging contaminants in landfills: a review. Curr Opin Environ Sci & Health 32:100430. https://doi.org/10.1016/j.coesh.2022.100430

OECD (2001) Organization for economic co-operation and development series on testing and assessment, number 33. Harmonised integrated classification system for human health and environmental hazards of chemical substances and mixtures.

Ogunbiyi OD, Ajiboye TO, Omotola EO, Oladoye PO, Olanrewaju CA, Quinete N (2023) Analytical approaches for screening of per- and poly fluoroalkyl substances in food items: a review of recent advances and improvements. Environ Pollut 329. https://doi.org/10.1016/j.envpol.2023.121705

Ort C, Lawrence MG, Rieckermann J, Joss A (2010) Sampling for Pharmaceuticals and Personal Care Products (PPCPs) and Illicit drugs in wastewater systems: are your conclusions valid? Critical review. Environ Sci Technol 44:6024–6035. https://doi.org/10.1021/es100779n

Park N, Choi Y, Kim D, Kim K, Jeon J (2018) Prioritization of highly exposable pharmaceuticals via a suspect/non-target screening approach: a case study for Yeongsan River, Korea. Sci Total Environ 639:570–579. https://doi.org/10.1016/j.scitotenv.2018.05.081

Paszkiewicz M, Godlewska K, Lis H, Caban M, Białk-Bielińska A, Stepnowski P (2022) Advances in suspect screening and non-target analysis of polar emerging contaminants in the environmental monitoring. TrAC - Trends Anal Chem 154. https://doi.org/10.1016/j.trac.2022.116671

Pinasseau L, Wiest L, Fildier A, Volatier L, Fones GR, Mills GA et al (2019) Use of passive sampling and high resolution mass spectrometry using a suspect screening approach to characterise emerging pollutants in contaminated groundwater and runoff. Sci Total Environ 672:253–263. https://doi.org/10.1016/j.scitotenv.2019.03.489

Place BJ, Ulrich EM, Challis JK, Chao A, Du B, Favela K, Feng YL, Fisher CM, Gardinali P, Hood A, Knolhoff AM, McEachran AD, Nason SL, Newton SR, Ng B, Nuñez J, Peter KT, Phillips AL, Quinete N, Renslow R, Sobus J, Sussman E, Warth B, Wickramasekara S, Williams AJ (2021) An introduction to the benchmarking and publications for non-targeted analysis working group. Anal Chem 93(49):16289–16296. https://doi.org/10.1021/acs.analchem.1c02660

Postigo C, Andersson A, Harir M, Bastviken D, Gonsior M, Schmitt-Kopplin P et al (2021) Unraveling the chemodiversity of halogenated disinfection by-products formed during drinking water treatment using target and non-target screening tools. J Hazard Mater 401. https://doi.org/10.1016/j.jhazmat.2020.123681

Postigo C, Cojocariu CI, Richardson SD, Silcock PJ, Barcelo D (2016) Characterization of iodinated disinfection by-products in chlorinated and chloraminated waters using Orbitrap based gas chromatography-mass spectrometry. Anal Bioanal Chem 408:3401–3411. https://doi.org/10.1007/s00216-016-9435-x

Puri M, Gandhi K, Kumar MS (2023) Emerging environmental contaminants: a global perspective on policies and regulations. J Environ Manag 332. https://doi.org/10.1016/j.jenvman.2023.117344

Purschke K, Zoell C, Leonhardt J, Weber M, Schmidt TC (2020) Identification of unknowns in industrial wastewater using offline 2D chromatography and non-target screening. Sci Total Environ 706. https://doi.org/10.1016/j.scitotenv.2019.135835

Rizzo L, Malato S, Antakyali D, Beretsou VG, Đolić MB, Gernjak W et al (2019) Consolidated vs new advanced treatment methods for the removal of contaminants of emerging concern from urban wastewater. Sci Total Environ 655:986–1008. https://doi.org/10.1016/j.scitotenv.2018. 11.265

Rosenfeld PE, Paul E, Feng LGH (2011) Risks of hazardous wastes. Elsevier/William Andrew. ISBN: 9781437778434. https://shop.elsevier.com/books/risks-of-hazardous-wastes/rosenfeld/ 978-1-4377-7842-7

Rout PR, Zhang TC, Bhunia P, Surampalli RY (2021) Treatment technologies for emerging contaminants in wastewater treatment plants: a review. Sci Total Environ 753:141990. https://doi.org/ 10.1016/j.scitotenv.2020.141990

Ruff M, Mueller MS, Loos M, Singer HP (2015) Quantitative target and systematic non-target analysis of polar organic micro-pollutants along the river Rhine using high-resolution mass-spectrometry - identification of unknown sources and compounds. Water Res 87:145–154. https://doi.org/10.1016/j.watres.2015.09.017

Ruttkies C, Neumann S, Posch S (2019) Improving MetFrag with statistical learning of fragment annotations. BMC Bioinformatics 20. https://doi.org/10.1186/s12859-019-2954-7

Samer M (2015) Biological and chemical wastewater treatment processes. Wastewater Treat Eng 150:61250. https://doi.org/10.5772/61250. InTech

Saravanan A, Senthil Kumar P, Jeevanantham S, Karishma S, Tajsabreen B, Yaashikaa PR et al (2021) Effective water/wastewater treatment methodologies for toxic pollutants removal: processes and applications towards sustainable development. Chemosphere 280:130595. https:// doi.org/10.1016/j.chemosphere.2021.130595

Schymanski EL, Jeon J, Gulde R, Fenner K, Ruff M, Singer HP et al (2014) Identifying small molecules via high resolution mass spectrometry: communicating confidence. Environ Sci Technol 48:2097–2098. https://doi.org/10.1021/es5002105

Schymanski EL, Singer HP, Slobodnik J, Ipolyi IM, Oswald P, Krauss M et al (2015) Non-target screening with high-resolution mass spectrometry: critical review using a collaborative trial on water analysis. Anal Bioanal Chem 407:6237–6255. https://doi.org/10.1007/s00216-015-8681-7

Segura PA, MacLeod SL, Lemoine P, Sauvé S, Gagnon C (2011) Quantification of carbamazepine and atrazine and screening of suspect organic contaminants in surface and drinking waters. Chemosphere 84:1085–1094. https://doi.org/10.1016/j.chemosphere.2011.04.056

Ševčík JGK (2005) Sampling|practice. In: Encyclopedia of analytical science, pp 198–204. Elsevier. https://doi.org/10.1016/B0-12-369397-7/00547-1

Sheng C, Nnanna AGA, Liu Y, Vargo JD (2016) Removal of trace pharmaceuticals from water using coagulation and powdered activated carbon as pretreatment to ultrafiltration membrane system. Sci Total Environ 550:1075–1083. https://doi.org/10.1016/j.scitotenv.2016.01.179

Sibiya I, Poma G, Cuykx M, Covaci A, Daso Adegbenro P, Okonkwo J (2019) Targeted and non-target screening of persistent organic pollutants and organophosphorus flame retardants in leachate and sediment from landfill sites in Gauteng Province, South Africa. Sci Total Environ 653:1231–1239. https://doi.org/10.1016/j.scitotenv.2018.10.356

Simon E, Van Velzen M, Brandsma SH, Lie E, Løken K, De Boer J, Bytingsvik J, Jenssen BM, Aars J, Hamers T, Lamoree MH (2013) Effect-directed analysis to explore the polar bear exposome: identification of thyroid hormone disrupting compounds in plasma. Environ Sci Technol 47:8902–8912. https://doi.org/10.1021/es401696u

Singh RR, Chao A, Phillips KA, Xia XR, Shea D, Sobus JR et al (2020) Expanded coverage of non-targeted LC-HRMS using atmospheric pressure chemical ionization: a case study with ENTACT mixtures. Anal Bioanal Chem 412:4931–4939. https://doi.org/10.1007/s00216-020-02716-3

Sjerps RMA, Vughs D, van Leerdam JA, ter Laak TL, van Wezel AP (2016) Data-driven prioritization of chemicals for various water types using suspect screening LC-HRMS. Water Res 93:254–264. https://doi.org/10.1016/j.watres.2016.02.034

Smith CA, Want EJ, O'Maille G, Abagyan R, Siuzdak G (2006) XCMS: processing mass spectrometry data for metabolite profiling using nonlinear peak alignment, matching, and identification. Anal Chem 78:779–787. https://doi.org/10.1021/ac051437y

Soulier C, Coureau C, Togola A (2016) Environmental forensics in groundwater coupling passive sampling and high resolution mass spectrometry for screening. Sci Total Environ 563–564:845–854. https://doi.org/10.1016/j.scitotenv.2016.01.056

Straube EA, Dekant W, Völkel W (2004) Comparison of electrospray ionization, atmospheric pressure chemical ionization, and atmospheric pressure photoionization for the analysis of dinitropyrene and aminonitropyrene LC-MS/MS. J Am Soc Mass Spectrom 15:1853–1862. https://doi.org/10.1016/j.jasms.2004.08.017

Stravs MA, Schymanski EL, Singer HP, Hollender J (2013) Automatic recalibration and processing of tandem mass spectra using formula annotation. J Mass Spectrom 48:89–99. https://doi.org/10.1002/jms.3131

Sultana T, Murray C, Kleywegt S, Metcalfe CD (2018) Neonicotinoid pesticides in drinking water in agricultural regions of southern Ontario, Canada. Chemosphere 202:506–513. https://doi.org/10.1016/j.chemosphere.2018.02.108

Tadić Đ, Manasfi R, Bertrand M et al (2022) Use of passive and grab sampling and high-resolution mass spectrometry for non-targeted analysis of emerging contaminants and their semi-quantification in water. Molecules 27. https://doi.org/10.3390/molecules27103167

Tian Z, Wark DA, Bogue K, James CA (2021) Suspect and non-target screening of contaminants of emerging concern in streams in agricultural watersheds. Sci Total Environ 795. https://doi.org/10.1016/j.scitotenv.2021.148826

Tröger R, Klöckner P, Ahrens L, Wiberg K (2018) Micropollutants in drinking water from source to tap - method development and application of a multiresidue screening method. Sci Total Environ 627:1404–1432. https://doi.org/10.1016/J.SCITOTENV.2018.01.277

Tröger R, Ren H, Yin D, Postigo C, Nguyen PD, Baduel C et al (2021) What's in the water? – Target and suspect screening of contaminants of emerging concern in raw water and drinking water from Europe and Asia. Water Res 198. https://doi.org/10.1016/j.watres.2021.117099.

Troxell K, Ng B, Zamora-Ley I, Gardinali P (2022) Detecting water constituents unique to septic tanks as a wastewater source in the environment by nontarget analysis: South Florida's deering estate rehydration project case study. Environ Toxicol Chem 5:1165–1178: https://doi.org/10.1002/etc.5309

Vaverková MD (2019) Landfill impacts on the environment—review. Geosciences 9(10):431. https://www.mdpi.com/2076-3263/9/10/431

Venkatesan AK, Halden RU (2014) Wastewater treatment plants as chemical observatories to forecast ecological and human health risks of manmade chemicals. Sci Rep 4:3731. https://doi.org/10.1038/srep03731

Wang S, Huo Z, Gu J, Xu G (2021) Benzophenones and synthetic progestin in wastewater and sediment from farms, WWTPs and receiving surface water: distribution, sources, and ecological risks. RSC Adv 11:31766–31775. https://doi.org/10.1039/D1RA05333G

Wang X, Yu N, Qian Y, Shi W, Zhang X, Geng J et al (2020a) Non-target and suspect screening of per- and polyfluoroalkyl substances in Chinese municipal wastewater treatment plants. Water Res 183. https://doi.org/10.1016/j.watres.2020.115989

Wang X, Yu N, Yang J, Jin L, Guo H, Shi W et al (2020b) Suspect and non-target screening of pesticides and pharmaceuticals transformation products in wastewater using QTOF-MS. Environ Int 137. https://doi.org/10.1016/j.envint.2020.105599

Wert EC, Rosario-Ortiz FL, Snyder SA (2009) Effect of ozone exposure on the oxidation of trace organic contaminants in wastewater. Water Res 43:1005–1014. https://doi.org/10.1016/j.watres.2008.11.050

Wood TP, Du Preez C, Steenkamp A, Duvenage C, Rohwer ER (2017) Database-driven screening of South African surface water and the targeted detection of pharmaceuticals using liquid chromatography - high resolution mass spectrometry. Environ Pollut 230:453–462. https://doi.org/10.1016/j.envpol.2017.06.043

Yue S, Ramsay BA, Brown RS, Wang J, Ramsay JA (2015) Identification of estrogenic compounds in oil sands process waters by effect directed analysis. Environ Sci Technol 49:570–577. https://doi.org/10.1021/es5039134

Part VI
Key Policy Messages and Recommendations

Chapter 21
Science-Based Policy Recommendations for Managing Emerging Pollutants: Protecting Water Quality for the Health of People and the Environment

Sarantuyaa Zandaryaa, Ali Fares, Gabriel Eckstein, Regina M. Buono, Mary Trudeau, James E. Nickum, Xinghui Xia, Atikur Rahman, Marijn Korndewal, Cassiana C. Montagner, Piero R. Gardinali, Robert Michael Di Filippo, and Anoop Veettil

Abstract Emerging water pollutants are a growing global concern due to their ubiquitous presence in water resources worldwide and their potential adverse effects on human health and ecosystems. Limited scientific understanding of sources of emerging pollutants' emissions to water bodies and their pathways, behaviour, and fate in aquatic environments, as well as human health and ecological effects, is a significant hindrance in managing emerging water pollutants. With exceptions concerning PFAS/PFOS and microplastic beads, there are few regulations for

S. Zandaryaa (✉)
Division of Water Sciences, Secretariat of the Intergovernmental Hydrological Programme, United Nations Educational, Scientific and Cultural Organization (UNESCO), Paris, France
e-mail: s.zandaryaa@unesco.org

A. Fares · A. Rahman · A. Veettil
Prairie View A&M University, Prairie View, TX, USA

G. Eckstein
Texas A&M University School of Law, Fort Worth, TX, USA

R. M. Buono
Environmental Law Institute, Washington, DC, USA

M. Trudeau
Envirings Inc, Ottawa, Canada

G. Eckstein · R. M. Buono · M. Trudeau · J. E. Nickum
International Water Resources Association, Paris, France

X. Xia
School of Environment, Beijing Normal University, Beijing, China

M. Korndewal
Organisation for Economic Co-operation and Development, Paris, France

C. C. Montagner
Institute of Chemistry, University of Campinas, Campinas, Brazil

© UNESCO 2025
S. Zandaryaa et al. (eds.), *Emerging Pollutants*, Advances in Water Security,
https://doi.org/10.1007/978-3-031-71758-1_21

emerging pollutants in national water and environmental policies, which results in a critical gap in safeguarding human health and aquatic ecosystems through effective prevention, reduction, and management strategies. This chapter presents a set of science-based policy recommendations for managing emerging water pollutants, particularly for the protection of aquatic ecosystems and groundwater resources, as well as through proper wastewater and waste management, including the circular economy approach and lifecycle management of pollutants. Policy recommendations are also proposed for managing priority emerging pollutants such as microplastics, nanomaterials, and trace chemicals. The policy recommendations emanate from key policy-relevant findings of research studies and scientific discussions presented at the UNESCO-IWRA International Conference on "Emerging Pollutants: Protecting Water Quality for the Health of People and Ecosystems," which took place online in January 2023, gathering over 170 state-of-the-art research studies on wide-ranging topics related to emerging water pollutants.

Keywords Emerging pollutants · Contaminants of emerging concern · Microplastics · PFAS · Water · Health · Environment

21.1   Emerging Pollutants: A Growing Threat to Aquatic Ecosystems

Emerging pollutants and their transformation products are ubiquitous in aquatic ecosystems, and many are toxic. Emerging pollutants have been detected in urban and ambient waters in different regions and can be considered a worldwide problem. The variety of pollutants is wide and growing due to the advancement of science and technology, which has led to the development of new substances for consumer products, healthcare and in the economy more broadly. The amount of pollutants generated depends on the specifics of product uses and human consumption patterns, but the quantity generally increases with population size. For example, pharmaceutical drug use has increased due to population growth and life expectancy.

Emerging pollutants detected in water bodies may cause ecological or human health impacts. Yet, the vast majority of emerging pollutants are not regulated under existing environmental laws. The concentrations of chemicals and other contaminants of emerging concern differ depending on their uses in each region and the level of wastewater treatment available or required. Pesticides, disinfectants, pharmaceuticals, microplastics, and engineered nanomaterials pose increasing water quality risks and human health threats.

P. R. Gardinali
Institute of Environment and Department of Chemistry and Biochemistry, Florida International University, Miami, FL, USA

R. M. Di Filippo
National Institute of Geological Sciences, University of the Philippines, Diliman, Philippines

Understanding the sources, pathways, and behaviour of emerging pollutants in water systems and aquatic environments is critical and the first step to addressing their ecological and human health impacts. Research is needed to trace specific pollution sources in groundwater and surface waters using physical and chemical tracers, improve understanding of pollutants' pathways, understand connections between water cycle components, and identify interactions of pollutants with biological processes in water. Moreover, research is needed on the effects of combinations of emerging pollutants and their transformation byproducts on human health and ecosystems. An improved scientific understanding of emerging pollutants is vital to efforts to limit the proliferation of emerging pollutants in the aquatic environment.

Some emerging pollutants and their transformation byproducts can be persistent, bioaccumulative, and toxic to humans and biota in various ways (Fig. 21.1). For example, per-and polyfluoroalkyl substances (PFAS), called "forever chemicals", can bioaccumulate and transfer from one trophic level to another in aquatic food webs, triggering multiple toxicity endpoints, such as endocrine, neurotoxic, carcinogenic, and immunotoxic effects, and impairing growth and reproduction of aquatic organisms. Another example is 6PPD-Q (2-anilino-5-[(4-methylpentan-2-yl)amino]-cyclohexa-2,5-diene-1,4-dione), which is a globally widespread tire rubber–derived oxidation product of 6PPD. The presence of 6PPD-Q in aquatic ecosystems induces the death of adult coho salmon when the salmon migrate to urban creeks contaminated with the substance, and can also induce acute mortality in juvenile coho salmon at trace levels (LC50 = 95 ng/L) (Greer et al. 2023).

Emerging pollutants also pose a global public health threat. Pharmaceutical waste residues from human and animal consumption of antibiotics are increasingly present in aquatic ecosystems worldwide, contributing to antimicrobial resistance—one of the most urgent public health concerns.

21.1.1   Challenges in Monitoring and Managing Emerging Pollutants in Aquatic Ecosystems

Most emerging pollutants are not monitored, and hazard information is still limited for some. This lack of information makes understanding the pollutants' environmental behaviour and assessing the risks they present very challenging, impeding decision-making and implementation of measures for their control and management by commercial sector, industries and government regulatory initiatives.

Fig. 21.1 Transformation process for organic pollutants. *Source* Authors

To protect and maintain the health of aquatic ecosystems, it is imperative to reduce the discharge of emerging pollutants in the environment. Although early-stage research efforts have enriched scientific understanding and knowledge of emerging pollutants, they are still quite limited. Three significant challenges confront monitoring and managing efforts for emerging pollutants:

1. *The lack of transparency around emerging pollutants and related products within their life cycles.* The production, use, and application of new materials and compounds are often protected, and thus information about their composition and properties is not publicly available. This makes it challenging to identify and understand their human health and ecological impacts after their intended uses once the compounds enter the environment as waste. Consequently, efforts for monitoring and managing emerging pollutants usually lag decades behind the discovery and use of new materials and compounds. For example, while liquid crystal monomers have been used in liquid crystal displays since the 1990s, environmental monitoring of liquid crystal monomers began only recently due to the confidential nature of their makeup among manufacturers. Further, the transportation and disposal of those new compounds are poorly documented, which may contribute to unintentional discharge and improper disposal, or accidental release. A lack of transparency further complicates the practical management

of emerging pollutants, research on their sources, pathways, and behaviour in aquatic ecosystems, and prevention and removal options.

2. *The difficulty in monitoring emerging pollutants.* Routine chemical analyses can identify a narrow portion of emerging pollutants. However, this approach is restricted to compounds with reference standards. When reference standards are not available, suspect screening (evaluating the presence of a list of suspected pollutants) and non-target analyses with high-resolution mass spectrometry (HRMS) provide extensive information (e.g., accurate mass and retention time) about known and unknown chemicals. However, only a fraction of HRMS information is well analyzed and interpreted, preventing the detection of some new emerging pollutants in water. Further, due to the properties of some emerging pollutants, the signals are masked, thereby preventing detection with HRMS.

3. *Identifying the hazard that emerging pollutants present.* Although identified as distinct compound classes (i.e., pollutant groups), the toxicities and ecological risks of different emerging pollutants for aquatic species have not been systematically studied, which impedes the enactment of regulations for emerging pollutants. Further, emerging pollutants often are present in the environment as mixtures, which makes their regulation even more challenging. The co-exposure of multiple emerging pollutants may cause adverse outcomes even when the concentrations of individual chemicals are very low or below analytical detection limits. Further, the transformation byproducts of emerging pollutants may be even more toxic than their parent compounds. Investigating the toxicity of thousands of emerging pollutants is time-, labour-, and cost-intensive, making it impractical to determine the multiple toxicity endpoints of each emerging pollutant.

21.1.2 Recommendations to Reduce and Manage Emerging Pollutants in Aquatic Ecosystems

Addressing the threats that emerging pollutants present for aquatic ecosystems across the globe is essential for protecting the health of aquatic ecosystems and ensuring the quality of urban waters, the sustainability of urbanization, and the achievement of United Nations Sustainable Development Goals (SDGs)—in particular, SDG 6: Clean water and sanitation, SDG 3: Good health and well-being, SDG 11: Sustainable cities and communities, and SDG 12: Responsible consumption and production.

Recommendations to improve the management and control of emerging pollutants in aquatic environments are given below:

1. *Increase and improve information available about emerging pollutants at each stage of their life cycles.* Better and more extensive information about the sources and synthesis of emerging pollutants, and the effluents in which they are present, would reduce their unintentional emissions and inadequate treatment, improve understanding of their physico-chemical properties and environmental

fate prediction, enable their replacement with less hazardous alternatives, and facilitate remediation of affected environments.

2. *Strengthen the analysis of emerging pollutants.* The integration of all viable methods for chemical analysis in a systematic approach would enable the detection of more emerging pollutants in aquatic environments and the subsequent understanding of their persistence, environmental fate, and levels of exposure (e.g., the transfer from one trophic level to another in aquatic food webs) in urban aquatic ecosystems. Meanwhile, developing data-mining programs, adequate sample pre-treatment methods, and advanced detection techniques would further decode emerging pollutants in complex aquatic samples.

3. *Accelerate toxicological testing for emerging pollutants.* High throughput tests, using robotics, high-speed computing, and other technological advances that allow a researcher to conduct millions of chemical, genetic, or pharmacological tests quickly, can be applied to carry out the toxicological profile of individual emerging pollutants and their mixtures, improving the understanding of their adverse effects and the quantitative structure–toxicity relationship for new emerging pollutants. The latter can be used to estimate the toxicity of untested emerging pollutants. It is known that the co-exposure to multiple contaminants may cause adverse outcomes even when the concentrations of individual pollutants are believed to be non-harmful. However, methods to assess the effects of mixtures of pollutants are currently inadequate and require further development. Effect-based methods, such as *in vitro* bioassays, are emerging in water quality testing and policies to assess the potential risk of chemicals in a water sample (OECD, 2023). Risk assessment of emerging pollutants and chemical mixtures in urban and ambient waters is only in the early stages of research and development.

4. *Establish national or regional regulations for emerging pollutants.* Where feasible, source control measures to eliminate or reduce the use and disposal of emerging pollutants are the most effective. The emissions of emerging pollutants in urban areas can differ depending on population size, human activities, urban development levels, and the chemicals and wastewater treatment facilities available. Therefore, in addition to the international conventions on chemicals and waste (e.g., the Basel, Rotterdam, Stockholm, and Minamata conventions), a series of national or regional regulations should be established, considering specific risk scenarios of exposure to emerging pollutants in water.

5. *Develop and implement other control measures when source controls are not feasible.* Collaborations in technology and knowledge transfer among less- and more-developed regions should be encouraged to mitigate the risks of emerging pollutants and to promote global health and equality. Emerging pollutants that can be removed in wastewater treatment technologies should be identified. For most countries around the globe, sanitation is still a big challenge. To improve this situation, efficient, low-cost sanitation and wastewater treatment options should be developed and urgently implemented to mitigate the risks of emerging

pollutants that compound those of old conventional contaminants (such as E. coli and nitrate).

Box 1. Emerging pollutants were found in Lake Guiers in Senegal

Lake Guiers, Senegal ©chelsealwood, CC BY SA 2.0. Source Wikimedia Commons (https://commons.wikimedia.org/wiki/File:Lac_de_Guiers_2017_01.jpg)

Growing anthropogenic pressures and climate change threaten the Lake Guiers Basin in Senegal. To understand the impact of human activities on the lake's water quality, researchers undertook a spatio-temporal evaluation of pesticide and land use in the basin (Saadi 2023). The study found increased pesticides in water samples, including organochlorines and organophosphates. It also demonstrated that both pesticides use and the occurrence of new compounds were increasing annually, although there was a decrease in agricultural land use during the COVID-19 pandemic period. High to moderate pollution levels were detected at some sites for most of the targeted parameters. The study highlighted that the environmental degradation of the lake basin requires the monitoring of chemicals and emerging pollutants while integrating management practices, involving multiple stakeholders, and increasing awareness to avoid the deterioration of the ecosystems.

21.2   Emerging Pollutants and Groundwater

Groundwater provides almost half of all drinking water sources worldwide and about 40% of the water used in irrigation (UNESCO 2023a, b, c, d, e). This underground resource is essential for health and food security in communities worldwide. Yet, monitoring groundwater quality can be expensive and requires specialized equipment, knowledge, and skills.

21.2.1   Emerging Pollutants in Groundwater

Groundwater protection is much more cost-effective than groundwater remediation after contamination. However, groundwater contamination can go undetected due to flows from unknown contamination sources (such as an underground chemical storage vessel), contamination from surface pollutant sources, or natural features of the soil or rock (for example, where arsenic or uranium naturally occur).

Emerging pollutants behave differently in groundwater than in surface water due to slower subsurface transport and higher residence times. Another reason emerging pollutants in groundwater are less known is the perception that groundwater is at a lower risk of contamination. A conceptual understanding and subsequent prioritization of emerging pollutants in groundwater requires knowledge of the mobility of specific pollutants in groundwater aquifers, environmental exposure, toxicity, and relative risks posed to groundwater. Regular monitoring and analytical testing are needed to improve the scientific evidence base to inform policy for groundwater protection, and, where available, existing monitoring data can be a basis for prioritization. Additional information is needed on contaminant source areas, occurrences, and concentration trends (e.g., the characteristics of a plume of contaminant in groundwater as it migrates from a pollution source), and degradation products as emerging pollutants may transform into other compounds of concern.

Further research is needed to identify and address priority contaminants in groundwater. A three-phase procedure can be used to understand emerging pollutants of concern in groundwater: (1) develop a conceptual understanding and approach; (2) prioritize specific emerging contaminants; (3) monitor and evaluate studies. More fundamentally, to manage emerging pollutants, there is an urgent need for a coherent and robust approach to protecting groundwater from contamination, including enacting more robust policies and changing how society views and uses the chemicals. Some regional initiatives are in place to improve the monitoring of priority emerging pollutants. For example, the European Groundwater Watch List is a voluntary process intended to produce high-quality Europe-wide monitoring data on substances that have not been routinely monitored and may pose a risk to groundwater.

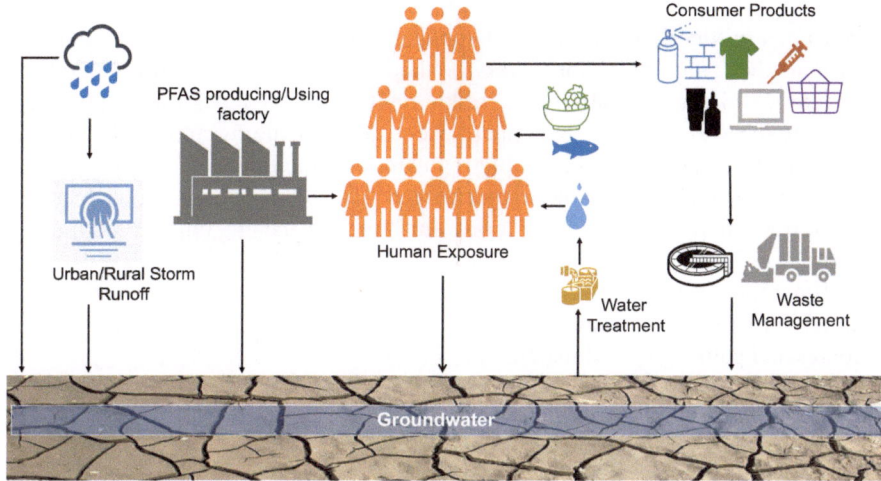

Fig. 21.2 Exposure routes to PFAS in groundwater. *Source* Authors

Contaminants of the greatest concern in groundwater

Six categories of contaminants of emerging concern have been identified in the literature on groundwater quality, which are also commonly detected in surface waters: (1) pharmaceuticals, (2) personal care and household cleaning products, (3) industrial chemicals, (4) flame retardants (which contain PFAS), (5) pesticides, (6) cyanotoxins, and (7) nanomaterials. Research also indicates that more sustained monitoring in groundwater should be directed at certain toxic pharmaceuticals and common toxic compounds in cosmetics and personal care products.

Box 2. Consumer products of emerging concern in groundwater

Emerging pollutants detected in groundwater originate from various human activities and sources through different routes and pathways. Figure 21.2 illustrates PFAS exposure routes to groundwater.

PFAS are a group of very diverse and complex chemicals that have been used in industry and consumer products worldwide over decades. Examples of PFAS-bearing consumer products include non-stick coatings (e.g., Teflon), textiles (e.g., Gore-Tex), stain-resistant carpets and furniture, and aqueous firefighting foams. Other consumer products with contaminants of emerging concern include cosmetics, deodorants, cleansers (soap, shampoo, toothpaste), moisturizers, perfumes, sunscreens, hair styling products, and shaving creams. Industrial chemicals with contaminants of emerging concern include pesticides, 1,4-Diozane (a chemical used in pharmaceutical and plastic production), and explosives. Nanoparticles are integral to hundreds of consumer products and biomedical applications, leveraging materials such as carbon fiber, aluminium, or silver for their antibacterial properties, optical displays, anti-static capabilities, wrinkle resistance, UV blocking, and more. Despite the widespread use of pharmaceuticals—including antibiotics, anti-inflammatories, pain relievers, lipid regulators, and antidiabetics—their presence in groundwater is seldom monitored. This lack of surveillance and monitoring leaves the concentrations and potential health impacts of these substances in groundwater unexplored.

Emerging pollutants resulting from these consumer products enter groundwater systems from a range of human activities and sources through diverse routes and pathways, as illustrated in Fig. 21.2.

21.2.2   Risk Assessment and Prioritization of Emerging Pollutants in Groundwater: Tools in Development

Risk assessment is a coherent, science-based decision-making process that also communicates chemical pollutant risk. Risk assessments contribute to governance processes for priority-setting for potential future scenarios of chemical exposure and environmental and human health analyses. Tools exist to conduct this kind of risk assessment. For example, Risk Assessment in the 21st Century (RISK21) (HESI 2024) provides a framework for screening and prioritizing chemicals to determine which warrant further assessment or additional data collection. This framework organizes all relevant information for an interactive and transparent evaluation of the available exposure and hazard data. Additionally, analytical tools based on Geographic Information System (GIS) technology assess the location and source areas of PFAS contamination. GIS tools also optimize the monitoring of contaminants of emerging concern.

Box 3. Emerging pollutants found in groundwater around the world

- Research studies indicate the presence of a variety of emerging pollutants in groundwater around the world, including: organic contaminants from sunscreens and antibiotics in the groundwater of the Yucatán peninsula as proxies to estimate human presence and related reduced water quality in recreational areas of Mexico (Leal-Bautista 2023);
- Organic contaminants in the groundwaters and recharge sources in Bengaluru City, Karnataka, India, include the ubiquitous detection of

sweeteners, which indicate groundwater age since their compounds were introduced only relatively recently (Brauns 2023);

- Various microplastics in urban soil and groundwater in Bauru, São Paulo, Brazil, impact biota and pose a risk to human health but are not well studied in groundwater (Lopez Ferreira et al. 2023);
- Antimicrobial resistance (AMR) in karst groundwater supplies in the Tampa Bay region of the United States of America was found to be disseminated throughout urban karst aquifers, replicating the results of a similar study in Kentucky, USA (Kaiser 2023).

21.2.3 Actions in Progress and Next Steps for Groundwater Protection from Emerging Pollutants

Some regional initiatives are in place to improve the monitoring of priority contaminants of emerging concerns.

In April 2024, the U.S. Environmental Protection Agency (2024) implemented a new National Primary Drinking Water Regulation that creates legally enforceable levels, called Maximum Contaminant Levels (MCLs), for PFOA, PFOS, PFHxS, PFNA, and HFPO-DA as contaminants with individual MCLs. It also created MCLs for PFAS mixtures containing at least two or more of PFHxS, PFNA, HFPO-DA, and PFBS using a Hazard Index MCL to account for the combined and co-occurring levels of these PFAS in drinking water. The USEPA also finalized health-based, non-enforceable Maximum Contaminant Level Goals (MCLGs) for these PFAS. In conjunction, the USEPA also finalized a new rule to reduce the presence of PFOA and PFOS in the aquatic and terrestrial environment.

The European Commission is reviewing proposed legislation to eliminate the widespread release of PFAS compounds in biosolids produced during biological wastewater treatment that are then used as soil amendments for agricultural fields and areas. PFAS compounds continue to leach into underlying groundwater from biosolid applications on soils. Researchers have recommended that other countries follow the European Commission's lead in managing biosolid use on agricultural soils. However, the issue is complex due to the potential nutrient supply from biosolids.

Legislation needs the support of the public for full and effective implementation. Therefore, public awareness of emerging pollutants in everyday consumer products should be part of a comprehensive approach for source control measures and mechanisms to reduce exposure to these chemicals. Researchers have also called on governments to require manufacturers to disclose information on the use of contaminants of emerging concern in their products.

21.3   Emerging Pollutants and Managing Wastewater and Waste

Widespread use of plastics, pharmaceuticals, and chemical pathogens has resulted in increasing levels of new and emerging pollutants found in the world's water resources. Wastewater treatment technologies are expensive, complex, and often ineffective in removing emerging pollutants from the water fraction. Pollutants removed from the water fraction are shifted to other media, such as sludge, biosolids, or air emissions.

21.3.1   Effectiveness of Wastewater Treatment Technologies in Removing Emerging Pollutants

Wastewater treatment technologies are vital in removing emerging pollutants from municipal and industrial effluents, consequently reducing their emissions into water bodies and mitigating their effects on aquatic ecosystems. However, there are limitations to the effectiveness of treatment processes in removing emerging pollutants from wastewater and freshwater resources. Moreover, some (advanced) treatment processes have high carbon emissions. Therefore, the most effective solution is to avoid or prevent the production of emerging pollutants at the source and their discharge into wastewater and freshwater bodies, reducing the need for expensive and complex wastewater treatment processes. Rigorous source control approaches can prevent many emerging pollutants from entering water bodies and protect ecosystems and human health from adverse effects. Where emissions of emerging pollutants cannot be avoided or reduced at the source, treatment of pollutants in wastewater is essential. Monitoring emerging pollutants in raw wastewater and wastewater treatment facility processes is highly complex, expensive, and challenging. Adequate sampling and analysis incur continuous costs and pose detection limitations.

Box 4. Pharmaceuticals in Brazil's water resources—data analytics of pharmaceuticals' sales

Water testing at a water treatment facility in Broken Bow, Oklahoma, USA ©Lance Cheung, US Department of Agriculture. *Source* Wiki media Commons (https://commons.wikimedia.org/wiki/File:Broken_Bow_Water_Treatment_Facility_water_testing.jpg)

To advance the understanding of emissions of emerging pollutants into Brazil's water resources, data analytics were applied to publicly available sources, including pharmaceutical sales data (Braz de Freitas Bijos 2023). A better understanding of the mass of pharmaceuticals sold and usage trends in major metropolitan areas can make monitoring of emerging pollutants in water bodies more targeted, thus reducing resource-intensive data collection. Additionally, study results on the concentrations and behavior of emerging pollutants before and after wastewater treatment were compiled from available literature, which helps further inform future monitoring programs in urban waterways. The comparative analysis demonstrated that open-source data significantly contributes to developing monitoring programs targeting specific emerging pollutants.

21.3.2 Changing Wastewater Treatment Strategies to Improve Performance: A Hopeful Note

Emerging pollutants occur in many combinations and have multifaceted origins and various chemical structures. Conventional biological wastewater treatment processes were not designed to remove these pollutants, although partial removals may occur. Adding a single tertiary treatment process will unlikely be sufficient to remove all emerging pollutants from wastewater. Further, there are many information and research gaps on emerging contaminants' fate in sludge, biosolids, and other wastewater treatment process wastes. Research on new and improved strategies for wastewater treatment to address emerging pollutants is developing, but further research is needed.

Three broad categories of wastewater treatment strategies for removing emerging pollutants are discussed below.

On-site treatment and removal at decentralized locations

New strategies for wastewater treatment can improve the performance of treatment processes. One option is to treat wastewater before it is released to the public sewerage system. As wastewater from different sources is aggregated into a centralized treatment facility, the effectiveness of pollutant removal decreases, particularly for substances in trace amounts, such as emerging pollutants. Implementing wastewater treatment systems tailored for specific pollutants at sites where the contaminants originate can provide promising results. For example, hospital wastewater discharge contains pharmaceutical residues and epidemiological vectors that pose potential human health risks. Treatment of hospital wastewater at hospital sites may lessen the release of these contaminants into municipal wastewater systems and local water bodies. Organic water treatment methods, for instance, when applied on a localized scale, are potentially more efficient and cost-effective than when they are part of a centralized wastewater treatment facility.

New methods and materials

Researchers are exploring various innovative treatment options and their effectiveness for removing various emerging pollutants from wastewater and freshwater, including using low-cost, locally available materials. Ceramic filters made of clay have been used for treating wastewater, stormwater, and drinking water. This method best suits lower water flow rates and increased process retention times. Studies have also shown that chemically activated carbon derived from bamboo sawdust has successfully removed specific pharmaceutical residues (paracetamol) from water. In India, a symbiotically functioning algal–bacterial community has been demonstrated to remove some emerging contaminants from hospital wastes. Other advanced treatment methods studied for their effectiveness in removing emerging pollutants include catalytic ozonation, activated carbon adsorption, Ultraviolet Advanced Oxidation Processes (UV-AOPs), and the heterogeneous photo Fenton process. These processes have the potential to remove emerging pollutants from wastewater. The European

Commission has proposed a revision of its Urban Wastewater Treatment Directive (2022) which requires advanced quaternary treatment of all urban wastewater in agglomerations larger than 10,000 population equivalent (PE) by 2045, with the aim of removing micropollutants. Further research is required to seek low-cost, sustainable technologies to remove emerging pollutants from wastewater.

Non-traditional solutions to wastewater treatment

Soil Aquifer Treatment System (SAT) is a powerful technology for water reuse and large-scale treatment using sorption barriers (compost, woodchips, biochar, clay, or zeolite that take up pollutants better than many soils alone). Trace organic contaminants (pharmaceuticals, personal care products, viruses, bacteria, and micro- and nano-plastics), still frequently found in treated wastewater, can be removed as water infiltrates soil supplemented with sorption barriers. SAT is especially desirable where wastewater can supplement groundwater resources through aquifer recharge. Yet, sampling and analysis of emerging pollutants in groundwater are costly.

Box 5. Research needs on emerging pollutants in wastewater

Increasing levels of emerging pollutants in wastewater urgently requires new regulatory and technological strategies, incorporating government regulations supported by robust research and increased public awareness. In particular, further research is needed to:

- Evaluate the effectiveness of wastewater treatment technologies in removing diverse types of emerging pollutants, especially those of priority concern.
- Determine the fate of emerging pollutants in sludge and wastewater treatment process byproducts to limit and reduce the re-introduction of emerging pollutants into the environment and their presence in crops and products through use and application of biosolids on agricultural soils as fertilizers and soil enhancement.
- Make risk assessments of the reuse of treated wastewater and sludge, especially for groundwater recharge and agricultural irrigation.

21.4 A Circular Economy Approach: Lifecycle Management of Emerging Pollutants

Managing pollutants successfully requires the consideration of the entire lifecycle of products that generate pollutants, meaning that not only the sources of the pollutants and their fate in water resources matter, but also product design, materials used to produce products, recyclability or biodegradability of materials used, disposal options after use, and any residual pollutants' afterlives. Such an approach starts with the identification of '*Who uses these materials before we mark them as pollutants?*'

and considers what alternatives are available that are less harmful to human health and ecosystems.

21.4.1 Adopting a Circular Economy Approach

The circular economy refers to an economic system based on redesigning, reusing, and regenerating materials or products and the processes that create them to improve sustainability. Essentially, the circular economy is a model of production and consumption that seeks to extend a product's lifecycle indefinitely through means such as redesigning, sharing, leasing, reusing, repairing, refurbishing, and recycling existing materials and products. The classical linear economy generates waste that eventually enters water bodies and the environment, causing environmental pollution and water quality degradation. In a circular economy approach, 'waste' created in every step of the production processes and uses becomes an input for further use, either for reuse as a used product or for recycling as a primary material in the production of other products needed in the economy, creating value. When further reuse, or recycling, is no longer feasible, the waste is disposed of safely or degraded into compost for biodegradable materials.

The concept of circular economy has received increasing attention as growing human populations and increased consumption levels place enormous pressure on natural resources around the globe, increase waste streams and generation, threaten water supplies, and pollute and harm the natural environment. The Organization for Economic Cooperation and Development (OECD) has noted that "*Through more efficient use of resources, eco-design, reuse, repurpose and remanufacturing the circular economy is an opportunity for a new way of thinking and an example of resilience in the face of future crises*" (OECD 2020). Further, SDG12, "Responsible consumption and production," sets forth objectives inherent in and served by the circular economy model. Water management based on circular approaches has great potential, as the natural hydrological cycle is inherently circular. Circular economy principles have been applied in technical and institutional aspects of water resource management for decades for water reuse and the recovery of energy, nutrients, and other byproducts from wastewater.

Box 6. Circular economy assessment of water and energy use in coffee production

Coffee berries ©Jmhullot CC BY SA 3.0. *Source* Wikimedia Commons (https://upload.wik imedia.org/wikipedia/commons/2/26/CoffeeBerry.jpg.)

Research and models are being developed to systematically consider the inputs and outcomes of supply chains, including food supply chains. One such model is the MICRON model, which provides a circular economy assessment framework aimed at helping companies assess the use of water and energy, the production and emissions of waste, and the durability of their products (Baratsas et al. 2022). The model seeks to provide insights into solutions and prioritization and to support the transition to a circular economy.

The circular economy assessment using the MICRON model is explained below through the example of the coffee production process. The assessment comprises the following:

(1) Identify and consider alternative production paths. Coffee is cultivated and processed using a "wet" method, which requires substantial amounts of energy and water, or a dry method that requires less water and energy but more time and space.
(2) Collect wastes from all supply chain stages and figure out how to use them. Coffee waste could be used for multiple purposes, including to produce biofuels, animal feed, activated carbon, composting material, and for use in particle board and dietary fiber.
(3) Develop network mapping pathways within the coffee supply chain.

(4) Develop a mathematical optimization model to minimize waste and GHG emissions, minimize use of natural resources, and maximize energy efficiency and profit of the system.

(5) Analyze demand scenarios to produce potential solutions that can inform decision making based on objectives. The production of instant coffee requires far more water and coffee cherries, and emits far more GHG than a mix of whole bean coffee and coffee beverages.

21.4.2 Applying a Circular Economy Concept to Emerging Pollutants

Policy interventions to reduce the risk of emerging pollutants accumulating in water resources should target all stages of the lifecycle of a product, including:

- Avoid and reduce pollution at the source during production processes. The supply chain of a product can be modeled to identify the part(s) of the process that can be improved to reduce the production of residuals, decrease emissions and the use of natural resources, increase the use of renewable energy, and extend the lifespan, repairability, and durability of products. Learning from modeling can help move toward sustainable development and decrease water pollution.
- Avoid or reduce pollution during product use. Proper storage of a product during use can help reduce pollution. For example, reduced packaging and reusable or biodegradable packages and containers reduce waste generated during product packaging, shipping, and storage.
- Minimize pollution at end-of-pipe destinations. Conventional wastewater treatment processes may remove some emerging pollutants from water with low to moderate efficiency but cannot remove most emerging pollutants. Even if removed from wastewater to a certain extent, the pollutants accumulate in sludge or biosolids. Residual pollutants are still present in treated wastewater, released to ambient water bodies, and potentially withdrawn as drinking water sources. Conventional wastewater and drinking water treatment processes were not designed to remove emerging, persistent contaminants. For example, in Brazil, antibiotics such as erythromycin and amoxicillin are commonly found in treated wastewater even after treatment because of the low efficiency of wastewater treatment processes for these substances and the societal overuse of antibiotics.
- Legislation and regulations to reduce pollutants at the source or to sustainably manage the use of products that cannot be eliminated (such as some pharmaceuticals) are essential to moving to a circular economy for managing emerging water pollutants.

21.4.3   Recovering and Generating Resources from Pollutants

A circular economy approach can be used to redesign products and repurpose residual "pollutants" to give them a new life and reintroduce them as recovered resources and materials in the economy. In addition to preventing and reducing pollution generation during the production and consumption processes, and consequently pollutants' emissions into the natural environment, the circular economy approach also provides opportunities for the recovery of valuable resources such as reclaimed water, energy (biogas), and nutrients (fertilizers), as well as other byproducts for further reuse in various supply and value chains. Examples of research on circular economy opportunities for managing emerging pollutants are described below.

- Lake Tegel in northwestern Berlin, Germany, is polluted with algae cyanobacteria and pharmaceutical micropollutants due to wastewater disposal and climate change. To remediate this problem using a circular economy approach, a start-up company is designing a process to separate the toxins generated by cyanobacteria from the nutrients (i.e., the microalgae) and then to use the microalgae as biofertilizer (Namba 2023).
- Colorants today represent a $68 billion industry based chiefly on synthetic colorants. Research studies explore opportunities to use fungi to produce natural colorants, which have the potential to make the industry sustainable and less polluting; however, while promising, the necessary energy demand of the process presents a challenge (de Oliveira et al. 2023).

21.4.4   The Importance of Proper Governance

Governments have a crucial role in supporting the development of circular economy approaches by creating enabling policy frameworks. Policy interventions are essential to enable more effective recycling and reuse programs, enact stricter regulatory standards to require compliance with production and pollution prevention rules, facilitate lifecycle approaches involving actors all along the supply chain, and encourage investments to stimulate innovation. Consultation between innovators (such as research laboratories and academic institutions) and regulators can help raise policymakers' awareness about the presence and effects of pollutants in water resources and develop and implement new, more effective policies and technologies. Multidisciplinary research can effectively bring knowledge together and implement novel policies and technological approaches to support a circular economy approach to managing emerging water pollutants.

21.5 Priority Emerging Pollutants: Microplastics, Nanomaterial, and Trace Chemicals

Emerging pollutants comprise a wide variety of contaminants found in water bodies in recent decades resulting from new or newly discovered materials thanks to advances in analytical technologies. Some emerging contaminants should receive priority in research and policy due to their toxicity, environmental persistence, or value as indicators of environmental conditions.

Priority topics for research on emerging pollutants include microplastics, PFAS, and pharmaceuticals and personal care products (such as shampoos, antibiotics, and pain reduction medications). In addition, nanomaterials' potential applications have benefits and potentially harmful effects for humans and aquatic organisms.

21.5.1 Microplastics and Nanoplastics

Among emerging water pollutants, microplastics have garnered significant attention from the research community. Microplastics are tiny pieces of plastic, typically 1–5 mm in size, which enter the environment through the release of consumer products (such as microbeads) and from clothes washing (microfibers), as well as from the breakdown of larger plastic debris and industrial waste. Research studies indicate the presence of microplastics in all types of freshwater bodies and ecosystems, from urban rivers and streams to freshwater lakes in remote locations and in the Arctic Ocean ice and water. Microplastics are considered priority emerging pollutants for multiple reasons, including their ubiquitous presence in the environment, the potential for ingestion by aquatic organisms, limited scientific data on their potential risks to human health and ecosystems, and the potential to accumulate and transport other pollutants, bacteria, and substances on microplastic surfaces, potentially increasing exposure of aquatic organisms to toxic chemicals and pollutants. Microplastic particles do not behave like traditional pollutants because, due to their characteristics, they interact with other substances and pollutants in water in ways that differ from other particles and pollutants. Key issues fuelling global research on microplastics include advancements in analytical techniques for detecting them in environmental media and organisms' tissues. However, further research is needed to better comprehend microplastic bioaccumulation (the buildup of chemicals in organisms), bioavailability (how easily a substance can be taken up by an organism), their transfer within the food web, potential adverse effects on human health and aquatic organisms, and the amplification of toxicity.

Microplastics enter aquatic systems through municipal and industrial wastewater effluent disposal and runoff from land during precipitation. Conventional wastewater treatment processes have been demonstrated effective in removing microplastics (except microfibres) from wastewater and consequently have been identified as a critical step to reducing the amount of microplastics that enter water bodies, particularly until other technologies and measures for source reduction can be developed and/or mature. However, microplastics removed from wastewater accumulate in sludge (or biosolids) from wastewater treatment processes, which are often used as fertilizer and soil enhancement. Accordingly, the reintroduction of microplastics into the environment through biosolids application on agricultural land requires special attention and should be researched. Proper disposal of sludge and biosolids containing microplastics is therefore important.

Nanoplastics in wastewater, freshwater resources, and drinking water are of special concern due to their particular properties and size. Nanoplastics are considered more toxicologically active than microplastics, primarily due to their smaller size and characteristics concerning their transport properties in the natural environment, the ability to penetrate tissues in human and animal organs, interactions with light and natural colloids, and bioavailability and diffusion times for the release of plastic additives (Gigualt et al. 2021). With research lacking on nanoplastics in water, there is a limited understanding of their effects on aquatic organisms and the potential health impacts of nanoplastic particles in drinking water. Conventional wastewater treatment processes are not effective for removing nanoplastics, as nanoparticles escape from treatment processes due to their size. A study (Devi et al. 2022) reviewed nanoplastics removal by various technologies, including advanced and new methods (microorganism-based degradation, membrane separation with a reactor, and photocatalysis) and traditional wastewater and drinking water treatment processes. It recommended combined methods (such as filtration with coagulation) as more effective for removing nanoplastics from water. Yet, the knowledge gap about the methods for nanoplastic removal should be addressed through more research. Even if removed from water, nanoplastic waste byproducts need to be managed so that they are not re-released into the environment.

Box 7. Microplastics in the aquatic environment

Microplastics © Oregon State University, USA, CC BY-SA 2.0. *Source* Wikimedia Commons (https://commons.wikimedia.org/wiki/File:Microplastic.jpg)

Microplastics can be a carrier of other pollutants such as heavy metals, chemicals, persistent organic compounds, and environmental pathogens because diverse types of contaminants sorb (or attach) to their surfaces. Sorption processes depend on the contaminant's characteristics, the microplastic's characteristics, such as the surface area, particle size, and polymer composition, and the characteristics of the environment, such as the pH or the presence of organic matter. Domingues et al. (2024, Chap. 2) examined the interaction of pesticides and biodegradation products in Brazil, specifically by studying fipronil, an insecticide of the phenyl pyrazole family commonly used for sugar cane and corn crops. Fipronil undergoes biotic and abiotic transformations in aquatic environments to create transformation products (fipronil sulphide, fipronil sulfone, and fipronil desulfinyl), considered more toxic and persistent than the original pesticide. The study evaluated the sorption of fipronil and two of its transformation products in ultrapure water and water from the Atibaia River, which contains high amounts of organic matter. All three of the compounds were highly sorbed by the microplastics in ultrapure water. In contrast, the sorption capacity observed in the river water was lower, partly due to the natural organic carbon in the water. The study underscored the importance of testing in environmental matrices to understand better how plastic pollution will interact with water in the environment.

21.5.2 PFAS: Long-Lasting "Forever" Pollutants

PFAS are a large group of manufactured chemicals widely used for decades in industry and consumer products, including household items. Typically designed to be strong and resistant to degradation, PFAS are long-lasting chemicals, the components of which break down very slowly over time in the environment. Accordingly, PFAS are considered "forever" chemicals because they persist in the environment, animals, and the human body.

Due to their extensive use and persistence in the environment, PFAS have been found in all environmental mediums (water, air, and soil) and in aquatic organisms and humans—including in the blood of people and animals—across the world. Humans are exposed to PFAS through contaminated food and drinking water. In contrast, the primary exposure pathway of aquatic organisms is water pollution by PFAS due to wastewater and industrial effluent discharges. PFAS are also present at low levels in a variety of food products, and studies have linked PFAS to harmful health effects in humans and animals. Because of the harmful human health effects of PFAS, some regions have started to put in place regulatory norms for PFAS in drinking water. For example, the recast EU Drinking Water Directive limits total PFAS in drinking water to 0.5 µg/l and for twenty specific PFAS compounds to 0.1 µg/l (EU 2020).

Research studies have explored various methods for removing PFAS from wastewater and freshwater resources used as drinking water, as conventional water treatment methods are ineffective. Granular activated carbon (GAC) and ion exchange treatments have been successful, whereas membrane systems are shown to be the most promising. Membranes can act in two ways, with both size exclusion (more giant molecules are strained by the membrane surface) and electrostatic interaction (a small electric charge increases contaminant removal) mechanisms.

Research in this field has been expanding in some regions. Still, there is a lack of studies on the extent of PFAS presence in water resources in developing regions, especially in Africa. This points to a need for further studies on the use of plastics and other PFAS-containing consumer products and their emissions to water resources. Moreover, there is a need for increased public awareness and regulatory action on PFAS worldwide, particularly in developing regions.

21.5.3 Pharmaceuticals and Personal Care Products

Pharmaceuticals and personal care products (PPCPs) are a group of emerging pollutants, including prescription and over-the-counter human and veterinary medications and various nonmedicinal personal care consumer chemicals. Personal care products include body care items, cosmetics, fragrances, ultraviolet blockers in sunscreens, and others. Among pharmaceutical residues found in wastewater and freshwater bodies, antibiotics are of significant concern because of their role in antimicrobial resistance (AMR). Antibiotics are increasingly present in wastewater effluents and treatment byproducts (e.g., sludge and biosolids).

Many PPCPs enter aquatic systems with untreated or insufficiently treated wastewater because conventional wastewater treatment processes are ineffective in removing them. Another primary source of PPCPs is runoff from agricultural land owing to the use of massive quantities of veterinary medications in livestock. This can present a problem for water reuse and recharge as treated wastewater is commonly used for water conservation. Diniz et al. (2024, Chap. 19) are in progress studying the effectiveness of various water treatment methods for removing various PPCPs in water intended for reuse and recharge in Brazil. The results of a multi-barrier process, incorporating reverse osmosis, oxidation, and activated carbon, met the Brazilian government's guidelines for water potability. PPCPs in wastewater treatment byproducts (i.e., biosolids) may have implications for agricultural practices, although wastewater biosolids contain essential nutrients for plant growth.

Researchers have been exploring alternative options for PPCP removal. Some pharmaceuticals may be removed from the aquatic environment using green degradation techniques, such as the application of microalgae. In Brazil, the biodegradation of sodium diclofenac (DCF) (a common anti-inflammatory chemical) was tested using *Tetraselmis sp.*, a marine microalga, as the degrading organism (Melegari 2023). The study demonstrated that the algae could degrade DCF, but the efficiency of the process appears low. While promising, more research is needed on alternative methods and modern technologies before they can be widely deployed.

Box 8. Residues of the pharmaceutical ciprofloxacin in the environment

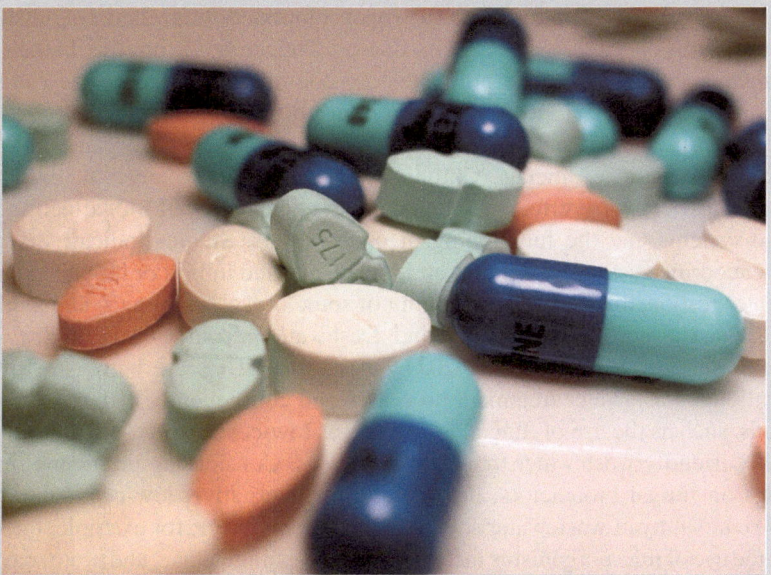

Pharmaceuticals. *Source* Wikimedia Commons, CC BY SA 3.0 (https://commons.wikime dia.org/wiki/File:VariousPills.jpg)

The metabolism of ciprofloxacin, an antibiotic used to treat a variety of bacterial infections, by humans and animals is limited. Approximately 70% of the antibiotic is excreted into waste discharge, ending up in wastewater effluents, wastewater treatment plant sludge biosolids, or animal manure. As wastewater treatment processes are not effective in removing it, ciprofloxacin residues are still present in treated wastewater and biosolids. Wastewater treatment byproducts and animal manure are commonly used as fertilizer in agricultural practices, and consequently, ciprofloxacin residues may be reintroduced into the environment with water reuse and biosolids applications. Ciprofloxacin residues might even persist after land application because of their persistence and low biodegradability. Ciprofloxacin has high sorption in soil, but it is immobile, persistent, and accumulative. A study evaluating the effect of ciprofloxacin on the soil invertebrate *Enchytraeus crypticus* over time found no significant difference in the reproduction rates of the first two generations but a significant decline in reproduction for the third (Ravanelli Martins 2023). The implications of this research are not fully understood, and more research is needed on trace antibiotic releases to the environment and their ecotoxicity.

21.5.4 Endocrine Disrupting Compounds

Among emerging water pollutants, a variety of compounds with endocrine-disrupting properties are of particular concern due to their effects on the human body and aquatic organisms. These pollutants are commonly referred to as endocrine-disrupting compounds (EDCs), which comprise diverse groups of emerging pollutants, including pharmaceuticals and personal care products, organic pollutants, plasticizers, and heavy metals. Chronic exposure to EDCs, even at low concentrations, causes adverse effects on the human body and animal life, affecting reproductive systems and the functioning of other organs. Studies indicate that EDCs affect endocrine systems in humans and organisms, and inhibit or imitate the natural hormones responsible for the functioning of some organs of humans and animals.

EDCs are found in wastewater and freshwater resources around the world. Research studies (Kasonga et al. 2021) stress that EDCs are barely removed by conventional wastewater treatment processes and consequently propose alternative methods, such as the use of fungal bioreactors as low-cost and eco-effective environmentally friendly wastewater treatment processes. Considering promising research findings on fungal bioreactors, the potential of alternative, low-cost solutions for EDCs removal from wastewater should be explored further; for example, the potential of the use of microorganisms including bacteria, fungi, algae, and protozoa, which have been found plausible to biodegrade these pollutants. Hence, further research is needed for the management and removal of EDCs from wastewater.

21.6 Key Policy Messages and Recommendations for Managing Emerging Pollutants in Water

Key policy messages and recommendations for managing emerging pollutants in freshwater resources and wastewater are summarized under five main themes based on the research findings and scientific discussions presented at the UNESCO-IWRA International Conference on "Emerging Pollutants: Protecting Water Quality for the Health of People and Ecosystems," which took place online in January 2023, gathering over 120 state-of-the-art research studies on wide-ranging topics related to emerging water pollutants. These policy messages and recommendations reflect the conclusions of ten thematic sessions and two high-level multi-stakeholder science-policy panel sessions, which brought together researchers and policymakers.

Theme 1. Emerging pollutants: a growing threat to aquatic ecosystems

- Emerging pollutants and their transformation products are ubiquitous in aquatic ecosystems, and many are toxic.
- Specifics differ from place to place, but pesticides, disinfectants, pharmaceuticals, microplastics, and engineered nanomaterials pose increasing risks worldwide to water quality and, thereby, to human health and ecological system integrity.

- Challenges for addressing these emerging pollutants include a lack of transparency on the pollutants' characteristics, limited scientific understanding, sparse data, difficulty in monitoring, and uncertainty regarding their risks.
- Water quality standards set for individual chemicals do not adequately account for the synergistic effects of multiple contaminants.
- Efficient control measures, such as source controls or specific wastewater treatment technologies, are needed to mitigate the risks posed by emerging pollutants to ecosystems and biodiversity.
- In less developed regions, emerging contaminants compound the risks posed by traditional pollutants such as *E. coli* or nutrients. Emerging pollutants typically are not yet regulated.
- Increasing economic growth and urbanization worldwide create opportunities, but also drive a vast and increasing consumption of chemicals directly or indirectly discharged into the environment.

Theme 2. Emerging pollutants and groundwater

- In principle, preventing groundwater contamination is less expensive than remediation, but the ability to monitor its contamination is limited.
- Risk assessment tools are available to assist decision-makers in prioritizing emerging substance contamination of groundwater and to improve groundwater quality monitoring programs.
- Emerging pollutants can enter groundwater from various sources such as wastewater, agricultural chemicals, veterinary medicines and solid waste. Cross-sectoral strategies are neccesary to protect groundwater resources from pollution by emerging contaminants.
- The behaviour and fate of emerging pollutants in groundwater are complex, posing significant health and environmental risks. A comprehensive approach is needed to investigate groundwater contamination and develop effective pollution prevention strategies.

Theme 3. Emerging pollutants and managing wastewater and waste

- The best treatment for emerging water pollutants is to avoid the discharge of the pollutants in the first place.
- Wastewater treatment technologies are vital in mitigating the effects of emerging pollutants, but there are limitations to their effectiveness.
- Sampling and analyzing emerging pollutants is expensive and limited, necessitating new, more effective analytical approaches.
- Potential shifts in strategy to improve wastewater treatment processes and technologies include: (1) decentralizing treatment processes, (2) developing and improving advanced approaches, (3) the innovative use of locally available materials, and (4) the use of non-traditional approaches such as soil aquifer treatment systems.
- Research should address: (1) the effectiveness of wastewater treatment technologies in removing emerging pollutants; (2) the fate of emerging pollutants in wastewater treatment process byproducts such as sludge and biosolids; (3)

health concerns about the reuse of treated wastewater and its byproducts; and (4) treated wastewater reuse for groundwater recharge and agricultural production.

Theme 4. A circular economy approach: lifecycle management of emerging pollutants

- Policy interventions should target all stages of a product's lifecycle to reduce the risk of emerging contaminants accumulating in water.
- With a circular economy approach, byproducts of production or from product use can be repurposed to create other products or to naturally biodegrade after use (e.g., colorants produced from fungi that can be dehydrated and composted after use).
- Governments play a crucial role in stimulating research and innovation to transform byproducts into resources and to redesign processes to avoid waste and pollutants that cannot be repurposed.
- Governments should require, where possible, the disclosure of contaminants of concern in consumer products to increase transparency and public awareness and to empower consumers to make informed consumption choices.
- Multidisciplinary teams, including academics and regulators, are an effective mechanism to bring knowledge together and identify novel policies and technological approaches to support a circular economy approach.

Theme 5. Priority emerging pollutants: microplastics, nanomaterial, PFAS, and trace chemicals

- Communities should prioritize research and policy to address emerging substances that are highly toxic, persist in the environment, or have value as indicators of water or environmental quality. This list will necessarily grow and evolve.
- PFAS (per- and polyfluoroalkyl substances) are of priority concern as they are known to be persistent in the environment and harmful to human health.
- Confirmed endocrine disruptors need to be regulated to mitigate exposure risks. Further research is needed on substances with suspected endocrine-disrupting properties.
- Plastic pollution is ubiquitous around the globe and adversely affects humans, wildlife, and the environment. Still, much remains unknown about the extent of these effects—especially microplastics, which are increasingly found in all environments.
- Technologies to improve our ability to identify and remove or degrade emergent pollutants are being developed, but their applicability and effectiveness on larger scales are still limited.
- Education and outreach activities are needed to raise public awareness about the sources of emerging pollutants, consumers' role in mitigating their release, and efforts and approaches to eliminate them from use, where possible.
- Manufacturers should be required, where possible, to disclose the use of priority emerging pollutants as part of a strategy to increase public awareness.
- Appropriate regulations need to be developed, based on the best available scientific data and knowledge, to mitigate risks of priority emerging pollutants.

Acknowledgements Policy recommendations presented in this chapter are drawn mainly from research findings presented at the UNESCO-IWRA Online Conference "Emerging Pollutants: Protecting Water Quality for the Health of People and Ecosystems," held on 17–19 January 2023, as well as key messages from the conference's thematic sessions and science-policy panel discussions. These findings were summarized in five UNESCO-IWRA Policy Briefs for the respective conference themes (UNESCO and IWRA 2023a, b, c, d, e). The authors also thank all participants of the conference. Their contributions to conference's discussions were insightful.

References

Baratsas SG, Pistikopoulos EN, Avraamidou S (2022) A quantitative and holistic circular economy assessment framework at the micro level. Comput Chem Eng 160. https://doi.org/10.1016/j.compchemeng.2022.107697

Brauns B (2023) Emerging contaminants in groundwaters and their relation to recharge sources in Bengaluru City, Karnataka, India. In: UNESCO/IWRA conference on emerging pollutants: protecting water quality for the health of people and the environment, 17–19 January 2023. https://iwra.org/proceedings/congress/resource/IWRA3rdOCJan2023_PPT025Brauns.pdf

Braz de Freitas Bijos JC (2023) Occurrence of antibiotics and psyquiatric drugs in Brazilian municipalities: a data analysis approach. In: UNESCO/IWRA conference on emerging pollutants: protecting water quality for the health of people and the environment, 17–19 January 2023. https://iwra.org/proceedings/congress/resource/IWRA3rdOCJan2023_PPTSess062Bijos.pdf

Devi MK, Karmegam N, Manikandan S, Subbaiya R, Song H, Kwon EE, Sarkar B, Bolan N, Kim W, Jörg Rinklebe J, Govarthanan M (2022) Removal of nanoplastics in water treatment processes: a review. Sci Total Environ 845:157168. https://doi.org/10.1016/j.scitotenv.2022.157168

Diniz V, Rohwedder JJR, Rath R (2024) Direct potable reuse: a prioritization of emerging contaminants for monitoring strategies and pilot-scale advanced treatment. In: Zandaryaa S, Fares A, Eckstein G (eds) Emerging pollutants: protecting water quality for the health of people and the environment. UNESO-Springer, New York

Diniz V, Reyes GM, Rath S, Cunha DGF (2020) Caffeine reduces the toxicity of albendazole and carbamazepine to the microalgae Raphidocelis subcapitata (Sphaeropleales, Chlorophyta). Int Rev Hydrobiol 105:151–161

Domingues, Dias RMA, Madeira CL, Starling MCVM, Neves TA, Montagner CC (2024) Occurrence of pesticides and emerging contaminants in the Pampulha Lake: anthropic pollution of a UNESCO heritage site. In Zandaryaa S, Fares A, Eckstein G (eds) Emerging pollutants: protecting water quality for the health of people and the environment. UNESO-Springer, New York

EU (2020) Directive (EU) 2020/2184 of the European Parliament and of the Council of 16 December 2020 on the quality of water intended for human consumption (recast) (OJ L 435, 23.12.2020), pp 1–62

European Commission (2022) Proposal for a directive of the European parliament and of the council concerning urban wastewater treatment. Recast. Brussels

Gigault J, Hadri HE, Nguyen B, Grassl B, Rowenczyk L, Tufenkji N, Feng S, Wiesner MR (2021) Nanoplastics are neither microplastics nor engineered nanoparticles. Nat Nanotechnol 16(5):501–507. https://doi.org/10.1038/s41565-021-00886-4

Greer JB, Dalsky EM, Lane RF, Hansen JD (2023) Tire-derived transformation product 6PPD-quinone induces mortality and transcriptionally disrupts vascular permeability pathways in developing Coho salmon. Environ Sci Technol 57(30):10940–10950. https://doi.org/10.1021/acs.est.3c01040

Health and Environmental Sciences Institute (HESI) (2024) Risk assessment in the 21st century (RISK21). https://hesiglobal.org/risk-assessment-in-the-21st-century-risk21/

Kaiser R (2023) Understanding the prevalence and occurrence of antimicrobial resistance in urban karst groundwater systems. In: UNESCO/IWRA conference on emerging pollutants: protecting water quality for the health of people and the environment, 17–19 January 2023. https://iwra. org/proceedings/congress/resource/IWRA3rdOCJan2023_PPT023Kaiser.pdf

Kasonga TK, Coetzee MAA, Kamika I, Ngole-Jeme VM, Momba MNB (2021) Endocrine-disruptive chemicals as contaminants of emerging concern in wastewater and surface water: a review. J Environ Manag 277:111485. https://doi.org/10.1016/j.jenvman.2020.111485

Leal-Bautista RM (2023) Assessment of emerging organic contaminants at the groundwater of Yucatán peninsula: recreational and water supply. In: UNESCO/IWRA conference on emerging pollutants: protecting water quality for the health of people and the environment, 17–19 January 2023. https://iwra.org/proceedings/congress/resource/IWRA3rdOCJan2023_PPT024LealBautista.pdf

Lopez Ferreira T et al (2023) Microplastics in urban soil and groundwater in the city of Bauru, São Paulo, Brazil. In: UNESCO/IWRA conference on emerging pollutants: protecting water quality for the health of people and the environment, 18th January 2023. https://iwra.org/proceedings/congress/resource/IWRA3rdOCJan2023_PPT027Ferreira.pdf

Melegari SP (2023) Toxicity and biodegradation of the pharmaceutical diclofenac employing the green marine microalga Tetraselmis sp.: a preliminary study. In: UNESCO/IWRA conference on emerging pollutants: protecting water quality for the health of people and the environment, 18th January 2023. https://iwra.org/proceedings/congress/resource/IWRA3rdOCJan2023_PPTSess104PedrosoMelegari.pdf

Namba K (2023) Bio-based and circular solutions for Harmful Algal Bloom (HAB) and water and climate change challenges in Berlin-Brandenburg. In: UNESCO/IWRA conference on emerging pollutants: protecting water quality for the health of people and the environment, 18th January 2023. https://iwra.org/proceedings/congress/resource/IWRA3rdOCJan2023_PPTSess063Namba.pdf

de Oliveira F et al (2023) Production and life cycle assessment of microbial colorants. In: UNESCO/IWRA conference on emerging pollutants: protecting water quality for the health of people and the environment, 18th January 2023

OECD (2020) The circular economy in cities and regions: Synthesis report, OECD urban studies, OECD publishing, Paris. https://doi.org/10.1787/10ac6ae4-en

OECD (2023) Endocrine disrupting chemicals in freshwater: monitoring and regulating water quality, OECD studies on water, OECD publishing, Paris. https://doi.org/10.1787/5696d960-en

Ravanelli Martins M (2023) Toxicity of ciprofloxacin through generations of the soil invertebrate Enchytraeus crypticus. In: UNESCO/IWRA conference on emerging pollutants: protecting water quality for the health of people and the environment, 18th January 2023. https://iwra. org/proceedings/congress/resource/IWRA3rdOCJan2023_PPTSess106RavanelliMartins.pdf

Saadi H (2023) Impacts of anthropogenic activities on water quality in the Guiers Lake basin (Senegal): spatio-temporal evolution of emerging pollutants. In: UNESCO/IWRA conference on emerging pollutants: protecting water quality for the health of people and the environment, 17–19 January 2023. https://iwra.org/proceedings/congress/resource/IWRA3rdOCJan2023_PPTSess052Saadi.pdf

UNESCO and IWRA (2023a) Theme I: emerging pollutants: a growing threat to aquatic ecosystems, policy brief no. 24, July 2023. https://iwraonlineconference.org/wp-content/uploads/2023/11/PB-24-July-2023-Web.pdf

UNESCO and IWRA (2023b) Theme II: emerging pollutants and groundwater, policy brief no. 25, July 2023. https://iwraonlineconference.org/wp-content/uploads/2023/11/PB-25-July-2023-Web.pdf

UNESCO and IWRA (2023c) Theme III: emerging pollutants and managing wastewater and waste, policy brief no. 26, July 2023. https://iwraonlineconference.org/wp-content/uploads/2023/11/PB-26-July-2023-Web.pdf

UNESCO and IWRA (2023d) Theme IV: a circular economy approach: lifecycle management of emerging pollutants, policy brief no. 27, July 2023. https://iwraonlineconference.org/wp-con tent/uploads/2023/11/PB-27-July-2023-Web.pdf

UNESCO and IWRA (2023e) Theme V: priority emerging pollutants: microplastics, nanomaterial, and trace chemicals, policy brief no. 28, July 2023. https://iwraonlineconference.org/wp-con tent/uploads/2023/11/PB-28-July-2023-Web.pdf

U.S. Environmental Protection Agency (2024) Per- and Polyfluoroalkyl Substances (PFAS): final PFAS national primary drinking water regulation. https://www.epa.gov/sdwa/and-polyfluoroal kyl-substances-pfas